Population Genetics in the Neotropics

Marcos Vinícius Bohrer Monteiro Siqueira
Enéas Ricardo Konzen
Pedro Manoel Galetti Junior
Editors

Population Genetics in the Neotropics

Springer

Editors
Marcos Vinícius Bohrer Monteiro Siqueira
Departamento de Ciências Agrárias e Biológicas (DCAB)
Universidade do Estado de Minas Gerais
Frutal, Brazil

Enéas Ricardo Konzen
Centro de Estudos Costeiros, Limnológicos e Marinhos
Universidade Federal do Rio Grande do Sul
Imbé, Brazil

Pedro Manoel Galetti Junior
Departamento de Genética e Evolução
Universidade Federal de São Carlos
São Carlos, Brazil

ISBN 978-3-031-83687-9 ISBN 978-3-031-83685-5 (eBook)
https://doi.org/10.1007/978-3-031-83685-5

This Springer imprint is published by the registered company Springer Nature Switzerland AG
The registered company address is: Gewerbestrasse 11, 6330 Cham, Switzerland

To the women of my life, Carmen Lúcia, Lilia Bohrer, and Mirian Cury
Marcos Vinícius Bohrer Monteiro Siqueira

To my beloved parents, whose wise thoughts came from experiencing the essence of nature
Enéas Ricardo Konzen

To my teachers and students
Pedro Manoel Galetti Junior

Preface

Welcome to "Population Genetics in the Neotropics," a collection of 18 chapters devoted to depict fundamental theoretical foundations of population genetics and their practical application with organisms of one of the most biodiverse biogeographic realms in our planet, the Neotropics. As you start reading this book, we invite you to think that you are traveling along the dense and huge forests of Amazon, with all those tall trees, vines, shrubs, and all the rodents, insects, birds, and other animals that move in between the trees, all of those with so many different sizes, shapes, and colors that you may have never seen before. Imagine yourself taking a sample of leaves, of water, or of soil and taking it to the lab and spreading it on a plate with a culture medium with a lot of nutrients and agar. After a few days, if you look at it on the microscope, you may see lots of microorganisms that you may never have thought that existed. If you were to sequence their DNA, with prior preparation of metagenomic libraries, multiple novel organisms would probably be detected after bioinformatic inspection.

You can now travel east and find similar trees to those you found in the Amazon, but way smaller and more sparse within fields with grasses and shrubs, the Brazilian Savanna or Cerrado. You can then go to the east Coast and find again that exuberant dense forests, or travel further south to find conifers mixed with tropical forests, and rocky fields amidst them. Going even more to the South, you will find less and less trees, different insects and animals, as you reach Patagonia. We could not forget Chaco, the west Pacific Coast, the Andes, part of Central America, the Caribbean, the Caatinga, and many other spots that make up the Neotropics.

Scientists have shown that the Amazon is a fundamental source of neotropical biodiversity, highlighting the complex interactions among several organisms that inhabit the Neotropics. However, it is not a surprise that this immense biodiversity is under threat, and humanity is to blame for that. Therefore, it is urgent that we seek strategies for conservation and restoration of degraded habitats. Moreover, the forests and fields of the Neotropics provide us with numerous food and housing supplies, so that conservation aligns with breeding desirable genetic materials as well. Conservation and breeding could not be done without the knowledge we have on population genetics. Population genetics is the field of biology that studies the allele frequencies of genes, their distribution within and among populations, and the factors that shape allele frequencies over generations (time) and among populations (space).

Until now, if you took a book on population genetics to study, you would frequently find a sequence of chapters with similar contents, starting with the basic assumptions and models, and being followed by the factors that shape allele frequencies in populations. To further understand the concepts and models, there is nothing better than having a lot of biological examples to guide you through your learning. Here is where we get at the keystone of our book. Up to date, there is no book that is primarily focused on examples of the neotropical realm. In general, there have been numerous population genetics studies going on in the Neotropics, but they have not been compiled in a book so far.

Given the neglected status of the neotropical realm in textbooks on population genetics, our goal is to provide the readers, whether living within the dominion of the Neotropics or not, with a sequence of chapters enriched with examples of animals, plants and other organisms that belong to the Neotropics. We hope that students and professionals interested in the field may acquire fundamental knowledge on population genetics processes and mechanisms by using examples that may applied to their own reality, or that could be contrasted with North America, Europe and other regions in the world.

For that purpose, we organized the present book as a sequence of chapters that may fit into a regular order of contents for population genetics, but aided by multiple examples of neotropical organisms that corroborate or even challenge basic concepts in the field. We start with an introductory chapter (Chap. 1) to go through the basic idea and applications of population genetics and a brief historical perspective on the main actors and actresses that moved this field of genetics forward ever since the beginning of the twentieth century. We also tried to provide information on the initial contributions of population geneticists to the study of neotropical populations.

Then we move forward to our Part I—Principles of Population Genetics, where a series of chapters conceptualize and provide basic theoretical foundations of population genetics. It is to be highlighted, though, that we start this section with a chapter devoted to defining biodiversity (Chap. 2), acknowledging the fundamental biodiversity hotspots in the Neotropics and the threats that have been imposed on them by human interference. The relation between population genetics and the crisis of biodiversity is highlighted. With that information, the reader can move toward a sequence of chapters to depict the concept of genetic variation and their components (Chap. 3), the concepts, basic theoretical foundations, applications, and case studies in the Neotropics for the topics of random genetic drift (Chap. 4), natural selection (Chap. 5), gene flow and migrations (Chap. 6), inbreeding and inbreeding depression (Chap. 7), and the quantitative nature of phenotypic variation, influenced either by genetic and environmental components of variances (Chap. 8).

After that theoretical appraisal, readers may proceed to Part II—Tools for Population Genetics. This section documents the most used methods applied into population genetic analyses. For starters, definitions and applications of the most important categories of molecular markers used for assessing intra- and interpopulational genetic variation are discussed (Chap. 9). Next, the revolution provided by constantly improving sequencing methods and their achievements in neotropical organisms so far are described (Chap. 10). Tools for integrating environmental data and prospect past, present, and potential future spanning ranges of species are at the core of discussion in Chap. 11. Moreover, how and whether the landscape influences the genetic makeup of populations is discussed on the theme of landscape genetics, in Chap. 12. The patterns of genetic structure through the geographic distribution of species and how evolutionary factors are implicated are the main subject of Chap. 13, on the field of phylogeography.

The third and last section, Part III—Population Genetics in Practice, puts the reader into the context of the fundamental strategies for the conservation of genetic resources of neotropical species. First, Chap. 14 provides the definitions and applications of ex situ and in situ approaches, among others that are discussed mainly for plants. Then, Chap. 15 deals with conservation strategies for the conservation of animals in the Neotropics, discussing important themes such as wildlife DNA forensics against illegal traffic of wild animals. The means by which humans have shaped and selected plants for desirable traits, that is, the domestication process, are described in Chap. 16. Domestication and breeding of animal models in the neotropical landscape are the matter of discussion of Chap. 17. Finally, we could not leave us, human beings, out of it! The main migrations and the genetic structure of human populations that inhabit the Neotropics are acknowledged by the final chapter (Chap. 18) of this collection.

With all that, we can only hope that readers navigate the field of population genetics in a distinctive manner, through one of the most biodiverse realms of the world. Let this book rehash the sentiment that we need to put in motion the conservation efforts described in this book or elsewhere to effectively preserve or conserve plants, animals, and other species that makeup the Neotropics and all the other realms in this planet, the only one that we have to live.

Frutal, Brazil — Marcos Vinícius Bohrer Monteiro Siqueira
Imbé, Brazil — Enéas Ricardo Konzen
São Carlos, Brazil — Pedro Manoel Galetti Junior

The original version of the book was inadvertently published without the acknowledgement. The acknowledgement has been included in this book.

Acknowledgments

We are grateful to Ricardo M. Degani, that produced scientific illustrations within the content of the following chapter numbers in our book: chapters 2, 3, 8, 9, 13 and 16.

Contents

Part II Tools for Population Genetics

Contributors

Alessandro Alves-Pereira Departamento de Genética, Escola Superior de Agricultura "Luiz de Queiroz", Universidade de São Paulo, Piracicaba, SP, Brazil

Patricia Amavet Laboratorio de Genética, Facultad de Humanidades y Ciencias, Universidad Nacional del Litoral-Consejo Nacional de Investigaciones Científicas y Técnicas, Santa Fe, Argentina

Larissa S. Arantes Leibniz Institute for Zoo and Wildlife Research (IZW), Berlin, Germany

Berlin Center for Genomics in Biodiversity Research, Berlin, Germany

Miklos Maximiliano Bajay Centro de Educação Superior da Região Sul, Universidade do Estado de Santa Catarina, Santa Catarina, SC, Brazil

Ossman Barrientos-Díaz Instituto de Biociências, Universidade Federal do Rio Grande do Sul, Porto Alegre, RS, Brazil

Marcelo Lattarulo Campos Instituto de Biociências, Universidade Federal de Mato Grosso, Cuiabá, MT, Brazil

Larissa Resende Carvalho Instituto de Ciências Biológicas, Universidade Federal de Goiás, Goiânia, GO, Brazil

Marcos Araújo Castro e Silva Departament de Medicina i Ciències de la Vida (MELIS-UPF), Universitat Pompeu Fabra, Barcelona, Spain

Charles R. Clement Instituto Nacional de Pesquisas da Amazônia, Manaus, Amazonas, AM, Brazil

Leonardo Carlos Jeronimo Corvalán Instituto de Ciências Biológicas, Universidade Federal de Goiás, Goiânia, GO, Brazil

Flaviane Malaquias Costa Departamento de Genética, Escola Superior de Agricultura "Luiz de Queiroz", Universidade de São Paulo, Piracicaba, SP, Brazil

Vanessa Paes da Cruz Departamento de Biologia Estrutural e Funcional, UNESP/IBB, Botucatu, SP, Brazil

Adelson Lemes da Silva Junior Departamento de Ciências Florestais, Universidade Federal de Lavras, Lavras, MG, Brazil

Gustavo Reis de Brito Departamento de Ciências Biológicas, Faculdade de Ciências e Letras, Campus de Assis, Universidade Estadual Paulista "Júlio de Mesquita Filho" - UNESP, Assis, SP, Brazil

Reinaldo Alves de Brito Departamento de Genética e Evolução, Universidade Federal de São Carlos, São Carlos, SP, Brazil

Dulcinéia de Carvalho Departamento de Ciências Florestais, Universidade Federal de Lavras, Lavras, MG, Brazil

Patrícia Domingues de Freitas Departamento de Genética e Evolução, Universidade Federal de São Carlos, São Carlos, SP, Brazil

Mario Luiz Teixeira de Moraes Departamento de Fitotecnia, Tecnologia de Alimentos e Sócio Economia, Faculdade de Engenharia—FEIS/UNESP, Ilha Solteira, SP, Brazil

Marcela Aparecida de Moraes Silvestre Departamento de Ciências Químicas e Biológicas, Instituto de Biociências—IBB/UNESP, Botucatu, SP, Brazil

Renata de Oliveira Dias Instituto de Ciências Biológicas, Universidade Federal de Goiás, Goiânia, GO, Brazil

Lucimara Cruz de Souza Departamento de Ciências Florestais, Universidade Federal de Lavras, Lavras, MG, Brazil

Pabila Stephanie de Souza Arruda Instituto de Biociências, Universidade Federal de Mato Grosso, Cuiabá, MT, Brazil

Wellington Ferreira do Nascimento Centro de Ciências de Chapadinha, Universidade Federal do Maranhão, Chapadinha, MA, Brazil

Marisol Domínguez Unit of Evolutionary Biology/Systematic Zoology, Institute for Biochemistry and Biology, University of Potsdam, Potsdam, Germany

Daniela Cristina Ferreira Instituto de Biociências, Universidade Federal de Mato Grosso, Cuiabá, MT, Brazil

Ana Flávia Francisconi Escola Superior de Agricultura "Luiz de Queiroz", Universidade de São Paulo, Piracicaba, SP, Brazil

Fernanda Amato Gaiotto Departamento de Ciências Biológicas, Universidade Estadual de Santa Cruz, Ilhéus, BA, Brazil

Pedro Manoel Galetti Junior Departamento de Genética e Evolução, Universidade Federal de São Carlos, São Carlos, Brazil

Tábita Hünemeier Departamento de Genética e Biologia Evolutiva, Universidade de São Paulo, São Paulo, SP, Brazil

Institute of Evolutionary Biology, IBE/CSIC-UPF, Barcelona, Spain

Adriana M. G. Ibelli Embrapa Pecuária Sudeste, São Carlos, São Paulo, SP, Brazil

Universidade Estadual do Centro-Oeste, Guarapuava, Paraná, PR, Brazil

Enéas Ricardo Konzen Centro de Estudos Costeiros, Limnológicos e Marinhos, Universidade Federal do Rio Grande do Sul, Imbé, Brazil

Thâmara Moura Lima Instituto de Ciências Agrárias, Universidade Federal Rural da Amazônia, Belém, PA, Brazil

Celso Luis Marino Departamento de Ciências Químicas e Biológicas, Instituto de Biociências—IBB/UNESP, Botucatu, SP, Brazil

Concepta McManus Universidade de Brasília (UNB), Instituto de Biologia, Brasília, DF, Brazil

Jonathan Morales-Marroquín Universidade Estadual de Campinas—Cidade Universitária Zeferino Vaz, Barão Geraldo, Campinas, Brazil

Alison Gonçalves Nazareno Instituto de Ciências Biológicas, Universidade Federal de Minas Gerais, Belo Horizonte, MG, Brazil

Kelly Nunes Departamento de Genética e Biologia Evolutiva, Universidade de São Paulo, São Paulo, SP, Brazil

Rhewter Nunes Universidade Estadual de Goiás, Iporá, GO, Brazil

Felipe Pimentel Universidade Anhembi Morumbi, São Paulo, SP, Brazil

Santiago Linorio Ferreyra Ramos Instituto de Ciências Exatas e Tecnologia, Universidade Federal do Amazonas, Itacoatiara, AM, Brazil

Luciana C. A. Regitano Embrapa Pecuária Sudeste, São Carlos, SP, Brazil

Universidade Federal de São Carlos, São Carlos, SP, Brazil

Héctor Antônio Assunção Romão Instituto de Ciências Biológicas, Universidade Federal de Goiás, Goiânia, GO, Brazil

Bruno Cesar Rossini Instituto de Biotecnologia – IBTEC/UNESP, Botucatu, SP, Brazil

Alesandro Souza Santos Departamento de Ciências Biológicas, Universidade Estadual de Santa Cruz, Ilhéus, BA, Brazil

Ana Lucia Anversa Segatto Universidade Federal de Santa Maria, Centro de Ciências Naturais e Exatas, Departamento de Bioquímica e Biologia Molecular, Santa Maria, RS, Brazil

Marcos Vinicius Bohrer Monteiro Siqueira Departamento de Ciências Agrárias e Biológicas (DCAB), Universidade do Estado de Minas Gerais, Frutal, Brazil

Evandro Vagner Tambarussi Departamento de Produção e Melhoramento Vegetal, Faculdade de Ciências Agronômicas, Universidade Estadual Paulista "Júlio de Mesquita Filho", Botucatu, SP, Brazil

Fernanda J. Trindade Instituto Tecnológico Vale Desenvolvimento Sustentável, Belém, PA, Brazil

Caroline Turchetto Instituto de Biociências, Universidade Federal do Rio Grande do Sul, Porto Alegre, RS, Brazil

Andreia Carina Turchetto-Zolet Universidade Federal do Rio Grande do Sul, Porto Alegre, RS, Brazil

Elizabeth Ann Veasey Departamento de Genética, Escola Superior de Agricultura "Luiz de Queiroz", Universidade de São Paulo, Piracicaba, SP, Brazil

Paulo Cesar Venere Instituto de Biociências, Universidade Federal de Mato Grosso, Cuiabá, MT, Brazil

Maria Imaculada Zucchi Escola Superior de Agricultura "Luiz de Queiroz", Universidade de São Paulo, Piracicaba, SP, Brazil

Agência Paulista de Tecnologia dos Agronegócios, Piracicaba, SP, Brazil

Chapter 1
Population Genetics in the Neotropics: Looking Back to See the Way Forward

Marcos Vinicius Bohrer Monteiro Siqueira, Enéas Ricardo Konzen, Pedro Manoel Galetti Junior, and Reinaldo Alves de Brito

Contents

Abstract Evolutionary biologists and geneticists have long recognized that genetic variation is a keystone for evolution to follow. That said, we cannot properly understand evolution if we do not know the levels and organization of the genetic variation that populations and species have. Population genetics is the field of genetics that deals with the genetic variation and their underlying factors within and among populations. Researchers in the field have amassed data on allele diversity and variation for countless species of animals, plants, fungi and other organisms to which has helped elucidate how much variation they have across their genomes. A brief history of the field discusses how population genetics theory was formulated based on genetic variation inferred from morphological traits and rapidly moved forward to addressing genetic variation at the molecular level. Technological inputs led a cascade of consecutive methods to detect and quantify allele frequencies, beginning

M. V. Bohrer Monteiro Siqueira (✉)
Departamento de Ciências Agrárias e Biológicas (DCAB), Universidade do Estado de Minas Gerais, Frutal, Brazil

E. R. Konzen
Centro de Estudos Costeiros, Limnológicos e Marinhos, Universidade Federal do Rio Grande do Sul, Imbé, RS, Brazil

P. M. G. Junior · R. A. de Brito
Departamento de Genética e Evolução, Universidade Federal de São Carlos, São Carlos, SP, Brazil

M. V. Bohrer Monteiro Siqueira et al. (eds.), *Population Genetics in the Neotropics*,
https://doi.org/10.1007/978-3-031-83685-5_1

with enzymes and their isoforms and culminating with studying variation at the DNA molecule itself. The so-called molecular markers have revealed singularities for the most diverse species across all continents, including organisms of the neotropical biogeographic realm, that includes South America and part of Central America. The increasing use of genome-based approaches should improve our knowledge on population genetic variation of all kinds of organisms, and how such variation is truly important for species adaptation in a changing world, particularly in the hyperdiverse Neotropics.

Keywords Genetic variation · Natural selection · Mutation · Polymorphisms

Learning Objectives
In this introductory chapter the reader is connected with the following concepts and ideas:

1. A brief historical perspective on the development of the field of population genetics;
2. A definition of population genetics and its importance in the conservation of biodiversity, among other goals;
3. The initial steps of an analysis and the most important types of molecular markers that have been used in population genetic studies in the Neotropics;
4. The most important applications and perspectives of population genetics.

1.1 A Short History on Evolutionary Theories and the Birth of Population Genetics

Variation, particularly, variation on fitness effects, is at the core of Charles Darwin's concept of natural selection. Despite that importance, at the time Darwin proposed his theory, very little was known on how variation segregated in populations, or even how it was passed on across generations, limiting the ability to explain how the environment influencing different organisms in the same population would effect change. As an attempt to understand these processes, Darwin considered a form of blending inheritance, pangenesis, although he emphasized the role of small changes accumulating through time as the main driver of the evolutionary process. After the publication of "On the Origin of Species", Galton and others published some quantitative work that suggested that the offspring tended to be the average of their parents, which was coined by Galton as the Law of the Regression to the Mediocrity (Galton 1886). He actually used this to form the basis of what he referred to as positive eugenics, aiming to improve human race, but also to suggest that some traits are more refractory to mixing, which led him to define nature and nurture. Regardless of the social implications of these results, and in part because of the problems with inheritance, some researchers at the time considered the unifying role of an individual and its development to suggest that very rarely mutations are small; rather, they may produce big phenotypic changes which would be in general detrimental, but eventually might provide for more opportunities to adaptation than small changes.

The central role of mutation in evolution is, therefore, one of the core points of the mutationist school, which purported that new species were produced rapidly through discontinuous transformations, not small changes, and the lingering qualms between these two points of view, to which we will return later, is a telling sign of how complex this question actually is, and how far we may be from solving it.

Even though Darwin's model for natural selection was in general readily accepted, several scholars questioned their dependence on small changes and considered that there must be different forms of inheritance that would explain evolution and inheritance. These topics remained a major point of contention for several more years, and that was not assuaged by the rediscovery of Mendel's work. In fact, when Mendel's studies were rediscovered at the turn of the twentieth century, the mutationist school considered Mendelian factors as a corroboration that changes were not as small as Darwin proposed, since they produced discontinuous phenotypes. Several researchers considered mutations to be generally large, and detrimental, and promoted the idea that there should be ways to expunge mutations and to prevent its culling, which reinforced the eugenics movement (van der Meer 2024). For instance, Morgan originally considered that Darwin's model of natural selection acting on small changes was clearly refuted, but was also skeptical of the generality of Mendelian factors (Shine and Wrobel 2014). Because visible mutants were the only foray into genetic variation, populations were considered to have little genetic variation, posing a problem for Darwin's model. To study the effect of mutations on variation and fitness, researchers such as Muller used x-ray in the lab, and showed that most variation produced was deleterious (Muller 1927). This was used as an argument to defend eugenics and suggest that medical attention to people with diseases would impact human evolution by increasing the number of deleterious alleles, or the genetic load (e.g., Muller 1950), as originally proposed by Muller.

It was only in 1918 that Fisher reconciled Mendelism and Darwinism in what may arguably be one of the most influential papers of all time. The article "The Correlation Between Relatives On The Supposition of Mendelian Inheritance" basically established the bases for animal crosses and epidemiology, modern statistics and, more to the point of this chapter, showed how Mendelism provides the genetic cornerstone for natural selection (Fisher 1919). Fisher reconciled Mendelian inheritance and quantitative variation by showing that continuous traits may be the result of a combination of many discrete Mendelian genes with small effects whose expression is affected by environmental effects, represented by reaction norms. More information on that is available from Chap. 8.

Fisher's studies, as well as papers and books by Wright (1943, 1951) and Haldane (1932, 1990), galvanized what has been dubbed as Neodarwinism, which can be summarized as the reinterpretation of the Darwinian theory considering genetic variation as the main driver for the origin of diversification and speciation and laid the mathematical foundations for how evolutionary forces could promote changes in populations. Several studies had shown, mostly on *Drosophila*, which was chosen as model species for historical and practical reasons, that there was little variation in nature. That came to be explained by Muller and Morgan, in what has been referred to as the classical school, by the direct impact of natural selection, which would quickly act upon any new variation, either to remove deleterious alleles, or lead the few advantageous to fixation. That was their reasoning to explain why there was so little variation in nature (Beatty 1987).

One name that is also associated with Neodarwinism, but not to the extent sometimes of its importance on the field, is Theodosius Dobzhansky, originally from Ukraine, but who studied in Russia and escaped to the United States to join Morgan to study natural populations. When he and associates studied karyotypes on populations of *Drosophila*, they found much more variation, particularly in the form of chromosomal inversions. Dobzhansky, in what has been referred to as the balanced polymorphism school, contended that these high levels of genetic variation at the population level was a consequence of heterozygote advantage, and therefore, there would be a lot of variation in nature. In fact, he proposed that there was so much variation that when there were homozygotes in a population, they most likely would be for different alleles. He found an association of several inversions in species of *Drosophila* to environmental cues in temperate climate and proposed that drift might have something to do with it, because populations would show large population size differences along the year, since they would increase in warmer months and dwindle during the winter, when populations would diapause and most of it would die (Dobzhansky 1949).

Dobzhansky proposed that there would be much less intrapopulation variation in the Tropics, because he expected population sizes there to be more stable through the year. He then took an opportunity presented by The Rockefeller group, which was trying to increase US soft power with its allies during and after World War II, to visit Brazil to study tropical populations of *Drosophila*. During several years, Dobzhansky visited Brazil and helped establish the evolutionary genetics community of scholars working on *Drosophila* in Brazil and South America, which is well covered in other publications (Pavan and Cunha 2003; de Carvalho 2020; Monte Sião and Pereira Martins 2020).

Furthermore, Dobzhansky was an integral part of the foundation of the Modern Synthesis, and his book Genetics and the Origin of Species (Dobzhansky 1937) was instrumental to expanding its theoretical underpinnings to a broader audience. He managed to explain the mathematical formulations of papers by Fisher, Wright and Haldane in broader biological principles, such as their impacts on population genetics and speciation which were much more amenable for biologists in general and had a great impact on naturalists and experimental biologists (Ayala and Fitch 1997). This was followed by other biologists, such as Simpson (1953), Stebbins (1950) and Mayr (1999), to provide a general framework that defined the Modern Synthesis.

For reasons that are still debated, several people considered that the main outcome of the Modern Synthesis was the overarching impact of natural selection on population variation, in what has later been referred to as the hardening of the Modern Synthesis (Gould and Lewontin 1979). At the time, though, the main conflict still stemmed from the fact that there was no effective way of measuring genetic variation, limiting our abilities to test hypotheses. That changed around the mid 1960s because of two important achievements. On one front, people were able to isolate and identify amino acid sequences from proteins of different organisms, even different species, and that led to an observation that there seemed to be a regular rate of evolution which led to what has been referred to an evolutionary clock. Because this seemed anathema to evolutionary processes that were known to that time, this was even referred to as "Non-Darwinian evolution" (King and Jukes 1969). On another front, as discussed below, new ways to investigate genetic variation on protein allozymes, pushed forth by one of Dobzhansky most famous students (Hubby and Lewontin 1966), led to the conclusion that there was a great deal of variation across populations and species.

The amount of variation across taxa found by allozyme studies was considered a vindication by the defenders of the balanced school but posed an important theoretical problem. Earlier theoretical work had estimated how much variation could be affected by selection in a population and derived an estimation of the cost of selection, that is, the cost of replacing an allele at the population level, considering the genetic load (Haldane 1957). These estimates indicated that the amount of variation at the population level was incompatible with an assumption that all this variation was being selected, or being maintained as balanced polymorphisms, suggesting, instead, that most variation segregating at the population level was neutral, or nearly neutral, and only a small amount was being affected by selection at any given time (Kimura and Crow 1964; Kimura 1968; Ohta 1973). Furthermore, if most variation is indeed neutral, it is easier to explain rates of evolution that are compatible with molecular clocks.

Interestingly, the neutral theory basically derived from the observation that all individuals are heterozygotes, for different alleles, which was exactly what Dobzhansky suggested over 40 years prior to that, but the main difference was that this was not maintained by selection, rather, it was neutral. We should note, however, that the discussion between the classical school, which defended that there was not much variation in nature, and the balanced school changed. As methods to survey molecular variation improved, it has become clear that there is a lot of variation, but we still struggle to understand how much of it is functional, how that variation is woven into new individuals in the context of different environments, and how these hierarchies are affected on populations by different evolutionary forces. Basically then, the emergence of the neutral theory changed the question from how much variation there is in nature, which is a lot, to how much of this variation is functional and perceived by selection. Much has been learned recently from the plethora of genetic data available and new methods, and this will be the focus of the remainder of this chapter, and this book.

1.2 Why Study Population Genetics? *The General Idea Behind Population Genetics and Genomics*

We started this chapter with a fascinating history depicting the fundamental events ever since the publication of Darwin on natural selection theory to the birth of population genetics. In that context, microevolutionary processes that constitute the basis of population genetics have been debated since the beginning of the past century. It is well accepted that population genetics provides the basic insights into the mechanisms of evolution (Charlesworth 2015). Michael Lynch even proposed an extension to Dobzhansky's well know phrase (Dobzhansky 1973) by stating that "nothing in evolution makes sense except in light of population genetics", and several researchers believe that if one wants to understand biological evolution, she or he should at least be slightly familiar with evolutionary change at the genomic level (e.g. Bromham 2009).

The use of molecular methods to study genetic polymorphisms has allowed populations to be studied across different species and habitat, and without the need for controlled crossings, mutant genes or any previous genetic studies (Hartl 2008). Numerous studies using genomic data have been produced making use of the revolution that new molecular markers and sequencing technologies have provided. Section 1.2 of our book depicts the fundamental molecular methods, sequencing technologies and recent statistical approaches to conducting population genetic studies.

In a detailed review describing the major questions and controversies that have concerned population geneticists in the recent past, Charlesworth and Charlesworth (2017) discuss how theoretical models and empirical data have shaped the understanding of key concepts such as genetic variation, selection, and drift. The authors also highlight the debates surrounding the relative importance of different evolutionary forces and the impact of molecular techniques on the field. Finally, they emphasize the ongoing challenges in integrating new technologies with traditional population genetics theory.

Three different fronts revolutionized population genetics (Hartl and Clark 2010). The first major revolution is related to the theory of coalescence. It employs a sample of individuals in a population to trace all alleles of a gene shared by all members of the population back to a single ancestral copy, known as the most recent common ancestor. This empirical revolution profited greatly from an independent revolution on methods to investigate genetic variation which culminated on the expansion of the field of genomics, that enabled the assembly of long DNA stretches using sequencing technologies that contributed in a never-before-seen way to the understanding of our position and that of other organisms in the tree of life. It is genomics and phylogenetic inferences that enabled us to predict even the last universal common ancestor (LUCA), that is, the putative first form of living cell that has arisen on Earth (Weiss et al. 2016). This was all made possible by the computational revolution, a third phase in this process, in which thousands of data are processed by increasingly smaller, faster and more accurate machines and statistical programs.

These revolutions happened at different times, places and purposes as population genetics matured (for review, Johnston et al. 2019; Okasaki et al. 2021).

As we move from the era of genetics to the era of genomics, we can ask ourselves, where are we going? What does genomics offer? Among other things, for instance, the population-genomics paradigm can facilitate biodiversity conservation through rapid biodiversity screening, identifying appropriate populations for translocations to rescue declining populations (Luikart et al. 2003). In Sect. 1.3 of this book, we will look at some practical studies that illustrate how genomics has allowed plants and animals to be conserved and their domestication process to be better understood in the Neotropics.

And when we think about the progress to promote global food security, let us state that the genomes of more than 100 crops have been sequenced to date. Plant genomic research has expanded across multiple fronts and the next few years promise to bring further advances spurred by the advent of new technologies and approaches—an advancing crop genomics from lab to field (Purugganan and Jackson 2021). With access to genomic data, a revolution in which molecular information is being used to enhance phenotypic responses to "genomic selection" is occurring. Hill (2014) discusses some of the history and population genetic issues as applied to the science of livestock improvement, which has had and continues to have major spin-offs into ideas and applications in other areas.

Scientific studies on population genetics and genomics have impacted our lives, particularly through medical advances, on our perception of ourselves and others, on our place in nature, and on our dreams for the present and future of humanity. Where did we come from, what are we now, and what will we become? Human population genomics leads to these and other questions. It is possible that in due time, population genetics and genomics may provide researchers with answers to specific questions, but frequently they actually drive more questions to be addressed.

1.3 When and How Have We Started to Study Population Genetics in the Neotropics?

The research in genetics in South America began in the early twentieth century and was greatly influenced by North American and European geneticists. From the first studies of insect chromosomes carried out by Francisco Sáez in 1925 in Argentina, the research on genetics in South America had gradually started in Brazil, Chile and Colombia, until approximately the first half of twentieth century (Cruz-Coke 1999).

The arrival of Dobzhansky in Brazil, as already discussed, was a decisive milestone for the beginning and growth of genetic studies in the country, with enormous repercussions in South America. Dobzhansky partnered up with some Brazilian researchers, especially with André Dreyfus (University of São Paulo), by studying chromosomal aberrations of populations of *Drosophilla ananassae* that occurred in

Brazil, resulting in their first co-authored publication (Dobzhansky and Dreyfus 1943). The studies on neotropical flies continued after that, resulting on the initial population genetic studies on the variation for chromosomal polymorphisms. These studies involved one of the main research goals of Dobzhansky, that is, deciphering the relationship between genetic variability with the forces of distinct environments that influence population survival, that is, the adaptive role from variation.

Advances in population genetics, though, were revolutionized in the mid 1960s after Hubby and Lewontin published their first article on allozyme variation in populations of *Drosophila* (Hubby and Lewontin 1966). After that, several researchers started to employ isozymes as genetic markers for depicting population-level allele frequencies and estimating levels of genetic diversity and structure. In the neotropical region, pioneering studies on population genetics used allozyme markers in a Neotropical species of mosquitoes, *Sabethes cyaneus* (Munstermann and Marchi 1986), and bees (Contel and Kerr, 1976). With plants, a review has described that the first studies on population genetics were published in the late 1980s, and no more than 10 papers per year with population genetics on plants were published until 2003 (Tinoco et al. 2015). Agricultural crops have also started being the targets of numerous population genetic studies with proteins and allozymes, such as the protein phaseolin and its variability on genotypes of common bean (*Phaseolus vulgaris*) based on SDS/PAGE (Gepts and Bliss 1986), and the patterns of variation at isozyme loci in three species of piper in Costa Rica (Heywood and Fleming 1986).

In general, among the distinct types of markers involved in population genetic studies, allozyme markers were favored in the 1990s and early 2000s, involving neotropical organisms. But it is DNA markers that paved the way for rapidly expanding population genetic studies on all sorts or species (further details on molecular markers are described in Chap. 9). Initially, restriction fragment length polymorphism (RFLP) markers were used to study populations of neotropically adapted honeybees (Muralidharan and Hall 1990). Earlier studies with random amplified polymorphic DNA markers (RAPD) were conducted with the neotropical genus *Leucaena* (Harris 1995) and cichlid fishes (Sultmann et al. 1995). RAPD markers were popular among researchers that worked with neotropical species until 2010. Amplified fragment length polymorphisms (AFLP) also took place in several population genetic studies in the end of the 1990s, such as with *Phaseolus lunatus* (Caicedo et al. 1999). Most studies with AFLP markers were published in the 2000s and early 2010s.

Microsatellites or SSRs (simple sequence repeats) and SNPs (single nucleotide polymorphisms) markers have been the most employed molecular tools for population genetics studies in the world (Ankita et al. 2022). Neotropical organisms were also the focus of numerous studies with such markers. Trinucleotide microsatellites in a neotropical wasp were studied and resulted in one of the earliest publications, in 1996 (Strassmann et al. 1997). Microsatellite markers are still very useful for population genetic studies, despite the fact that sequencing technologies have enabled the development of studies based on SNP markers. Inter-simple sequence repeats (ISSR), markers adjacent to common SSR throughout plant and animal genomes, were also used in several studies with neotropical species, and the earliest

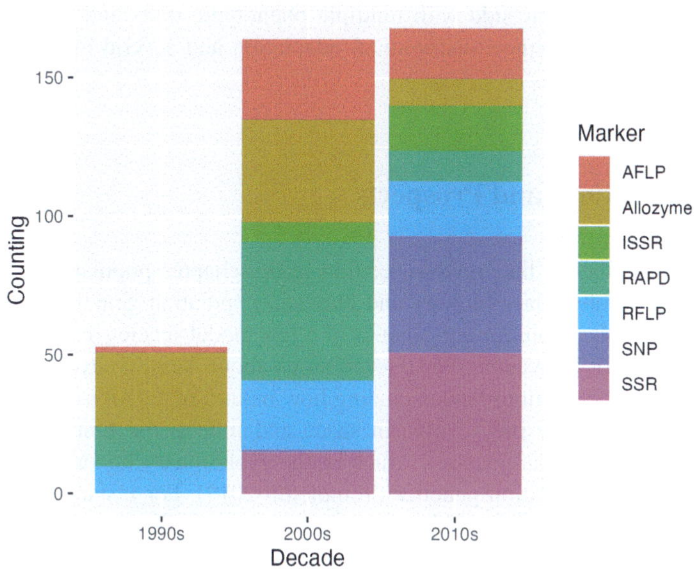

Fig. 1.1 A simple panorama of the distribution of population genetic studies per decade and type of marker, as retrieved from Web of Science (Clarivate) by searching two terms: marker (one word among: allozyme, RFLP, RAPD, AFLP, SSR, ISSR or SNP) + neotropical

records date from 2005. Chap. 9 discusses the main molecular methods for conducting population genetic studies using some of these methods.

Overall, a search for different types of molecular markers accompanied by the word Neotropical in the Web of Science platform (May 2024) showed a rapid change of the marker choice over the years (Fig. 1.1). Let us clarify that the numbers shown in the figure are absolute when it comes to the specific search for the pair of the following words: marker (allozyme OR RFLP OR RAPD OR AFLP OR SSR OR ISSR OR SNP) AND Neotropical. However, those numbers clearly do not account for the numerous studies that have been published so far, but simply the absolute number of articles that have both the marker type and Neotropical as terms indicated by the publications. Those numbers would have been much larger with the use of alternative phrases. Therefore, they give a brief picture of the main markers used in the 1990s, 2000s, 2010s, the three last decades in which records appear. It is also worth noting that studies with RFLP markers, especially in the 2000s and 2010s, are mainly PCR-derived, that is, using restriction enzymes to digest PCR products directly, therefore, not the RFLP as the original method that used X-rays and DNA probes in 1980s and early 1990s. In the 2020s, most studies in population genetics in the neotropical region involve SNP markers, whether using reduced representation of genomes, with methods such as genotyping-by-sequencing, or even whole-genome sequencing comparisons (see Chap. 10). In the years to come, we certainly will see numerous publications coming forth with these high-throughput methods to unravel the specific patterns of genetic diversity and structure of neotropical species, alongside with multiple phenotypic traits and environmental variables for modeling their distribution, adaptation and several other objectives of study.

1.4 Applications and Prospects

As already described in the previous sections of this chapter, population genetics is at the core of evolutionary biology and classical population genetic studies have long focused on the evolutionary genetics of a few model species (e.g. *Drosophila*, Dobzhansky 1946; Polivanov 1964), teaching us about the microevolutionary factors guiding their evolution. Understanding how natural populations belong to different species and how they evolve in space and time is the first and foremost application of population genetics. Much of the evolutionary history of humanity has been told by population genetics (Templeton 2019). For instance, population genetics was used to show that Philippine islands were populated by multiple waves of human migration during the last 50,000 years, contributing to our understanding of how these waves of migration changed the landscape genetics of the Asia-Pacific region (Larena et al. 2021). In the Neotropics, studies carried out in Amerindian populations reported high frequency of the human alpha-actinin skeletal muscle isoform 3 gene which could be associated to their high quality muscular constitution, and polymorphisms of genes related to the tumor protein 53 (TP53) pathway are related to adaptation of the Andean populations living in high altitude (Salzano 2016). And so, population genetics has given important clues as to how natural populations of all types of organisms have diverged in response to the presence of barriers, demographic, climatic or even anthropogenic effects, consolidating the importance of evolutionary genetics as the basis for all other applications of population genetics.

Population genetics easily facilitated revealed the manipulation of domesticated populations in the improvement of plants and animals, another great application of the field (see Chap. 17 for more details and examples). Although little is still understood about which genes can be linked to a given phenotype, and how they interact with one another and their environmental, neutral markers-based population genetics underlies a range of subareas, in human and animal medicine, for instance, from paternity studies (based on knowledge of allele frequencies) to immunogenetics, pharmacogenetics and, more recently, pharmacogenomics. A better understanding of the association between adaptive genetic variation and individual responses to different drugs, for example, is a window of great importance for general health improvement in the near future and has its theoretical and methodological basis in population genetics. The same happens in agriculture, where the evolution of genetic variation in different plant populations explains, for example, the development of resistance to agricultural pesticides (e.g. Brown 1996).

Today, as we experience a true biodiversity crisis around the world, population genetics emerges as the origin of conservation genetics to bring important contributions to conservation biology. The recent book Conservation Genetics in the Neotropics (Galetti Jr 2023), presented several studies in which conservation genetics provided excellent contributions to our understanding on how animals and plants respond to human-promoted environmental changes that have been threatening biodiversity in the Neotropics, such as habitat degradation, loss and fragmentation, pollution, hunting, and climate change. For instance, in the emblematic maned wolf (*Chrysocyon brachyurus*), population genetic analyses were able to reveal a reduction of gene flow among populations and recent bottleneck signatures that could be consequence of the huge human landscape modification, mainly for agricultural purposes along the past century in the Brazilian savanna. It was suggested that the population structuring and recent bottleneck observed in the maned wolf may compromise the maintenance of genetic variation at the local populations level and impair the long-term species survival (Rodriguez-Castro et al. 2022). It is important to highlight that much of what is known about the population genetics of neotropical biodiversity is still based on mitochondrial genetic markers or few nuclear ones (e.g. microsatellites). Although there has been a marked expansion of genomic data in the last two decades, the use of genomic tools is still incipient in population studies in the Neotropics, with still a handful of studies even using reduced-genome strategies, such as RADseq, and ddRADseq (e.g. Tomé et al. 2021) (see Chap. 10).

The reduction in the cost of large-scale sequencing methods and the increased use of such approaches in low- and medium-income countries, which constitute the majority of the neotropical region, should foster population genomics studies, increasing knowledge on population attributes such as genetic diversity, inbreeding, gene flow and population structuring, as well as allowing the assessment of genetic signatures of adaptive significance for populations in the Neotropics.

1.5 Final Remarks

Studying population genetics in the Neotropics is crucial due to the extraordinary biodiversity and ecological complexity of this region. The Neotropics, which include tropical rainforests, savannas, and other species-rich ecosystems, are home to a vast array of organisms, many of which are endemic and exhibit unique ecological interactions. Population genetics in this region allows for a better understanding of how genetic diversity is maintained, how populations adapt to heterogeneous environments, and how evolutionary processes shape biodiversity. This knowledge is essential for species conservation and the sustainable management of ecosystems.

Moreover, the Neotropics are frequently subjected to anthropogenic pressures such as deforestation, land-use changes, and climate change, which can drastically alter population dynamics. Population genetics offers tools to detect genetic changes that may indicate loss of diversity, population fragmentation, or adaptation to new stressors. Knowledge derived from these tools are vital for the implementation of conservation strategies aimed at mitigating the effects of these pressures and promoting ecosystem resilience.

Finally, studying population genetics in the Neotropics also contributes to the global understanding of evolutionary processes. The region provides a natural laboratory for exploring fundamental questions such as speciation, coevolution, and adaptation in extremely diverse environments. Research in these ecosystems can offer models for biodiversity conservation in other tropical regions and may reveal new species and ecological interactions, broadening our knowledge of the evolution of life on Earth.

Key Points

- Population genetics deals with alleles frequencies in populations and how they are shaped by a diverse range of factors;
- Ronald Fisher reconciled Mendelian genetics and biometrics by showing that continuous traits may be the result of a combination of many discrete Mendelian genes with small effects whose expression is affected by environmental effects;
- Neodarwinism emerged as a combination of the contributions of Fisher, Haldane, Wright and other geneticists, leading to further development of population genetics;
- Using organisms from the Neotropics, the pioneer population genetic studies were published in the late 1980s, using biochemical (isozymes) and DNA-based markers (mostly RFLP at the time);
- The Neotropics hold high levels of biodiversity, but are severely threatened by human action, a fact that demands population genetics studies and effective conservation and breeding efforts.

Questions

1. Why was Ronald Fisher important in the early phase of the development of population genetics?
2. How did Dobzhansky contribute to the development and expansion of population genetics in South America?
3. Which molecular markers best reflected population genetics in the last three decades?
4. What are the main applications of population genetics? Describe the importance of the field into conservation and breeding in the context of the Neotropics.

References

Ankita M, Pramod KS, Abhishek B, Vinay S, Vikas S, Pradeep S, Joy R, Himanshu S (2022) Chapter 8 - Analysis of SSR and SNP markers, Editor(s): Dev Bukhsh Singh, Rajesh Kumar Pathak, Bioinformatics, Academic Press, Pages 131—144, https://doi.org/10.1016/B978-0-323-89775-4.00017-1

Ayala FJ, Fitch WM (1997) Genetics and the origin of species: an introduction. Proc Natl Acad Sci USA 94:7691–7697. https://doi.org/10.1073/pnas.94.15.7691

Beatty J (1987) Weighing the risks: stalemate in the classical/balance controversy. J Hist Biol 20:289–319. https://doi.org/10.1007/BF00139457

Bromham L (2009) Does nothing in evolution make sense except in the light of population genetics? Biol Philos 24:387–403. https://doi.org/10.1007/s10539-008-9146-6
Brown TM (1996) Applications of molecular genetics in combatting pesticide resistance. An overview. In: Brown TM (ed) Molecular genetics and evolution of pesticide resistance, vol 645. ACS Symposium Series, American Chemical Society, pp 1–8
Caicedo AL, Gaitán E, Duque MC, Chica OT, Debouck DG, Tohme J (1999) AFLP fingerprinting of Phaseolus lunatus L. and related wild species from South America. Crop Sci 39(5):1497–1507. https://doi.org/10.2135/cropsci1999.3951497x
Charlesworth B (2015) What use is population genetics? Genetics 200:667–669. https://doi.org/10.1534/genetics.115.178426
Charlesworth B, Charlesworth D (2017) Population genetics from 1966 to 2016. Heredity 118:2–9. https://doi.org/10.1038/hdy.2016.55
Contel EPB, Kerr WE (1976) Esterase polymorphism in bees. Biochemical Genetics, 14(3-4):379–387
Cruz-Coke R (1999) The history of genetics in Latin American countries during the twentieth century. Revista Medica de Chile 127(12):1524–1532
de Carvalho TB (2020) Modern evolutionary biology and Brazilian population genetics: Theodosius Dobzhansky at the University of São Paulo. Perspect Sci 28:223–243. https://doi.org/10.1162/posc_a_00339
Dobzhansky T (1937) Genetics and the origin of species. Columbia University, New York
Dobzhansky T (1946) Genetics of natural populations; recombination and variability in populations of *Drosophila pseudoobscura*. Genetics 31(3):269–290
Dobzhansky T (1949) Observations and experiments on natural selection in Drosophila. Hereditas 35:210–224. https://doi.org/10.1111/j.1601-5223.1949.tb03334.x
Dobzhansky T (1973) Nothing in biology makes sense except in the light of evolution. Am Biol Teach 35(3):125–129. https://doi.org/10.1093/biosci/biaa170
Dobzhansky T, Dreyfus A (1943) Chromosomal aberrations in Brazilian Drosophila ananassae. Proc Natl Acad Sci USA 29(10):301–305. https://doi.org/10.1073/pnas.29.10.30
Fisher RA (1919) XV.—the correlation between relatives on the supposition of Mendelian inheritance. Trans Roy Soc Edinburgh 52:399–433. https://doi.org/10.1017/S0080456800012163
Galetti PM Jr (2023) Conservation genetics in the Neotropics. Springer. Cham, p 583. https://doi.org/10.1007/978-3-031-34854-9
Galton F (1886) Regression towards mediocrity in hereditary stature. J Anthropol Inst G B Irel 15:246–263. https://doi.org/10.2307/2841583
Gepts P, Bliss FA (1986) Phaseolin variability among wild and cultivated common beans (*Phaseolus vulgaris*) from Colombia. Econ Bot 40(4):469–478. https://doi.org/10.1007/BF02859660
Gould SJ, Lewontin RC (1979) The spandrels of San Marco and the Panglossian paradigm: a critique of the adaptationist programme. Proc Royal Soc Lond B 205:581–598
Haldane JBS (1932) The causes of evolution, Longmans. Green and Co., Ltd, London, New York and Toronto
Haldane JBS (1957) The cost of natural selection. J Genet 55:511–524
Haldane JBS (1990) A mathematical theory of natural and artificial selection—I. Bull Math Biol 52:209–240. https://doi.org/10.1007/BF02459574
Harris SA (1995) Systematics and randomly amplified polymorphic DNA in the genus Leucaena (Leguminosae, Mimosoideae). Plant Syst Evol 197:195–208. https://doi.org/10.1007/BF00984639
Hartl DL (2008) Princípios de Genética de População. Editora Funpec, p 217
Hartl DL, Clark AG (2010) Princípios de Genética de Populações. Editora Artmed, p 660
Heywood JS, Fleming TH (1986) Patterns of allozyme variation in three Costa Rican species of Piper. Biotropica 18:208–213
Hill WG (2014) Applications of population genetics to animal breeding, from wright, fisher and lush to genomic prediction. Genetics 196(1):1–16. https://doi.org/10.1534/genetics.112.147850

Hubby JL, Lewontin RC (1966) A molecular approach to the study of genic heterozygosity in natural populations. I. The number of alleles at different loci in Drosophila pseudoobscura. Genetics 54(2):577–594. https://doi.org/10.1093/genetics/54.2.577

Johnston HR, Keats BJB, Sherman SL (2019) Population genetics. In: Pyeritz RE, Korf BR, Grody WW (eds) Emery and Rimoin's principles and practice of medical genetics and genomics. Academic, pp 359–373. https://doi.org/10.1016/B978-0-12-812537-3.00012-3

Kimura M (1968) Evolutionary rate at the molecular level. Nature 217:624–626

Kimura M, Crow JF (1964) The number of alleles that can be maintained in a finite population. Genetics 49:725–738

King JL, Jukes TH (1969) Non-Darwinian evolution. Science 164:788–798

Larena M, Sanchez-Quinto F, Sjödin P et al (2021) Multiple migrations to The Philippines during the last 50,000 years. PNAS 118(13):e2026132118

Luikart et al (2003) The power and promise of population genomics: from genotyping to genome typing. Nat Rev Genet 4:981–994. https://doi.org/10.1038/nrg1226

Mayr E (1999) Systematics and the origin of species, from the viewpoint of a zoologist. Harvard University Press

Monte Sião JF, Pereira Martins LA-C (2020) Dobzhansky and Dreyfus's group: the introduction of natural population genetics studies in Brazil (1943–1960). Perspect Sci 28:244–276. https://doi.org/10.1162/posc_a_00340

Muller HJ (1927) Artificial transmutation of a gene. Science 66:84–87. https://doi.org/10.1126/science.66.1699.84

Muller HJ (1950) Our load of mutations. Am J Hum Genet 2:111–176

Munstermann LE, Marchi A (1986) Cytogenetic and isozyme profile of Sabethes cyaneus, a mosquito of the neotropical canopy. J Hered 77:241–248. https://doi.org/10.1093/oxfordjournals.jhered.a110229

Muralidharan K, Hall HG (1990) Prevalence of African DNA RELP alleles in neotropical African honeybees. Arch Insect Biochem Physiol 15(4):229–236. https://doi.org/10.1002/arch.940150404

Ohta T (1973) Slightly deleterious mutant substitutions in evolution. Nature 246:96–98

Okazaki A, Yamazaki S, Inoue I et al (2021) Population genetics: past, present, and future. Hum Genet 140:231–240. https://doi.org/10.1007/s00439-020-02208-5

Pavan C, Cunha AB (2003) Theodosius Dobzhansky and the development of genetics in Brazil. Genet Mol Biol 26:387–395. https://doi.org/10.1590/S1415-47572003000300027

Polivanov S (1964) Selection in experimental populations of *Drosophila melanogaster* with different genetic backgrounds. Genetics 50:81–100. https://doi.org/10.1093/genetics/50.1.81

Purugganan MD, Jackson AS (2021) Advancing crop genomics from lab to field. Nat Genet 53:595–601. https://doi.org/10.1038/s41588-021-00866-3

Rodriguez-Castro KG, Lemos FG, Azevedo FC et al (2022) Human highly modified landscapes restrict gene ow of the largest neotropical canid, the maned wolf. Biodivers Conserv 31:1229–1247. https://doi.org/10.1007/s10531-022-02385-x

Salzano FM (2016) The role of natural selection in human evolution—insights from Latin America. Genet Mol Biol 39(3):302–311. https://doi.org/10.1590/1678-4685-GMB-2016-0020

Shine I, Wrobel S (2014) Thomas hunt Morgan: pioneer of genetics. University Press of Kentucky, Lexington

Simpson GG (1953) The major features of evolution. Columbia University Press, New York

Stebbins GL (1950) Variation and evolution in plants. Columbia University Press

Strassmann JE, Peters JM, Barefield K, Solís CR, Hughes CR, Queller DC (1997) Trinucleotide microsatellite loci and increased heterozygosity in cross-species applications in the social wasp, Polistes. Biochem Genet 35:273–279. https://doi.org/10.1023/A:1021849100346

Sültmann H, Mayer WE, Figueroa F, Tichy H, Klein J (1995) Phylogenetic analysis of cichlid fishes using nuclear DNA markers. Mol Biol Evol 12(6):1033–1047. https://doi.org/10.1093/oxfordjournals.molbev.a040279

Templeton AL (2019) Human population genetics/genomics and society. In: Templeton ALE (ed) Human population genetics and genomics. Academic, p 498. https://doi.org/10.1016/B978-0-12-386025-5.00014-2

Thomé MTC, Carstens BC, Rodrigues MT et al (2021) A role of asynchrony of seasons in explaining genetic differentiation in a Neotropical toad. Heredity 127:363–372. https://doi.org/10.1038/s41437-021-00460-7

Tinoco CF, Lima NE, Lima-Ribeiro MS, Collevatti RG (2015) Research and partnerships in studies on population genetics of Neotropical plants: a scientometric evaluation. Biochem Syst Ecol 61:357–365. https://doi.org/10.1016/j.bse.2015.07.00

van der Meer M (2024) Sown without care: Dutch eugenicists and their call for optimising developmental conditions, 1919–1939. Soc Hist Med hkae002:473–493. https://doi.org/10.1093/shm/hkae002

Weiss MC, Sousa FL, Mrnjavac N, Neukirchen S, Roettger M, Nelson-Sathi S, Martin WF (2016) The physiology and habitat of the last universal common ancestor. Nat Microbiol 1(9):1–8

Wright S (1943) Isolation by distance. Genetics 28(2):29114–29138. https://doi.org/10.1093/genetics/28.2.114

Wright S (1951) The genetical structure of populations. Ann Eugenics 15:323–352. https://doi.org/10.1111/j.1469-1809.1949.tb02451.x

Part I
Principles of Population Genetics

Chapter 2
Why Study Biodiversity in the Neotropics?

Patrícia Domingues de Freitas and Pedro Manoel Galetti Junior

Contents

Abstract First coined in the 1980s to refer to the biological diversity existing on Earth, the term "biodiversity" is currently defined at three different levels: genetics, species and ecosystems. Biological diversity exists because genetic changes made it possible for organisms to adapt in the most diverse environments. The biodiversity on Earth is distributed across biogeographic regions, which include the Neotropics, comprising both Central and South America. The Neotropics have the highest biodiversity richness in the world, and are characterized by their tropical rainforests, including the Amazon Forest, which is home to the greatest species diversity on Earth. Despite this huge biodiversity, the Neotropics as well as other regions on the planet have been dramatically affected by human activities, and biodiversity loss is one of the most pressing concerns in the Anthropocene. Populations adapted to very specific conditions are being locally extinct, depleting species diversity and increasing the risk of species extinction. By assessing the frequency of alleles in populations and their changes associated with mutation, genetic drift, natural selection, and gene flow, population genetics is a powerful tool to evaluate the effects of the Anthropocene on the long-term persistence of species. Nevertheless, neotropical

P. D. de Freitas (✉) · P. M. G. Junior
Departamento de Genética e Evolução, Centro de Ciências Biológicas e da Saúde, Universidade Federal de São Carlos, São Carlos, SP, Brazil
e-mail: patdf@ufscar.br

M. V. Bohrer Monteiro Siqueira et al. (eds.), *Population Genetics in the Neotropics*,
https://doi.org/10.1007/978-3-031-83685-5_2

species, and their populations and diversification process are still poorly studied. This chapter highlights key questions about the importance of studying and conserving neotropical biodiversity, indicating that population genetics can provide essential information for biological conservation, contributing to the sustainability and maintenance of life on Earth, one of the greatest challenges of contemporary human society.

Keywords Conservation biology · Biogeographical regions · Anthropocene · Population genetics

Learning Objectives

This chapter presents a brief overview of biodiversity, with special emphasis on the neotropical region. The core structure is centered on three key questions that allow us to understand biodiversity in a broader sense, including humans as an integral part of it:

1. What is biodiversity and how does it fit into the different fields of life sciences and human society, standing out as a valuable and fundamental component for sustaining life on Earth?
2. How is biodiversity distributed on our planet and what are the main threats to its conservation?
3. How can population genetics help us to expand knowledge about biodiversity and promote its long-term viability?

Following these questions, this chapter should aid the reader with a better understanding of biological diversity and related concepts, including the relevance of studying neotropical biodiversity at different levels.

2.1 Levels of Biodiversity

In the preface to the book Biodiversity (Wilson and Peter 1988), which brought together the studies presented at the National Forum on BioDiversity in Washington D.C., in September 1986, under the sponsorship of the National Academy of Sciences and the Smithsonian Institution, Edward O. Wilson teaches us that the term "BioDiversity" was written for the first time by Walter G. Rosen, representative of the National Academy of Sciences on the organizing committee of the aforementioned forum. This new word, derived from a contraction of the terms Biological and Diversity, was formerly written with capital B and D, but immediately came into common use and began to be written as we know it today.

To theoretically conceptualize biodiversity, we must bring together different points of view of science in a multi and transdisciplinary sense. From this vein, the comprehension of biodiversity has been expanding, especially after gathering deeper knowledge of ecosystem functionalities and the rapid advances in molecular

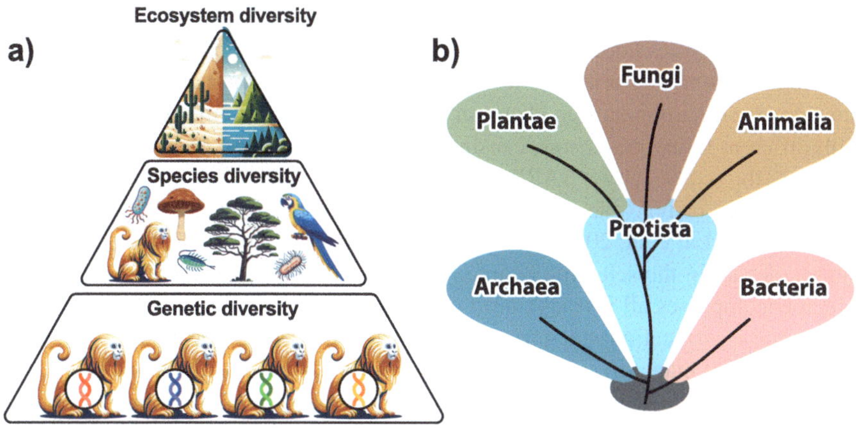

Fig. 2.1 (**a**) Levels of biodiversity, from genetic to ecosystem diversity and (**b**) a tree of life, indicating major *taxa*

biology technologies in recent times (Wilson and Peter 1988; Ponge 2021; Antonelli and Sanmartín 2011). As we currently understand, biodiversity is defined by different dimensions, from DNA to individuals, populations and species and their interactions in communities and ecosystems. Therefore, biodiversity encompasses genetic, species and ecosystem diversity. These three levels can be schematically represented in a pyramid according to their potential amount of diversity, in which the base is the genetic diversity, the middle is the variety of species; and the top is the diversity of ecosystems (Fig. 2.1a).

Moving from the top to the bottom, ecosystems are composed of biotic and abiotic components, including the species that occupy the environment and perform important functions for its maintenance. In ecosystems, communities of organisms of various species and their populations, and environmental elements, form a series of chains of interdependent relationships (Ponge 2021). Indeed, ecosystems encompass all living organisms (plants, animals, fungi and microorganisms) existing on Earth and the interactions among them and with non-living components (e.g. water and soil minerals).

The set of all ecosystems forms the Biosphere that represents the variety of environments that make life possible on the planet (Levin 1998). In summary, we recognize two main types of ecosystems: the terrestrials, represented by forests, deserts, mountains, prairies and pastures; and the aquatic, which include marine (seas and oceans) and freshwater (lakes, mangroves and rivers) environments and their communities.

The second level of biodiversity refers to species, a fundamental concept in biology generally defined as a group of individuals that share specific morpho-physiological characters and are capable of reproducing and generating viable offspring (Mayr 1970). This biological concept designates the basic unit of Linnaeus' taxonomic system, proposed by the Swedish naturalist Carl Linnaeus in the eighteenth century, which is often used in the scientific classification of species, including eukaryotes (animals, plants, fungi and protists) and prokaryotes (bacteria, cyanobacteria and archaea). It is worth to mention that although viruses occupy the most diverse environments and have an enormous capacity to multiply and interact with different living organisms, causing damage to the health of many of them, – such as the coronavirus SARS-CoV-2 (Severe Acute Respiratory Syndrome CoronaVirus), responsible for the COVID-19 pandemic that occurred in the the earlier 2020s (Lai et al. 2020)—, they are not considered living beings and despite intense debate in the scientific community, viruses remain outside the tree of life, which encompasses all species (multi and unicellular), at their respective levels of taxonomic classification recognized by science (Fig. 2.1b). Worthy of note is also the existence of giant viruses, such as the *Mimivirus*, which has DNA and RNA, and the ability to produce its own proteins and repair its own genome, characteristics not observed in other virus to date, even though they still need to host a cell to display those functions (La Scola et al. 2003).

Defining and classifying species is definitely not a simple task, once there are multiple concepts from different perspectives, such as phenetic, ecological, genetic or phylogenetic, among others (Queiroz 2005; Aldhebiani 2018). In this sense, there are several ways to identify a species, whether through single or integrated approaches, including analyses of morphological traits, behavior (e.g. courtship to mating), geographic distribution or area of occurrence, habitat and niche occupied, as well as DNA assessments. Integrated approaches in general tend to be broader and more reliable since they are based on a combination of traits (O'Neill et al. 2003; Queiroz 2005; Aldhebiani 2018).

Regardless of the methods used to their definition, species are constituted of a set of individuals and their individual genes and genotypes, distributed in single or multiple populations, sustained by their genetic variability, the third level of biodiversity. Because of their genetic variability, species can evolve and adapt to the most diverse environments (see more in Chap. 3). In short, genes and their interaction with the environment determine the biological functions and adaptation responses of organisms. The genetic diversity of a single species can be distributed in different populations, showing different allele frequencies, depending on stochastic factors, being genetic or environmental, and on selection and adaptation (Wilson and Peter 1988; O'Neill et al. 2003; Queiroz 2005; Aldhebiani 2018). This is why genetic diversity has been considered the base of the pyramid of biological diversity.

The recent development of robust technologies to assess genetic diversity has allowed the production of an enormous amount of relevant information about biodiversity. Part of these data have enabled the identification of a hidden biodiversity, often representing populations or species still unknown by science. Molecular operational taxonomic units (MOTUs), which represent unique evolutionary lineages, were massively described in the past decade, greatly increasing the potential number of species in a variety of ecosystems and biogeographic regions (Ryberg 2015; Milanese et al. 2019; Rouhan and Gaudeul 2021; Xavier et al. 2023), including the Neotropics (e.g. Pires et al. 2017; Silva-Santos et al. 2018; Lopes et al. 2020; Ramirez et al. 2021). In some cases, a few lineages may be highly genetically

Table 2.1 Number of eukaryote and prokaryote *taxa* registered on the Encyclopedia of Life (EOL 2024)

Taxon	N species	N genera	N families
Eukaryotes	2,003,399	183,878	12,017
Prokaryotes (excluding Archaea)	24,510	4127	548

divergent, even at the species level, but do not present enough diagnostic morphological traits capable of pointing to distinct species, those being considered cryptic species (Silva-Santos et al. 2023).

Overall, in a comprehensive way, biodiversity covers the entire range of population and genetic variation of lineages and/or species of all terrestrial and aquatic ecosystems. According to the Encyclopedia of Life (EOL) up to the time of publication of this book there are approximately 2,000,000 registered species of eukaryotes (Table 2.1), which are classified into approximately 184,000 genera and 12,000 families (EOL 2024). In fact, science has been more concerned and/or assertive in accessing eukaryotic biodiversity, and it is believed that the number of known species is still underestimated, since the literature has been showing an expressive increase in the number of new descriptions from virtually all continents and all taxonomic groups. This picture has become more evident with the use of taxonomy integrated by morphological, chromosome and molecular data (e.g. Padial et al. 2010; Ramirez et al. 2017; Turon et al. 2003).

Among eukaryotes, protists (protozoa and algae) exhibit high genetic diversity and biomass and contain many evolutionary innovations. They are morphologically diverse and mostly unicellular and microscopic living beings, and are found in diverse environments as free-living organisms, such as in soil and mainly in oceans, but also as parasites of other organisms, including humans and other non-human primates (Katz 2012; Mann et al. 2020; Morard et al. 2022). To date it is believed that approximately 70,000 protist species exist (Table 2.2), with many unidentified taxa so far (Pawlowski et al. 2012). Molecular data from microscopic protist lineages have substantially improved our understanding about the origin and diversification of eukaryotes, by providing insights on the nature of the genome of the last eukaryotic common ancestor (LECA) and evidences for mitochondria acquisition by eukaryotic cells. Fungi are also very abundant among eukaryotes and are widely spread throughout the world, being found in the air, soil, water and inside plants and animals. They present a huge variation in shapes and size and can also be microscopic. According to Antonelli et al. (2023), there are approximately 2.5 million species of fungi, of which only 155,000 have been identified by science (Table 2.2). Therefore, to date, more than 90% of them remain unknown.

Most assuredly researchers have been quite efficient in counting some groups of animals, such as mammals, of which there are over 5500 described species, and plants, such as trees, of which there are over 60,000 named species. However, various animals and plants are still expected to be undiscovered (Table 2.2). Approximately 8.7 million species are believed to be arthropods, of which, about 5.5 million are insects and 1.5 million are beetles. Regarding plants, about 300,000

Table 2.2 Number of species (N) known by science within each major eukaryote group

Eukaryotes	N	References
Protists	70,000	Pawlowski et al. (2012)
Fungi	155,000	Antonelli et al. (2023)
Animals		
Vertebrates	74,962	IUCN (2024)
Invertebrates	1,480,351	IUCN (2024)
Plants	425,035	IUCN (2024)

species of flowering plants are recognized thus far, but the true number may be closer to 370,000, with around 2000 species being discovered each year (Britannica 2024; Christenhusz and Byng 2016).

Although the macroscopic eukaryotes (plants, animals, and some fungi species) are perceived by us as dominating all ecosystem landscapes, prokaryotes are considered the most abundant and metabolically and phylogenetically diverse group on Earth (Oren 2004). Until recently our knowledge on prokaryotes was limited by our ability to cultivate bacteria, and previous studies have reported approximately 6200 (Oren 2004) or even 10,000 or up bacterial species (Ward 2002), highlighting that most are not cultivable and cannot be analyzed for accurate classification. Nevertheless, modern technologies for culturing bacteria and archaea, and recent genomic approaches, such as metabarcoding, have allowed the assessment of a hidden biodiversity of prokaryotes, although new species-level descriptions are still limited (Lewis et al. 2021; Wang et al. 2021). Today, about 24,510 prokaryotic species, distributed in 4127 genera and 548 families (Table 2.1), excluding archaea, are recognized (EOL 2024). The microbial biodiversity forms the microbiotas, which are communities of microorganisms that live together and perform distinct physiochemical functions (Whipps et al. 1988).

Microbiotas may colonize a variety of environments, occurring in soil, water, air, and also in organs and tissues of other living beings, including human and other animals and plants. They comprise a huge variety of microorganisms, from prokaryotes (e.g. bacteria) to eukaryotes (e.g. fungi), in addition to virus (virome), and interact with various components of the environment and/or host species (holobiont), resulting in what we known as microbiome. Therefore, microbiomes consist of a bunch of microorganisms that interact among them, and with environment and/or host elements, performing important roles for the balance of ecosystems and species with which they are associated (Berg et al. 2020). In turn, a holobiome is formed by a host species and the many other species living in or around it, while the combined genomes and the gene interactions of all those species is the hologenome (Bordenstein and Theis 2015), which has been assumed to be an evolutionary unit, as it integrates genetic and metabolic relationships, performing relevant functions, so that organisms can better adapt to the environment (Morris 2018; Berg et al. 2020).

The study of microbiomes and hologenomes has enabled the description of important ecological and biological functions, and a better understanding of issues related to the functioning of biodiversity, as an assembly of species that interact with each other in different ways, due to their genomes and genetic diversity (Berg et al. 2020). The recent advances of next generation sequencing approaches has enabled access both micro and macro-biodiversity, through the development of technologies such as metabarcoding, genotyping-by-sequencing (GBS) and whole genome assemblies (WGS), among others. Thus, a significant number of new species and their genetic diversity and biological functions have emerged (e.g. Berg et al. 2020; Mann et al. 2020).

2.2 Biodiversity Values

Biodiversity is high, complex and interconnected, also presenting interdependent relationships with the environment, playing essential roles that allow the ecosystem balance and the existence of life on Earth in a self-sustainable synergy. Prokaryotes, per instance, are relevant for the fixation of nutrients, participating in the nitrogen and carbon cycles, and in the decomposition process. They also have an important role in the digestion system of some large eukaryotes, such as ruminants and humans. As a natural source for biotechnology, studying prokaryotes has allowed numerous advances in the development of medicines and food, and also as agents in decomposition of pollutants, such as crude-oil, petroleum and derivatives (Dusenbery 1996; Whitman et al. 1998; Cohen 2002).

Regarding eukaryotes, several protozoa and also algae are important for maintaining ecological balance (Dunlap 2001). Protists, for instance, are relevant for the global food network and the carbon cycle; therefore, predicting their response to projections of future climate change is essential for modeling ecosystem dynamics in a changing world (Morard et al. 2022). Among these, protozoa are consumers, participating in the process of decomposition of organic matter and in the recycling of substances from the environment, including decomposition of environmental pollutants (Grimstone 1966; Pratt and Cairns 1985). Algae are responsible for producing up to 90% of the planet's oxygen. Furthermore, some species capture carbon, others are food sources for other organisms, including those that live in the aquatic environment. On the other hand, fungi can produce bactericidal components useful for controlling human diseases and have been used in the pharmaceutical industry for producing antibiotics; in addition, several fungi species are edible, such as shitake and shimeji, or are used in the production of food and beverages such as cheese, bread, wine and beer (Grimstone 1966; Pratt and Cairns 1985; CBD 2011; Alho 2008). Therefore, such biodiversity has been exploited to obtain several consumables for human needs, such as food, beverages, medicines, fuels, fertilizers, and cosmetics (CBD 2011; Alho 2008).

The value of all species to the maintenance of others, including humans (e.g. metabolic and physiological functions in immunity, digestion, nutrition, metabolism, etc.), as well as for the ecosystem balance (e.g. production of humus, nutrient cycling, degradation of organic matter and xenobiotics, bioremediation of impacted environments, and others) is undeniable. Nevertheless, the importance of plant and animal biodiversity for the maintenance of ecosystems and life on Earth is enormous and stands out for being unique in many cases, for example, for seed dispersal and pollination performed by specific taxa, including rare and/or threatened species (Hanninen et al. 2010; Antonelli and Sanmartín 2011; Mugnai et al. 2021; Barrere et al. 2024).

Plants and animals are vital components of ecosystems and their functions are often interconnected and co-dependent (Antonelli and Sanmartín 2011; Jordano 2016; Barrere et al. 2024). Plants are primary producers (autotrophic), that is, they are able to produce their own food through photosynthesis. This process converts sunlight, carbon dioxide and water into organic compounds (e.g. glucose), providing the basis for the trophic chain. Without plants, most of the living species would have no food and/or energy sources available. On the other hand, several animals are primarily consumers (heterotrophic), that is, they rely on plants and/or other animals for energy and nutrition. They perform relevant functions in the food trophic chain, breaking down organic matter and recycling nutrients (Hanninen et al. 2010; Jordano 2016). Therefore, fauna and flora diversity interact and form complex food webs (producers, consumers and decomposers), participating in nutrient cycling (Fig. 2.2).

Plants and animals are also important bioindicators for ecosystem health (Mugnai et al. 2021). Animal species play unique roles in nature, such as seed dispersal (e.g. primate, birds and bats) and pollination (e.g. bees), contributing to the balance of the ecosystems. Herbivory, for example, impairs some plants from dominating environments; conversely, pollination and seed dispersal promote flowering and growth/

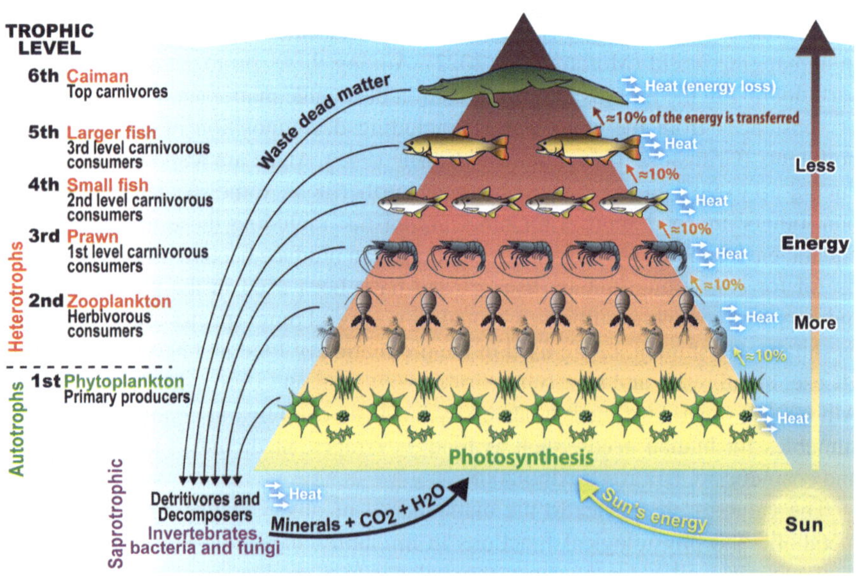

Fig. 2.2 A representation of a freshwater trophic web, indicating trophic levels (producers, consumers, decomposers), energy flow and major nutrient cycle

germination, assisting in the maintenance of plant diversity. Disbalances on ecological relationships of plants and animals may negatively impact populations, changing communities and the environment as a whole (e.g., Hanninen et al. 2010; Antonelli and Sanmartín 2011; Barrere et al. 2024).

Overall, from an anthropogenic perspective, the value of biodiversity can be summarized into some main interrelated categories: environmental-ecosystem, producer-consumer, social-economic, aesthetic-cultural, ethical-moral (Moran and Bann 2000; Alho 2008; CBD 2011). Certainly, such categorization arises mostly from human impressions and needs, however, they also denote the relevance of all biodiversity for itself, since the complex interdependence of species is at least represented by the environmental-ecosystem and consumer-producer values.

In the last centuries, several plants and animals have suffered the negative impact of human exploitation. Natural resources (soil, rocks, minerals/ores, water, etc.), in addition to genetic resources (all species and their populations varieties) have been massively overexploited (CBD 2011). In aquatic ecosystems, for example, overexploitation of natural stocks of fishing resources (e.g. fishes, shrimps, crabs, wales, lobsters, oysters), for food, but also for obtaining sub-products like oil and cosmetics, is negatively impacting aquatic communities and ecosystems. In terrestrial ecosystems, forests have been devastated to expand urban centers, and agricultural practices such as cattle grazing, and cultivation of large areas with monocultures, wood trade, and the construction of roads and dams, among others, threatening species and increasing the risk of extinction of many plants and animals (Eckert et al. 2023; Barrere et al. 2024). Thus, one of the greatest challenges of modern human societies has been to reconcile their activities and the sustainable use of natural resources, including biodiversity.

For the purpose of biological diversity conservation, in a broader sense, people need awareness that we are an integral part of it and must live in harmony with nature. In the book entitled "Ideas to postpone the end of the world", Ailton Krenak, a Brazilian environmentalist and philosopher, devises important reflections on the current myth of "sustainability", in which forests, mountains and rivers, and their associated biodiversity have been devoured. According to Krenak, the idea of humanity being dissociated from the Earth, also an organism by his definition, of which humans are just a component, denies the plurality of forms of life, existence and habits, leaving aside the value of the traditional human societies (Krenak 2020).

In addition to the undeniable value of fauna and flora for the existence of life on Earth, from now on we bring to light one of the main reasons why the scientific community has been paying special attention to plant and animal biodiversity: the current crisis of biodiversity (Eckert et al. 2023). It is also for this reason that we dedicate a key topic in this chapter to address such relevant issues and how population genetics can help us to understand biodiversity and explore it in a sustainable way, promoting the maintenance of its evolutionary potential and its conservation (Barrere et al. 2024). For now, let's focus on the distribution patterns of plant and animal species across the world and the global importance of the neotropical region, namely tropical America, considered the region with the highest species richness on Earth (Antonelli and Sanmartín 2011).

2.3 Biodiversity Centers and the Neotropics

Faced with a catastrophic scenario of unsustainable use of natural resources, the concept of 'biodiversity hotspots' emerged at the end of 1990s, as a strategy to protect areas with high rates of habitat loss and exceptional richness of plant species, mainly endemic ones (Myers 1988). First coined in 1988 by Norman Myers, a British environmentalist, the term "hotspot", in the biological context, was established to denote a biogeographic region that is home to a high diversity of endemic species (i.e. species that occur just in a strict geographic area) of vascular plants (Pteridophytes, Gymnosperms and Angiosperms), which may be threatened due to a significative habitat loss (e.g. at least ¾ of vegetation cover), mainly due to deforestation (Myers et al. 2000). Ten biodiversity hotspots were first established in tropical forest regions worldwide (Myers 1988). Later, 25 areas were identified as hotspots (Myers et al. 2000), and currently this number has expanded to 36 regions (Fig. 2.3) across the globe (Conservation International 2024).

Despite the initial concept of biodiversity hotspots being restricted to define areas of high diversity and endemism of plant species, subsequent studies have taken into account animal diversity of vertebrate (e.g. Myers et al. 2000; Jenkins et al. 2013) and invertebrate (e.g. Mumladze et al. 2020) species. In this sense, biogeography and population genetics stand out as relevant disciplines to study the geographic distribution patterns of species and their populations, in addition to the factors that determine such patterns, increasing knowledge about how plant and animal diversity is spread worldwide (King et al. 2006; Landguth and Cushman 2010).

Indeed, biodiversity is not homogeneously distributed on the planet, but is concentrated in specific geographical areas, that present particular geo-climate features

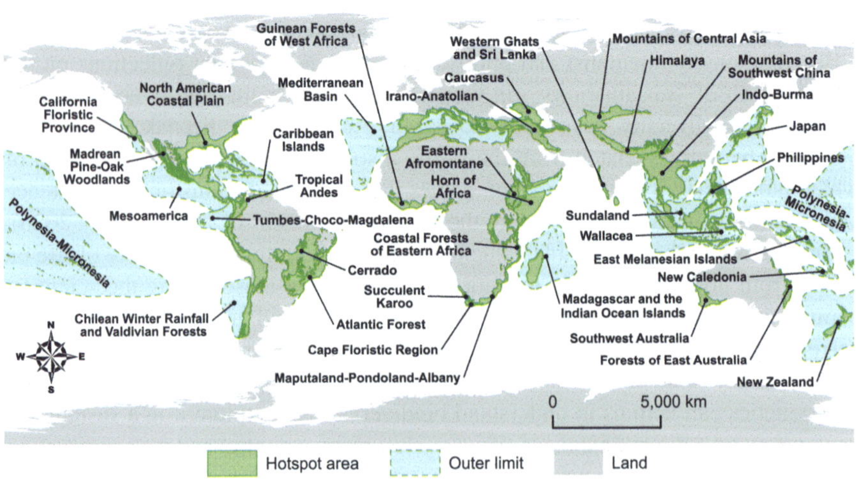

Fig. 2.3 Biodiversity hotspots worldwide, defining regions where biodiversity conservation is most urgent because of high levels of endemism and human threat. (Adapted from Hoffman et al. 2016)

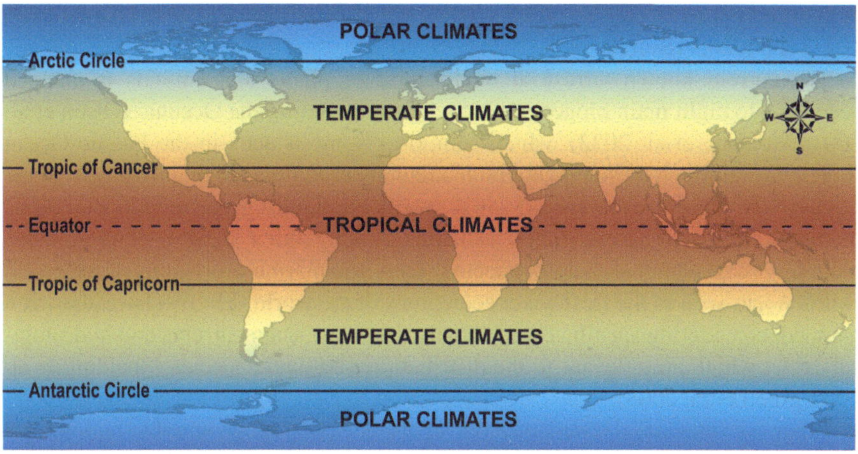

Fig. 2.4 World climate zones. (Adapted from Natural Earth, a public domain map dataset (https://www.naturalearthdata.com))

(e.g., type of soil, temperature, solar incidence, humidity, among others) that contribute to species living there or not (Myers et al. 2000; Mittermeier et al. 2011). Due to the curvature and inclination of the Earth on its axis in relation to the Sun, different regions of the planet receive different amounts of solar irradiation throughout the year, influencing the climate and hydrological cycle, and defining five major climate zones (Fig. 2.4), according to the Köppen climate classification, the most widely used geographical and climatological system across the world (Alvares et al. 2013; Beck et al. 2018).

Tropical zones are undoubtedly those with the greatest biodiversity on the planet, especially the tropical forests. Approximately 44% of all vascular plant species and 35% of all species from four vertebrate groups are confined to 25 hotspots, comprising just 1.4% of Earth's land surface (Myers et al. 2000; Brooks et al. 2006). Climate conditions (drier, wetter, colder and/or warmer), combined with hydrological and geological features, are primarily responsible for the biodiversity distribution across the globe. However, the dispersal capacity of species, along with vicariant events (e.g. geographic barriers), that may limit dispersion and gene flow, are also responsible for the biogeographical patterns found over time (Antonelli and Sanmartín 2011).

One of the first biogeographic maps for animal species was proposed by Phillip Lutley Sclater, an English ornithologist, who presented the bird diversity distribution pattern (Sclater 1858). Alfred Russel Wallace, considered the father of biogeography, expanded the study to other vertebrate species (Wallace 1876). The geographic distribution of plant species had already been mapped in 1805, by Lamarck and Candolle, who depicted floristic provinces in France (Ebach and Goujet 2006). However, it was zoo-biogeography that actually boosted this important field of study during the second half of the nineteenth century, increasing knowledge about biodiversity and its distribution patterns (Sclater 1858; Wallace 1876). Since then, records for biogeographic maps have increased, especially for vertebrates (Procheş and Ramdhani 2012), but also for invertebrates (Mugnai et al. 2021) and plants (Biurrun et al. 2021).

The current known terrestrial biodiversity of plants and animals is mapped according to eight main biogeographic regions, including the Oceanic (Olson et al. 2001; Morrone et al. 2022), which comprises some islands in the Pacific Ocean (Fig. 2.5). Although these biogeographic regions or realms are mainly based on the biodiversity of terrestrial species, it is worth highlighting that continental areas include systems of rivers, lakes, estuaries, among other aquatic environments, including seas, which are interconnected. Thus, the marine ecoregions of the world (MEOW), as well as the freshwater ecoregions of the world (FEOW) have also been mapped, considering specific characteristics, such as ocean currents and hydrological cycle, which are relevant to the distribution of aquatic species (Spalding et al. 2007; Abell et al. 2008).

When we integrate datasets of biodiversity hotspots, biogeographic regions and climatic zones, we realize that the species number strongly declines from the tropics towards the temperate and polar zone north and south of the equator, indicating that, though biodiversity is abundant in regions rich in orographic structure or high in geo-diversity regions, such as mountains and coastal areas, they are mainly

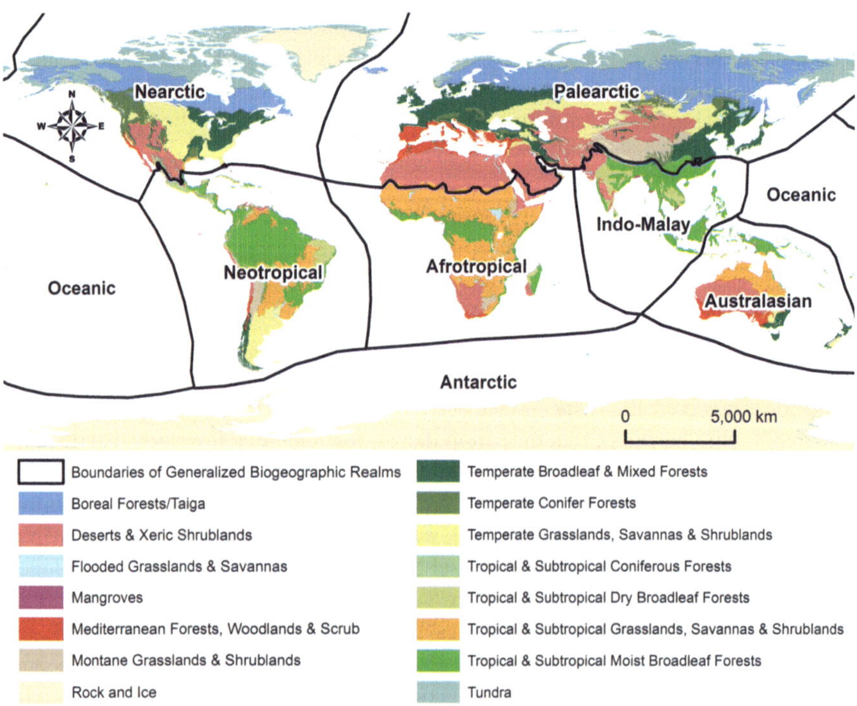

Fig. 2.5 A global map showing the eight biogeographical realms of the world, obtained from Olson et al. (2001), highlighting the fourteen terrestrial biomes as obtained from RESOLVE Ecoregions dataset (Dinerstein et al. 2017)

associated with warm and humid tropical conditions, pointing to the global importance of these climatic/ecological zones (Forseth 2010; Olson et al. 2001), that comprise the most species-rich biomes (savanna and rain forests). Although there are different prepositions for classifying biomes (e.g. Whittaker 1962, 1970), in summary, we can point to fourteen terrestrial biomes (Dinerstein et al. 2017) distributed throughout the main climatic-ecological zones (Fig. 2.5).

The tropical forests present the highest biodiversity worldwide. This feature is due to the combination of a set of particular environmental conditions, including temperature, rainfall, soil nature and topography and terrain ruggedness (Liang et al. 2022; Sabatini et al. 2022). Whilst the most biodiverse temperate forest shows up 50 tree species per hectare, a tropical forest can present four times more species per hectare. On the other hand, Central European forests have even less species diversity when compared to similar latitudes in North and South America and Asia (Global Forest Resources Assessment 2005; Leuschner and Ellenberg 2017; Liang et al. 2022; Sabatini et al. 2022). Tropical rainforests occur closer to the Equator, where the days are long and the light lasts around 12 h, precipitation is greater than in any other region and temperatures are high, averaging 20–25 °C, with little seasonal variation (Beck et al. 2005; Forseth 2010; Alvares et al. 2013; CCKP 2024). The most biodiversified tropical forests are found in the Neotropics (Olson et al. 2001).

The neotropical region extends from central Mexico to southern Brazil, including Central America, the Caribbean islands (Antilles) and almost all of South America. It stands out for having the largest number of plants and animals in the world, including some of the current biodiversity hotspots (Fig. 2.3) recognized to date (Morrone 2014; Morrone et al. 2022). According to the Brazilian Ministry of Environment (MMA), Brazil is the neotropical country with the greatest biodiversity worldwide, with more than 116,000 and 46,000 animal and plant species, respectively. This huge biodiversity is spread across its entire territory that occupies almost half of South America and has both the most diverse biomes in the world: tropical forests and savannas (MMA 2024).

Savannas are dominated by a mixture of grasses and small trees, generally deciduous, and represent a transition from tropical forests to deserts. They are located to the north and south of tropical forests, and they are in regions with lower annual precipitation and consequently longer dry seasons. About 60% and 23% of Africa and Australia, respectively, are covered by various types of savannas, which can be distinguished by the relative abundance of trees and grass (Neldner et al. 2001; Forseth 2010). Actually, there are different types of savannas, such as temperate, mediterranean and mountainous, however, the savanna with the greatest biodiversity is within the Neotropics. The Cerrado, known as Brazilian Savanna, is considered the most species-rich savanna in the world, and a biodiversity hotspot (Fig. 2.3). Cerrado extends mainly in Brazil, slightly on the border of Bolivia and Paraguay, and is home to 11,000 plant species, 837 birds, 120 reptiles, 150 amphibians, 1200 fishes, 90,000 insects and 199 mammals; comprising 5% of the total species on Earth, of which 32% are endemic to this biome (WWF 2015).

Despite the high biodiversity and endemism of savannas, indubitably, the most biodiverse biomes worldwide are the tropical rainforests, which are estimated to contain more than half of Earth's terrestrial species. From around 40–75% of the whole biodiversity are natives to tropical forests, which harbors approximately 50% of the known species of animals and plants, and 70% of flowering plants. In addition to having the greatest biodiversity, tropical forests present the highest primary productivity among terrestrial biomes, being both of the aspects possibly related to the combination of environmental and climate conditions, such as temperature, rainfall, soil composition and nature, terrain ruggedness, among others (Forseth 2010; Antonelli and Sanmartín 2011; Liang et al. 2022; Sabatini et al. 2022). In turn, the most species-rich tropical forests are in the Neotropics. In addition to harboring the largest number of plants and animals, the neotropical rainforests also present a high rate of endemic species.

The largest rainforest in the world is the Amazon forest, representing over 50% of Earth's rainforests, covering most of the Amazon basin, including nine countries of South America, of which about 60% of the Amazon forest is in Brazil, 13% in Peru, 10% in Colombia, and the remaining ones in Bolivia, Ecuador, French Guiana, Guyana, Suriname, and Venezuela. The Amazon rainforest also shows the greatest availability of freshwater in the world. This enormous rainforest houses approximately 10% of the whole biodiversity known to date, including endemic and endangered plant and animal species. This biome is home to about 20,000–50,000 tree species, reaching up to more than 200 tree species per hectare. Regarding animal species there are about 300 species of mammals, 370 reptiles, 420 amphibians, 1300 birds, and 2400 fish, in addition to 1800 butterflies and 3000 species of bees (Liang et al. 2022; Sabatini et al. 2022; MMA 2024). Likewise, the Atlantic forest is home to a huge biodiversity, being considered one of the richest forests in the world in terms of tree diversity, with more than 20,000 plant species, of which approximately 40% are endemic (WWF 2024).

The Atlantic rainforest extends from northeast to south of Brazil along the Atlantic coastline, and inland into northeast Argentina and eastern Paraguay border, and also harbors a great variability of animal species, with more than 2000 vertebrates (30% endemic), of which about 1023 are birds (20% endemic), 475 are amphibians (60% endemic), 306 are reptiles (30% endemic), and 300 are mammals (30% endemic), including 24 species of primates (80% endemic). This rainforest is among the world's top global biodiversity hotspots (Fig. 2.3), housing one in 14 of the Earth's total plant species and one of 20 vertebrate species (WWF 2024). Despite this magnitude, since the initial period of colonization of Brazil, at the beginning of the sixteenth century, to the present day, the Atlantic Forest has been reduced to only about 12% of its original coverage (WWF 2024), this tropical forest is considered one of the most threatened biomes in the world due to the high rate of deforestation (Ribeiro et al. 2009, 2011).

Indeed, an expressive number of animal and plant species have been dramatically affected by human activities, and the biological diversity loss is one of the biggest concerns in the Anthropocene (Dirzo et al. 2014; Pievani 2014; Malhi 2017). The global growing rate of species extinction is currently estimated to be between 1000 and 10,000 times higher than the natural rate, evidencing a true crisis of biodiversity, in which the loss of species and populations can lead to ecosystems collapse and ecological processes (Dirzo et al. 2014; Turvey and Crees 2019; Eckert et al. 2023). Nevertheless, because of the highest diversity in the Neotropics, many species and their populations are still poorly studied.

2.4 Biodiversity Crisis and Population Genetics

Life on Earth probably arose around 3.7 billion years ago, that is, approximately 0.8 billion years after the planet's emergence. Since then, different forms of life have emerged and evolved, but have also become extinct, mainly due to extreme phenomena, such as asteroid fall, continent movements, glaciations and severe warming events, which promoted the five great mass extinctions and the loss of most species that existed in previous geological periods from the Paleozoic to the Cretaceous eras (Fig. 2.6) (Phillips 1860; Alvarez et al. 1980; Turvey and Crees 2019).

In addition to eras and periods, geological time is also divided into epochs. The current Cenozoic era presents three periods (Paleogene and Neogene—anteriorly named as Tertiary -, and Quaternary) and seven main epochs (Paleocene, Eocene, Oligocene, Miocene, Pliocene, Pleistocene and Holocene). Both Pleistocene and Holocene are epochs from the current Quaternary period (Fig. 2.6). The Pleistocene is the longest epoch (2.5 million) and the Holocene is the shorter one (11,700 years) to date (Phillips 1860; Cohen et al. 2013). One of the most known mass extinctions occurred 65 million years ago, during the Cretaceous-Paleogene, when dinosaurs

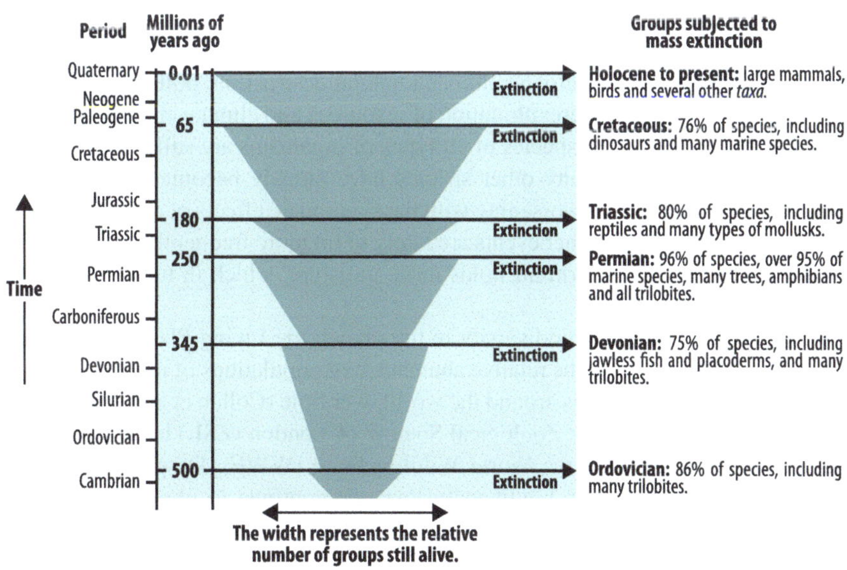

Fig. 2.6 Mass extinctions timeline, including the sixth mass extinction (Holocene to present) which characterize the Anthropocene

and many marine species disappeared from Earth. However, during the Quaternary, large mammals (e.g. mammoths, mastodons, and the saber-toothed tiger), from the megafauna (animals weighing more than 44 kg), also became extinct due to longer periods of glaciations (50,000 years) intercalated with shorter warming periods, that occurred 650,000 years ago, at the late Pleistocene (Fig. 2.6), period also known as The Great Ice Age or Quaternary Glaciation (Phillips 1860; Ehlers and Gibbard 2011; Turvey and Crees 2019).

From the late Pleistocene to the beginning of the Holocene, several species extinctions have occurred, especially from the megafauna. Such extinctions resulted in a decrease in fauna density and diversity worldwide and are different from previous ones due to the lack of ecological succession to replace the extinct species and the change in the regime of faunal relationships and previously established habitats as a consequence (Stuart 1999; Sandom et al. 2014; Pires 2024). After the Quaternary Glaciation, several species continue to become extinct. However, no extreme natural events have occurred to explain the contemporary extinctions other than human interference (Malhi 2017; Turvey and Crees 2019; Spalding and Hull 2021).

In the early 2000s, the term "Anthropocene" was proposed by Paul J. Crutzen, a Dutch scientist Nobel laureate, to designate a new geological epoch characterized by environmental and climatic changes caused mainly by anthropogenic activities (Crutzen 2002). According to Crutzen, "the Anthropocene could be said to have started in the late 18th century, when analyses of air trapped in polar ice showed the beginning of growing global concentrations of carbon dioxide and methane". Although this term is controversial, it has become popular and has been widely used in various contexts, including to explain the contemporary biodiversity crisis, considered the sixth ongoing extinction (Crutzen 2002; Dirzo et al. 2014; Spalding and Hull 2021; Eckert et al. 2023).

Indeed, it is common sense that the Earth is experiencing a true biodiversity loss. Many factors causing this biodiversity crisis are directly or indirectly related to human activities, such as deforestation and habitat destruction, pollution, introduction of invasive species, overexploitation of resources and climate change (Spalding and Hull 2021). Countless species of all types of organisms are suffering increased population decline and many other species have already become extinct (IUCN 2024), and this situation is worsening with the synergistic effects of ongoing climate changes. The extreme weather events are likely to hit more frequently as the climate crisis goes on, because warm air holds more moisture, which in turn fuels storms (Barrere et al. 2024).

To evaluate the loss of biodiversity in the planet, the Living Planet Index (LPI) was proposed for tracking the relative abundance of populations of mammals, birds, fish, reptiles and amphibians around the world over time (Collen et al. 2009; McRae et al. 2017). Since 2016 the Zoological Society of London (ZSL) has managed the index in cooperation with the World Wildlife Fund (WWF). Despite 30 years of policy interventions to stop biodiversity loss, we continue to observe population declines. The 2022 global LPI shows a 69% average decline in populations monitored between 1970 and 2018 (range: −63% to −75%).

Changes in biodiversity loss vary in different parts of the world. While LPI in North America showed a decline equivalent to 20% between 1970 and 2018, this index declined 18% in Europe and Central Asia, 55% in Asia and the Pacific, 66% in Africa and 94% in Latin America and the Caribbean (WWF 2022). Precisely, these estimates are showing that the most biodiverse regions of the world are also losing their species in greater numbers. Additionally, if we consider the diversity of populations adapted to very particular conditions of neotropical biomes and hotspots, such as Cerrado and Atlantic forest, the loss of biodiversity is even greater. In this sense, population genetics may contribute to assess the population diversity, which is diversifying at slower paces in contrast to rapid environmental and climate changes caused by human activity and unsustainable use of natural resources.

Populations adapted to very particular conditions are being locally extinct, depleting species diversity and increasing the risk of species extinction. Although not all genetic diversity promotes an adaptive advantage, it is well accepted that by assessing the frequency of alleles in different populations and changes associated with evolutionary factors, such as mutation, genetic drift, natural selection, and gene flow, population genetics is as powerful tool to inferring the synergistic effects of human activities and climate changes on the long persistence of species, especially if small and threatened populations are a target.

2.5 Final Remarks

Biodiversity is critical for the range of services it provides to ecosystems and human well-being. Biodiversity owns the genetic resources which represent the basis of the biodiversity pyramid (genetic diversity), that which ensures the diversification and adaptation of the species that interact and evolve in the ecosystems. Genetic resources are responsible for the biological basis for both global environmental health and human needs, such as food security. The increased human interference in the ecosystems has promoted a true biodiversity crisis which has been accentuated by the synergistic effects of climate changes.

The Neotropics harbor one of the highest biodiversity areas worldwide, but also several hotspots where the loss of biodiversity is quite prominent, resulting in increased risk of species extinction and ecosystem collapse. For all we know, it is unquestionably important to preserve biodiversity in terms of ecosystem and species diversity and genetic diversity, especially in light of climate change predictions and increasing global warming. Therefore, preserving maximum genetic diversity means ensuring biodiversity and genetic resources for the future of our planet. In this sense, population genetic studies can provide essential information to saving species at extinction risk, mainly in geographic regions with high biodiversity. Moreover, population genetics may also diagnose the effects of increased environmental modifications promoted by humans in the last centuries, as is the case of the Neotropics.

Key Points

- Biodiversity comprises genetic variation, species and ecosystems diversity.
- Biodiversity has a number of direct and indirect values.
- Hotspots of biodiversity are concerns of conservation.
- The Neotropics own a huge biodiversity and part is still unknown by science.
- Population genetics is a powerful tool for identifying hidden biodiversity and assessing population response to anthropogenic habitat modification.

Questions

1. What are the differences between species diversity and biodiversity?
2. What are the values of biodiversity?
3. What is hidden biodiversity and how is genetics useful for its identification?
4. How can population genetics assist in protecting species under risk of extinction?

Acknowledgments PDF and PMGJ are thankful to the Brazilian agencies Conselho Nacional de Desenvolvimento Científico e Tecnológico (CNPq, 317345/2021-4, 303524/2019-7) and Fundação de Amparo à Pesquisa do Estado de São Paulo (2017/23548-2, 2021/06668-0, 2021/10639-5).

References

Abell RM, Thieme ML, Revenga C et al (2008) Freshwater ecoregions of the world: a new map of biogeographic units for freshwater biodiversity conservation. Bioscience 58:403–414. https://doi.org/10.1641/B580507

Aldhebiani AY (2018) Species concept and speciation. Saudi J Biol Sci 25(3):437–440. https://doi.org/10.1016/j.sjbs.2017.04.013

Alho CJR (2008) The value of biodiversity. Braz. J Biol 68(4, Suppl):1115–1118. https://doi.org/10.1590/s1519-69842008000500018

Alvares CA, Stape JL, Sentelhas PC et al (2013) Köppen's climate classification map for Brazil. Meteorol Z 22(6):711–728. https://doi.org/10.1127/0941-2948/2013/0507

Alvarez LW, Alvarez W, Asaro F et al (1980) Extraterrestrial cause for the cretaceous-tertiary extinction – experimental results and theoretical interpretation. Science 208:1095–1108. https://doi.org/10.1126/science.208.4448.109

Antonelli A, Sanmartín I (2011) Why are there so many plant species in the Neotropics? Taxon 60:403–414

Antonelli A, Fry C, Smith RJ et al (2023) State of the world's plants and fungi. Royal Botanic Gardens, Kew, p 92

Barrere J, Reineking B, Jaunatre M et al (2024) Forest storm resilience depends on the interplay between functional composition and climate Insights from European-scale simulations. Funct Ecol 38(3):500–516. https://doi.org/10.1002/tax.602010

Beck C, Grieser J, Rudolf B (2005) A new monthly precipitation climatology for the global land areas for the period 1951 to 2000, Climate status report. German Weather Service, Offenbach

Beck H, Zimmermann N, McVicar T et al (2018) Present and future Köppen-Geiger climate classification maps at 1-km resolution. Sci Data 5:180214. https://doi.org/10.1038/sdata.2018.214

Berg G, Rybakova D, Fischer D et al (2020) Microbiome definition re-visited: old concepts and new challenges. Microbiome 8:103. https://doi.org/10.1186/s40168-020-00875-0

Biurrun I, Pielech R, Dembicz I et al (2021) Benchmarking plant diversity of Palaearctic grasslands and other open habitats. J Veg Sci 32:e13050. https://doi.org/10.1111/jvs.13050

Bordenstein SR, Theis KR (2015) Host biology in light of the microbiome: ten principles of holobionts and hologenomes. PLoS Biol 13(8):e1002226. https://doi.org/10.1371/journal.pbio.1002226
Britannica (2024) https://www.britannica.com
Brooks TM, Mittermeier RA, da Fonseca GA et al (2006) Global biodiversity conservation priorities. Science 313(5783):58–61. https://doi.org/10.1126/science.1127609
CBD (2011) Convention on biological diversity, secretariat of the convention on biological diversity. United Nations Environment Programme, Montreal. https://www.cbd.int
CCKP (2024) Climate Change Knowledge Portal. https://climateknowledgeportal.worldbank.org
Christenhusz MJM, Byng JW (2016) The number of known plants species in the world and its annual increase. Phytotaxa 261(3):201–217. https://doi.org/10.11646/phytotaxa.261.3.1
Cohen Y (2002) Bioremediation of oil by marine microbial mats. Int Microbiol 5(4):189–193. https://doi.org/10.1007/s10123-002-0089-5
Cohen KM, Finney SC, Gibbard PL (2013) The ICS international chronostratigraphic chart. Episodes 36(3):199–204
Collen B, Loh J, Whitmee S et al (2009) Monitoring change in vertebrate abundance: the living planet index. Conserv Biol 23(2):317–327. https://doi.org/10.1111/j.1523-1739.2008.01117.x
Conservation International (2024) https://www.conservation.org/priorities/biodiversity-hotspots
Crutzen P (2002) Geology of mankind. Nature 415(23). https://doi.org/10.1038/415023a
Dinerstein E, Olson D, Joshi A et al (2017) An ecoregion-based approach to protecting half the terrestrial realm. Bioscience 67(6):534–545. https://doi.org/10.1093/biosci/bix014
Dirzo R, Young HS, Galetti M et al (2014) Defaunation in the Anthropocene. Science 345:401–406. https://doi.org/10.1126/science.125181
Dunlap PV (2001) Microbial diversity. In: Simon AL (ed) Encyclopedia of biodiversity, 2nd edn. Academic, pp 280–291
Dusenbery DB (1996) Life on a small scale: the behavior of microbes. Scientific American Library, WH Freeman and Company, New York, p 214
Ebach MC, Goujet DF (2006) The first biogeographical map. J Biogeogr 33:761–769. https://doi.org/10.1111/j.1365-2699.2006.01477.x
Eckert I, Brown A, Caron D et al (2023) 30×30 biodiversity gains rely on national coordination. Nat Commun 1(7750). https://doi.org/10.1038/s41467-023-42737-x
Ehlers J, Gibbard P (2011) Quaternary glaciation. In: Singh VP, Singh P, Haritashya UK (eds) Encyclopedia of snow, ice and glaciers, Encyclopedia of earth sciences series. Springer, Dordrecht, pp 873–882
EOL (2024) Encyclopedia of life, Eukaryotes. https://eol.org/pages/2908256
Forseth I (2010) Terrestrial biomes. Nat Educ Knowl 3(10):11
Global Forest Resources Assessment (2005) Biological diversity chapter. FAO. Food and Agriculture Organization. https://www.fao.org/3/A0400E/A0400E04.pdf
Grimstone AV (1966) Structure and function in protozoa. Ann Rev Microbiol 20(1):131–150. https://doi.org/10.1146/annurev.mi.20.100166.001023
Hanninen OOP, Mustafa A, Mansourian BP et al (2010) Medical and health sciences. Encyclopedia of biological, physiological and health sciences in the global Encyclopedia of Life Support Systems (EOLSS) 468 pp.
Hoffman M, Koenig K, Bunting G et al (2016) Biodiversity Hotspots (version 2016.1) (2016.1) [Data set]. Zenodo. https://doi.org/10.5281/zenodo.3261807
IUCN (2024) International Union for Conservation of Nature. https://www.iucnredlist.org
Jenkins CN, Pimm SL, Joppa LN (2013) Global patterns of terrestrial vertebrate diversity and conservation. Proc Natl Acad Sci 110:e2602–e2610
Jordano P (2016) Sampling networks of ecological interactions. Funct Ecol 30:1883–1893. https://doi.org/10.1111/1365-2435.12763
Katz LA (2012) Origin and diversification of Eukaryotes. Ann Rev Microbiol 66(1):411–427. https://doi.org/10.1146/annurev-micro-090110-102808
King RC, Stansfield WD, Mulligan PK (2006) A dictionary of genetics. Oxford University Press, Oxford

Krenak A (2020) Ideias para adiar o fim do mundo, segunda edição. Companhia das Letras, São Paulo

La Scola B, Audic S, Robert C et al (2003) A giant virus in amoebae. Science 299(5615):2033. https://doi.org/10.1126/science.108186

Lai CC, Shih TP, Ko WC et al (2020) Severe acute respiratory syndrome coronavirus 2 (SARS-CoV-2) and coronavirus disease-2019 (COVID-19): the epidemic and the challenges. Int J Antimicrob Agents 55(3):105924. https://doi.org/10.1016/j.ijantimicag.2020.105924

Landguth EL, Cushman SA (2010) CDPOP: a spatially explicit cost distance population genetics program. Mol Ecol Resour 10:156–161. https://doi.org/10.1111/j.1755-0998.2009.02719.x

Leuschner C, Ellenberg H (2017) Ecology of central European forests. In: Vegetation ecology of Central Europe, volume 1. Springer Nature, 972 p

Levin S (1998) Ecosystems and the biosphere as complex adaptive systems. Ecosystems 1:431–436. https://doi.org/10.1007/s100219900037

Lewis WH, Tahon G, Geesink P et al (2021) Innovations to culturing the uncultured microbial majority. Nat Rev Microbiol 19:225–240. https://doi.org/10.1038/s41579-020-00458-8

Liang J, Gamarra JGP, Picard N et al (2022) Co-limitation towards lower latitudes shapes global forest diversity gradients. Nat Ecol Evol 6:1423–1437. https://doi.org/10.1038/s41559-022-01831-x

Lopes U, Galetti PM Jr, Freitas PD (2020) Hidden diversity in Prochilodus nigricans: a new genetic lineage within the Tapajós River basin. PLoS One 15:e0237916. https://doi.org/10.1371/journal.pone.0237916

Malhi Y (2017) The concept of the Anthropocene. Annu Rev Environ Resour 42(1):77–104. https://doi.org/10.1146/annurev-environ-102016-060854

Mann AE, Mazel F, Lemay MA et al (2020) Biodiversity of protists and nematodes in the wild nonhuman primate gut. ISME J 14:609–622. https://doi.org/10.1038/s41396-019-0551-4

Mayr E (1970) Populations, species and evolution. Harvard University Press, Cambridge, MA, 485 p

McRae L, Deinet S, Freeman R (2017) The diversity-weighted living planet index: controlling for taxonomic bias in a global biodiversity indicator. PLoS One 12(1):e0169156. https://doi.org/10.1371/journal.pone.0169156

Milanese A, Mende DR, Paoli L et al (2019) Microbial abundance, activity and population genomic profiling with mOTUs2. Nat Commun 10(1014). https://doi.org/10.1038/s41467-019-08844-4

MMA (2024) Ministry of environmental. Federal Government, Distrito Federal. https://www.gov.br/mma/pt-br/assuntos/biodiversidade

Moran D, Bann C (2000) The valuation of biological diversity for National Biodiversity Action Plans and strategies: a guide for trainers. United Nations Environment Program (UNEP)

Morard R, Weinkauf MF, Brombacher A et al (2022) Protists as model ecological and evolutionary study systems: emerging methodologies of the 2st century. Front Mar Sci 9:1014238. https://doi.org/10.3389/fmars.2022.1014238

Morris JJ (2018) What is the hologenome concept of evolution? F1000Research 19(7). https://doi.org/10.12688/f1000research.14385.1

Morrone JJ (2014) Biogeographical regionalisation of the Neotropical region. Zootaxa 3782:1–110. https://doi.org/10.11646/zootaxa.3782.1.1

Morrone JJ, Escalante T, Rodríguez-Tapia G et al (2022) Biogeographic regionalization of the Neotropical region: new map and shape file. An Braz Acad Sci 94(1):e20211167. https://doi.org/10.1590/0001-3765202220211167

Mugnai F, Meglécz E, Abbiati M et al (2021) Are well-studied marine biodiversity hotspots still blackspots for animal barcoding? Glob Ecol Conserv 32:e01909. https://doi.org/10.1016/j.gecco.2021.e01909

Mumladze L, Japoshvili B, Anderson EP (2020) Faunal biodiversity research in the republic of Georgia: a short review of trends, gaps, and needs in the Caucasus biodiversity hotspot. Biologia 75:1385–1397. https://doi.org/10.2478/s11756-019-00398-6

Myers N (1988) Threatened biotas: hot spots in tropical forests. Environmentalist 8:187–208. https://doi.org/10.1007/BF02240252

Myers N, Mittermeier R, Mittermeier C et al (2000) Biodiversity hotspots for conservation priorities. Nature 403:853–858. https://doi.org/10.1038/35002501

Neldner VJ, Fox ID, Wilson GW et al (2001) The vegetation of the Australian tropical savannas. Conference: standalone publication. In: The Brisbane volume, Queensland Herbarium Publication, Australia

O'Neill R, Snowdon R, Köhler W (2003) Population genetics: aspects of biodiversity. In: Esser K, Lüttge U, Beyschlag W, Hellwig F (eds) Progress in botany, vol 64. Springer, Berlin/Heidelberg. https://doi.org/10.1007/978-3-642-55819-1_7

Olson DM, Dinerstein E, Wikramanayake ED et al (2001) Terrestrial ecoregions of the world: a new map of life on Earth. Bioscience 51(11):933–938. https://doi.org/10.1641/0006-3568(2001)051[0933:TEOTWA]2.0.CO;2

Oren A (2004) Prokaryote diversity and taxonomy: current status and future challenges. Philos Trans R Soc Lond Ser B Biol Sci 359(1444):623–638. https://doi.org/10.1098/rstb.2003.1458

Padial JM, Miralles A, De la Riva I et al (2010) The integrative future of taxonomy. Front Zool 7(16). https://doi.org/10.1186/1742-9994-7-16

Pawlowski J, Audic S, Adl S et al (2012) CBOL Protist working group: barcoding eukaryotic richness beyond the animal, plant, and fungal kingdoms. PLoS Biol 10(11):e1001419. https://doi.org/10.1371/journal.pbio.1001419

Phillips J (1860) Life on earth: its origin and succession. Macmillan and Company/Oxford University, 224 p

Pievani T (2014) The sixth mass extinction: Anthropocene and the human impact on biodiversity. Rend Fis Acc Lincei 25:85–93

Pires MM (2024) The restructuring of ecological networks by the Pleistocene Extinction. Annu Rev Earth Planet Sci 52(1):133–158. https://doi.org/10.1146/annurev-earth-040722-104845

Pires AA, Ramirez JL, Galetti PM Jr et al (2017) Molecular analysis reveals hidden diversity in Zungaro (Siluriformes: Pimelodidae): a genus of giant South American catfish. Genetica 1:1–5. https://doi.org/10.1007/s10709-017-9968-8

Ponge JF (2021) Communities, ecosystem engineers, and functional domains. Ecol Res 36(5):766–777. https://doi.org/10.1111/1440-1703.12247

Pratt JR, Cairns JJ (1985) Functional groups in the Protozoa: roles in differing ecosystems. J Protozool 32:415–423. https://doi.org/10.1111/j.1550-7408.1985.tb04037.x

Procheş S, Ramdhani S (2012) The world's zoogeographical regions confirmed by cross-taxon analyses. Bioscience 62(3):260–270. https://doi.org/10.1525/bio.2012.62.3.7

Queiroz K (2005) Ernst Mayr and the modern concept of species. PNAS 25(102):6600 6607. https://doi.org/10.1073/pnas.0502030102

Ramirez JL, Simbini L, Marques CG et al (2021) DNA barcoding of Penaeidae (Decapoda; Crustacea): non-distance-based species delimitation of the most economically important shrimp family. Diversity 13:460

Ribeiro MC, Metzger JP, Martensen AC et al (2009) The Brazilian Atlantic Forest: How much is left, and how is the remaining forest distributed? Implications for conservation. Biol Conserv 142(6):1141–1153. https://doi.org/10.1016/j.biocon.2009.02.021

Ribeiro MC, Martensen AC, Metzger JP et al (2011) The Brazilian Atlantic Forest: a shrinking biodiversity hotspot. In: Zachos FE, Habel JC (eds) Biodiversity hotspots. Springer, Heidelberg/Berlin

Rouhan G, Gaudeul M (2021) Plant taxonomy: a historical perspective, current challenges, and perspectives. In: Besse P (ed) Molecular plant taxonomy. Methods in molecular biology, volume 2222. Humana, New York

Ryberg M (2015) Molecular operational taxonomic units as approximations of species in the light of evolutionary models and empirical data from Fungi. Mol Ecol 24(23):5770–5777. https://doi.org/10.1111/mec.13444

Sabatini FM, Jiménez-Alfaro B, Jandt U et al (2022) Global patterns of vascular plant alpha diversity. Nat Commun 13:4683. https://doi.org/10.1038/s41467-022-32063-z

Sandom C, Faurby S, Sandel B et al (2014) Global late quaternary megafauna extinctions linked to humans, not climate change. Proc R Soc Lond B Biol Sci 281(1787):20133254. https://doi.org/10.1098/rspb.2013.3254

Sclater PL (1858) On the general geographical distribution of the members of the class Aves. Zool J Linnean Soc 2:130–145. https://doi.org/10.1111/j.1096-3642.1858.tb02549.x
Silva-Santos RS, Ramirez JL, Galetti PM Jr et al (2018) Molecular evidences of a hidden complex scenario in *Leporinus* cf. *friderici*. Front Genet 9:1–9. https://doi.org/10.3389/fgene.2018.00047
Silva-Santos RS, Machado CB, Camelier P et al (2023) Molecular characterization of *Astyanax* species (Characiformes: Characidae) from the upper Paraguaçu River basin, a hydrographic system with high endemism. Neotrop Ichthyol 21:e230032. https://doi.org/10.1590/1982-0224-2023-0032
Spalding C, Hull PM (2021) Towards quantifying the mass extinction debt of the Anthropocene. Proc Biol Sci 288(1949):20202332. https://doi.org/10.1098/rspb.2020.2332
Spalding MD, Fox HE, Allen GR et al (2007) Marine ecoregions of the world: a bioregionalization of coastal and shelf areas. Bioscience 57(7):573–583. https://doi.org/10.1641/B570707
Stuart AJ (1999) Late pleistocene megafaunal extinctions. In: MacPhee RDE (ed) Extinctions in near time. Advances in vertebrate paleobiology, volume 2. Springer, Boston, pp 257–269
Turon X, Zarcero J, Antich AW et al (2003) Metabarcoding the eukaryotic community of a threatened, iconic Mediterranean habitat: Posidonia oceanica seagrass meadows. Front Mar Sci 10:1145883. https://doi.org/10.3389/fmars.2023.1145883
Turvey ST, Crees JJ (2019) Extinction in the Anthropocene. Curr Biol 29(19):R982–R986. https://doi.org/10.1016/j.cub.2019.07.040
Wallace AR (1876) The geographical distribution of animals, volume 1. Harper and Brothers/Cambridge University Press
Wang F, Li M, Huang L et al (2021) Cultivation of uncultured marine microorganisms. Mar Life Sci Technol 3:117–120. https://doi.org/10.1007/s42995-021-00093-z
Ward BB (2002) How many species of prokaryotes are there? PNAS 99(16):10234–10236. https://doi.org/10.1073/pnas.162359199
Whipps J, Lewis K, Cooke R (1988) Mycoparasitism and plant disease control. In: Burge M (ed) Fungi in biological control systems. Manchester University Press, pp 161–187
Whitman WB, Coleman DC, Wiebe WJ (1998) Prokaryotes: the unseen majority. Proc Natl Acad Sci USA 95(12):6578–6583. https://doi.org/10.1073/pnas.95.12.6578
Whittaker RH (1962) Classification of natural communities. Bot Rev 28(1):1–239. https://doi.org/10.1007/BF02860872
Whittaker RH (1970) Communities and ecosystems. MacMillan Publishing Co, New York, 162 p
Wilson EO, Peter FM (1988) Biodiversity. National Academy Press, Washington, DC, p 521
WWF (2015) The big five of the Cerrado. https://www.wwf.org.br/?50242/The-Big-Five-of-the-Cerrado. Accessed 17 July 2024
WWF (2022) Living Planet Report 2022. In: Almond REA, Grooten M, Bignoli D et al (eds) Building a nature-positive society. WWF, Gland
WWF (2024) Atlantic forest. https://www.worldwildlife.org/places/atlantic-forest. Accessed 19 Aug 2024

Chapter 3
What Is and Why Is Genetic Variation Important?

Fernanda Amato Gaiotto, Thâmara Moura Lima, and Alesandro Souza Santos

Contents

Abstract Most of the books and texts we read during our learning on topics related to genetics and evolution describe genetic variation as "the raw material for evolution". This statement is accurate; however, in this chapter, we will explore this topic and its definitions in greater depth to better understand its applications. We will explore a few key questions: why is genetic diversity necessary for a neotropical species? What is the importance of genetic variation for the use and domestication of genetic resources? Why do we need to conserve existing genetic diversity? Genetic diversity is generally higher in neotropical regions than in other areas on our planet. This makes sense as in this region we also have a wide range of environments variation (altitudes, temperatures, soil types, salinity, among others). The interaction between genetic and environmental variation results in more adapted phenotypes or phenotypes with greater potential for use by humans. However, other aspects of genetic variation, such as epigenetics, influence rapid adaptation, a process that we will address in this chapter.

F. A. Gaiotto (✉) · A. S. Santos
Departamento de Ciências Biológicas, Universidade Estadual de Santa Cruz, Ilhéus, Bahia, BA, Brazil
e-mail: gaiotto@uesc.br

T. M. Lima
Instituto de Ciências Agrárias, Universidade Federal Rural da Amazônia, Terra Firme – Belém, Pará, PA, Brazil

M. V. Bohrer Monteiro Siqueira et al. (eds.), *Population Genetics in the Neotropics*,
https://doi.org/10.1007/978-3-031-83685-5_3

Keywords Genetic variability · Adaptation · Domestication · Conservation

Learning Objectives
This chapter should help readers understand the following:

1. Basic definitions and fundamental theories of population genetics;
2. Understand the origin of genetic variability;
3. The definition and differences between neutral alleles and those under selection;
4. The importance of genetic diversity for adaptation to modified environments and landscapes;
5. The uses of genetic variation for domestication and pre-breeding.

3.1 Introduction

The perception of genetic variability becomes straightforward for humans when we begin by observing other individuals of our species and noticing the diversity among us. If we analyze faces, hair types, skin color, nose shape, and even mouth size, we perceive a wide range of human characteristics, hereafter frequently referred to as traits. This phenotypic and, therefore, genetic variability is easily observed when we do this among individuals of our own species (Fig. 3.1a).

Understanding genetic variability in other species, such as cashew trees, bromeliads, sea turtles, or small wild rodents, poses a greater challenge as the human eye isn't trained to detect such differences. Even in well-known species like potatoes, beans, or rice, which we encounter daily, distinguishing variation among individuals in the same species is quite challenging. Few species have discernible characteristics for assessing genetic diversity from our naked eye. In Fig. 3.1b, various corn cobs are presented to illustrate this perception, as corn serves as a didactic example. We can observe different colors and shapes of grains, some rounder, others longer, as well as variations in ear size, some shorter and robust, others longer and tapered.

Fig. 3.1 Examples of observable phenotypic variation in humans (**a**) and corn (**b**)

However, in most cases, when morphological (phenotypic) variation is absent (at least to our perception), detection is routinely accomplished through molecular markers.

The entire phenotypic diversity observed in both groups in Fig. 3.1 arises from the genotype (genes) and their interactions with the environment. For genetic variability to occur, there needs to be a diversity of expressed genes. However, the unexpressed portion of the genome also exhibits variability, which tends to be proportional to the variability of expressed alleles (Singh et al. 2023). This unexpressed variability can be detected through neutral molecular markers, meaning it is genetic diversity related to traits that may or may not be expressed but are not subject to selection pressure. To understand the extent of genetic variability within a species typically composed of various populations, we turn to population genetics. Hence, genetic markers play a crucial role in this process.

3.2 Challenges of Neotropical Species to Evolve and the Importance of Genetic Variability in This Process

Threats arising from human activities, such as the introduction of invasive alien species, air pollution, climate change, defaunation, and deforestation, threaten global biodiversity. In particular, deforestation resulting from the expansion of human activities has increased dramatically in recent years across the world, conditioning biota to survive and play their functional roles in anthropic environments with hostile conditions (Díaz et al. 2019). The conversion of natural environments into anthropic landscapes has caused the loss of species and ecological processes, as well as the reduction in fundamental ecosystem functions, such as carbon storage and nutrient cycling (Faria et al. 2023). For these reasons, the conversion of natural systems into anthropic ones is the main concern and target of studies by conservationists of this century, who seek to understand how these changes impact different levels of biodiversity (i.e., species diversity, genetic diversity, and of ecosystems). This negative impact is even more serious in tropical forests, which are globally recognized for their high biodiversity and account for more than 90% of global deforestation considering the period from 2000 to 2018 (FAO 2020). As a result, the biodiversity associated with these environments is at serious risk of being lost. Within neotropical regions in particular, the scenario of forest destruction is especially worrying for the Brazilian Atlantic Forest, which has high species richness and endemism, considered a Biodiversity hotspot (Martini et al. 2007). Sources differ in their estimates, but SOS Mata Atlântica and INPE (2020) estimate that the current forest coverage is approximately 28% of the original, mostly distributed in small and isolated fragments, with great impoverishment of species (SOS Mata Atlântica and INPE 2020).

Taking into account that the various human actions mentioned above are rapidly modifying the neotropical environments, it is essential to understand how species will respond to such changes and especially whether they will be able to adapt to these modified habitats. In this sense, assessing the genetic variability of species is extremely important from an ecological, evolutionary, and conservationist point of view, since this approach makes it possible to detect the fragility of populations in face of threats in a scenario of anthropogenic disturbances (Schmidt et al. 2023). In particular, the field of conservation genetics can predict the adaptive responses of species to the challenges of surviving in a modified environment, providing essential information to support efficient conservation strategies. For example, a recent study highlighted the need to develop and evaluate genetic diversity metrics to inform conservation policies, such as the IUCN (International Union for Conservation of Nature) red list and recent policy adopted by the UN's Convention on Biological Diversity Kunming-Montreal Global Biodiversity Framework (Schmidt et al. 2023).

The most used means of accessing genetic variability for plants, animals, and microorganisms ever since the 1980s has been the use of molecular markers (here you can read about them in the context of their application, but Chap. 9 should outline much more details on the specific matter of molecular tools for population genetics). This approach to collecting genetic information made major advances in the late 1980s, with the development of PCR (polymerase chain reaction), which allowed rapid amplification (multiplication) of target regions of DNA with greater ease and precision (Mullis and Faloona 1987). The emergence of PCR promoted a rapid increase in the use of molecular markers focused on conservation genetics, especially microsatellites. The great popularity of microsatellites was mainly because they are codominant markers and have a high degree of polymorphism, which allows natural populations to be better characterized for their genetic variation. Moreover, microsatellites are excellent tools for accessing genetic variability arising from uni or biparental inheritance. This last aspect is interesting because markers based on biparental inheritance generally have higher mutation rates and are recombinant, while uniparental markers are non-recombinant and have lower rates of genetic mutations. Therefore, the use of markers of uni or biparental inheritance is complementary to each other, making it possible to evaluate how older events (uniparental, for example, microsatellites based on chloroplast DNA) or more recent (biparental, microsatellites based on nuclear DNA) influenced or influence genetic diversity (Santos et al. 2019). As an example, in a study that combined this complementary approach of uni and biparental inheritance, Santos and collaborators (2019) highlighted the sharing of the contemporary gene pool, genetic bottleneck, and founder effect in subpopulations of a tree species known as Biriba (*Eschweilera ovata* (Cambess.) Mart. ex Miers) inserted in anthropized landscapes of the Atlantic Forest of southern Bahia, Brazil.

Although microsatellite markers are efficient for studies aimed at genetically characterizing populations of a given species, they are considered neutral, meaning they are not directly influenced by natural selection. Therefore, these markers are recommended to understand and characterize microevolutionary mechanisms related to gene flow and genetic drift, revealing dispersal mechanisms and events, demographic history, and the spatial and temporal dynamics of populations of interest. For example, in the neotropical region, several studies have used microsatellite markers to uncover the effects of deforestation and defaunation on plant species native to the Brazilian Atlantic Forest. In studies carried out with a species of palm tree threatened with extinction and known as Juçara palm (*Euterpe edulis* Martius), prior research has revealed that deforestation and defaunation have negatively influenced this species, reducing its genetic diversity and contemporary gene flow distances (Santos et al. 2016; Carvalho et al. 2021). Thus, these disturbances represent a threat to this species of great ecological relevance (food resource for approximately 58 species of birds and 20 species of mammals) and an important genetic resource (extraction of the apical meristem for human consumption).

However, the negative impacts associated with human actions are not always noticeable in a short time and do not affect all species equally (González et al. 2020). In a recent study based on 10 species of trees economically exploited in the Brazilian Atlantic Forest, Santana and collaborators (2023) showed that the populations maintained high genetic variability in general. The authors attribute this apparent genetic resilience of trees to the longevity of the species, the short time they have been exploited, and the recent history of intensified deforestation in the biome (1980s). However, the authors also highlight that the ability to maintain genetic diversity in this disturbed environment may vary according to the different types of economic exploitation and the reproductive biology of each species.

For studies that aim to evaluate microevolutionary mechanisms related to natural selection and, therefore, the use of markers considered neutral (e.g., microsatellites) would not be appropriate. For that purpose, one of the most recommended molecular tools are SNP (single nucleotide polymorphism) markers. SNP markers are widely distributed and with high frequency in the genome of any organisms of interest, and may be located in neutral or functional regions of the genome. With this type of marker, it is possible to use neutral regions of the genome, that is, they do not directly influence individual fitness, to make inferences related to demographic processes and genetic drift. On the other hand, processes related to natural selection and local adaptations can be evaluated based on functional regions of the genome, that is, those that can directly influence the level of individual fitness (Manel and Holderegger 2013). Typically, the use of this molecular tool makes it possible to simultaneously evaluate thousands of points in the genome (locus) in a single analysis of the target species, using a genomic approach. Certainly, this broad sampling at the genomic level and the discrimination of neutral versus non-neutral regions increases the inferential power of the results obtained and enables a greater understanding of the genetic status of target populations, as well as the evolutionary mechanisms that are shaping the genetic variation of such species (Ćalić et al. 2016).

Unfortunately, studies with a genomic approach are incipient for the socioeconomic context of the neotropical region (Santos and Gaiotto 2020). In one of the rare studies using this approach, Santos et al. (2023) used SNPs inserted into transcribed regions of the genome of heart-of-palm to investigate patterns of variability and genetic structure in anthropogenic landscapes of the Brazilian Atlantic Forest.

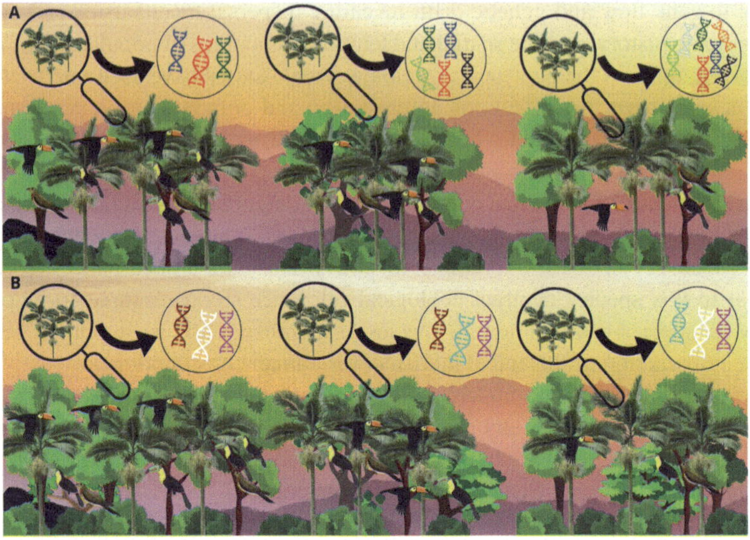

Fig. 3.2 Graphic representation of the increase in the number of alleles inserted into gene regions of the genome and supposedly under natural selection in three populations of the Juçara palm, in a more deforested landscape and with different levels of abundance of seed-dispersing birds in the Atlantic Forest of southern Bahia, Brazil (**a**) and representation of the lack of relationship in the number of this type of alleles in three populations of the same species, located in a more forested landscape and with different abundances of birds (**b**). The colors and number of double helices (i.e. DNA molecules) represent the SNP alleles. (For more details on this study, see Santos et al. (2023))

These authors showed that populations of this threatened species have patterns of structure and genetic variability that differ between regions with different levels of deforestation and biogeographical histories. Furthermore, they suggest that deforestation associated with the defaunation process may drive the levels of genetic variability related to the local adaptation of this neotropical species. This research reveals that landscapes with greater intensity of anthropogenic disturbances have a greater number of gene regions under selection pressure, due to greater biotic and abiotic stresses in these highly modified environments, leading to the detection in genetic analysis of a greater number of alleles supposedly under the action of selective forces (Fig. 3.2). On the other hand, more forested landscapes are supposedly under less selection pressure related to anthropogenic disturbances, leading to a smaller number of alleles supposedly under selection identified in the analyses. With this, the authors suggest that populations inserted in environments with greater intensity of disturbances must deal with increasingly greater adversities and environmental variation, which could require different adaptation mechanisms. The major problem in this case is that, initially, populations may be experiencing a transient increase in genetic variability, as postulated by the positive selection model. However, once the selected allele reaches fixation in response to conditions that are acting as an environmental filter, that locus becomes monomorphic, and genetic variation is lost, which reduces the ability of these populations to respond adaptively in the future, given the large environmental fluctuations that normally occur in highly deforested landscapes. In this context, population genomics approaches can contribute to understanding the genetic adaptation and conservation of neotropical species, which are inserted in an environment that is undergoing rapid changes induced by human activities (Santos and Gaiotto 2020).

Typically, studies that aim to understand the adaptation of populations to their local habitats relate genetic variation from non-neutral regions of the genome to environmental variables (e.g., temperature, precipitation, etc.), in an approach called landscape genomics (i.e., describing how genetic variation is associated with environmental characteristics) (Ahrens et al. 2018). Thus, it is possible to investigate the environmental factors that are influencing specific gene regions and, ultimately, influencing population fitness. In this context, considering the great environmental heterogeneity in the neotropical region, whether due to its own natural dynamics or resulting from serious changes induced by humans, it is necessary to unveil the evolutionary potential of the species that inhabit this part of the planet. In this sense, we need to consider that the evolutionary potential of species depends on genetic variation that directly influences survival traits and reproduction of individuals (that is, fitness). The greater the variation among individuals, the better the chances of a population to respond to environmental changes over time. Natural selection acts and selects phenotypic traits (and the genotypes associated) that confer some adaptive advantage according to the environment in question, yielding greater fitness on individuals carrying such traits over generations, thus leading to the adaptive evolution of populations. In this way, natural selection shapes the population's genetic composition over time, according to the pressures of local biotic and abiotic factors.

Generally speaking, populations are subject to three basic types of natural selection: directional selection, disruptive selection, and stabilizing selection (further details in Chap. 5). Directional selection can occur in two different ways: (1) positive directional selection, when a genetic variant increases the fitness of the carrying individual and increases its frequency over generations and (2) purifying directional selection when certain heritable characteristics reduce the individual fitness and its frequency is reduced or even eliminated from the population over generations. In stabilizing selection, the intermediate phenotype and its associated genotype will be favored and increase in frequency while the extremes are diminished. Finally, in disruptive selection, the frequency of two or more phenotypes that are fitter than the intermediate phenotypes increases in frequency in the population over time. Thus, although genetic variation is fundamental and functions as the raw material upon which natural selection acts (Orr 2005), it is expected that this process, which allows populations to adapt to their environments, will lead to less genetic variability over time.

3.3 Genetic Resource and Domestication (Breeding)

Genetic resources are defined as genetic materials of factual or potential value, with valuable genetic diversity for enabling ecosystem functions as well as for acknowledging current or future economic and social needs. Thus, many species can be considered genetic resources due to the potential benefits they may offer from various perspectives (Fenzi and Bonneuil 2016). We have already described that genetic diversity is crucial for species adaptation within different ecosystems (Hoban et al. 2021). Additionally, genetic diversity is the foundation for human activities aimed at improving plant and animal attributes, thereby enhancing their utility. This has been observed since the beginning of domestication, through selection, cultivation, or breeding of genotypes (individuals) with desirable traits over others (Moyers et al. 2018). Domestication is understood as an evolutionary process mediated by artificial selection to meet human needs, which basically alters the genetic and phenotypic diversity of selected populations (Miller and Gross 2011).

The evolution of genetic resources in response to the pressure of artificial selection has provided the foundation for agricultural production, promoting targeted changes in gene and allele frequencies in populations over generations, and consequently in the phenotypes of interest (Wedger et al. 2021). Among the changes in phenotypic characteristics, examples include plant species producing seeds with lower dormancy rates, and grains with higher yields of oil or carbohydrates (Munguía-Rosas et al. 2019). Generally, populations of species subjected to artificial selection exhibit higher productivity, ease of management, and fruits with various colors, flavors, and textures (Clement et al. 2021). This set of changes associated with human preferences is termed the domestication syndrome, which can also encompass traits related to the reproductive biology and phenology of plant species (Ferrer et al. 2020). The intensity of artificial selection pressure varies, determining phenotypic patterns and alterations in the genetic structure of populations (Francisconi et al. 2021). This results in a continuum of genetic variation from wild-type populations, through incipiently domesticated ones, to those unable to survive without human intervention (Lins Neto et al. 2014). Typically, the first two types of populations have greater genetic diversity, providing the potential for the selection of traits or genes of interest to humans. In contrast, cultivated populations exhibit a narrow genetic base, with little expectation of high genetic diversity.

The domestication of plant species in neotropical regions occurred within the context of artificial selection in agroecological systems, where indigenous peoples managed the environment without necessarily implementing intensive food production systems (Lowe et al. 2018). In the Neotropics, domestication was shaped by the peculiar human integration with natural ecosystems through local cultural and ethnic factors. For example, ethnobotanical and evolutionary research reveals that domestication in the Amazon region was influenced by indigenous knowledge of management practices that promoted the establishment of patches of useful plant communities in forest ecosystems. Thus, considering the superficial selection on the genetic diversity of different species, it is possible to observe the domestication not only of plants but also of landscapes (Levis et al. 2018).

The relationship between indigenous peoples and natural ecosystems is further illustrated in recent research on the domestication of cupuaçu (*Theobroma grandiflorum* (Willd. ex Spreng.) K. Schu), a fruit found in the Amazon Rainforest. Although naturally occurring in this neotropical biome and previously classified as native, recent scientific findings have revealed that it has been domesticated by indigenous peoples for over 5000 years and is a variant of another species, cupuí, which has smaller fruits. Over time, human-directed crossing between genotypes with larger fruits of cupuí, including backcrossing with the original parents, led to the transfer of genes expressing the larger fruit size of cupuaçu (Colli-Silva et al. 2023).

In modern domestication practices, genetic improvement is a field in which humans employ various tools to increase productivity and resistance/tolerance to biotic (e.g., pests and diseases) and abiotic (e.g., salinity and water stress) factors (Turner-Hissong et al. 2020). The provision of products and services through genetic improvement contributes to the development of agriculture, forestry, aquaculture, and other sectors (e.g., the pharmaceutical industry). In this sense, genetic improvement, or also genetic breeding, serves as the link between genetic resources and their economic potential. Its success depends on genetic diversity among the genotypes within populations, allowing for the recombination of genes and alleles that express desirable traits. The genotypes available for such purposes are referred to as germplasm, representing a collection of accessions with characteristics aligned with the goals of the breeding program (Migicovsky et al. 2019). Figure 3.3 represents a germplasm bank from the perspective of the genetic diversity of a tree species with timber value.

Different types of germplasm collections consolidate ex-situ genetic conservation, safeguarding genetic variation through accessions derived from breeding (current and historical commercial cultivars/clones), as well as native varieties and wild

Fig. 3.3 Illustration of a germplasm bank representing genetic diversity among accessions of a timber genetic resource, based on color differentiation

relatives (Priyanka et al. 2021). Diversity characterization is performed through evaluation of phenotypic (morphological) and genetic (molecular) descriptors (Bhandari et al. 2017). Traits are validated as descriptors when they exhibit a high heritability coefficient, indicating a strong likelihood that the trait will be observed in subsequent generations due to greater genetic control (Timaeus et al. 2021). Identifying genetic diversity among accessions is crucial information for breeders to promote crosses with hybrid vigor (heterosis), ensuring that progeny exhibit superior traits compared to their parents. This enhances selection efficiency and genetic gain (Labroo et al. 2021).

Genetic mapping is another strategy in breeding for identifying genes associated with desired phenotypes (Migicovsky and Myles 2017). The success of this approach relies on the extensive genetic diversity within the germplasm, which includes multiple recombination events, improving the ability to accurately predict phenotypes based on gene/allele composition. This allows breeders to select parents and test progenies for their performance using genetic information. Materials in germplasm collections can include ovules, embryos, pollen, seeds, entire plants or specific sources of propagules. The choice of conservation method depends on the type of material and can include in vitro preservation, cryopreservation, field storage, or maintenance in controlled environmental conditions (Priyanka et al. 2021).

Due to the higher levels of genetic diversity in populations with wild genotypes, especially those which experience reduced artificial selection pressure, they become important targets for on-site conservation. This strategy, known as in situ genetic conservation, allows populations to evolve in their natural habitat, ensuring the preservation of the evolutionary potential of genetic resources (Chacón-Labella et al. 2019). Given that increased domestication intensity reduces genetic diversity, in situ conservation is essential for maintaining the source of genes and alleles of genetic resources available for use. Ex situ conservation strategies depend on the reproductive system of the species, effective population size, and knowledge of the level of genetic diversity, among other factors. These distinct modes of genetic conservation are understood as complementary approaches (Zonneveld et al. 2018). Figure 3.4 illustrates the types of conservation, where the different natural occurrence sites of populations must be considered for in situ genetic conservation.

Fig. 3.4 Complementarity of Genetic Resources Conservation strategies, illustrating the genetic variation (different colors) contained in natural populations from several locations (*In Situ*), safeguarded in germplasm (*Ex Situ*)

Conversely, ex situ genetic conservation should capture part of this variation to effectively maintain the germplasm and make accessions available for use.

Genetic diversity is the foundation for domestication and the development of breeding activities, which ensure the production of food and other essential products and services for humans. Therefore, in situ and ex situ genetic conservation is a crucial strategy for food security and the stability of globally important production sectors. In the context of negative impacts due to the unsustainable exploitation of various genetic resources over time, safeguarding genetic diversity ensures the rational use of this component of biodiversity.

3.4 Conservation

As biodiversity conservation is much needed, genetic variation becomes crucial as it enables species to adapt to environmental changes and stresses, such as climate change, habitat loss, and new pests and diseases (Stange et al. 2021). Also, it provides the raw material for natural selection to act upon, allowing populations to evolve and survive in face of changing conditions.

Genetic variation within species contributes to the diversity and stability of ecological communities and ecosystems, as it enhances ecosystem functioning by supporting processes like nutrient cycling, pollination, and seed dispersal carried out by diverse groups of species. Therefore, maintaining genetic variation is essential for the long-term survival of endangered species, as small populations are susceptible to genetic drift and inbreeding, which can lead to reduced genetic diversity, increasing disease susceptibility, for example. Conservation efforts often focus on increasing genetic diversity through techniques like environment restoration, genetic rescue, and reintroduction programs (Alves et al. 2023).

Genetic diversity is crucial for silviculture, agriculture, and food security too. Crops with low genetic diversity are more vulnerable to pests, diseases, and environmental stresses. Maintaining the genetic diversity of plant species guarantees the sustainable production of food, bioactives, fibers, resins, and wood products. Thus, genome-wide genetic variation generally improves population viability by preventing inbreeding depression and maintaining adaptive potential (Kardos et al. 2021). Conserving genome-wide variation is usually the best approach to prevent extinction, as focusing only on functional variation is often misleading and counterproductive.

3.5 Final Remarks

Genetic variation represents all the variation contained in the DNA molecule, which ultimately originates from mutations combined with recombinations of genes and alleles, promoting the variation within and between populations. This variation is manifested through the phenotypic differences observed in all species, which are

Box 3.1 A Few and Basic Concepts to Keep in Mind After This Chapter

Phenotype is the set of observable morphological and functional characteristics of an organism, resulting from the interaction of genes with each other and with the environment.

Genotype refers to the specific set of alleles (variants of genes) present in an organism. It is the genetic makeup of an individual, focusing on particular genes or genetic loci that determine specific traits or characteristics.

Genome is the complete set of an organism's genetic material, including all of its genes and non-coding sequences. It encompasses the entire DNA (or RNA, in some viruses) within an organism's cell.

Alleles are different versions of the same gene that arise through mutation and are found at the same place (locus) on a chromosome.

Germplasm is the living genetic material, such as seeds, pollen, spores, or tissue cultures, that is preserved for the purpose of breeding, research, and conservation of genetic diversity in plants or animals. It contains the genetic information necessary for the propagation and survival of a species.

Domestication syndrome refers to a suite of traits or features that tend to be consistently observed in domesticated plants and animals compared to their wild ancestors. These traits often include changes in morphology, behavior, physiology, and life history, which have been inadvertently or intentionally selected for during the process of domestication.

Hybrid vigor, also known as heterosis, is the phenomenon where offspring of genetically diverse parents exhibit superior traits or performance compared to either of their parents. It is commonly observed in the first generation (F1) hybrids resulting from crosses between two genetically distinct parental lines. The phenomenon is widely used in agriculture, where hybrid crops and livestock are bred to capitalize on the benefits of heterosis, resulting in improved yields, health, and other desirable traits.

Genetic mapping is a process used to determine the relative positions of genes or other DNA sequences on chromosomes. It involves identifying and recording the locations of genes or genetic markers, such as single nucleotide polymorphisms (SNPs) or microsatellites, along the chromosomes of an organism. Genetic maps provide valuable information about the organization and inheritance patterns of genes within a genome. They are essential for various genetic studies, including gene identification, linkage analysis, and marker-assisted breeding in plants and animals.

Heritability coefficient, often denoted as H^2, represents the proportion of phenotypic variation in a population that is attributable to genetic variation among individuals.

Microevolution refers to the evolutionary changes that occur within a population or species over a relatively short time. These changes involve alterations in the frequencies of alleles within the gene pool of a population due to processes such as natural selection, genetic drift, mutation, and gene flow. Microevolutionary processes can lead to observable changes in traits, frequencies of genetic variants, and adaptations within a population, ultimately shaping its genetic composition and phenotypic diversity over time.

Molecular markers are identifiable and heritable variations in DNA sequences that can be used to track genetic differences among individuals, populations, or species. These markers serve as molecular signposts along the genome, allowing researchers to detect genetic diversity, map genes, study genetic relationships, and identify genomic regions associated with specific traits.

Adaptation refers to the evolutionary process by which organisms develop or optimize traits that improve their survival and reproduction in a particular environment. These traits are often shaped by natural selection, as individuals with advantageous traits are more likely to survive and pass on their genes to the next generation.

Box 3.2 Relationship Between Neutral Genetic Variability and Genes Under Selection

Genetic variability can be categorized into neutral variability and variability under selection. We use molecular markers, simulations, and population genetics to classify this variability. In simple terms, neutral loci are genomic regions that are (i) unexpressed or located in intergenic regions, (ii) transcribed but provide no selective advantage, or (iii) unlinked (in linkage equilibrium) to loci under selection.

On the other hand, selective loci can be detected by molecular markers that are either linked to or located within genes that confer some capacity for longer survival (longevity) or greater reproductive success, thus ensuring that more individuals with this allele are present in the next generation. This provides a selective advantage compared to other individuals in the population who do not possess the selective variant.

The origin of both types of loci is the same: mutations. However, neutral variation arises solely from random mutations, while variation under selection arises due to natural selection. Both types can be perpetuated across generations: selective alleles because they confer an advantage, and neutral alleles because changes in DNA sequences do not affect the fitness of organisms and are located in non-transcribed regions.

Box 3.3 Epigenetics and Its Role in Rapid Adaptation

Epigenetics refers to changes in gene expression that do not involve changes in the DNA sequence. Epigenetic modifications occur in regions surrounding the DNA (such as methylation of the DNA, histone methylation or acetylation, or miRNA), yet they necessarily result in changes in gene expression and, consequently, the corresponding phenotypes. These changes can be heritable and influenced by the environment. Epigenetic mechanisms, by impacting genetic activity without altering the DNA sequence itself, play a crucial role in the regulation of gene expression, cellular differentiation, organism development, and responses to environmental stimuli. Therefore, the interdisciplinary nature of epigenetics is essential for understanding mechanisms related to adaptive plasticity that are transmitted across generations (Herman and Sultan 2016).

Epigenetic marks play a crucial role in the adaptation of living organisms to their environment, influencing traits related to fitness and contributing to adaptive variation. Additionally, epigenetic inheritance can modulate characteristics that result in phenotypic variation independent of genetic factors, highlighting the importance of considering epigenetic mechanisms for a comprehensive understanding of adaptation. The heritability of epigenetic variation is influenced by various factors, including DNA methylation patterns and environmental gradients. The study conducted by Whipple and Holeski (2016) identified epigenetic quantitative trait loci (eQTLs) that account for a significant portion of heritability in ecologically relevant traits. Additionally, natural epigenetic variation has been associated with environmental gradients in plant populations, indicating a genetic basis for DNA methylation and its potential role in local adaptation.

Additionally, studies on epigenetic inheritance at the landscape level provide valuable insights into adaptive variation, elucidating the relative contributions of genetic variation and phenotypic plasticity. By integrating epigenetics with genomics, modeling, and experimentation, researchers can achieve a comprehensive understanding of how epigenomic variation influences evolution at the landscape scale (Whipple and Holeski 2016).

Therefore, in the context of rapid adaptation to environmental pressures such as pathogens or changing conditions, epigenetic modifications can lead to phenotypic changes that enhance an organism's survival and reproductive success. Some mechanisms such as (i) rapid response, (ii) reversible changes, (iii) phenotypic plasticity, (iv) interaction with genetics, (v) gene regulation, and (vi) adaptive phenotypes are key points regarding the role of epigenetics in rapid adaptation (Gawra et al. 2023).

essential for their survival since it serves as raw material for natural selection, promoting adaptations to different environmental conditions. It also contributes to the stability of ecological communities and ecosystems, maximizing the chances of maintaining high species diversity in the Neotropics, especially in the context of global climate change.

Furthermore, genetic variation represents the basis for the domestication of species in genetic breeding programs, those that are intended to increase productivity and resistance to biotic and abiotic factors, crucial for forestry and agriculture. In this context, maintaining genetic variation is fundamental from the perspective of environmental conservation and the use of genetic resources. This highlights the relevance of studies that assess the genetic variation of native species, particularly Neotropical species that are still so little studied.

Advancing scientific knowledge in this area is important to support ecological restoration programs in the decade of restoration (2021–2030). With genetic and molecular knowledge, we can promote the efficient and long-term rescue of populations threatened with extinction, with reintroduction programs for locally extinct species, as well as with germplasm banks. Certainly, these initiatives can effectively contribute to the in situ and ex situ genetic conservation of populations of target species.

Key Points

1. Genetic variability refers to the differences in DNA sequences among individuals. This variation is crucial for species to adapt to changing environmental conditions, such as climate change, habitat loss, and the emergence of new pests and diseases. It provides the raw material for natural selection, allowing populations to evolve and survive in the face of environmental challenges. The interaction between genetic variation and environmental factors results in more adapted phenotypes with greater potential for survival and reproduction in specific habitats.
2. The perception of genetic variability becomes evident when observing the wide range of characteristics among individuals within a species. To facilitate this observation, we use molecular tools. In neotropical regions, where genetic diversity is generally higher, the interaction between genetic and environmental variation result in well-adapted phenotypes to their specific habitats.
3. Genetic diversity is crucial for domestication and breeding of plants and animals, enabling selection for traits such as disease resistance, yield improvement, and environmental adaptation. It supports the development of improved varieties for agriculture, forestry, and conservation. Additionally, conserving genetic diversity in wild populations provides a reservoir for future breeding and ecosystem restoration efforts.
4. Conserving genetic diversity is essential for long-term species survival, especially in the face of increasing environmental pressures and human-induced disturbances. Small populations are particularly vulnerable to genetic drift and inbreeding, which can lead to reduced genetic diversity and increased susceptibility to diseases and environmental stresses. Conservation efforts often focus on increasing genetic diversity through habitat restoration, genetic rescue, or reintroduction programs.

Questions

1. Create a mind map about the main applications of genetic variability, considering nature conservation and food production.
2. Imagine that you are a decision-maker and need to guide public policies for forest restoration in a certain region. What are your main recommendations considering the importance of genetic variability for the longevity of the restored populations?

3. What are the main adaptive advantages when epigenetic changes occur instead of variations at the genomic level?
4. Each genomic region can confer advantages or interesting morphologies for genetic improvement. Your task is to review the literature and describe examples where genetic variability has been used for the domestication of species useful to humans.
5. Considering that a new species has been discovered, describe best practices for their genetic conservation.

Acknowledgments We would like to thank the Conservation Genetics group, the Center for Biotechnology and Genetics (CBG), and the Applied Ecology and Conservation Laboratory (LEAC) at the State University of Santa Cruz, BA, Brazil, for the valuable discussions that led to the writing of this chapter. We are also grateful to the funding agencies that support our research in the conservation genetics of tree species (Capes, CNPq, Fapesb, Finep, Funbio, and the Rufford Foundation).

References

Ahrens CW, Rymer PD, Stow A et al (2018) The search for loci under selection: trends, biases and progress. Mol Ecol 27(6):1342–1356. https://doi.org/10.1111/mec.14549

Alves F, Banks SC, Edworthy M et al (2023) Using conservation genetics to prioritise management options for an endangered songbird. Heredity 130(5):289–301. https://doi.org/10.1038/s41437-023-00609-6

Bhandari HR, Bhanu AN, Srivastava K, Singh MN, Shreya, Hemantaranjan A (2017) Assessment of genetic diversity in crop plants - an overview. Adv Plants Agric Res 7:279–286. https://doi.org/10.15406/apar.2017.07.00255

Ćalić I, Bussotti F, Martínez-García PJ, Neale DB (2016) Recent landscape genomics studies in forest trees—what can we believe? Tree Genet Genomes 12(1):3. https://doi.org/10.1007/s11295-015-0960-0

Carvalho CDS, Lucas MS, Côrtes MC (2021) Rescuing intraspecific variation in human-impacted environments. J Appl Ecol 58(2):350–359. https://doi.org/10.1111/1365-2664.13764

Chacón-Labella J, García Palacios P, Matesanz S et al (2019) Plant domestication disrupts biodiversity effects across major crop types. Ecol Lett 22(9):1472–1482. https://doi.org/10.1111/ele.13336

Clement CR, Casas A, Parra-Rondinel FA et al (2021) Disentangling domestication from food production systems in the Neotropics. Quaternary 28;4(1):4. https://doi.org/10.3390/quat4010004

Colli-Silva M, Richardson JE, Neves EG et al (2023) Domestication of the Amazonian fruit tree cupuaçu may have stretched over the past 8000 years. Commun Earth Environ 1;4(1):401. https://doi.org/10.1038/s43247-023-01066-z

Díaz S, Settele J, Brondízio ES et al (2019) Pervasive human-driven decline of life on Earth points to the need for transformative change. Science 13;366(6471):eaax3100. https://doi.org/10.1126/science.aax310

FAO (2020) Global Forest resources assessment 2020—key findings, Rome

Faria D, Morante-Filho JC, Baumgarten J et al (2023) The breakdown of ecosystem functionality driven by deforestation in a global biodiversity hotspot. Biol Conserv 1(283):110126. https://doi.org/10.1016/j.biocon.2023.110126

Fenzi M, Bonneuil C (2016) From "genetic resources" to "ecosystems services": a century of science and global policies for crop diversity conservation. Cult Agric Food Environ 38:72–83. https://doi.org/10.1111/cuag.12072

Ferrer MM, Montañez-Escalante PI et al (2020) Growing out of the tropical forests: domestication syndrome of native Mesoamerican trees in Mayan homegardens. Cult Agric Food Environ 67:587–604. https://doi.org/10.1007/s10722-019-00833-2

Francisconi AF, Alves RP, Clement CR et al (2021) Genetic structure and diversity identify incipient domestication of Piquiá [*Caryocar villosum* (Aubl.) pers.] along the lower Tapajós River, Brazilian Amazonia. Genet Resour Crop Evol 68:1487–1501. https://doi.org/10.1007/s10722-020-01078-0

Gawra J, Valdivieso A, Roux F et al (2023) Epigenetic variations are more substantial than genetic variations in rapid adaptation of oyster to Pacific oyster mortality syndrome Sci Adv 8;9(36):eadh8990. https://doi.org/10.1126/sciadv.adh8990

González AV, Gómez-Silva V, Ramírez MJ et al (2020) Meta-analysis of the differential effects of habitat fragmentation and degradation on plant genetic diversity. Conserv Biol 34(3):711–720. https://doi.org/10.1111/cobi.13422

Herman JJ, Sultan SE (2016) DNA methylation mediates genetic variation for adaptive transgenerational plasticity. Proc R Soc B Biol Sci 283(1838):20160988. https://doi.org/10.1098/rspb.2016.0988

Hoban S, Bruford MW, Funk WC et al (2021) Global commitments to conserving and monitoring genetic diversity are now necessary and feasible. Bioscience 71:964–976. https://doi.org/10.1093/biosci/biab054

Kardos M, Armstrong EE, Fitzpatrick SW et al (2021) The crucial role of genome-wide genetic variation in conservation. Proc Natl Acad Sci U S A 30;118(48):e2104642118. https://doi.org/10.1073/pnas.2104642118

Labroo MR, Studer AJ, Rutkoski JE (2021) Heterosis and hybrid crop breeding: a multidisciplinary review. Front Genet 12:1–19. https://doi.org/10.3389/fgene.2021.643761

Levis C, Flores BM, Moreira PA et al (2018) How people domesticated Amazonian forests. Front Ecol Evol 5:171. https://doi.org/10.3389/fevo.2017.00171

Lins Neto EMDF, Peroni N, Casas A et al (2014) Brazilian and Mexican experiences in the study of incipient domestication. J Ethnobiol Ethnomed 10:1–12. https://doi.org/10.1186/1746-4269-10-33

Lowe AJ, Breed MF, Caron H et al (2018) Standardized genetic diversity-life history correlates for improved genetic resource management of Neotropical trees. Divers Distrib 24:730–741. https://doi.org/10.1111/ddi.12716

Manel S, Holderegger R (2013) Ten years of landscape genetics. Trends Ecol Evol 28(10):614–621. https://doi.org/10.1016/j.tree.2013.05.012

Martini AM, Fiaschi P, Amorim AM, Paixão JL (2007) A hot point within a hot-spot: a high diversity site in Brazil's Atlantic Forest. Biodivers Conserv 16(11):3111–3128. https://doi.org/10.1007/s10531-007-9166-6

Migicovsky Z, Myles S (2017) Exploiting wild relatives for genomics-assisted breeding of perennial crops. Front Plant Sci 8:1–16. https://doi.org/10.3389/fpls.2017.00460

Migicovsky Z, Warschefsky E, Klein LL, Miller AJ (2019) Using living germplasm collections to characterize, improve, and conserve woody perennials. Crop Sci 59:2365–2380. https://doi.org/10.2135/cropsci2019.05.0353

Miller AJ, Gross BL (2011) From forest to field: perennial fruit cropdomestication. Am J Bot 98:1389–1414. https://doi.org/10.3732/ajb.1000522

Moyers BT, Morrell PL, McKay JK (2018) Genetic costs of domestication and improvement. J Hered 109:103–116. https://doi.org/10.1093/jhered/esx069

Mullis KB, Faloona FA (1987) Specific synthesis of DNA in vitro via a polymerase-catalyzed chain reaction. In: Methods in enzymology (eds) Academic Press, pp 335–350

Munguía-Rosas MA, Jácome-Flores ME, Bello-Bedoy R et al (2019) Morphological divergence between wild and cultivated chaya (*Cnidoscolus aconitifolius*)(Mill.) IM Johnst. Genet Resour Crop Evol 66:1389–1398. https://doi.org/10.1007/s10722-019-00790-w

Orr HA (2005) The genetic theory of adaptation: a brief history. Nat Rev Genet 6(2):119–127. https://doi.org/10.1038/nrg1523

Priyanka V, Kumar R, Dhaliwal I, Kaushik P (2021) Germplasm conservation: instrumental in agricultural biodiversity—a review. Sustain For 13:1–18. https://doi.org/10.3390/su13126743

Santana NS, Santos AS, Borges DB et al (2023) Genetic resilience of Atlantic forest trees to impacts of biome loss and fragmentation. Eur J For Res 142(1):161–174. https://doi.org/10.1007/s10342-022-01516-x

Santos AS, Gaiotto FA (2020) Knowledge status and sampling strategies to maximize cost-benefit ratio of studies in landscape genomics of wild plants. Sci Rep 10(1):3706. https://doi.org/10.1038/s41598-020-60788-8

Santos AS, Cazetta E, Dodonov P et al (2016) Landscape-scale deforestation decreases gene flow distance of a keystone tropical palm, *Euterpe edulis* Mart (Arecaceae). Ecol Evol 6(18):6586–6598. https://doi.org/10.1002/ece3.2341

Santos AS, Borges DB, Vivas CV et al (2019) Gene pool sharing and genetic bottleneck effects in subpopulations of *Eschweilera ovata* (Cambess.) Mart. ex Miers (Lecythidaceae) in the Atlantic Forest of southern Bahia. Brazil Genet Mol Biol 42:655–665. https://doi.org/10.1590/1678-4685-GMB-2018-0140

Santos AS Cazetta E, Faria D, et al (2023) Tropical forest loss and geographic location drive the functional genomic diversity of an endangered palm tree. Evolutionary Applications, 16(7):1257–1273. https://doi.org/10.1111/eva.13525

Schmidt C, Hoban S, Hunter M et al (2023) Genetic diversity and IUCN Red List status. Conserv Biol 37(4):e14064. https://doi.org/10.1111/cobi.14064

Singh A, Singh A, Barb J, Mahama AA (2023) Genetic variation and germplasm usage. In: Suza WP, Lamkey KR (eds) Crop improvement. Iowa State University Digital Press

SOS Mata Atlântica, INPE (2020) Atlas dos remanescentes florestais da Mata Atlântica: período 2020–2021

Stange M, Barrett RD, Hendry AP (2021) The importance of genomic variation for biodiversity, ecosystems and people. Nat Rev Genet 22:89–105. https://doi.org/10.1038/s41576-020-00288-7

Timaeus J, Weedon OD, Finckh MR (2021) Combining genetic gain and diversity in plant breeding: heritability of root selection in wheat populations. Sustain For 13. https://doi.org/10.3390/su132212778

Turner-Hissong SD, Mabry ME, Beissinger TM et al (2020) Evolutionary insights into plant breeding. Curr Opin Plant Biol 54:93–100. https://doi.org/10.1016/j.pbi.2020.03.003

Wedger MJ, Schumann AC, Gross BL (2021) Candidate genes and signatures of directional selection on fruit quality traits during apple domestication. Am J Bot 108:616–627. https://doi.org/10.1002/ajb2.1636

Whipple AV, Holeski LM (2016) Epigenetic inheritance across the landscape. Front Genet 7:220280. https://doi.org/10.3389/fgene.2016.00189

Zonneveld M, Thomas E, Castañeda-Álvarez NP et al (2018) Tree genetic resources at risk in South America: a spatial threat assessment to prioritize populations for conservation. Divers Distrib 24:718–729. https://doi.org/10.1111/ddi.12724

Chapter 4
What Is the Meaning of Genetic Drift?

Miklos Maximiliano Bajay

Contents

Abstract This chapter begins by stating the core principle for any genetic analysis, the Hardy-Weinberg equilibrium. Any deviation from the model implicates that one or more evolutionary mechanisms are acting in the populations. We start with the first, random genetic drift. Genetic drift is the random fluctuation in allele frequencies due to sampling error in finite populations. All populations are finite; therefore genetic drift occur in every population continuously. However, the magnitude and speed of allele frequency changes throughout the genome over time are influenced by population size. Effective population size refers to the theoretical size of an idealized population that experiences the same amount of genetic drift as the actual population under consideration. Small or isolated populations are susceptible to genetic drift, and as a result, alleles may become lost over time, especially rare ones, increasing the genetic differentiation between populations. The neotropical region contains many endemic species that are found in specific habitats or geographic locations. Therefore, genetic drift can be particularly important due to the high levels of biological diversity and regional isolation of these populations. Understanding genetic drift and effective popula-

M. M. Bajay (✉)
Centro de Educação Superior da Região Sul, Universidade do Estado de Santa Catarina, Laguna, Santa Catarina, SC, Brazil
e-mail: miklos.bajay@udesc.br

M. V. Bohrer Monteiro Siqueira et al. (eds.), *Population Genetics in the Neotropics*,
https://doi.org/10.1007/978-3-031-83685-5_4

tion size has important implications for conservation efforts, as loss of genetic diversity can reduce the adaptive potential of a population, making it less able to respond to environmental change.

Keywords Hardy-Weinberg equilibrium · Evolutionary mechanisms · Population differentiation · Neutral evolution

Learning Objectives

- Describe the core model used in population genetics, the Hardy-Weinberg equilibrium, and argue on the deviations of that principle, placing genetic drift as the first mechanism to be debated.
- Define genetic drift and explain how it influences allele frequencies in a population.
- Understand the difference between genetic drift and natural selection.
- Identify factors that can increase the likelihood of genetic drift occurring in a population.
- Describe the effects of genetic drift on genetic diversity within a population.
- Analyze real-life examples of genetic drift in action, such as founder effects and population bottlenecks.
- Critically evaluate the implications of genetic drift for conservation biology and the management of endangered species in the Neotropics.

4.1 The Most Basic Concepts and a Core Model to Keep in Mind

If you are a beginner on population genetics, there are a few basic concepts that you need to understand clearly before you go on with the main subject of this chapter. Chapter 3 disclosed a few of the most elementary concepts to keep in mind for this book, such as gene, allele, and genetic variation. Here, for any endeavors toward comprehending and interpreting population genetic mechanisms such as genetic drift, some initial and basic mathematical foundations of an essential statistical model are presented: the Hardy-Weinberg equilibrium principle.

Just to remind the reader, a gene is a segment along a chromosome, that is, a chain of nucleotides that can be transcribed to messenger RNA, and ultimately be translated into one or more proteins. Alleles are different variants of the same gene produced by mutations. In a biological sense, alleles are any difference observable in a trait due to nucleotide changes in the same gene. The combination of two alleles, whether the same (homozygous) or different (heterozygote) in a diploid organism, make up the *genotype* for that gene. The observable manifestation of that genotype is the *phenotype*.

Taking one of the traits that Gregor Mendel worked with peas, seed shape, there are usually two discernable phenotypes: round or wrinkled. Based on Mendel's experiments, seed shape has been suggested as a trait controlled by a single gene, that is a monogenic trait. By that logic and assuming that only two phenotypes are distinguished (round or wrinkled), we can represent the gene controlling seed shape as *R* (for round), with two alleles for the gene, *R* for round and *r* for wrinkled. *R* has complete dominance over *r*, that is, the heterozygote *Rr* produces round seeds. Only the homozygous recessive form *rr* produces wrinkled seeds.

Considering two parental pures lines (*RR* and *rr*) being crossed, 100% of the F_1 generation should have round seeds. After a generation of selfing, though, 75% of the seeds produced are round and 25% of them are wrinkled. This is what we expected under the assumptions of allele segregation following Mendelian expectations in the following Punnet square:

	R	*r*
R	*RR*	*Rr*
r	*Rr*	*rr*

And we can further analyze this example and provide novel details to other important concepts in population genetics:

- *allele frequencies*: the proportion that each allele has in the population or sampled materials;
- *genotypic frequencies*: the proportion of each genotype;
- *phenotypic frequencies*: the proportion of each phenotype.

For the example under consideration above:

- allele frequencies: *R* is 50%, and *r* is 50%;
- genotypic frequencies: 25% is *RR*, 50% is *Rr* and 25% is *rr*;
- phenotypic frequencies: 75% of seeds are round, and 25% of seeds are wrinkled.

For further information, the difference between round and wrinkled peas has been elucidated at the molecular level as a major polymorphism within the gene controlling the production of the starch-branching enzyme (*SBEI*) (Bhattacharyya et al. 1990). *RR* embryos normally produce the starch-branching enzyme and go through normal synthesis of starch, lipids, and proteins that produce round seeds. On the other hand, *rr* embryos produce an aberrant form of the protein, with deep consequences in the metabolism, ultimately leading to wrinkled peas. The *rr* peas have a *SBEI* gene interrupted by an insertion of a 0.8 kb transposable element similar to the *Ac/Ds* family in maize (Bhattacharyya et al. 1990).

Calculations of allele frequencies are paramount in population genetics. With the allele frequencies in hand, population geneticists can compute numerous statistics that lead to valuable interpretations of the mechanisms shaping populations. But those statistics need a reference model to be based on. Models are important to place comparisons with any situations or mechanisms that diverge from them. The difference between the expected and the observed is referred to as deviation, and in population genetics, we always want to know how much deviation we have in relation to an ideal model. That assists geneticists in reaching conclusions over population genetic mechanisms.

Two papers published independently in 1908 by G. H. Hardy and W. Weinberg stated what we know today as the core model for population genetic analyses. Decades passed until scientists noted that the two researchers published similar conclusions on how allele frequencies behave in consecutive generations (Bacaer 2011). While the paper of Hardy was published in English, in Science (Hardy 1908), the manuscript of Weinberg (1908) was written and published in German. Collectively, the foundations that were independently provided were recognized as the Hardy-Weinberg equilibrium.

The basic assumptions of the Hardy-Weinberg equilibrium are that in the absence of any factors that could alter allele frequencies in an infinitive large and panmictic population, such as the absence of mutations, recombinations, natural selection, random genetic drift, and migrations, the allele frequencies should remain the same along consecutive generations. One generation of crossings should be sufficient to stabilize the allele frequencies as by the model. By *panmictic* we mean a population, in that crosses among individuals are totally random.

Instead of using the previous example, let's consider an important gene of a neotropical legume. The neotropical species *Phaseolus vulgaris* L., the common bean, produces edible grains that are widely consumed in Latin America and other regions in the world. A seed coat trait that is worth mentioning for this species is the presence or absence of the seed coat glossiness. Seed coat glossiness is a trait mainly controlled by the *Asper* (abbreviated as *Asp*) gene, but it is also influenced by the *Joker* (*J*) locus in common bean. Konzen and Tsai (2014) have shown that distinct cultivars and landraces of common bean segregated for two alleles of *Asp*: *Asp*, that is the dominant allele, conferring a glossy aspect to the seed coat; and *asp*, conferring a dull aspect to the seed coat. Using F_1, F_2 and F_3 progenies of a set of crosses and their reciprocals among parental lines differing for glossiness (glossy versus dull seed coats), the authors confirmed the expected ratios for Mendelian segregation, i.e., the monogenic pattern of inheritance. The phenotypic differences are easily distinguishable at the ultrastructural level as an irregular surface on the seed coat as well as a thinner palisade epidermis on dull seed coats (Konzen and Tsai 2018).

Taking the latter trait as an example, let's consider five genotypes varying for seed coat glossiness, such as:

$$Asp\,Asp, Asp\,asp, Asp\,asp, asp\,asp, asp\,asp$$

For this case:

- the allele frequencies are $Asp = 0.40$ or 40%; and $asp = 0.60$ or 60%;
- the genotypic frequencies are $Asp\,Asp = 0.20$ or 20%; $Asp\,asp = 0.40$ or 40%; and $asp\,asp = 0.40$ or 40%;
- the phenotypic frequencies are 60% of seeds with glossiness; and 40% of seeds with dull seed coat.

Now, let's consider a cross between two pure parental lines for *Asp Asp* and *asp asp*. The F_1 generation produced 100% of seeds with a glossy seed coat. By selfing, F_2 produces 75% of lines with glossy seed coat, while 25% have dull seed coat. The expected genotypic frequencies are: 25% *Asp Asp*, 50% *Asp asp*, and 25% *asp asp*.

Usually, we represent one allele (in this case, the dominant) as p, and the alternative allele as q. That considered, we calculate the probability by chance that the p allele (*Asp*) will be transferred to the next generation. Having p and q equal chances of being transmitted, with no differences in fitness or any other factor (fundamental premises of the Hardy-Weinberg principle), the probability of a given genotype having the *Asp* allele in the next generation is 0.5. By means of chance, the probability of the other allele of the diploid genotype being q (asp) is also 0.5. Also by random, the probability that an individual is *Asp Asp* or *Asp asp* or *asp asp* in F_2 is the product between the probabilities of each allele, that is 0.5 × 0.5 = 0.25. Therefore, by designing a Punnet square with all the genotype combinations, we have:

	Asp	*asp*
Asp	$\boldsymbol{p^2}$ *Asp Asp*	$\boldsymbol{pq}$ *Asp asp*
asp	$\boldsymbol{pq}$ *Asp asp*	$\boldsymbol{q^2}$ *asp asp*

Note that we distributed the total probability of 1 (100%) in four genotypes:

- *Asp Asp* – 50% of probability for each allele, $p \times p = p^2 = 0.5 \times 0.5 = 0.25$;
- *Asp asp* – repeated two times, so $2pq = 2 \times 0.5 \times 0.5 = 0.5$;
- *asp asp* – 50% of probability for each allele, $q \times q = q^2 = 0.5 \times 0.5 = 0.25$.

Summing up, we have:

$$25\%\left(p^2\right)+50\%\left(2pq\right)+25\%\left(q^2\right)=100\%$$

That provides the general equation or model:

$$p^2+2pq+q^2=1$$

The basic algebraic expression that mathematically represents what the Hardy-Weinberg equilibrium states. A deeper examination of the mathematical foundations can be read at Hartl and Clark (2007).

Deviations from the equilibrium can be due to a diverse range of factors. Any deviations can be determined through quite simple statistical approaches, such a cui-square test. Significant deviation suggests that one or more evolutionary mechanisms are shaping populations under analysis. In this chapter, we start with one of them, random genetic drift.

4.2 Genetic Drift

Evolution is the change in allele frequency over time. Evolution is only possible with mutations and creations of novel alleles, thereby, with changes in allele frequencies due to multiple other factors over time. *Genetic drift* is the random change in allele frequency due to sampling error in a finite population and it can induce evolution. Unlike natural selection, genetic drift outcomes are unpredictable, however, it is an evolutionary force that can alter the population's gene pool through time.

The number of individuals and their gametes that contribute to the next generation within a population can affect the impact of genetic drift, which is a random process. The frequency of each allele is equally likely to change from generation to generation due to genetic drift. Even if we cannot predict the direction of the change, we can characterize the expected magnitude of the allelic frequency shift.

When a population is small, fluctuations in their allele frequencies are more frequent and the probability that one allele will be fixed or lost at a locus is higher. In Fig. 4.1, we simulated several replicates of the initial population, with two alleles for a given locus. Note that the ancestral gene pool is the same ($p = 0.5$ for one of the alleles and, therefore, 1–0.5 for the other allele) in all the populations simulated and genetic drift is the only evolutionary mechanism at work in populations; then sampling error causes allele frequencies to fluctuate randomly. Now suppose that no gametes are exchanged between the subpopulations (that is, genetically isolated from one another). Therefore, Fig. 4.1 allows us to investigate the role of genetic drift upon subpopulations.

With increasing time, more subpopulations become fixed for one allele or the other, having lost their genetic variation at the locus. Ultimately, all populations are expected to become fixed, that is, all copies of the remaining allele have to be identical-by-descent.

A population of N individuals contains a total of *2 N* gene copies, for diploid species into consideration. If every one of these gene copies is a unique allele and drift is the only mechanism of evolution at work, one of the *2 N* alleles will drift to fixation, and all the others will be lost. Each allele must have an equal chance of being the one that drifts to fixation; that is what we meant when we assumed that drift is the only mechanism of evolution at work. So we have *2 N* alleles, each with an equal probability of becoming fixed. Each allele's chance must therefore be *1/2 N*.

Considering a gene with 2 alleles in a finite population in which A_l is the allele frequency, there is a A_l chance that the allele drifts to fixation and will be A_l and $1 - A_l$ chance that it will be the other allele drifts to fixation.

Sewall Wright (1931) showed that, averaged across many populations, heterozygosity is predicted by the equation:

$$H_{t+1} = H_t \left(1 - 1/2N\right)$$

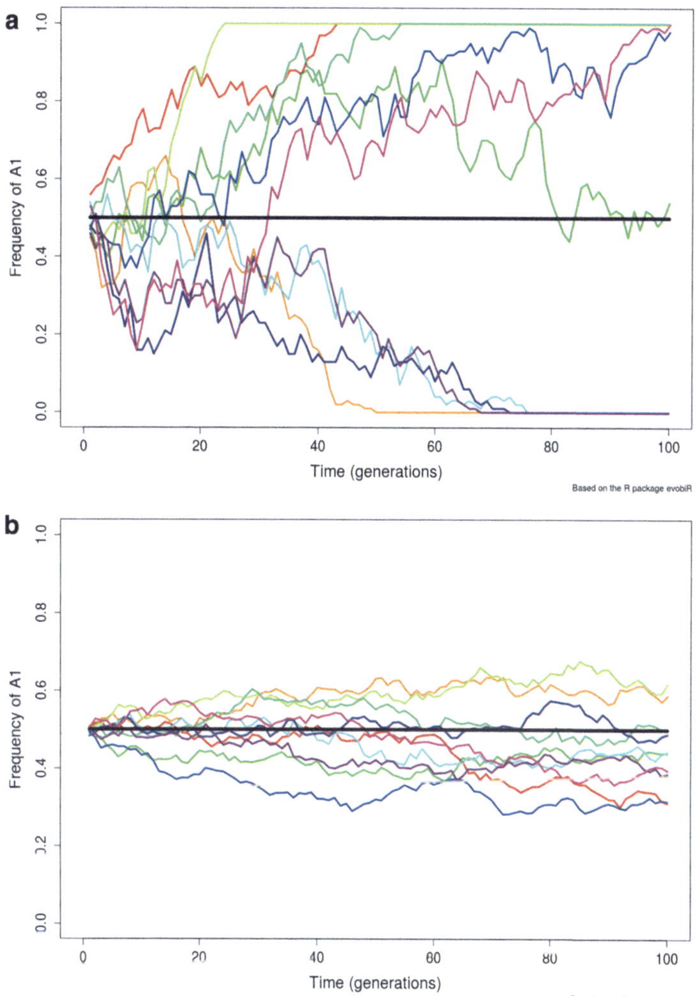

Fig. 4.1 The graph represents a computer simulation of genetic drift in 10 replicate populations starting with an initial allele frequency of 0.5 over a period of 100 generations. The population size is kept constant at 10 individuals (**a**) and 1000 individuals (**b**) every generation. The black line represents the initial frequency kept constant over generations. The simulations were performed using the R package *learnPopGen* (Revell 2019)

where H_{t+1} is the heterozygosity in the next generation, and the decline in heterozygosity depends on the heterozygosity in the current generation (H_t) in a population of N diploid individuals.

As allele frequencies in a finite population drift toward fixation or loss, the frequency of heterozygotes decreases. Figure 4.2 shows the decline in the average frequency of heterozygotes in simulated populations. The frequency of heterozygotes indeed tends to rapidly fall in small populations. The contrary should be

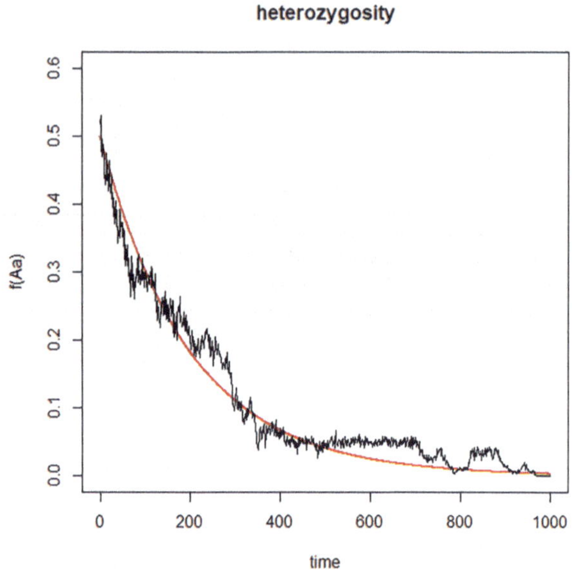

Fig. 4.2 A computer simulation of genetic drift in 10 replicate populations starting with an initial frequency of heterozygotes of 0.5 over a period of 1000 generations. The population size is kept constant at 100 individuals every generation. The red curve represents the rate of decline predicted. The simulations were performed using the R package *learnPopGen* (Revell 2019)

expected in large populations. Eventually one allele or the other will become fixed in every population, and the average frequency of heterozygotes will fall to 0.

It is well established that population extinction risk due to environmental, demographic or genetic stochastic processes is linked to population size. Genetic stochasticity may increase the probability of extinction of small populations due to the enhancement of deleterious effects of genetic drift and inbreeding and loss of genetic variability may lead to a reduction in individual fitness and population evolutionary potential (Willi and Hoffmann 2009).

The probability that a given allele will be the one that drifts to fixation is equal to that allele's initial frequency. Wright's model of genetic drift provides an intuitive explanation of fixation probabilities.

4.3 Effective Population Size

The *effective population size* (*Ne*) can be defined as the size of an ideal population experiencing the same magnitude of random genetic drift over time as the real population under consideration (Wright 1931), that is, a sample of the population has allele frequencies that are subject to genetic drift in the same magnitude as if all individuals of that population were considered.

Because real populations deviate from the idealized models, effective sizes are typically much smaller than census sizes, and this genetic drift is generally measured as a loss of heterozygosity, increase of identity by descent, or change in allele frequency over time.

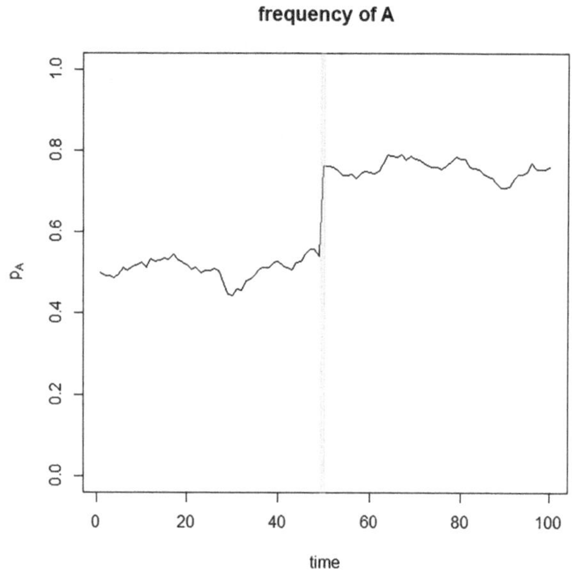

Fig. 4.3 A computer simulation of genetic drift in a population starting with an initial allele frequency of 0.5 over a period of 100 generations. The population size is kept constant at 1000 individuals every generation, until generation 50, at which point the population size is reduced to 2 individuals. The population size then rebounds to 1000 individuals at generation 51 and remains at 1000 for the remainder of the simulation to simulate a bottleneck effect. The simulations were performed using the R package learnPopGen (Revell 2019)

The increased likelihood of stochastic genetic processes can either be a consequence of bottlenecks and founder effect or can be an independent process. In cases where a drastic reduction of population size happens in situ, the bottleneck may be either instantaneous (one generation, as shown in Fig. 4.3) or gradual and take years (generations) to become effective. Founder effect occurs when the population was derived from a small number of founding individuals drawn from a large ancestral population. The most obvious difference between a founder event and a bottleneck event in situ is the spatial movement in the first case and the permanence in the same area in the second.

While drift causes changes in allele frequencies, it also changes multi-locus gamete frequencies. Genetic drift leans towards creating linkage disequilibrium (LD) and stochastic associations between loci. Multi-loci changes in allele frequencies, that is, gamete frequencies changes have potentially stronger evolutionary effects than changes at a single locus. Founder and bottleneck effects are predominantly effective in producing LD and these stochastic associations can persist for many generations. Only recombination can break LD created by genetic drift.

The effectiveness of gene flow in counteracting genetic stochasticity and fostering population genetic connectivity is strongly correlated to *Ne*, especially when populations are subdivided (Palstra and Ruzzante 2008).

Although it has been recognized that very small numbers of migrants per generation may be enough to counteract the deleterious effects of genetic drift (Slatkin, 1985), population structure is expected even in not isolated populations due to genetic stochasticity (Lenormand et al. 2009; Lowe and Allendorf 2010). Therefore, effective population size is a crucial constraint to evaluate population persistence probability and to understand species evolution (Luikart et al. 2010).

Habitat fragmentation influences evolution by changing patterns of genetic drift and gene flow among populations, which is usually referred to as *nonadaptive evolution* (Schulz et al. 2022). Because of the genetic isolation under fragmentation, drift will operate independently in each subpopulation. Because of the randomness of the evolutionary direction of drift, it is unlikely that all the independent subpopulations will evolve in the same direction. The chances of any particular population deviating from 0.5 and the magnitude of that deviation increase with each generation. At the population level, heterozygosity levels among populations are measured by Wright's F_{ST} statistic (Wright 1965):

$$F_{ST} = \left(H_T - H_S\right) / H_T$$

Where H_T is the expected heterozygosity under Hardy–Weinberg equilibrium in a total population created by combining all of our separate populations, and H_S is the average across separate populations (also known as subpopulations) in their expected heterozygosities. F_{ST} is also called the *fixation index.*

Insular populations are characterized by smaller effective population sizes than mainland populations and low genetic diversity due to founder effects (Robinson et al. 2016; Mayr 1967; Frankham 1997). The size of an island, its distance from the mainland, and the time elapsed since its isolation will affect the genetic structure of its communities, and the likelihood of local extinction (Jaenike 1973). However, the estimates of genetic variation in neotropical island populations do not always have lower genetic diversity in large islands that only recently separated from the mainland (Francisco et al. 2016).

4.4 Neutral Evolution in the Neotropical Landscape

The geological history of landscapes is strongly associated with the evolution of their resident communities; therefore, the complex Neotropical landscape provides a natural source for population structure.

Understanding which geographic barriers promote and maintain population divergence is critical to conserving genetic diversity and endemism within the region. Geographic barriers are associated with a wide span of divergence times, representing a protracted history of species distribution postdating barrier development. The Rio Negro, a major tributary of the Amazon River, has played a key role in driving allopatric speciation for various animal taxa. Similarly, the gene flow for the *Buchenavia oxycarpa* Amazon tree population is restricted by riverine barriers (Nazareno et al. 2019).

It has been proposed that the climate of the Neotropics, with a more constant temperature and reduced seasonal variation, has given rise to amplified ecological specialization, increased unevenness of species distributions, and less migration rate among populations (Terborgh 1980). Neotropical populations that are historically separated by large rivers, mountain chains, or islands, tend to be relatively distinct genetically, and even morphologically (Marks et al. 2002). Dellinger and collaborators (2022) found consistently high genetic differentiation among populations of neotropical plant species differing in pollination strategy.

Neotropical forests are the most widespread and the most biodiverse ecosystems on Earth, yet only 24% remains intact. Indeed, the Neotropics are undergoing intense economic development. It certainly puts a tension on natural resources, foremost on forests, which have been cut for timber, pasture and for croplands (Lewis et al. 2015).

In recent years, urbanization of Neotropical biomes is predicted to strongly influence genetic drift, which produces stochastic changes in allele frequencies between generations. Urbanization can cause such impacts in multiple ways, including primarily the loss of natural habitat caused by fragmentation, founder effects and bottlenecks associated with response to urban expansion. These circumstances are predicted to result in both a loss of genetic diversity within populations and amplified population structure (Ordóñez-Delgado et al. 2022).

In fact, spatial isolation of urban populations may restrict connectivity, leading to low levels of gene flow between fragments and subsequent higher genetic structure and lower genetic diversity in isolated populations (Honnay and Jacquemyn 2007). Urbanization limits gene flow and has an effect in the genetic structure of some neotropical species, but barely affects others. Such differences are associated with distinct levels of tolerance to urban environments and the resources that these landscapes offer to each species (Rodríguez-Bardía et al. 2022).

The impact of habitat fragmentation in natural populations can significantly disturb the movement of animals, pollen, and seeds (Tewksbury et al. 2002), which can change population genetic structure (Sebbenn et al. 2011) and decrease genetic diversity (Aguilar et al. 2008).

To avoid the loss of genetic diversity in neotropical regions by genetic drift, it is imperative to protect populations from fragmentation within each region of endemism, warranting the continued persistence of evolutionarily distinct lineages within species and preserve suitable tracts of natural habitat within each region of endemism. Special emphasis should be placed on endemism regions separated by lowland barriers and deep intermontane river valleys (Weir 2009).

However, population genetic structure does not always lead to extinction (Nosil 2012). For instance, geographically isolated populations diverge and persist in a continuous process, until the evolution of reproductive isolation and ecological discrepancy finishes the allopatric speciation (Allmon 1992). Population structure per se is an indispensable step of diversification, implying that differentiation and diversification should be associated over a deep evolutionary time (Harvey et al. 2017).

Diversification in the Neotropical genus *Mabuya*, endemic to the coastal mountain range of Venezuela, started approximately in the Middle Miocene (15.5–13.4 Ma) and seems to be a good example to demonstrate the relevance of bottleneck effects for speciation. Furthermore, the split between *M. bistriata* and *M. altamazonica* and between the occidental and (meridional + oriental) clades of *M. nigropunctata* fits very well with the biogeographic split between the eastern and western Amazon basins reported for several other taxa (Miralles et al. 2010).

Another case is the isolation, and continuous drift occurred in the three populations of the ancestral *M. bistriata* found in the Hylaea, the northern Atlantic Forest, and the Cariri enclave. Nevertheless, significant divergence occurred only in the smaller population. When the small size of the geographical areas of isolation and differentiation from which one might infer that the effective population sizes were also small (Loebmann and Haddad 2010).

The habitat fragmentation of plants and animals in neotropical refuges may have resulted in differentiation, subspeciation, or speciation, while extinction was the consequence in other cases. Nevertheless, genetic drift is the major force responsible for changes in populations allele frequencies across the genome over time (Lynch et al. 2011).

When genetic drift is the only evolutionary force at work, the rate of substitution is equivalent to the mutation rate. This is right regardless of the population size, since more mutations arise in a larger population, nevertheless, in a large population each new mutation has a smaller probability of drifting to fixation.

Under genetic drift, large populations generate and maintain more genetic variation than small populations, but populations of all sizes accumulate substitutions at the same rate. However, there are certainly other evolutionary forces than genetic drift that are frequently at work. Considering natural selection, most mutations are deleterious while others are selectively neutral or even advantageous (at a lower frequency).The deleterious mutations will be eliminated by natural selection and will never become fixed. The rate of substitution will then be equal to the rate at which neutral and advantageous mutations occur. Nonetheless, advantageous mutations are exceedingly rare and that most alleles of most genes are selectively neutral. In the most part of the genome, the rate of evolution will be equal to the neutral mutation rate.

The evolution would be determined by the selective force against deleterious mutations and the capability of the population to overcome genetic drift, which is determined by effective population size.

4.5 Key Notes

- The Hardy-Weinberg equilibrium states that allele frequencies remain equal through consecutive generations in a large, infinite and panmictic population, unless random genetic drift, natural selection, mutations, recombination,

migration and other factors cause the frequencies to differ from their previous generation.
- Genetic drift is a random process that can have a significant impact on the genetic diversity of populations, particularly in small and isolated populations.
- Small population sizes and habitat fragmentation can lead to increased levels of genetic drift, potentially resulting in reduced genetic diversity and increased inbreeding.
- Genetic drift plays a role in genetic structure, as populations that become isolated from each other may accumulate genetic differences over time.
- In the Neotropics, genetic drift may be particularly important due to the high levels of species diversity and the fragmentation of habitats.
- Conservation efforts in the Neotropics should take genetic drift into account, as it can have important implications for the long-term viability of populations and species.

4.6 Final Remarks

Genetic drift is a fundamental process in evolution, and it plays a significant role in shaping the genetic diversity of populations. Neotropical populations are dynamic and respond to their changing environments in complex ways. Genetic drift shape the genome makeup of these diverse ecosystems.

Questions

1. How does genetic drift affect Neotropical populations?
2. Why does genetic drift tend to have a greater impact on neotropical populations compared to temperate populations?

References

Aguilar R, Quesada M, Ashworth L et al (2008) Genetic consequences of habitat fragmentation in plant populations: susceptible signals in plant traits and methodological approaches. Mol Ecol 14:1811–1820. https://doi.org/10.1111/j.1365-294X.2008.03971.x

Allmon WD (1992) A causal analysis of stages in allopatric speciation. Oxf Surv Evol Biol 8:219–257. https://doi.org/10.1093/oso/9780195062892.003.0007

Bacaër N (2011) A short history of mathematical population dynamics. Springer-Verlag, London. https://doi.org/10.1007/978-0-85729-115-8

Bhattacharyya MK, Smith AM, Ellis THN et al (1990) The wrinkled-seed character of pea described by Mendel is caused by a transposable element insertion in a gene encoding starch-branching enzyme. Cell 60(1):115–122. https://doi.org/10.1016/0092-8674(90)90722-J

Dellinger AS, Paun O, Baar J et al (2022) Population structure in Neotropical plants: integrating pollination biology, topography and climatic niches. Mol Ecol 31:2264–2280. https://doi.org/10.1111/mec.16403

Francisco FO, Santiago LR, Mizusawa YM et al (2016) Genetic structure of Island and mainland populations of a Neotropical bumble bee species. J Insect Conserv 20:383–394. https://doi.org/10.1007/s10841-016-9872-z

Frankham R (1997) Do Island populations have less genetic variation than mainland populations? Heredity (Edinb) 78:311–327. https://doi.org/10.1038/hdy.1997.46

Hardy GH (1908) Mendelian proportions in a mixed population. Science 28(706):49–50. https://doi.org/10.1126/science.28.706.49

Hartl DL, Clark AG (2007) Principles of population genetics. Sinauer Associates

Harvey MG, Seeholzer GF, Smith BT et al (2017) Positive association between population genetic differentiation and speciation rates in New World birds. Proc Natl Acad Sci USA 114:6328–6333. https://doi.org/10.1073/pnas.1617397114

Honnay O, Jacquemyn H (2007) Susceptibility of common and rare plant species to the genetic consequences of habitat fragmentation. Conserv Biol 21:823–831. https://doi.org/10.1111/j.1523-1739.2006.00646.x

Jaenike JR (1973) A steady state model of genetic polymorphism on islands. Am Nat 107:793–795. https://doi.org/10.1086/282878

Konzen ER, Tsai SM (2014) Seed coat shininess in Phaseolus vulgaris: rescuing a neglected trait by its screening on commercial lines and landraces. J Agric Sci 6(8):113

Konzen ER, Tsai SM (2018) Genetic variation of landraces of common bean varying for seed coat glossiness and disease resistance: valuable resources for conservation and breeding. In: Rediscovery of landraces as a resource for the future. IntechOpen, London, pp 177–193

Lenormand T, Roze D, Rousset F (2009) Stochasticity in evolution. Trends Ecol Evol 24:157–165. https://doi.org/10.1016/j.tree.2008.09.014

Lewis SL, Edwards DP, Galbraith D (2015) Increasing human dominance of tropical forests. Science 349:827–832. https://doi.org/10.1126/science.aaa9

Loebmann D, Haddad CFB (2010) Amphibians and reptiles from a highly diverse area of the Caatinga domain: composition and conservation implications. Biota Neotropica 10(3):227–256. https://doi.org/10.1590/S1676-06032010000300026

Lowe WH, Allendorf FW (2010) What can genetics tell us about population connectivity? Mol Ecol 19:3038–3051. https://doi.org/10.1111/j.1365-294X.2010.04688.x

Luikart G, Ryman N, Tallmon DA et al (2010) Estimation of census and effective population sizes: the increasing usefulness of DNA-based approaches. Conserv Genet 11:355–373. https://doi.org/10.1007/s10592-010-0050-7

Lynch M, Bobay LM, Catania F et al (2011) The repatterning of eukaryotic genomes by random genetic drift. Annu Rev Genomics Hum Genet 12:347–366. https://doi.org/10.1146/annurev-genom-082410-101412

Marks BD, Hackett SJ, Capparella AP (2002) Historical relationships among Neotropical lowland forest areas of endemism as determined by mitochondrial DNA sequence variation within the Wedge-billed Woodcreeper (Aves: Dendrocolapitidae: *Glyphorynchus spirurus*). Mol Phyl Evol 24:153–167. https://doi.org/10.1016/S1055-7903(02)00233-6

Mayr E (1967) The challenge of Island faunas. Aust Nat Hist 15:369–374

Miralles A, Carranza S (2010) Systematics and biogeography of the Neotropical genus Mabuya, with special emphasis on the Amazonian skink *Mabuya nigropunctata* (Reptilia, Scincidae). Mol Phyl Evol 54:857–869. https://doi.org/10.1016/j.ympev.2009.10.016

Nazareno AG, Dick CW, Lohmann LGA (2019) Biogeographic barrier test reveals a strong genetic structure for a canopy-emergent Amazon tree species. Sci Rep 9(18602):18602. https://doi.org/10.1038/s41598-019-55147-1

Nosil P (2012) Ecological speciation. Oxford; New York: Oxford University Press.

Ordóñez-Delgado L, Iñiguez-Armijos C, Díaz M et al (2022) The good, the bad, and the ugly of urbanization: response of a Bird Community in the Neotropical Andes. Front Ecol Evol 10(844944):1–11. https://doi.org/10.3389/fevo.2022.844944

Palstra FP, Ruzzante DE (2008) Genetic estimates of contemporary effective population size: what can they tell us about the importance of genetic stochasticity for wild population persistence? Mol Ecol 17:3428–3447. https://doi.org/10.1111/j.1365-294x.2008.03842.x

Revell LJ (2019) learnPopGen. An R package for population genetic simulation and numerical analysis. Ecol Evol 9(14):7896–7902. https://doi.org/10.1002/ece3.5412

Robinson JA et al (2016) Genomic flatlining in the endangered Island fox. Curr Biol 26:1183–1189. https://doi.org/10.1111/j.1365-294X.2008.03842.x

Rodríguez-Bardía M, Fuchs EJ, Barrantes G et al (2022) Genetic structure in neotropical birds with different tolerance to urbanization. Sci Rep 12:6054. https://doi.org/10.1038/s41598-022-09961-9

Schulz L, Sendker FL, Hochberg GKA (2022) Non-adaptive complexity and biochemical function. Curr Opin Struct Biol 73:102339. https://doi.org/10.1016/j.sbi.2022.102339

Sebbenn AM, Carvalho ACM, Freitas MLM et al (2011) Low levels of realized seed and pollen gene flow and strong spatial genetic structure in a small, isolated and fragmented population of the tropical tree *Copaifera langsdorffii* Desf. Heredity 106:134–145. https://doi.org/10.1038/hdy.2010.33

Slatkin M (1985) Gene flow in natural populations. Annu Rev Ecol Evol Syst 16:393–430. https://doi.org/10.1016/j.sbi.2022.102339

Terborgh J (1980) Causes of tropical species diversity. Proc Int Ornithol Congr 17:955–961

Tewksbury JJ, Levey DJ, Haddad NM et al (2002) Corridors affect plants, animals, and their interactions in fragmented landscapes. Proc Natl Acad Sci 99:12923–12926

Weinberg W (1908) Über den Nachweis der Vererbung beim Menschen. Jahreshefte des Vereins für Vaterländische Naturkunde in Württemberg 64:368–382

Weir JT (2009) Implications of genetic differentiation in neotropical montane forest birds. Ann Missouri Bot Gard 96(3):410–433. https://doi.org/10.3417/2008011

Willi Y, Hoffmann AA (2009) Demographic factors and genetic variation influence population persistence under environmental change. J Evol Biol 22:124–133. https://doi.org/10.1111/j.1420-9101.2008.01631.x

Wright S (1931) Evolution in Mendelian populations. Genetics 16:97–159. https://doi.org/10.1093/genetics/16.2.97

Wright S (1965) The interpretation of population structure by F-statistics with special regard to systems of mating. Evolution 19:395–420. https://doi.org/10.1111/j.1558-5646.1965.tb01731.x

Chapter 5
How Do Natural Selection and Human Interactions Modulate Genetic Variation?

Flaviane Malaquias Costa, Ana Flávia Francisconi, Jonathan Morales-Marroquín, and Maria Imaculada Zucchi

Contents

Abstract The theory of natural selection is a crucial topic that any student focusing on population genetics should understand. Here, we provide a broad and simple definition of natural selection, accompanied by examples from the neotropical region. Moreover, the processes and mechanisms by which natural selection alters genetic variation are briefly described. Signatures of selection in living organisms such as plants refer to the observable patterns in the genetic variation of populations that may indicate the action of natural selection. These patterns are the result of specific alleles or genetic variants being favored or disfavored by selective pressures

F. M. Costa · A. F. Francisconi
Escola Superior de Agricultura "Luiz de Queiroz", Universidade de São Paulo, Piracicaba, SP, Brazil

J. Morales-Marroquín
Universidade Estadual de Campinas – Cidade Universitária Zeferino Vaz – Barão Geraldo, Campinas, Brazil

M. I. Zucchi (✉)
Escola Superior de Agricultura "Luiz de Queiroz", Universidade de São Paulo, Piracicaba, SP, Brazil

Agência Paulista de Tecnologia dos Agronegócios, Piracicaba, SP, Brazil

M. V. Bohrer Monteiro Siqueira et al. (eds.), *Population Genetics in the Neotropics*,
https://doi.org/10.1007/978-3-031-83685-5_5

in the environment. Several methods are used by researchers to detect and study these kinds of signatures. The results can provide useful insights into how organisms adapt and evolve in response to their ecological surroundings, as well as providing valuable insights into their response to changes in their environment. Consequently, this can help us comprehend the mechanisms that underlie evolution and ultimately contribute to efforts in conservation and ecosystem recovery. In this chapter, we describe the results obtained from genetic analyses using different molecular markers and genomic sequencing methods, and present case studies involving the neotropical species *Euterpe precatoria*, *Acrocomia aculeata* and *Bertholletia excelsa*. This knowledge is critical for comprehending how these species respond to environmental changes, and for developing effective conservation strategies to preserve their genetic diversity and ecological roles in different ecosystems.

Keywords SNPs · SSR · Whole genome sequencing · Outliers markers · Adaptation

Learning Objectives
This chapter explores the following topics:

- Explain how natural selection acts on genetic variation in plant populations and identify the genetics/genomic methods used to detect selection pressures.
- Introduce the concept of natural selection, its different modes, and how they help to explain adaptation.
- Determine how sequencing methods, such as genotyping-by-sequencing (GBS) and whole genome sequencing (WGS) contribute to the identification of selection signatures in different Neotropical species.
- Evaluate how selection promotes adaptation to specific environments and how studies using SNPs and SSR markers are used to comprehend adaptations.
- Identify the main differences between neutral markers and outlier markers in genomic analyses and how such differences affect the interpretation of genomic data.
- Apply the signatures of selection detected in genomic studies to predict responses of species to environmental changes and to support the development of conservation strategies.

5.1 Introduction

Natural selection is the process by which advantageous characteristics become more frequent in a population over time, due to the higher reproductive success of the individuals (Futuyma 2005; Freeman and Herron 2007). In a population, there is natural variation, which is hereditary and results from genetic differences between individuals. In a competitive environment, individuals with characteristics that provide an adaptive advantage are more capable to survive and reproduce. Survival of the fittest is the mechanism by which natural selection acts, favoring individuals who have characteristics that best adapt them to their environment. Therefore, natural selection is the process that results in the adaptation of species (Futuyma 2005; Freeman and Herron 2007).

The relationship between phenotype and fitness can be complex, since there are different modes of selection (Futuyma 2005; Freeman and Herron 2007) which can act on populations: directional selection, stabilizing selection, disruptive selection, purifying selection and artificial selection (more and related information in Chap. 3 as well as in the following Box 5.1).

Box 5.1 What Are the Different Modes of Selection?

Directional selection: fitness increases (or decreases) in proportion to the value of a trait. Directional selection on a continuous trait modifies the average value of the trait in the population.

Stabilizing selection: individuals with intermediate values of a trait have higher fitness. Stabilizing selection on a continuous trait does not change the average value of the trait in the population. However, it reduces variation by removing the extremes of the trait's distribution.

Disruptive selection: individuals with extreme values of a trait have higher fitness. Disruptive selection on a continuous trait does not change the average value of the trait in the population.

Purifying selection: removes deleterious genetic variants. Purifying selection on a continuous trait modifies the mean value of the trait in the population.

Artificial selection: corresponds to human directed selection for desirable characteristics in the population. Artificial directional selection on a continuous trait modifies the average value of the trait in the population. For example, selecting maize varieties that produce more kg per hectare than the average of the population.

Genetic variation in plants, animals and other organisms associated with evolutionary processes and adaptation has been characterized through genome sequencing (Meyer and Purugganan, 2013; Kistler et al. 2020), and also by evidence of natural selection and as well as artificial selection due human interference (Beaumont and Nichols 1996; Vigouroux et al. 2002; Da Fonseca et al. 2015; Alves-Pereira et al. 2020, 2022). By studying population genomics, biogeography, phylogenetics, and the distribution of neotropical species, we can reveal the historical patterns and processes that have shaped their current distribution and evolutionary processes. Some of the key factors influencing organisms diversification and adaptation are the historical dispersal events, geographical barriers, and geological activity (Eiserhardt et al. 2013). In addition to geographical factors, environmental conditions such as climate, biome and region characteristics also play a significant role in shaping species distribution.

In plants, candidate genes for selection were identified in maize (Meyer and Purugganan 2013) and cassava genomes (Alves-Pereira et al. 2020, 2022), which allowed us to better understand the selection forces that acted on the evolutionary process of such neotropical species. Moreover, this can provide valuable information for predicting the potential impacts of future environmental and anthropogenic changes. Overall, studying the selective signatures on species diversity in the Neotropics is not only important to understand the factors that have shaped diversity, but also for conducting conservation strategies and deepening our knowledge on tropical species ecology.

Molecular markers used in genetic and genomic analyses can provide different types of information (Avise 2004), and have been useful in evolutionary studies of species in the Neotropical region (Clement et al. 2010; Cristo-Araújo et al. 2013; Sujii et al. 2015; Alves-Pereira et al. 2018, 2020, 2022). DNA polymorphism based on simple sequence repeats (SSR) or microsatellites are considered the most important molecular markers for estimating population genetic parameters, as they are highly informative due to their codominance and because they have multi-allele patterns and high heterozygosity (Collevatti et al. 1999). Although microsatellite markers are valuable for population studies, it has been shown that a small fraction of single nucleotide polymorphism (SNP) markers can have very high information content for population structure analysis, outperforming microsatellites (Luikart et al. 2003; Helyar et al. 2011). With the advent of next generation sequencing (NGS), it has become possible to identify thousands of single nucleotide polymorphism (SNP) markers, even in non-model organisms. Genotyping-by-sequencing (GBS) is one of the approaches used to obtain SNP markers. GBS uses restriction enzyme-based complexity reduction together with DNA barcode adapters, which allows the construction of multiplex libraries (NGS; Poland and Rife 2012). SNP markers are widely distributed throughout the genome. Ever since the development of GBS, it has been possible to sample at least 6% of the total genome (Poland and Rife 2012). For further information on molecular markers, read Chap. 9.

In addition to determining genomic diversity and structure based on SNP markers alone, under the assumption of neutrality, this new approach enables us to identify loci that show signatures of selection (Etter and Johnson 2012; Zucchi et al. 2019). This means that neutral SNP markers found throughout the genome will be similarly affected by the demography and evolutionary history of populations. At the same time, SNP markers under selection should behave differently. They usually exhibit "outliers" patterns of variation (Luikart et al. 2003), indicating putative adaptation or functional role. Outlier SNP-containing DNA sequences can be annotated with gene ontology (GO) terms to explore information about cellular components, molecular functions, and biological processes (Blake 2013). Finally, the combined use of neutral SNP markers with outliers can also add a new dimension to population genetics (Luikart et al. 2003, 2018), now called population genomics. By combining these two types of SNP, it is possible to delineate differentiated populations and indicate priority sites for conservation and management initiatives within the species (Funk et al. 2012; Luikart et al. 2018; Breed et al. 2019).

With advances in NGS, whole genome sequencing projects have become much easier to accomplish. Whole genome sequencing (WGS) is used to determine the complete genome sequence of a species, including all genetic information contained in their genome, from chromosomes to the smallest fragments of DNA (Ekblom and Wolf 2014; Fuentes-Pardo and Ruzzante 2017). WGS can be broadly applied to study evolution, genetic diversity, taxonomy, genetic mutations, identification of genes, regulatory elements, potential selection marks in the genome, interactions between genes and the environment, as well as other genomic elements. The application of WGS in population genetics allows us to provide a more comprehensive and detailed view of the genetic variability present in a given population, as it enables the detection of diversity and population structure on an unprecedented scale in terms of the number of loci, at a scale of millions of molecular markers (Ekblom and Wolf 2014; Fuentes-Pardo and Ruzzante 2017; Kistler et al. 2018, 2020; Sandercock et al. 2022). Despite technological advances, WGS can still be challenging and expensive, especially for large or large-scale genomes. However, the costs associated with sequencing have decreased over time, making WGS much more affordable.

NGS techniques have allowed the transition from the classical genetic approach, focused on more specific regions of the genome, to whole genome scale analyses (Allendorf et al. 2010; Ekblom and Wolf 2014; Lewin et al. 2018). In this way, by accessing the partial or total genome of organisms, it is possible to perform evolutionary studies, based on more robust data, and expand our understanding of the adaptive potential of populations in responses to environmental changes, in order to support conservation strategies (Allendorf et al. 2010; Ekblom and Wolf 2014; Lewin et al. 2018).

The main objective of this chapter was to explore how natural selection and human interactions influence genetic variation in various species, especially in the Neotropical region. The chapter aims to describe the processes and mechanisms by which selection affects genetic diversity, using specific examples of plant species such as *Euterpe precatoria*, *Acrocomia aculeata* and *Bertholletia excelsa*. It emphasizes the importance of understanding genetic variation for conservation strategies, ecological studies and the adaptation of species to environmental changes. The reader should be reminded that although plants are the focus of examples here, similar approaches can be applied to animals as well.

5.2 Natural Selection and Anthropogenic Action as Shapers in the Genetic Diversity and Structure of Açaí Palms

Palm diversity in the Neotropical region has long fascinated researchers, as it represents one of the most diverse and iconic plant groups in these tropical areas. Finding selection signatures on palm genomes is essential for unraveling the complex patterns and mechanisms driving their distribution and evolution. Such insights can aid valuable information to conservation initiatives, as well as contribute to our understanding of the broader field of tropical plant ecology and evolution. *Euterpe precatoria* Mart., popularly known as "açaí", is a single-stemmed and solitary palm that can reach 23 m in height (Isaza et al. 2017). *E. precatoria* is widely distributed in Central and South America (Henderson and Galeano 1996), being the most abundant plant species in the Amazon (Ter Steege et al. 2013, Fig. 5.1). Within the Amazon, the species can occur in areas of terra-firme (upland areas), non-floodable areas and várzeas (lowland areas; Bernardes Filho 2021). In recent years, *E. precatoria* has become the most collected species for non-timber forest products (NTFPs; Lopes et al. 2022), due to the production of açaí fruit (Fig. 5.1c), as Brazil has experienced a significant increase in the yield of this fruit in the last decade (80.32%, Instituto Brasileiro de Geografia e Estatística, IBGE 2021). The extractivism of *E. precatoria* is the main factor that has made the state of Amazonas the second largest producer of açaí fruit in Brazil (around 31% of Brazilian production; Melo et al. 2021; Ramos et al. 2021). This palm is vital to traditional river dwellers and contributes to the economy, society, and culture of 6 million households (Lopes et al. 2022).

Euterpe precatoria has been the subject of recent population genetic studies, mainly using simple sequence repeat (SSR) molecular markers. Ramos et al. (2016)

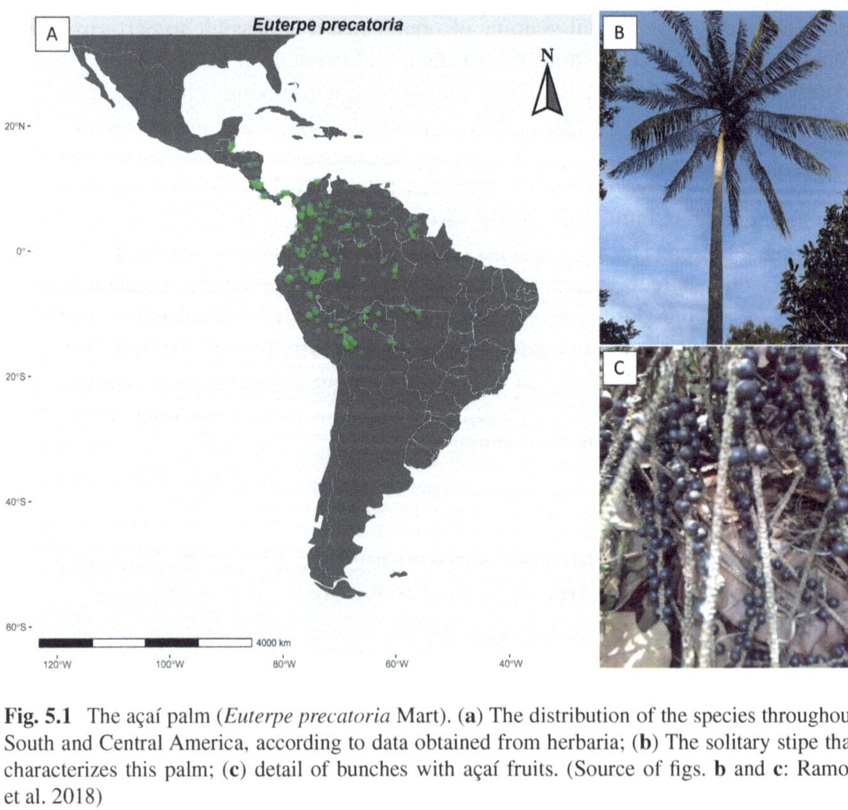

Fig. 5.1 The açaí palm (*Euterpe precatoria* Mart). (**a**) The distribution of the species throughout South and Central America, according to data obtained from herbaria; (**b**) The solitary stipe that characterizes this palm; (**c**) detail of bunches with açaí fruits. (Source of figs. **b** and **c**: Ramos et al. 2018)

developed microsatellite markers for the species and characterized their reproductive system, genetic diversity and population structure, pollen and seed flow in the city of Parintins, Amazonas (Ramos et al. 2018, 2019). More recently, they also characterized the genetic diversity and structure of 19 sampling sites in different regions of Brazilian Amazonia (Ramos et al. 2021). Other population studies have also been conducted for this species with microsatellite markers transferred from *E. edulis* (Gaiotto et al. 2001). These were carried out by da Cruz et al. (2022), who analyzed individuals from five municipalities located in northern Mato Grosso state, where deforestation pressure is high. Finally, the last population-based assessment of *E. precatoria*, using SSRs transferred from *E. edulis*, included three municipalities in the Amazonas state to determine population diversity and structure, as well as their mating system using adult individuals and their seedlings (Perrut-Lima et al. 2023).

These studies characterize the species as being predominantly allogamous (Ramos et al. 2018, 2019; Perrut-Lima et al. 2023). *E. precatoria* has an entomophilous pollination, made by beetles and bees (Küchmeister et al. 1997) and revealed a pollen flow from 50–1500 m (Ramos et al. 2018). The seed dispersal varies from 4–1400 m (mean of 400 m), with most seeds distributed near the maternal palms (Ramos et al. 2018). This pattern is influenced by primary and secondary seed dispersal mechanisms of the species. Primary dispersal occurs by precipitation. Secondary dispersal is mainly performed by rodents and birds over short distances (Zona and Henderson 1989) and by human management over long distances (Ramos et al. 2018, 2021). In addition, dispersal can also occur according to the river flow (Bovi and Castro 1993).

Among the results obtained by Ramos et al. (2018) and Perrut-Lima et al. (2023), mature palms tend to present greater genetic diversity, represented by higher heterozygosities, than seedlings. The species has factors that contribute to gene flow and pollen availability, such as being highly abundant (Ter Steege et al. 2013; Perrut-Lima et al. 2023), and reproduction practically through crossing (Ramos et al. 2019; Perrut-Lima et al. 2023). However, the short primary and secondary dispersal of seeds, by precipitation and small animals, leads to restricted gene flow and interbreeding between relatives (Ramos et al. 2018; Perrut-Lima et al. 2023). This factor has been suggested as the cause for the lower genetic diversity in seedlings. As higher values were detected in mature palms (Perrut-Lima et al. 2023, Ramos et al. 2021), natural selection may operate in guaranteeing higher heterozygosities in natural populations of the species.

As previously mentioned, human management is also one of the secondary factors in the dispersal of *E. precatoria*. Human action on açaí palms, as well as their domestication process, may have been occurring since the time of the Amerindians (Clement 2001). This action has also influenced the pattern of their population in large aggregations, called açaizais in the Amazon, which may be one of the reasons for their abundance in the biome (Ter Steege et al. 2013). The effects of recent human actions are clearly identified in the study by Cruz et al. (2022), where low heterozygosity and high inbreeding were found, contrary to the other population genetic results (Ramos et al. 2018; Perrut-Lima et al. 2023). Fragmentation due to deforestation in the state of Mato Grosso has been identified as one of the main

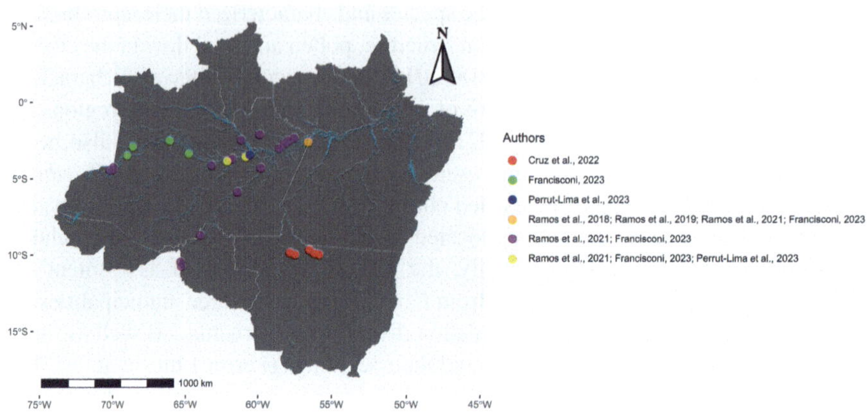

Fig. 5.2 Locations of all collection sites involved in genetic and genomic studies of populations with the *E. precatoria*. In Ramos et al. (2018, 2019, 2021) the populations were analyzed using single sequence repeat (SSR) markers from *E. precatoria*. In da Cruz et al. (2022) and Perrut-Lima et al. (2023), SSR markers transferred from *E. edulis* were used. In Francisconi (2023), single nucleotide polymorphism (SNP) were used, divided into putatively neutral and outliers

contributors to this result, as the mosaic created among fragments obstructs the action of dispersal agents (da Cruz et al. 2022). Thus, in addition to natural selection modulating the diversity of adult palms, human intervention cannot be ruled out for this species. Besides ancient management, humans have widely consumed açaí, dispersed its seeds and fragmented its environment, thus interfering with the genetic diversity and population structuring process of *E. precatoria*.

Efforts have been made to re-evaluate and expand the sampling areas of *E. precatoria* in the Brazilian Amazon. By sampling 23 sites, a new population genetic study with SNPs markers in *E. precatoria* was carried out (Francisconi 2023, Fig. 5.2). Based on different tests for deviations from neutrality, the set of markers was divided into putatively neutral and under-selection SNPs (outliers). GO terms were identified in the DNA sequences with outlier SNP markers, and genomic diversity and population structure were estimated using both neutral and outlier markers (Francisconi 2023). As a result, in sequences with outlier SNPs, functions such as tolerance to heat stress and increased photosynthetic capacity were found. Increased resistance to fungal infections and response to jasmonic acid were also identified, as well as the control of root growth by auxin and root morphogenesis. These findings suggest selective pressure favoring increased resistance and photosynthetic capacity, resulting in greater energy efficiency and, consequently, higher fruit production.

When comparing neutral and outlier markers, the most notable difference appeared in the population structure analysis, with a decrease in the number of genomic groups. In the same analysis, discriminant analysis of principal components (DAPC; Jombart et al., 2010; Fig. 5.3a, b), the number of genomic groups decreased from 12 with neutral SNPs to nine with outlier SNPs (Francisconi 2023). Thus, the use of outlier SNPs indicated selective pressure. Given the

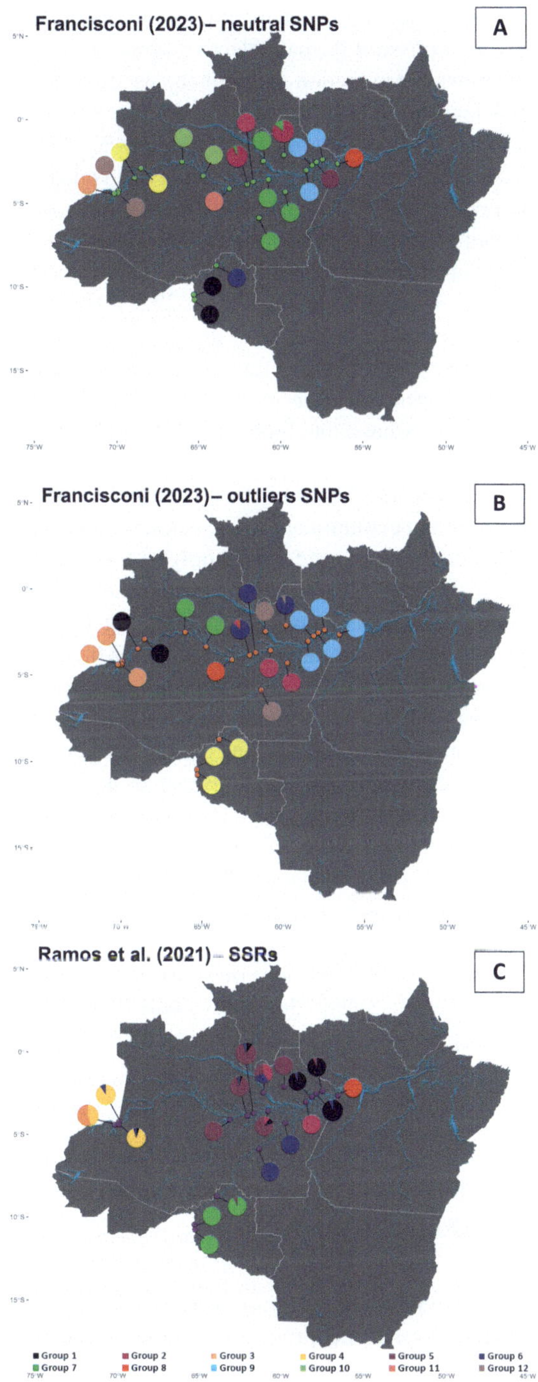

Fig. 5.3 Results of population structure analyses in *E. precatoria* by Francisconi (2023) and Ramos et al. (2021). The results are based on discriminant analysis of principal components (DAPC), where each color in the circles of the collection sites represents a specific genetic/genomic group. In (**a**) Neutral SNP markers were used in 23 collection sites in the Brazilian Amazon, in which 12 genomic groups were identified (Francisconi 2023). (**b**) SNP markers under selection (outliers) were used in the same 23 locations in the Amazon, in which nine genomic groups were identified (Francisconi 2023). (**c**) SSR markers were used to characterize 19 collection sites, determining eight genetic groups (Ramos et al. 2021)

significant demand for açaí fruit, as well as the results identified by the GO terms, there is a possibility that selective pressure is being directed towards palms with higher fruit production per bunch, as this represents a significant productivity advantage for *E. precatoria* (Cartaxo et al. 2020). However, another noteworthy finding was the similarity observed between the neutral and outlier SNP datasets, which is the influence of river dispersal syndromes (Fig. 5.3a, b). When a specific site does not consist of an exclusive genomic group, neighboring sites tend to form a genomic cluster, following the flow pattern of the rivers. This result was first observed in the population structure determined by DAPC in the 19 populations analyzed with SSRs (Ramos et al. 2021). However, in this case only eight genetic groups were determined, showing the relative lower discriminatory power of SSRs compared to SNPs (Fig. 5.3a, c). The same pattern of distribution of genetic groupings by riverine location was observed for *Mauritia flexuosa* L.f., another abundant palm in the Amazonian biome (Ter Steege et al. 2013; Sander et al. 2018).

Obtaining large-scale results by combining the two types of SNPs is essential for optimal decisions about species management and conservation, since they can provide different evolutionary information (Funk et al. 2012). Deforestation in Brazilian Amazonia has occurred particularly in the states of Pará, Mato Grosso, Rondônia and Amazonas (Albert et al. 2023). The sustainable use of biodiversity and its ecosystem services is one way of encouraging and reversing this scenario (Albert et al. 2023). As the most abundant species in the biome and a prominent source of NTFPs, *E. precatoria* becomes an important resource for restoration, conservation, and management programmes (Francisconi 2023). Using accessions from a particular region is more effective for conserving the species gene pool and increasing its resilience in present and future scenarios (Breed et al. 2019). Identifying the distribution of genomic groups through structural analyses such as DAPC and other methods using the two types of SNPs will, therefore, guide the appropriate locations for species different efforts (Francisconi 2023). For example, using the DAPC results with both SNPs, it is possible to define at least 12 genomic groups that should be sampled between the collection sites (Fig. 5.3a, b). If the groups are shared between locations, those with the greatest diversity can be chosen to conserve or provide progeny for restoration, since adult individuals of *E. precatoria* are naturally selected for greater genetic diversity in the species (Perrut-Lima et al. 2023).

5.3 From Natural Populations to Agricultural Systems: The Case of *Acrocomia aculeata*, an Incipiently Domesticated Palm

Acrocomia palms are a genus of Neotropic plants that hold great potential for being a commodity for oil production through crop improvement. *Acrocomia aculeata* (Arecaceae: Arecoideae: Cocoseae: Bactridinae subtribe) is the species with the largest geographical distribution of the genus, occurring from central Mexico throughout Central America to northern Argentina (except Ecuador, Peru and Chile) and most of Brazil (de Lima et al. 2018; Díaz-Hernández et al. 2023), preferably in open areas associated with savannas, grasslands, dry broadleaf forests, xeric shrublands, and rain forest transition areas to dry ones (Fig. 5.4a, b). The species can be easily recognized by the large number of thorns on the stipe (steam), the frequent presence of remnants of the sheath of fallen leaves, the presence of senile leaves and coccoid yellow fruits with great oil content (Fig. 5.4b, c) (Lorenzi 2010; de Lima et al. 2018). Withing the genus, also *A. totai* Mart. and *A. intumescens* Drude. show promising potential for oil production and bioeconomy. The extraction of oil from the fruit pulp of *A. aculeata* from populations in the Brazilian state of Minas Gerais yielded 65% w/w on a dry basis (Colombo et al. 2018).

The species exhibits potential for plant breeding, but it currently exists in its wild form and is considered an incipiently domesticated crop (Clement 1999a, b). Domesticated plant species are those that have been genetically altered through artificial selection such that phenotypic characteristics distinguish them from wild progenitors. An incipiently domesticated population exhibits both the mean and variance of a selected trait within the range of variation of the species, displaying a subtle domestication syndrome (Anderson 1952; Piperno 2011; Clement et al.

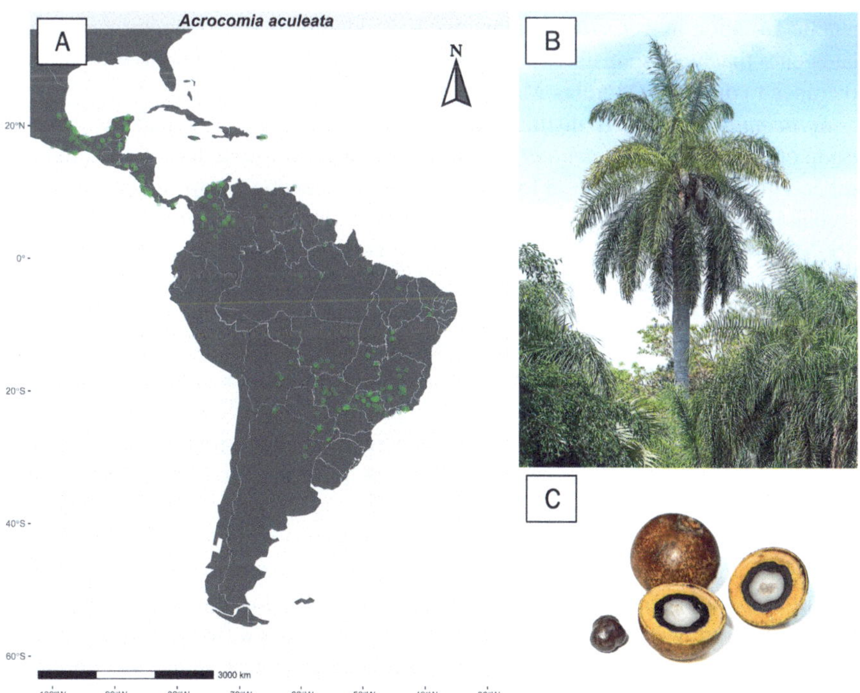

Fig. 5.4 Characteristics of the *Acrocomia aculeata*. (**a**) Distribution of the species throughout the Neotropic, according to data obtained from herbaria. (**b**) *A. aculeata* palm. (**c**) *A. aculeata* fruits and seeds. (Photos: Giuseppe Mazza)

2021). Due to its wide geographical distribution, it has been used in different ways. In Mexico and some Central American countries, the sap from the stipe is used to make a fermented drink. In Brazil, the fruit is commercially exploited to extract oil from the pulp for biofuels and the kernel for pharmaceutical products and the cosmetics industry (Aires and de Carvalho Junior 2023).

Natural populations of *Acrocomia* palms have been found to exhibit genetic and phenotypic variation that is associated with adaptation to various environmental conditions. *Acrocomia aculeata* (Jacq.) Lodd. ex Mart. presents the largest distribution of the neotropical palms, occurring in six of the seven biogeographic regions proposed by Eiserhardt et al. (2011). Such variation can contribute to the development of improved crop varieties with enhanced traits such as drought tolerance, disease resistance, and high oil yield.

For genetic studies on crop breeding and domestication, a high-quality reference genome sequence and chromosomal annotation of plant species are necessary. Since *Acrocomia* is a novel crop and it does not have a reference genome, population genomics techniques such as GBS or RAD-Seq can be used to explore the evolutionary mechanisms that underlie its diversity and variety. Understanding the genetic diversity in *Acrocomia* palms is crucial for identifying valuable traits and selecting suitable breeding materials. Genetic diversity plays a key role in the adaptation and resilience of plant species to changing environmental conditions. By studying the genetic variation in *Acrocomia* palms, we have identified and characterized candidate genes that are responsible for important traits such as adaptation to drought and tolerance to diseases (Morales-Marroquín et al. 2024).

Acrocomia aculeata is distinguished at the genus level as a separate genetic group or taxon. Within *A. aculeata*, a distinct substructure is visible using both neutral and under natural selection loci, displaying two genetic groups that correlate to a break in the continent's gene pools—one from South America and the other from North America (Fig. 5.5). The existence of two distinct groups, could be the consequence of separate evolution and reproductive isolation brought about by the Amazon Rainforest functioning as a physical barrier to restrict gene flow between them (Díaz et al. 2021; Morales-Marroquín et al. 2024). There is a sub-structure in those gene pools, for example, Central America presents 3 genetic groups that occurs in different bioregions.

Acrocomia totai exhibits distinct genetic structure among populations from the Brazilian Cerrado, Pantanal, Chaco, and Espinal. In contrast, earlier research using SNPs and microsatellites to exclusively target neutral loci revealed less or almost none genetic structure (Fig. 5.5). This suggests ongoing speciation and recent diversification within the populations of *A. totai* (Morales-Marroquín et al. 2024). In the case of *A. intumescens*, rather than being influenced by differences in biomes, the genetic structure was shaped by allopatric isolation across different phytophysiognomies within the Caatinga. Specifically, significant genetic structure was observed between groups in the North Sertaneja Depression and those in the transition from the Borborema Highlands to the South Sertaneja Depression within the Caatinga biome (Fig. 5.5) (Morales-Marroquín et al. 2024). Ecological interactions have altered the phenotype of the three species of Acrocomia by influencing the evolution of certain features necessary for its survival.

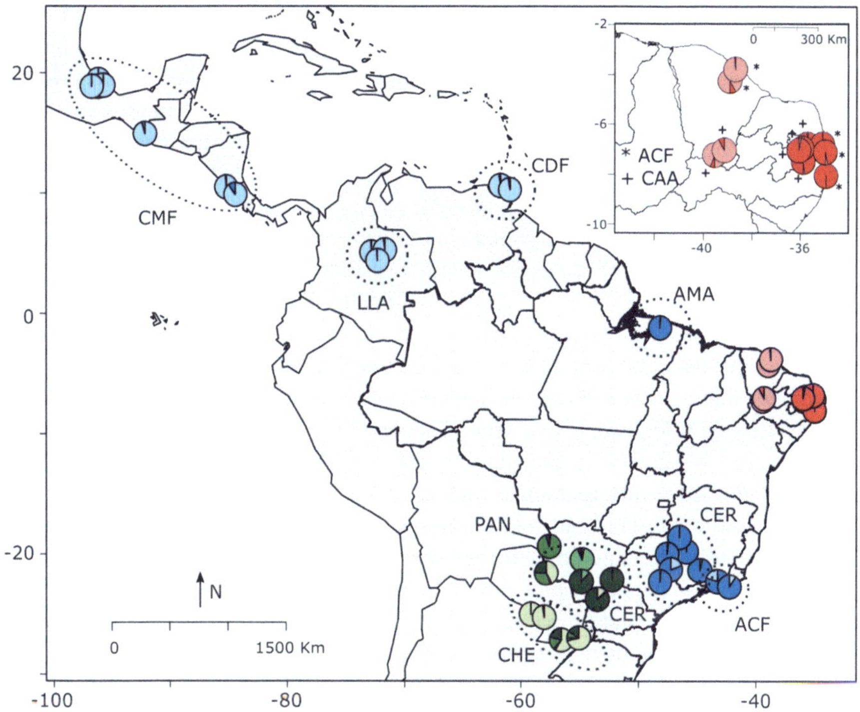

Fig. 5.5 Map showing sampling locations for Acrocomia aculeata (blueish), A. totai (greenish) and A. intumescens (redish), and the major genetic divergence among biogeographic groups as suggested by sparse non-negative matrix factorization (sNMF) analyses. Pie charts show the average sNMF ancestry coefficients across the genetic clusters represented by different color shades. Acronyms represent the macrobiome where the species occurs: CMF: Central America Moist Forest; CDF: Caribbean Dry Forest; LLA: Llanos; AMA: Amazon; ACF: Atlantic Coastal Forest; CAA: Caatinga; CER: Brazilian Cerrado; PAN: Pantanal; CHE: Chaco Espinal.

The relative abundance of palms was more strongly associated with local conditions, specifically higher mean annual precipitation (Muscarella et al. 2020). We have also found outlier SNPs related with the biological processes of metabolic process, cellular process, and the molecular functions in drought environment adaptations and oil yield production (Couto et al. 2024; Morales-Marroquín et al. 2024). This could be one of the main drivers in *A. aculeata* diversification. Is still unclear where the species originated. On the other hand, the most widely recognized theory on *A. aculeata* indicates that the species originated in northern Brazil (State of Pará) at 11,200 MY and was then distributed by humans to Central America based on the dates of archeological records.

Given that *A. aculeata* disperses throughout South, Central, and North America, it is possible that a founder effect contributed to the lower genetic diversity values for the species found in the northern group. Humans have a significant association with the species. It is widely used in several American countries, despite being regarded as a species that is only incipiently domesticated. Also, animal–plant interactions, which include antagonistic connections like herbivory and seed predation as well as mutualism, impact the fitness of individual plants. As suggested by the seed-dispersal syndrome hypothesis, the morphology of *A. aculeata* fruit, including its size, color, and content, appears to have undergone correlated evolution prior to the start of its domestication process through very recent human interaction (~10,000 BP) (Rodríguez and Aschero 2005; de Lima et al. 2018; Nascimento et al. 2020). Therefore, patterns of genetic diversity and structure can also be the result of the Great American Biotic Interchange (GABI).

Considering the regional differences in frequency distributions for stress response and the identification of accessions with resistance to specific stresses, further research on the genetic variation in *Acrocomia* palms can also facilitate the development of crop varieties that are tailored to specific regions and their unique environmental conditions (Couvreur et al. 2011). By understanding the genetic variation in *Acrocomia* palms, researchers can identify valuable traits for crop improvement such as drought tolerance, disease resistance, and high yield potential. This knowledge can then be applied in breeding programs to develop improved varieties that are better adapted to different environmental conditions and have increased productivity and resilience. This can lead to more sustainable and productive agriculture, benefiting both farmers and the broader society.

Palm species in the neotropics exhibit high genetic variation, which enables them to adapt to various niches within their habitat. This is evident in the case of *Acrocomia* species in the neotropics, where genetic diversity allows for niche adaptation and the ability to thrive in different environmental conditions (Díaz-Hernández et al. 2023; Couto et al. 2024). The genetic diversity of palm species in the neotropics is not only essential for their ability to adapt to different environmental conditions but also contributes to the overall resilience of the ecosystem. Furthermore, the genetic diversity found in neotropical palms serves as a reservoir of valuable traits that can be important for future breeding and conservation efforts.

In addition to their ecological significance, palm species in the neotropics also hold cultural and economic importance for local communities. The diverse genetic traits present in *Acrocomia* palms may offer opportunities for the development of new crops or the improvement of existing economically valuable species, providing sustainable livelihoods for local populations. Overall, the genetic diversity of palm species in the neotropics is a fascinating and crucial aspect of their biology and adaptation strategies.

5.4 Whole Genome Sequencing (WGS) of Brazil Nut: A Million Loci Approach to Detect Signatures of Selection

The Brazilian nut (*Bertholletia excelsa* Humb. & Bonpl., Lecythidaceae) is an important genetic resource for local human populations in Amazonia and has high potential for the market (Fig. 5.6). In 2020, the total production in the country was

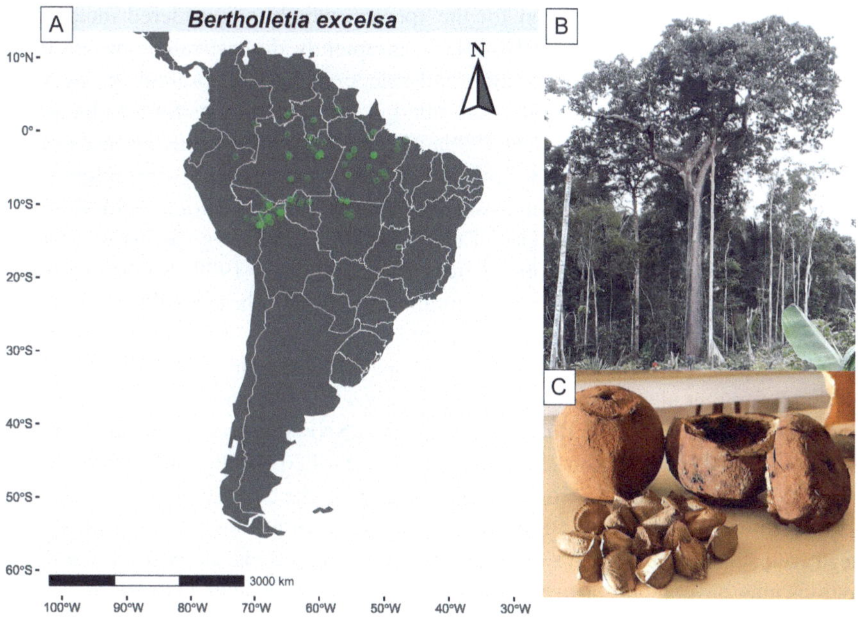

Fig. 5.6 Characteristics of the Brazilian nut (*Berthollletia excelsa*). (**a**) The distribution of the species throughout Amazonia, according to data obtained from herbaria. (**b**) Brazilian nut tree. (**c**) Brazilian nut fruits and seeds. (Photos **b** and **c**: Flaviane Malaquias Costa)

33,118 tons and the revenues achieved R$ 98.551 million (IBGE 2020). The species is distributed in the Amazon of Brazil, Bolivia, Peru, Colombia, Venezuela and the Guianas (Fig. 5.6). The trees are distributed in populations with around 50–100 individuals, separated by areas where the species is completely absent (Peres and Baider 1997), which may suggest that these forests were planted by ancient indigenous populations (Posey 1985; Clement et al. 2010; Shepardjr and Ramirez 2011).

The association between Brazil nut and Amazonian archaeological sites has been frequently observed (Pärssinen et al. 2009, 2020), demonstrating the interface between the species and ancient human populations, which acted in the dispersal and domestication of the species (Shepardjr and Ramirez 2011). The species has nutritious seeds that are rich in oils, which can be consumed *in natura*, roasted or pressed (aimed at obtaining Brazil nut milk) (Shepardjr and Ramirez 2011), that being considered an important trait for selection in the domestication of the species (Fig. 5.6c). In Acre, an archaeological study corroborates this hypothesis, by recording archaeological Brazil nut fruits dating 2000 years before the present, with an average fruit diameter of 5.7 cm (Pärssinen et al. 2020). Current Brazilian nut fruits measure around 11–15 cm, with higher counts and larger seeds (Shepardjr and Ramirez 2011).

The center of domestication of the Brazil nut has not yet been confirmed (Clement et al. 2010, 2015). Some hypotheses suggest that Northern Amazonia may involve a potential center of domestication for the species, which is considered incipiently domesticated (Clement et al. 2010, 2015). Incipiently domesticated species have been modified by human intervention and selection, but still present the average phenotype within the range of variation found in wild populations for the characteristics subject to selection (Clement 1999a, b). Studies performed using microsatellite markers (Buckley et al. 1988; Sujii et al. 2015; Cabral et al. 2017; Baldoni et al. 2020), one involving cpDNA microsatellite markers in populations widely separated by geographic distance (Gribel et al. 2007), found lower levels of genetic diversity compared to other tropical trees (Buckley et al. 1988; Kanashiro et al. 1997) and did not reveal significant geographic structure. These results suggest that the evolutionary history of the Brazilian nut is recent and their dispersal was faster, which is typical of incipiently domesticated species (Clement et al. 2010).

Sandercock et al. (2022) used WGS to analyze the genetic diversity of populations of the Brazilian nut. Based on 7,000,000 SNP markers retrieved from the analyses, potential selection signatures were identified throughout their genome using Tajima's *D* index (see Box 5.2) (Tajima 1989). The analyses detected genomic regions under potential selection located on two chromosomes, which presented negative peaks of the Tajima's *D* index for a few populations, suggesting that these regions may have undergone recent selection for environmental adaptation.

Box 5.2 Analytical Approaches Used to Detect Selection Pressures in Evolutionary Studies of Neotropical Species

Outlier SNP markers are identified as putatively "outliers" because they do not follow the normal (neutral) distribution observed for most SNPs in the population. Outlier SNP can indicate regions of the genome that are under natural selection or other types of selection, where genetic variation is being significantly advantageous or quite the opposite. Identifying these SNP markers can help to understand the mechanisms of adaptation and evolution. The detection of outlier SNP markers usually involves statistical and computational methods, such as positive selection tests, F_{ST} analysis (a measure of genetic differentiation between populations), and other approaches that compare the expected allele frequency with the observed one. DNA sequences with putatively outlier SNPs can be annotated according to gene ontology (GO) terms to explore information about cellular components, molecular functions, and biological processes.

Tajima's *D* index is a statistical measure that focuses on detecting deviations from the neutral model of evolution in a population. It compares segregation diversity (variation in allelic composition) with nucleotide diversity (variation in DNA sequences) to identify whether events such as selection and population expansion/reduction have recently occurred. Tajima's *D* is used to identify deviations from the neutral model of evolution, and positive or negative values can indicate recent natural selection.

Genomic analyses involving nine populations and 76 individuals of the Brazil nut were performed using WGS, based on 27,000,000 SNP markers. Tajima's *D* index was computed throughout the 17 chromosomes of the species, within which we can detect positive and negative peaks that may indicate potential signatures of selection in the whole genome of the different Brazil nut populations, represented by different colors in the graphs (Fig. 5.7). Discrepant peaks in opposite directions such as observed on chromosome 2 may suggest that potential positive and negative selection processes involve this region of the genome for the different populations analyzed. The analyses indicate patterns in population 4 (yellow) in which two negative Tajima's *D* peaks on chromosome 8 and three positive Tajima's *D* peaks on chromosome 17 suggest potential signature of selection in these regions. Further analyses of genome annotation should be able to identify functions associated with these regions of the genome, in order to elucidate evolutionary processes of the Brazil nut and prospect genes under selection.

5.5 Final Remarks

Selective pressures, such as natural selection, refer to the processes by which specific alleles or genetic variants are favored due to environmental changes and anthropogenic actions, resulting in the adaptation of species to their environment. Genetic variation associated with evolutionary processes and adaptation is characterized through genomic sequencing and the molecular markers obtained are used as tools to understand genetic/genomic diversity. Using examples of neotropical species, we explore how historical factors, geographical barriers and environmental conditions have shaped the genetic diversity of populations of *Euterpe precatoria*, *Acrocomia aculeata* and *Bertholletia excelsa*. We discussed mechanisms and processes by which natural selection has altered genetic variation in populations, including genomic signatures and the methods for detecting them. We demonstrated the implications of these studies for the adaptation, evolution and conservation of these species. In *Euterpe precatoria*, for example, the comparison between neutral and outlier markers contributed to discuss how selective pressures can influence the structuring of populations. The widely distributed species *Acrocomia aculeata* has been structured into two distinct genetic groups between South and North America, possibly due to physical barriers and reproductive isolation. The presence of outlier SNP markers involved in metabolic and stress response processes were identified, with potential for crop improvement. As also outlined here, a genomic approach using WGS and the Tajima's *D* index identified regions of the genome under potential selection in populations of *Bertholletia excelsa*. Positive and negative peaks on different chromosomes may indicate recent environmental adaptation or anthropogenic pressure. Analysis of the signatures of selection and genomic annotation associated with these regions should assist to understand the evolutionary and adaptive processes and the potential genes under selection. In this chapter we highlight the importance of combining different types of molecular markers (SSR, neutral and

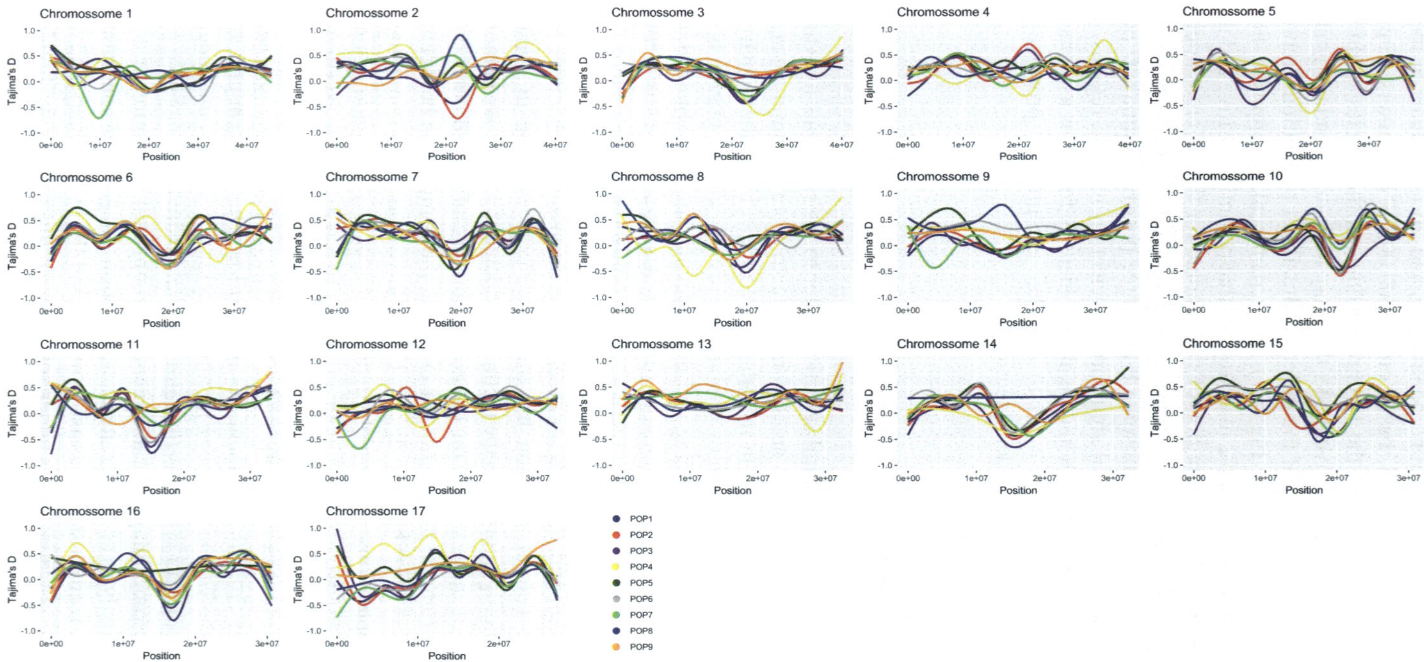

Fig. 5.7 Tajima's D index in the Brazil nut genome of nine populations based on 27,000,000 SNPs obtained by Whole Genome Sequencing. Positive and negative variation of D suggest that mechanisms of selection may be operating on specific genomics regions

outlier SNP), to obtain a comprehensive view of genetic/genomic diversity and selective pressures. New modern genome sequencing methods, such as WGS, offer new perspectives on species adaptation and will be crucial for more efficient conservation and management strategies for native species in the face of climate change in the twenty-first century.

Key Points

- Natural selection and human interactions can shape the genetic variation of species. Genomic and molecular studies contribute to understanding how these selective pressures affect genetic diversity.
- Genotyping approaches such as genotyping-by-sequencing (GBS) and whole genome sequencing (WGS) are used to identify genomic selection signatures and putatively outlier markers. These methods help detect genetic variations that may be under selective pressures.
- *Euterpe precatoria*: the genetic diversity of this species is influenced by both natural factors and human activities, such as habitat fragmentation and management. The use of SNP markers and the comparison between neutral and outlier markers contributed to identifying selective pressures acting on the structure of *Euterpe precatoria* populations.
- *Acrocomia aculeata*: this species shows genetic variation associated with adaptation to different environmental conditions and has potential for crop improvement. The identification of SNP markers outside the neutral pattern suggests adaptation to environmental stresses and potential for developing improved varieties.
- *Bertholletia excelsa*: studies based on WGS, on a scale of millions of loci, identified potential selection signatures in genomic regions of the Brazil nut. Tajima's *D* was used to detect deviations from neutral expectation, indicating selection pressures in the genome of populations.
- Understanding genetic variation and selection signatures supports the development of more effective conservation strategies and more sustainable management of natural resources, considering both natural and anthropogenic pressures. Modern genomics and sequencing techniques are essential for understanding these processes.

Questions

1. How does natural selection affect genetic variation in plant populations, and what genomic methods can detect this selection?
2. What are the main components of the theory of natural selection proposed by Charles Darwin, and how are they related to explain the adaptation of species?
3. How do modern sequencing techniques, such as GBS and WGS, contribute to the detection of selection signatures in plant species?
4. What are the main differences between neutral markers and outlier markers in genomic analysis, and how do these differences influence the interpretation of genetic data?
5. How does selection promote adaptation to specific environments and how do studies using SNPs help to understand these adaptations?

6. How are the signatures of selection detected in genomic studies used to predict the response of species to environmental changes and support conservation strategies?
7. How does natural selection affect genetic variability within a population and what are the mechanisms by which advantageous genetic variants become more abundant?

References

Aires GCM, de Carvalho Junior RN (2023) Potential of supercritical Acrocomia aculeata oil and its technology trends. Appl Sci 13(15):8594. Multidisciplinary Digital Publishing Institute. https://doi.org/10.3390/APP13158594

Albert JS, Carnaval A, Flantua SG et al (2023) Human impacts outpace natural processes in the Amazon. Science 379(6630):eabo5003. https://doi.org/10.1126/science.abo5003

Allendorf FW, Hohenlohe PA, Luikart G (2010) Genomics and the future of conservation genetics. Nat Rev Genet 11(10):697–709. https://doi.org/10.1038/nrg2844

Alves-Pereira A, Clement CR, Picanço-Rodrigues D et al (2018) Patterns of nuclear and chloroplast genetic diversity and structure of manioc along major Brazilian Amazonian rivers. Ann Bot 121(4):625–639. https://doi.org/10.1093/aob/mcx190

Alves-Pereira A, Clement CR, Picanço-Rodrigues D et al (2020) A population genomics appraisal suggests independent dispersals for bitter and sweet manioc in Brazilian Amazonia. Evol Appl 13(2):342–361. https://doi.org/10.1111/eva.12873

Alves-Pereira A, Zucchi MI, Clement CR et al (2022) Selective signatures and high genome-wide diversity in traditional Brazilian manioc (Manihot esculenta Crantz) varieties. Sci Rep, Nature Publishing Group 12(1):1–14. https://doi.org/10.1038/s41598-022-05160-8

Anderson E (1952) Plants, man and life, 1st edn. University of California Press, Los Angeles. https://doi.org/10.2307/jj.8085357

Avise JC (2004) Molecular markers, natural history, and evolution. Sinauer Associates. Sinauer Associates, Sunderland

Baldoni AB, Ribeiro Teodoro LP, Eduardo Teodoro P et al (2020) Genetic diversity of Brazil nut tree (Bertholletia excelsa Bonpl.) in southern Brazilian Amazon. For Ecol Manag 458:117795. https://doi.org/10.1016/j.foreco.2019.117795

Beaumont M, Nichols R (1996) Evaluating loci for use in the genetic analysis of population structure. Proc R Soc B Biol Sci 263:1619–1626

Bernardes LA (2021) Perfil Econômico e Análise Da Qualidade Do Fruto e Polpa de Açaí (Euterpe Precatoria Mart.) Comercializados Em Manaus, Amazonas. Instituto Nacional de Pesquisas da Amzônia

Blake JA (2013) Ten quick tips for using the gene ontology. PLoS Comput Biol 9(11):9–10. https://doi.org/10.1093/nar/gks1050.PLOS

Bovi MLA, de Castro A (1993) Assaí. In: Clay JW, Charles Clement R (eds) Selected species and strategies to enhance income generation from Amazonian forests. FAO, Rome, pp 58–67

Breed MF, Harrison P, Blyth C et al (2019) The potential of genomics for restoring ecosystems and biodiversity. Nat Rev Genet 20(10):615–628. https://doi.org/10.1038/s41576-019-0152-0

Buckley DP, O'malley DM, Apsit V, Prance GT, Bawa KS (1988). Genetics of Brazil nut (Bertholletia excelsa Humb. & Bonpl.: Lecythidaceae) 1. Genetic variation in natural populations

Cabral JC, Baldoni AB, Tonini H et al (2017) Diversity and genetic structure of the native Brazil nut tree (Bertholletia excelsa bonpl.) population. Genet Mol Res 16(3). https://doi.org/10.4238/gmr16039702

Cartaxo C, de Marçal Vasconcelos MA, Almeida Papa D et al (2020) Euterpe Precatoria Mart.: Boas Práticas de Produção Na Coleta e Pós-Coleta de Açaí-Solteiro. Rio Branco: Embrapa Acre

Clement CR (1999a) 1492 and the loss of amazonian crop genetic resources. I. The relation between domestication and human population decline. Econ Bot 53(2):188–202. https://doi.org/10.1007/BF02866498

Clement CRC (1999b) 1492 and the loss of Amazonian crop genetic resources. II. Crop biogeography at contact. In economic botany (Vol. 53, Issue 2)

Clement CR (2001) Melhoramento de Espécies Nativas. In: Nass LL, Valois ACC, Melo IS et al (eds) Recursos Genéticos & Melhoramento-Plantas. Fundação de Apoio à Pesquisa Agropecuária de Mato Grosso-Fundação MT, Rondonópolis

Clement CR, de Cristo-Araújo M, d'Eeckenbrugge GC et al (2010) Origin and domestication of native Amazonian crops. Diversity 2(1):72–106. https://doi.org/10.3390/d2010072

Clement CR, Denevan WM, Heckenberger MJ et al (2015) The domestication of Amazonia before european conquest. Proc R Soc B 282(1812). Royal Society of London. https://doi.org/10.1098/rspb.2015.0813

Clement CR, Casas A, Parra-Rondinel FA et al (2021) Disentangling domestication from food production Systems in the Neotropics. Quaternary, Multidisciplinary Digital Publishing Institute 4(1):4. https://doi.org/10.3390/QUAT4010004

Collevatti RG, Brondani RV, Grattapaglia D (1999) Development and characterization of microsatellite markers for genetic analysis of a Brazilian endangered tree species Caryocar brasiliense. Heredity 83(6):748–756

Colombo CA, Chorfi Berton LH, Diaz BG et al (2018) Macauba: a promising tropical palm for the production of vegetable oil. OCL 25:D108. https://doi.org/10.1051/ocl/2017038

Couvreur TL, Forest F, Baker WJ (2011) Origin and global diversification patterns of tropical rain forests: inferences from a complete genus-level phylogeny of palms. BMC Biol 9(1):44. https://doi.org/10.1186/1741-7007-9-44

Cristo-Araújo M, Reis VM, Rodrigues DP et al (2013) Domestication of peach palm in Southwestern Amazonia. Tipití 11(2):74–80

Cruz JSC, Freitas J, Souza Rodrigues A et al (2022) Diversity and genetic structure of natural populations of the palm tree Euterpe Precatoria (Arecaceae). Revista de Biologia Tropical 70(1):213–221. https://doi.org/10.15517/rev.biol.trop..v70i1.42942

Da Fonseca RR, Smith BD, Wales N et al (2015) The origin and evolution of maize in the southwestern United States. Nat Plants 1. https://doi.org/10.1038/nplants.2014.3

de Couto EGO, Morales-Marroquín J, Alves-Pereira A et al (2024) Genome-wide association insights into the genomic regions controlling oil production traits in Acrocomia aculeata (neotropical native palm). bioRxiv. Cold Spring Harbor Laboratory, p 2024.01.17.576138. https://doi.org/10.1101/2024.01.17.576138

de Lima NE, Carvalho AA, Meerow AW, et al. (2018) A review of the palm genus Acrocomia: Neotropical green gold. In: Organisms diversity & evolution. Springer Berlin Heidelberg, pp 1–11. https://doi.org/10.1007/s13127-018-0362-x

Díaz BG, Zucchi MI, Alves-Pereira A et al (2021) Genome-wide SNP analysis to assess the genetic population structure and diversity of Acrocomia species. PLoS ONE 16(7 July):1–24. https://doi.org/10.1371/journal.pone.0241025

Díaz-Hernández BG, Colombo CA, Morales-Marroquín JA et al (2023) Assessing the genetic vulnerability of Macaúba palm [Acrocomia aculeata (Jacq.) Lodd. ex Mart.] through the mating system and genetic diversity of open-pollinated progenies. Ann Appl Biol. John Wiley & Sons, Ltd. https://doi.org/10.1111/AAB.12879

do Nascimento LF, Guimarães PR, Onstein RE et al (2020) Associated evolution of fruit size, fruit colour and spines in Neotropical palms. J Evol Biol Oxford Academic 33(6):858–868. https://doi.org/10.1111/JEB.13619

Eiserhardt WL, Pintaud JC, Asmussen-Lange C et al (2011) Phylogeny and divergence times of Bactridinae (Arecaceae, Palmae) based on plastid and nuclear DNA sequences. Taxon 60(2):485–498

Eiserhardt WL, Svenning J-C, Baker WJ et al (2013) Dispersal and niche evolution jointly shape the geographic turnover of phylogenetic clades across continents. Sci Rep 3(1):1164. https://doi.org/10.1038/srep01164

Ekblom R, Wolf JBW (2014) A field guide to whole-genome sequencing, assembly and annotation. In: Evolutionary Applications 7(9):1026–1042. Wiley-Blackwell. https://doi.org/10.1111/eva.12178

Etter PD, Johnson E (2012) RAD paired-end sequencing for local De Novo assembly and SNP discovery in non-model organisms. In: Pompanon F, Bonin A (eds) Data production and analysis in population genomics: methods and protocols, vol 888. Springer Science+Business Media New, New York, pp 135–151. https://doi.org/10.1007/978-1-61779-870-2

Francisconi AF (2023) Genomics and ecological niche modeling of palms from the genus Euterpe. Universidade Estadual de Campinas

Freeman S, Herron JC (2007) Evolutionary analysis, 4th edn. Hardcover

Fuentes-Pardo AP, Ruzzante DE (2017) Whole-genome sequencing approaches for conservation biology: advantages, limitations and practical recommendations. Mol Ecol 26(20):5369–5406. https://doi.org/10.1111/mec.14264

Funk C, McKay J, Hohenlohe PA et al (2012) Harnessing genomics for delineating conservation units. Trends Ecol Evol 27(9):489–496. https://doi.org/10.1016/j.tree.2012.05.012

Futuyma DJ (2005) Evolution, 3rd edn. Sinauer Associates Inc

Gaiotto FA, Brondani RPV, Grattapaglia D (2001) Microsatellite markers for heart of palm - Euterpe Edulis and E. Oleracea Mart. (Arecaceae). Mol Ecol Notes 1(1–2):86–88. https://doi.org/10.1046/j.1471-8278.2001.00036.x

Gribel R, Lemes MR, Bernardes LG et al (2007) Phylogeography of the Brazil-nut tree (Bertholletia excelsa, Lecythidaceae): evidence of human influence on the species distribution. Association for Tropical Biology and Conservation

Helyar SJ, Hemmer-Hansen J, Bekkevold D et al (2011) Application of SNPs for population genetics of nonmodel organisms: new opportunities and challenges. Mol Ecol Resour 11(SUPPL. 1):123–136. https://doi.org/10.1111/j.1755-0998.2010.02943.x

Henderson A, Galeano G (1996) Euterpe, Prestoea, and Neonicholsonia (Palmae), 72. New York: Published for Organization for Flora Neotropica by the New York Botanical Gard. https://www.jstor.org/stable/4393873

IBGE, Instituto Brasileiro de Geografia e Estatística (2020) Produção da Extração Vegetal e da Silvicultura. https://www.ibge.gov.br/estatisticas/economicas/agricultura-e-pecuaria/9105-producao-da-extracao-vegetal-e-da-silvicultura.html?=&t=resultados

IBGE, Instituto Brasileiro de Geografia e Estatística (2021) Produção de Açaí (Cultivo). https://www.ibge.gov.br/explica/producao-agropecuaria/acai-cultivo/br

Isaza C, Bernal R, Galeano G et al (2017) Demography of Euterpe Precatoria and Mauritia Flexuosa in the Amazon: application of integral projection models for their harvest. Biotropica 49(5):653–664. https://doi.org/10.1111/btp.12424

Jombart, T., Devillard, S., & Balloux, F. (2010). Discriminant analysis of principal components: a new method for the analysis of genetically structured populations. *BMC genetics, 11*, 1–15.

Kanashiro M, Harris SA, Simons A (1997) RAPD diversity in Brazil nut (Bertholletia excelsa Humb. & Bonpl.: Lecythidaceae). Silvae Genet 46(4):219–223

Kistler L, Yoshi Maezumi S, De Souza JG et al (2018) Multiproxy evidence highlights a complex evolutionary legacy of maize in South America. Science 362(6420):1309–1313. https://doi.org/10.1126/science.aav0207

Kistler L, Thakar HB, VanDerwarker AM et al (2020) Archaeological Central American maize genomes suggest ancient gene flow from South America. Proc Natl Acad Sci USA 117(52):33124–33129. https://doi.org/10.1073/PNAS.2015560117

Küchmeister H, Silberbauer-Gottsberger I, Gottsberger G (1997) Flowering, pollination, nectar standing crop, and nectaries ofEuterpe precatoria (*Arecaceae*), an Amazonian rain forest palm. Plant Syst Evol 206:71–97. https://doi.org/10.1007/bf00987942

Lewin HA, Robinson GE, Kress WJ et al (2018) Earth BioGenome Project: sequencing life for the future of life. Proc Natl Acad Sci USA 115(17):4325–4333. https://doi.org/10.1073/pnas.1720115115

Lopes R, Vieira Da Cunha RN, dos Santos Tavares M et al (2022) Seasonality of fruit production of Euterpe Oleracea and E. Precatoria Açaí Palm Trees Cultivated in the Metropolitan Region of Manaus (AM). Revista Agro@Mbiente On-Line 16:1–14. https://doi.org/10.18227/1982-8470ragro.v16i0.7282

Lorenzi H (2010) Flora Brasileira Arecaceae (Palmeiras), Instituto Plantarum De Estudos Da Flora Ltda. Instituto Plantarum, Nova Odessa

Luikart G, England PR, Tallmon D et al (2003) The power and promise of population genomics: from genotyping to genome typing. Nat Rev Genet 4(12):981–994. https://doi.org/10.1038/nrg1226

Luikart G, Kardos M, Hand BK et al (2018) Population genomics: advancing understanding of nature. March:3–79. https://doi.org/10.1007/13836_2018_60

Melo G, Souza Costa F, da Silva LC (2021) O Cenário Da Produção Do Açaí (Euterpe Spp.) No Estado Do Amazonas/ The production scenario of Açaí (Euterpe Spp.) in the State of Amazonas. Brazil J Dev 7(7):71536–71549. https://doi.org/10.34117/bjdv7n7-365

Meyer RS, Purugganan MD (2013) Evolution of crop species: genetics of domestication and diversification. Nat Rev Genet 14(12):840–852. https://doi.org/10.1038/nrg3605

Morales-Marroquín JA (2024) Genetic variations associated with adaptation processes in Acrocomia palms: a comparative study across the Neotropic for future crop improvement, in review process

Muscarella R, Emilio T, Phillips OL et al (2020) The global abundance of tree palms. Glob Ecol Biogeogr, John Wiley & Sons, Ltd 29(9):1495–1514. https://doi.org/10.1111/GEB.13123

Pärssinen M, Schaan D, Ranzi A (2009) Pre-Columbian geometric earthworks in the upper Purús: a complex society in western Amazonia

Pärssinen M, Ferreira E, Virtanen PK et al (2020) Domestication in motion: macrofossils of pre-colonial brazilian nuts, palms and other Amazonian planted tree species found in the upper Purus. Environ Archaeol 26:309–322. https://doi.org/10.1080/14614103.2020.1765295

Peres CA, Baider C (1997) Seed dispersal, spatial distribution and population structure of Brazilnut trees (Bertholletia excelsa) in southeastern Amazonia. J Trop Ecol 13(4):595–616. https://doi.org/10.1017/S0266467400010749

Perrut-Lima P, Sebbenn AM, Francisconi AF et al (2023) Genetic diversity and mating system of Euterpe Precatoria in three localities along the lower Solimões River in Central Amazonia. Silvae Genet 72(1):81–91. https://doi.org/10.2478/sg-2023-0008

Piperno DR (2011) The origins of plant cultivation and domestication in the New World Tropics patterns, process, and new developments. Curr Anthropol 52(SUPPL. 4):S453–S470. https://doi.org/10.1086/659998

Poland J, Rife TW (2012) Genotyping-by-sequencing for plant breeding and genetics. The plant. Genome 5(3). https://doi.org/10.3835/plantgenome2012.05.0005

Posey DA (1985) Indigenous management of tropical forest ecosystems: the case of the Kayapo Indians of the Brazilian Amazon I

Ramos S, Dequigiovanni G, Sebbenn AM et al (2016) Spatial genetic structure, genetic diversity and pollen dispersal in a harvested population of Astrocaryum Aculeatum in the Brazilian Amazon. BMC Genet 17(1):1–10. https://doi.org/10.1186/s12863-016-0371-8

Ramos S, Dequigiovanni G, Sebbenn AM et al (2018) Paternity analysis, pollen flow, and spatial genetic structure of a natural population of Euterpe Precatoria in the Brazilian Amazon. Ecol Evol 8(22):11143–11157. https://doi.org/10.1002/ece3.4582

Ramos S, Gomes Lopes MT, Lopes R et al (2019) Mating system analysis of Açaí-Do-Amazonas (Euterpe Precatoria Mart.) using molecular markers. Crop Breed Appl Biotechnol 19(1):126–130. https://doi.org/10.1590/1984-70332019v19n1n17

Ramos S, Dequigiovanni G, Gomes Lopes MT et al (2021) Genetic structure in populations of Euterpe Precatoria Mart. In the Brazilian Amazon. Front Ecol Evol 8(January):1–11. https://doi.org/10.3389/fevo.2020.603448

Rodríguez MF, Aschero CA (2005) Acrocomia chunta (Arecaceae) raw material for cord making in the Argentinean Puna. J Archaeol Science Academic Press 32(10):1534–1542. https://doi.org/10.1016/J.JAS.2005.04.016

Sander NL, Pérez-Zavala F, Da Silva CJ et al (2018) Rivers shape population genetic structure in Mauritia Flexuosa (Arecaceae). Ecol Evol 8(13):6589–6598. https://doi.org/10.1002/ece3.4142

Sandercock AM, Westbrook JW, Zhang Q et al (2022) Frozen in time: Rangewide genomic diversity, structure, and demographic history of relict American chestnut populations. Mol Ecol 31(18):4640–4655. https://doi.org/10.1111/mec.16629

Shepardjr GH, Ramirez H (2011) Made in Brazil: human dispersal of the Brazil nut (Bertholletia excelsa, Lecythidaceae). in Ancient Amazonia 1

Sujii PS, Martins K, Wadt LH et al (2015) Genetic structure of Bertholletia excelsa populations from the Amazon at different spatial scales. Conserv Genet 16(4):955–964. https://doi.org/10.1007/s10592-015-0714-4

Tajima F (1989) Statistical method for testing the neutral mutation hypothesis by DNA polymorphism. Genet Soc Am 123:585–595

Ter Steege H, Pitman NC, Sabatier D, Baraloto C, Salomão RP, Guevara JE, Silman MR (2013) Hyperdominance in the Amazonian tree flora. Science 342(6156):1243092

Vigouroux Y, McMullen M, Hittinger CT et al (2002) Identifying genes of agronomic importance in maize by screening microsatellites for evidence of selection during domestication. Proc Natl Acad Sci USA 99(15):9650–9655. https://doi.org/10.1073/pnas.112324299

Zona S, Henderson A (1989) A review of animal-mediated seed dispersal of palms. Selbyana 11:6–21

Zucchi MI, Cordeiro E, Allen C et al (2019) Patterns of genome-wide variation, population differentiation and SNP discovery of the red banded stink bug (Piezodorus Guildinii). Sci Rep 9(1):1–11. https://doi.org/10.1038/s41598-019-50999-z

Chapter 6
How Do Gene Flow and Migration Affect Populations?

Mario Luiz Teixeira de Moraes, Marcela Aparecida de Moraes Silvestre, Bruno Cesar Rossini, and Celso Luis Marino

Contents

Abstract Gene flow is the movement of genes among and within populations of a species, which can occur in time and space. The process by which individuals (alleles) enter or leave a population is known as migration and is more common in animals. Thus, these evolutionary factors are determinants of the genetic diversity within and among populations. In this chapter, we introduce basic concepts of populations related to the genetic basis and the importance of gene flow in the survival and reproduction of plant species. The influence of gene flow and migration on the genetic diversity of these populations and their consequences are addressed. In neo tropical trees, gene flow occurs through the dispersal of pollen (contemporary) and seeds (historical) dispersal, carried out by animals or by the action of physical factors such as wind and water. An effective gene flow depends on the number of genes

M. L. T. de Moraes (✉)
Departamento de Fitotecnia, Tecnologia de Alimentos e Sócio Economia, Faculdade de Engenharia – FEIS/UNESP, Ilha Solteira, SP, Brazil
e-mail: mario.moraes@unesp.br

M. A. de Moraes Silvestre · C. L. Marino
Departamento de Ciências Químicas e Biológicas, Instituto de Biociências – IBB/UNESP, Botucatu, SP, Brazil

B. C. Rossini
Instituto de Biotecnologia – IBTEC/UNESP, Botucatu, SP, Brazil

M. V. Bohrer Monteiro Siqueira et al. (eds.), *Population Genetics in the Neotropics*, https://doi.org/10.1007/978-3-031-83685-5_6

and the genetic divergence of the participating populations, as well as on the dispersal agents. Migration is the change in the constitution of the population and the ecosystem. Migration is more noticeable in animal populations compared to plants, which require a longer time for such changes to become evident. As examples, we briefly discuss the knowledge about gene fluctuations in case studies of two types of populations: (*i*) natural populations; (*ii*) *aroeira* genebanks (AGB). In the first case, there is no anthropic interference, and it allows for a better understanding of how gene fluctuations occur in natural populations. In AGB there is a higher frequency of certain (favorable) alleles, in other words, a direction in the gene flow has been directed through human intervention. The importance of migration in the biodiversity of biomes and colonization of some plant species are also discussed. Perspectives on the usefulness of gene flow and migration studies are also discussed.

Keywords Colonization · Dispersal agents · Evolutionary factors · Genetic basis · Genetic diversity

Learning Objectives:
This chapter will help readers to understand the following:

- The role of gene flow and migration among populations.
- The importance of pollinators and seed dispersers in maintaining genetic diversity.
- Methods for estimating gene flow.
- Apply knowledge of gene flow in practice, such as establishing strategies for seed collecting for *ex situ* conservation.
- Impacts of changes in migratory routes on animals on the local ecosystem and society.
- How migration was essential for the constitution of biomes and the evolutionary history of organisms.

6.1 Introduction

Since the beginning, the Earth has undergone countless changes in the landscape and their inhabitants, after all, nature is nothing short of dynamic. Unfortunately, we are limited to mainly caring about the planet from prehistory to the present day, but this period is very short compared to the age of the Earth. Willis and Mcelwain (2014) reported that the first photosynthesizer fossils date back to around 3.5 billion years ago. Plants began colonizing our planet around 450 Ma. At around 75 Ma before the present, trees with 11 to 60 meters in height appeared, which demonstrates the fantastic evolution of life on our planet.

Landscape changes caused mainly by weather conditions have driven changes in organisms through the evolutionary process over geological time. If there was no evolution (Box 6.1) acting on populations, it would be difficult for them to survive in the past. Thus, we can suggest that evolution is the primary cause of species adaptation.

Box 6.1 Basic Concepts to Keep in Mind

Evolution: set of adaptive modifications causing changes in the gene frequencies of the population, as a result of selective pressure, among other factors, in a gradual process.

Population: a group of individuals that have the same genetic base, not necessarily the same species, and that share the same reproductive system in which they can exchange genes with each other, living in the same geographic area and time interval.

Genetic basis: refers to the total genetic variability of a given population.

Evolutionary factors: drift, mutation, selection and migration/gene flow.

Gene flow: movement of genes between and within populations that can occur in time and space.

Migration: corresponds to the entry or exit of individuals (alleles) from a population.

Linkage disequilibrium: non-random association of alleles among two or more loci. Loci can be physically linked along the same chromosome, but linkage disequilibrium can also occur between loci at different chromosomes.

Source: based on Darwin (2018).

Another important aspect concerns how the evolutionary process affects the population. This is influenced by the genetic basis (Box 6.1) of the population and the intensity by which evolutionary factors are acting isolated or in interaction. In this occasion, it is significant to highlight that the concept of migration is related to the movement of alleles or individuals between populations. They are followed by crossings, which then characterize gene flow, which will alter the gene or allele frequency of the population. It is also important to mention that migration between populations generates linkage disequilibrium (Resende 2015) (see Box 6.1 for a definition). In this chapter, gene flow and migration will guide the discussion along with case studies involving neotropical species.

Consider a plant species that has a mating system which requires outcrossing (allogamy) and that makes up two populations (A and B). Population A is located in an ideal habitat, that is, suitable weather and soil, presence of pollinators, among other favorable conditions. In this case, the pollinator has a large supply of flowers from different individuals, which allows it to exchange genes from several flowers and travel great distances throughout the range that area spans. In contrast, population B is the opposite. It is not in a favorable area and it does not have a large number of pollinators available. Thus, the food supply will limit the pollinator to visit a smaller number of flowers and travel lower distances, as it should save energy to survive. Thus, population A will have descendants with greater genetic variability, that is, many genes make up the genetic basis from different combinations in the reproduction process over the years. This contribution of genes increases the chances for population survival in the event of weather extremes. Population B tends towards reduced genetic variation, that is, as pollination is restricted to a few individuals, their descendants will be similar, causing a narrowing of the genetic basis and limiting adaptive fitness.

Degen and Sebbenn (2014) reviewed the literature on the subject and reported that the main vector of pollen dispersal in most temperate tree species is the wind, and that most neotropical tree species are pollinated by animals. Seeds can be dispersed in both temperate and neotropical trees by wind and animals and, in some cases, water also serves as a dispersal vector. In neotropical species, 72% of immigrant pollen was found for continuous populations, 70% for fragmented populations and 10% for isolated trees in pastures. Therefore, the authors state that pollination by animals is efficient in promoting gene exchange between populations even under different conditions.

About gene flow via seeds, efficiency is determined by the size and shape of the seeds and the type of vector (Box 6.2). Seed dispersal generally occurs in different stages. After a random or clustered primary distribution, a second dispersal further affects seed establishment. The location where seeds are deposited depends on the mode of seed dispersal. The success of seed dispersal in the genetic composition of populations is determined by the survival of the seed to reproductive age. Seed establishment depends, in many cases, on seed density and dispersal distance. It also is controlled by predation and the environment in which seeds are established.

Box 6.2 Modes of Seed Dispersal

Wind: generally present in species that have light and winged seeds.

Gravity: in species that have large, heavy seeds that simply fall from the tree to the ground.

Water: rivers and streams, as in the case of Amazon rainforest.

Animal dispersal: most neotropical trees have fruits adapted to dispersal by animals such as birds, bats, primates and rodents.

Source: Degen and Sebbenn (2014), Correa et al. (2015).

Migration is most studied in animals. It is very common for media outlets to report the migration of birds, fish, whales, penguins, etc. These animals leave their natural habitat during the reproductive period or due to unfavorable local conditions, such as lack of food and adverse weather conditions. Duponchelle et al. (2021) reported the importance of fish migration from the Amazon basin. Most of these fish have key ecological roles as apex predators, ecological engineers, or seed dispersal species. However, reducing their population sizes could induce cascading effects with implications for ecosystem stability and associated services. The fish movements in the Amazon basin are often related to the direction of their migrations, which can be longitudinal (long and medium distance) or lateral (short distance). Longitudinal migrations occur within the upstream-downstream gradient of rivers in both directions and can connect different parts of the river (upstream with downstream) or even the estuary or the main river channels with the headwaters or smaller tributaries. As an example, the species *Brachyplatystoma rousseauxii* performs a trans-Amazonian round-trip migration of up 12,000 km between its hatching areas in the Andean Piedmont of Bolivia, Colombia, Ecuador and Peru and its nursery in the lower Amazon and estuary in Brazil. In contrast, lateral migrations occur among habitats of river floodplains, including the main river channels and lakes, connecting channels, and various vegetated habitats, such as flooded forests. Osteoglossiformes, a ray-finned fish order, such as *Arapaima*, have been shown to move synchronically through a total of eight distinct river floodplain habitats over the course of the flood pulse closely following river water levels. However, the main threats to freshwater fish in the Amazon to migration routes are overexploitation, flow and habitat modification by hydropower development, deforestation, and climate change.

Migrations may have profound consequences for the population dynamics of various organisms. Back to the example of migratory fishes, they may influence the trophic chain of rivers, by strongly impacting the habitat structure and local communities through their feeding and breeding activities. As a few examples, such fish are able to remove the largest *Brachyplatystoma* species. More so, their activity may decrease the biomass of local invertebrates and conversely, increase the biomass of macrophytes (aquatic plants). Such changes can cause drastic changes in the structure and functioning of food webs in the Amazon basin.

Other crucial ecological aspects of various Amazonian migratory fishes are their frugivory and ability to disperse seeds and improve germination rates, taking part in the spatial distribution and diversity of riparian forests. It is believed that fishes may have been the first vertebrate dispersers in the Neotropics, as of 70 million years ago. As most tree species produce their fruits during the high-water season, when fish invade the flooded forest to feed, seed dispersal is likely to result from a long process of mutually beneficial co-adaptation. In this context, the removal of thousands of fish species can have a major impact on the ecosystem, causing irreversible damage. This information is overwhelming since the Amazon basin hosts the Earth's highest diversity of freshwater fish.

Gene flow and mutation can provide new genotypes in local populations that are not adapted to these conditions. Stearns and Hoekstra (2000) report that in southern France a small bird (blue tit) breeds in downy (deciduous) and holm (non-deciduous) oaks. This bird is adapted to make their nests in the downy oak. However, as the gene flow is very intense, it leads blue tits to build nests in holm oaks as well, where the peak of insects available as their food occurs later, so that they are poorly adapted to the latter.

In relation to plant species, migration is more noticeable over a long period of time, such as the process of colonization of these organisms through our planet. Considering human action as a vector of migration, it is possible to mention what happened in 1876. British Henry Wickham took 70 thousand seeds of the rubber tree from the Brazilian Amazon rainforest to England and later expanded their cultivation to Asia.

All that considered, this chapter aims to demonstrate how gene flow and migration affect populations. For gene flow we will report, based on field observations, how this factor alters the genetic basis in the long-lived tree known in Brazil as *aroeira* (*Astronium urundeuva* (M. Allemão) Engl. – Anacardiaceae) in its natural habitat. A comparison between the latter and the so called *Aroeira* Genebank (AGB), a progeny test, is also discussed within the context of gene flow. For migration, we report how some plant species colonized neotropical regions, with an emphasis on Brazilian biomes, millions of years ago.

6.2 *Astronium urundeuva*: Case Study of the Interference of Gene Flow in the Genetic Structure of a Population

On the course of plant evolution, angiosperms succeeded in relation to gymnosperms due to some factors such as: (*i*) their diploid phase is much greater than the haploid phase in their life cycle; (*ii*) pollen transport occurs by means other than exclusively by wind, which was fundamental for efficient gene flow in the neotropical landscape; (*iii*) double fertilization: embryo (2n) and endosperm (3n); and (*iv*) seeds are protected by a fruit (Paterniani 1974).

Astronium urundeuva (*aroeira*) is a tree species typical of South America and occurs in the following biomes: Chaco (Argentina, Paraguay, Bolivia and Brazil), Pantanal, Cerrado, Atlantic Forest and Caatinga (Brazil). It is a dioecious tree species, with some records of monoecy in the literature. Flowering occurs during dry winters; the species is deciduous, which facilitates pollination by bees and dispersion of seeds by the wind. It is also worth noting that macaws disperse the seeds of *A. urundeuva* over long distances, hence its name in Tupi-Guarani (indigenous Brazilian language) "*arara* tree" (Carvalho 2003). This behavior leads *A. urundeuva* to be considered, on the scale of ecological succession, as an anthropic pioneer. It makes up almost pure forests in open areas, and as a late secondary in vegetated areas, two contrasting landscapes that produce distinct effects on gene flow.

The aforementioned ecological aspects of *A. urundeuva* along with an ethnobotanical and an economic significance led the São Paulo State University (UNESP) to create a germplasm collection (*aroeira* genebank or AGB) on an experimental farm located in Selvíria, MS, Brazil. Genetic materials were collected from several locations throughout Brazil, including Caatinga, Cerrado, the Atlantic Forest/Cerrado ecotone, and a Semidecidous Seasonal Forest within the Atlantic Forest. In total, 15 populations were sampled for seeds from 398 open pollinated trees, leveling up to an effective population size (N_e) of 1592. Thus, the AGB collection has a considerable genetic base, which is fundamental in an *ex situ* genetic conservation program and gene recombination between and within populations from different biomes (Fig. 6.1).

Fig. 6.1 Progeny test of *Astronium urundeuva* with the presence of typical Cerrado fauna. (Photo credit: Francieli Alves Caldeira Saul)

The AGB experimental site was installed as a progeny test, which allows their transformation, over time, into seedling seed orchards. In this way, gene flow within and between populations has been evaluated using quantitative traits (phenotypic) as well as molecular markers.

To further illustrate the differences in gene flow behavior, we will report some studies and field observations carried out by our group with populations of *A. urundeuva*. In the first scenario, *A. urundeuva* occupies a pasture field with the behavior of an anthropic pioneer. Therefore, gene flow should be intense among related trees. In the second case, a natural population in a forest fragment with a density of one plant per hectare, in which gene flow occurs among unrelated plants (Fig. 6.2).

For the progeny tests, we discuss the provenances of Aquidauana/PT-AQ (Pantanal) with five years of cultivation, Itarumã/PT-IT (Cerrado) with 11 years, Paulo de Faria/PT-PF (Cerrado/Atlantic Forest) with 18 years, and Seridó/PT-SE (Caatinga) with 18 years. All populations are planted at the same experimental location. Gene flow estimates were obtained with microsatellite markers developed for the species. Microsatellite markers are molecular markers used to estimate the genetic differences between individuals from a population and will be discussed later (see Chaps. 5 and 9). From the analyses of the microsatellite genotypes, we were able to calculate several estimates related to gene flow, such as the effective number of pollen donors and the number of mother trees. The effective number of pollen donors were counted to 5 for PT-AQ and PT-IT, 12 for PT-PF and 7 for PT-SE. The number of mother trees for seed collection were 55 for PT-AQ, 53 for PT-IT, 48 for PT-PF and 52 for PT-SE.

Fig. 6.2 (**a**) *Astronium urundeuva* behaving as an anthropic pioneer. (**b**) Identification of a mother tree of *A. urundeuva* in a natural population. (Photo credit: Mario Luiz Teixeira de Moraes)

Based on the allele frequencies in each population, we were also able to estimated the average difference among populations, estimated as the coefficient $G'_{ST} = 0.33$. This estimate indicates that 33% of the variation is found among populations. That means that the remaining 67% of the genetic diversity is distributed within the populations. These results suggest that the species has genes adapted to specific environments, in this case biomes, possibly due to the colonization process and evolutionary factors, such as gene flow. It can be observed that the presence of pollinators interferes with the recombination mode of these genes and is reflected in seed collection planning for adequate sampling of the genetic basis.

In the case of natural populations, the ecological station Paulo de Faria (PFES), located in the northwest of São Paulo state, was sampled. For this study a census was carried out in the PFES where 467 adult-trees identified were sampled, mapped and their diameter at breast height measured. Among the adult individuals, 175 were female, 268 were male and 24 were not sexed due to the absence of flowers. The effective number of pollen donors was 11, comprising an effective neighbor pollination area of 22.1 ha. The average pollen dispersal distance was of 138 m and the pollen pool comprises 97.3% within the fragment. The variance of effective size was 3.14, and it is necessary to collect seeds from 50 mother trees. This population is similar to PT-PF, witch indicates that seed collection was adequate and the present progeny test represents the genetic basis of the natural population.

Another interesting study was carried out by Caetano et al. (2008). The authors sampled 53 natural populations of *A. urundeuva* distributed in Argentina (nine), Bolivia (six), Brazil (22) and Paraguay (16). These areas were classified as Caatinga, Cerrado, Chaco and transitional or non-specified vegetation. With nine microsatellite markers, two large clusters were detected: (*i*) populations of Argentina, Bolivia and Paraguay; and (*ii*) populations of Brazil. Brazilian sites could be separated into two subgroups: (*i*) Caatinga; (*ii*) other sampled areas. Populations from Caatinga had lower genetic diversity, suggesting that in the past some barriers interfered with gene flow and led to genetic drift in this population, making them more specialized in adapting to this biome. The authors also emphasize that due to the lack of paleobotanical information from eastern South America, the colonization process of *A. urundeuva* and subsequent inferences about the history of the formation of dry forests should be treated with caution. However, the most accepted hypothesis is that the species expanded from northeast to other locations southwards. That was suggested, as divergent alleles were detected between Caatinga and Cerrado populations, suggesting that some barriers were not as robust and allowed long-distance dispersal. Paleobotanical knowledge is important for better understanding the process of colonization and gene dispersion of organisms, as is briefly accounted for in Sect. 6.4 of this chapter.

It is important to highlight that gene flow presents different behavior depending on the biology of the species, pollinator colonization and type of habitat. Therefore, more studies are needed on the natural populations that provided seeds to compose the progeny test in discussion here to further clarify how pollen flow can interfere with the adaptation of *A. urundeuva*. Readers are advised to further readings, such as the studies of Gaino et al. (2010), Souza et al. (2017) and Manoel et al. (2021), in order to check more details of what is reported in this section. The evolutionary relationships among other genera within the Anacardiaceae family, including *A. urundeuva* can be found in Rossini et al. (2021).

6.3 How to Estimate Gene Flow?

Gene flow is the transfer of alleles between populations. Now consider a population where plants with purple flowers are more present than white flowers. In another population, there are more white flowers than purple and even red. This means that we have different frequencies of each phenotype in each population. If these populations come into contact, as occurs in species with long pollen dispersal, when individuals from these populations exchange genetic material, then alleles are transferred from one population to another, causing allele frequency to change in this new generation. In this way, by evaluating the frequency of a trait, whether phenotypic or based on DNA sequences, it is possible to estimate gene flow between populations. A few other cases are worth discussing:

(a) *Based on population density and distances between mother trees and seed dispersal*

In the literature, several expressions are found to estimate gene flow between populations, one of them was used by Kageyama (1987), in which the gene flow index (I) is expressed by: $I = \frac{\underline{D}}{\underline{d}}$; where: $\underline{D}$ is the average seed dispersal distance from the mother tree and $\underline{d}$ is the average distance between the mother trees. The author cites two examples: (*i*) *Ceiba speciosa:* there is 1 tree in 5 ha, $\underline{D} = 200$ m and $\underline{d} = 200$ m. Thus, $I = 1.0$; (*ii*) *Euterpe edulis:* 100 trees in 1 ha, $\underline{D} = 60$ m and $\underline{d} = 10$ m, which corresponds to $I = 6.0$. In this index, the cause-and-effect relationship in population density in the population is fundamental in the expression of a greater or lesser index of gene flow between populations.

(b) *Entry of individuals or pollen into a population by migration*

Migration refers to the entry (immigration) or exit (emigration) of alleles or individuals between populations followed by crossings (recombination), which leads to changes in the allele frequencies of the reference population. For illustration purposes, *e.g.*: suppose that the reference population has the A_1 allele with a frequency q_0, in the G_0 generation. If in the G_1 generation the reference population is not influenced by any evolutionary factor, in the case of the example without gene flow, the frequency of A_1 will continue to be q_0, that is, $\Delta_q = q_1 - q_0 \therefore \Delta_q = 0$. However, if "m" immigrants enter the RP population, with a frequency of q_m, it will be verified in the G_1 generation, after recombination, the frequency of A_1 will be:

$$\Delta_q = q_1 - q_0 \therefore \Delta_q = \left[(1-m)q_0 + mq_m\right] - q_0 \therefore \Delta_q = m(q_m - q_0)$$

Therefore, for the migratory process (Δ_q) to be effective, the allelic divergence and immigration rate must be relevant. This process is more evident in reference populations with a restricted genetic basis (Resende and Mauro 2003). When it comes to *ex situ* genetic conservation programs, collecting samples from different populations and placing them in the same site is a strategy for expanding the genetic variability of a species. This is considered as a way of migration since plants do not move as animals (Resende 2015).

(c) *The pollinator is known as is the phenology of the species*

One of the most popular species from Brazilian Cerrado is the *pequi* (*Caryocar brasiliense* – Caryocaraceae), which has a complete flower with male and female organs in the same flower, which allows self-fertilization. Collevatti et al. (2010) reported bats as pollinators and the record of inbred crosses, as revealed by molecular markers. They also highlight that the number of seeds per fruit can range from 1 to 4. That verified, we can make an "assumption" as an exercise, e.g.: if three self-fertilizations and one cross-fertilization occurred, one seed is formed in the fruit only. However, if the fruit produced two seeds, we have two self and two cross-fertilization events and so on. Based on these findings, two populations of *C. brasiliense* were studied: (*i*) in the first, individuals from the population are found in a legal reserve on a farm in the region of Três Lagoas-MS/Brazil (POP-LR); and (*ii*) the individuals are randomly distributed in a 110 ha or an *Eucalyptus* plantation (POP-EUCA). In these populations, the authors found a trend for producing seeds (1 to 4) that behaved as a positive asymmetric type curve in which the mode (most frequent value, in this case 1 seed) is lower than the median and is lower than the average.

In this context, by knowing the type of flower of the species, the pollinator agent, the distribution of the species, the morphology of the produced fruits, and also the phenology of the plant. We can make inferences regarding the gene flow that is occurring in these populations. Thus, in POP-RL and POP-EUCA, the formation of 1 seed predominated, but no fruitless trees occurred in POP-RL. However, the same no longer occurred in POP-EUCA where all trees flowered, but 25% of them did not produce fruits due to the absence of cross-pollination. This characterizes a lower efficiency of gene flow within this population due to the greater distance between the *C. brasiliense* trees and the barriers imposed on the pollinator by planting the eucalyptus trees.

(d) *Gene flow in populations located in forests and Cerrado fragments*

A model developed by Chybicki et al. (2021) in relation to pollen donors was used by Zaruma et al. (2024) to verify the effects of self-fertilization and biparental. According to the results, the crossings were not random due to the pollen dispersion pattern, which ranged from 3 to 184 m. Thus, two dispersal models were found depending on the foraging behavior of the bat, the main pollinator: the first indicates a high occurrence of crossings between nearby trees and the second, on a smaller scale, a dispersal occurring over longer distances. It is interesting to mention that from an evolutionary point of view, Futuyma (1998) mentions that many species present gene flow over long distances. However, the average level of gene flow is restricted to short distances, which implies the divergence of local populations through genetic drift or local adaptation. An important aspect of gene flow is related to speciation, which can be observed in the study carried out by Pinto et al. (2023) in 35 populations of *H. stigonocarpa* complex in central Brazil, separating two species, *H. stigonocarpa* var. *brevipetiolata* and *H. stigonocarpa* var. *stigonocarpa*.

Cerrado, Amazon and Atlantic Forest are under habitat fragmentations at an accelerated pace due to human interference. Kubota et al. (2024), in an ecotone region between the Cerrado and the Atlantic Forest (448.2 ha), studied gene flow between four subpopulations of *Cariniana estrellensis*: Sp1 (4.9 ha), Sp2 (55.1 ha), Sp3 (63.8 ha) and Sp4 (17.66 ha) using nine microsatellite loci and following the approaches of Adams et al. (1997) and Slavov et al. (2004). The authors observed that the subpopulations are not isolated, receiving 18.3% of pollen outside the fragment and the gene exchange between the subpopulations varies from 16.1% to 31.3%. They also found that pollen dispersion via gene flow reaches 3.5 km. Therefore, for the purpose of collecting seeds from mother trees for ex situ conservation, the recommendation is that the mother trees should be more than 200 m away, to avoid inbreeding. At the stage of fragmentation that these populations are, information of genetic variation based on molecular markers, geographic locations of parental trees, distance of pollen dispersal, and agents for pollination and seed dispersal are crucial for appropriate design of conservation and management of populations. Promoting the connectivity between subpopulations and others that are external to those forest fragments advised for preventing reproductive isolation among populations.

6.4 Influence of Migration on the Evolutionary History of Neotropical Species

To illustrate this point, we report how some plant species colonized South America. Dick and Pennington (2019) addressed the history of the diversification of neotropical tree species in the last 100 million years, which is worth discussing in this section. In their study, four neotropical biomes were investigated: (*i*) lowland rain forests; (*ii*) lower to upper montane forests; (*iii*) seasonally dry forests; (*iv*) savannahs. During this period, South America began to separate from Africa and Laurasia, and it was biotically connected across North America and Eurasia.

The first South American rainforest appeared 58 Ma ago, within the Paleogene (Paleocene + Eocene), where there was a global expansion of forests dominated by angiosperms at midlatitudes and the equatorial region. In contrast, the first modern dry forests and savannahs emerged later, in the Miocene (22–11 Ma) and Pliocene (11–2 Ma).

The emergence of these biomes can be explained by four migratory routes that occurred in the Eocene (50 Ma): (*i*) Beringia (from Alaska to North America); (*ii*) North Atlantic Land Bridge (from the European continent to North America); (*iii*) Antillean Land Bridge (bridge connecting North America to South America); and (*iv*) Magellan Land Bridge (from Antarctica to Chile).

In the past, Antarctica was partially forested and had a mixture of temperate and Neotropical plant lineages. The Magellan Land Bridge allowed the migration of taxa such as Cunoniaceae, Proteaceae, Myrtaceae and some groups of Lauraceae; the conifers Podocarpaceae and Araucariaceae; and tree ferns (*e.g.*, Cyatheaceae) to the Southern Hemisphere.

Another example of migration was found in a case study carried out in the Yasuní Forest (lowland rain forest/ Amazon rainforest) in Ecuador. The results suggest that around 50% of tree species belonging to that area descend from immigrant lineages that colonized South America during the Cenozoic period. Some examples of these taxa are: 37 genera and 114 species of Fabaceae; 12 genera and 75 species of Lauraceae; 7 genera and 53 species of Sapotaceae; 13 genera and 39 species of Annonaceae; 5 genera and 37 species of Meliaceae; 5 genera and 13 species of Lecythidoideae. These populations probably used the Antillean Land Bridge route and dispersed over long distances using sea currents to reach South America. Phylogenetic studies show that populations of tree species within the Amazon basin have a random phylogenetic structure and are brought together by widespread dispersal, that being an evidence of the hypothesis of migration of several genera and species mentioned above, and which makes up the great biodiversity of this biome.

It is supposed that glacial and interglacial periods paved the way to drastic changes in the terrestrial landscape, which led to the fragmentation of the Amazon rainforest. These events divided the biome into lowland rain forests and seasonally dry forests. Later on, some regions became savannahs. Evidence of that was found by the representation of 72 out of 90 dominant woody genera in the Brazilian Cerrado in the Amazon. In the same biome, four woody endemic clades were shown based on the evolutionary transitions from surrounding biomes, promoting adaptations parallel to frequent fires, such as corky bark and the ability of the roots to sprout.

A community phylogenetic approach found that populations of tree species from seasonally dry forests are more strongly phylogenetically clustered than those found in rainforests and savannah populations. This information suggests that seasonally dry forests had strong adaptive barriers that difficulted migrants to settle over there, possibly due to erratic rainfall or different soil conditions.

The review of the migratory history of some plant species to neotropical biomes made by Dick and Pennington (2019) is based on phylogenetic studies in specific periods and with populations of tree species that were the most abundant in the biomes. However, there are several gaps to be clarified and require more in-depth studies on the matter. We encourage readers to delve deeper into these studies by reading their work.

6.5 Final Remarks

In neotropical tree species, gene flow occurs through the dispersal of pollen and seeds, carried out by animals or through the action of physical factors such as wind and water. However, efficiency depends on the number of genes, genetic divergence of the involved populations, dispersal agents and environmental conditions. We discuss the use of knowledge on genetic fluctuations in a case study of *A. urundeuva* in two ways: (*i*) natural populations; (*ii*) *aroeira* genebanks (AGB) as a progeny test. In the first case, there is no anthropogenic interference and it allows a better understanding of how genetic fluctuations occur in natural populations. In AGB there is a higher frequency of certain (favorable) alleles, that is, there is a direction in gene flow made by human action. However, the formation of AGB is directly related to seed collection.

As mentioned above, molecular markers are widely used in estimating gene flow in neotropical species, of which microsatellite markers have been involved in multiple studies. Given a genetic marker, a region of DNA that can be transcribed or not and that has polymorphism, we can estimate genetic differences between individuals. We can also measure whether there is a flow of alleles over generations between different populations based on allele frequencies. These molecular markers identify differences between individuals or their geographic distribution. Currently, these markers are widely used to measure differences between populations, as well as estimate the number of migrants between populations, due to the great variability of polymorphisms that do not directly reflect the phenotype.

However, we report that gene flow interferes with the number of mother trees for seed collection. Brazil has great biodiversity in its biomes, but it is still far from knowing the dynamics of the pollen and seed pools of its tree species. Therefore, it is necessary to continue research on this subject to generate a database that allows the establishment of better public policies in relation to seed collection for the *ex situ* conservation of these species. We believe that with the advances of molecular tools, such as more robust markers that allow access to the genome as described by Rossini et al. (2021) for *A. urundeuva*, these questions could be clarified in the near future.

Migration can occur naturally, caused by the movement of organisms from one location to another. Animals are driven by the search for food and reproduction, playing a fundamental role in maintaining ecosystems. However, humans have played a big role in animal migrations, especially due to economic uses. For plants, there was a leading role at the time of the expansions of great civilizations, by land or sea routes, in which they exchanged vegetal material with each other and also biopiracy.

In this chapter the objective was to show the participation of migration in the evolutionary process of some neotropical tree genera and species. However, as reported in item 6.4, phylogenetic studies carried out on samples dated 100 Ma and current samples are still inconclusive. However, with the advent of technology, especially in the genomic area and the availability of resources and qualified researchers, it will be possible to carry out more detailed analyses on the species already studied and expand the number of species. Therefore, these studies might go to the individual and population levels, making it possible to clarify some gaps left in the evolutionary history of neotropical tree species and strengthening the hypotheses reported in this chapter.

Key Points

- Gene flow can in many cases enhance or neutralize the action of other evolutionary factors such as natural selection in the reference population.
- When a uniform natural population goes through a process of fragmentation, several subpopulations arise, and higher homozygosity is a consequence. However, if gene flow is reestablished, homozygosity tends to decrease.
- Migration is an evolutionary process that allows natural populations to survive and sustain ecosystems.
- Migration probably contributed to the formation of neotropical biomes and biodiversity composition.

Questions

1. What is the relation between gene flow and natural selection?
2. What are the consequences of changes in migratory routes in animals?
3. How important is gene flow in the evolutionary process of neotropical trees?
4. Discuss the importance of the migration of cultivated plant seeds made by great navigators among continents.

Acknowledgements We thank the financial support for our research from FAPESP n° 2022/03320-5. Celso L. Marino (306.974/2023-1) and Mario L.T. Moraes (311.164/2021-8) are supported by CNPq research fellowships and Marcela A.M. Silvestre (Unesp PROPE n° 4372 – PD III fellowship).

References

Adams WT, Hipkins VD, Burczyk J et al (1997) Pollen contamination trends in a maturing Douglas-fir seed orchard. Can J For Res 27:131–1344. https://doi.org/10.1139/cjfr-271-131

Caetano S, Prado D, Pennington RT et al (2008) The history of Seasonally Dry Tropical Forests in eastern South America: inferences from the genetic structure of the tree *Astronium urundeuva* (Anacardiaceae). Mol Ecol 17:3147–3159. https://doi.org/10.1111/j.1365-294X.2008.03817.x

Carvalho PER (2003) Espécies arbóreas brasileiras. (Coleção Espécies Arbóreas Brasileiras, v.1). Brasília: Embrapa Informação Tecnológica; Colombo: Embrapa Florestas

Chybicki IJ, Oleksa A, Dering M (2021) Identification of determinants of pollen donor fecundity using the hierarchical neighborhood model. Mol Ecol Resour 21:781–800. https://doi.org/10.1111/1755-0998.13307

Collevatti RG, Estolano R, Garcia SF et al (2010) Short-distance pollen dispersal and high self-pollination in a bat-pollinated neotropical tree. Tree Genet Genomes 6:555–564. https://doi.org/10.1007/s11295-010-0271-4

Correa SB, Costa-Pereira R, Fleming T et al (2015) Neotropical fish–fruit interactions: eco-evolutionary dynamics and conservation. Biol Rev 90:1263–1278. https://doi.org/10.1111/brv.12153

Darwin C (2018) The origin of species (6th edn). Independently published Michigan, 592p

Degen B, Sebbenn AM (2014) Genetics and tropical forests. In: Köhl M, Pancel L (eds) Tropical forestry handbook. Springer, pp 1–30. https://doi.org/10.1007/978-3-642-41554-8_75-1

Dick CW, Pennington RT (2019) History and geography of neotropical tree diversity. Annu Rev Ecol Evol Syst 50:279–301. https://doi.org/10.1146/annurev-ecolsys-110617-062314

Duponchelle F, Isaac VJ, Doria CRC et al (2021) Conservation of migratory fishes in the Amazon basin. Aquatic Conserv 31:1087–1105. https://doi.org/10.1002/aqc.3550

Futuyma D (1998) Evolutionary biology, 3th edn. Sinauer Associates, Inc

Gaino APSC, Silva AM, Moraes MA et al (2010) Understanding the effects of isolation on seed and pollen flow, spatial genetic structure and effective population size of the dioecious tropical tree species *Myracrodruon urundeuva*. Conserv Genet 11:1631–1643. https://doi.org/10.1007/s10592-010-0046-3

Kageyama PY (1987) Conservação "*In Situ*" de recursos genéticos de plantas. IPEF 16(35):7–35

Kubota TYK, Hallsworth JE, Silva AM et al (2024) Pollen dispersal and mating patterns determine resilience for a large-yet-fragmented population *Cariniana estrellensis*. Conserv Genet 25:117–132. https://doi.org/10.1007/s10592-023-01557-8

Manoel RO, Rossini BC, Cornacini MR et al (2021) Landscape barriers to pollen and seed flow in the dioecious tropical tree *Astronium fraxinifolium* in Brazilian savannah. PLoS One 16(8):e0255275. https://doi.org/10.1371/journal.pone.0255275

Paterniani E (1974) Evolução dos sistemas reprodutivos nos vegetais. Cienc Cul 26(5):476–492

Pinto RB, Ferreira RSB, Telles MPC et al (2023) Untangling the *Hymenaea stigonocarpa* complex (Fabaceae) using population genetics and morphology. Bot J Linn Soc 202(1):134–147. https://doi.org/10.1093/botlinnean/boac044

Resende MDV (2015) Genética quantitativa e de populações. Suprema, Viçosa

Resende MDV, Mauro RA (2003) Genética de populações e conservação de animais silvestres. In: Costa RB (ed) Fragmentação florestal e alternativas de desenvolvimento rural na Região Centro-Oeste. UCDB, Campo Grande, pp 75–112

Rossini BC, Moraes MLT, Marino CL (2021) Complete chloroplast genome of *Myracrodruon urundeuva* and its phylogenetics relationships in Anacardiaceae family. Physiol Mol Biol Plants 27:801–814. https://doi.org/10.1007/s12298-021-00989-1

Slavov GT, Howe GT, Gyaourova AV et al (2004) Pollen Flow (PFL): a computer program for estimating pollen flow using highly variable markers and paternity exclusion. PNWTIRC Publications

Souza DCL, Rossini BC, Souza FB et al (2017) Development of microsatellite markers for *Myracrodruon urundeuva* (F.F. & M.F. Allemão), a highly endangered species from tropical

forest based on next-generation sequencing. Mol Biol Rep 45:71–75. https://doi.org/10.1007/s11033-017-4142-z
Stearns SC, Hoekstra RF (2000) Evolution: an introduction. Oxford University Press
Willis KJ, Mcelwain JC (2014) The evolution of plants, 2nd edn. Oxford University Press, New York, p 424p
Zaruma DUG, Sebbenn AM, Rossini BC et al (2024) Mating system and inbreeding depression in *Hymenaea stigonocarpa*. Tree Genet Genomes 20:6. https://doi.org/10.1007/s11295-024-01639-9

Chapter 7
What Are the Genetic Consequences of Inbreeding?

Daniela Cristina Ferreira, Pabila Stephanie de Souza Arruda, Marcelo Lattarulo Campos, and Paulo Cesar Venere

Contents

Abstract The term inbreeding refers to any form of mating that involves genetically similar or related organisms, whether intentional or not. While inbreeding does not change the frequency of the alleles in a population, it does redistribute genotypes, and increase consanguinity and homozygosity, ultimately decreasing levels of genetic diversity in a population. One of the principal consequences of inbreeding is inbreeding depression, which is a reduction in the evolutionary fitness of the individuals that make up a population. Although inbreeding may often lead to a decrease in the reproductive success of a group of organisms, it also represents a potentially useful tool for the isolation and retention of a desired trait in breeding programs of both plants and animals. Inbreeding can occur naturally in the wild, and will contribute to the propagation of a species, although its effects can vary considerably between large, randomly-sorted populations and small, isolated populations. In small populations, inbreeding can have drastic consequences, such as the rapid manifestation of undesirable traits, including the expression of deleterious alleles or the acquisition of traits that decrease reproductive success. This chapter addresses the causes and effects of inbreeding, and discusses biological concepts, such as the coefficient of endogamy, inbreeding depression, and the consequences of reduced heterozygosity, using populations of neotropical organisms as examples.

D. C. Ferreira (✉) · P. S. de Souza Arruda · M. L. Campos · P. C. Venere
Instituto de Biociências, Universidade Federal de Mato Grosso, Cuiabá, MT, Brazil

M. V. Bohrer Monteiro Siqueira et al. (eds.), *Population Genetics in the Neotropics*,
https://doi.org/10.1007/978-3-031-83685-5_7

Keywords Breeding systems · Consanguinity · Homozygosity · Inbreeding coefficient · Inbreeding depression

Learning Objectives
At the end of this chapter, the reader should be able to:

- Define and explain the importance of inbreeding;
- Understand the methods employed to quantify and analyze inbreeding;
- Understand the implications of inbreeding for the genetic "wellbeing" of both natural and domesticated populations.

7.1 What Is Inbreeding?

Inbreeding, or endogamy (from the Greek *endon*—within, and *gamos*—sexual union or mating), refers to any reproductive system in which closely-related or genetically similar organisms mate among themselves, intentionally or otherwise. Inbreeding applies to any organism that reproduces sexually, including both plants and animals. Genetic traits are determined by alternative forms of a gene (alleles) that vary in their nucleotide composition, but occupy the same position on a chromosome. These alleles may be manifested in either the homozygous (two identical copies of the same gene) or the heterozygous condition, that is, two distinct alleles (Snustad and Simmons 2015, Fig. 7.1). When individuals share a common ancestor, they tend to inherit the same alleles of a given gene, and if they mate together, they tend to pass these alleles on to their progeny. The most extreme example of endogamy is the self-fertilization observed in some plant species, while in animals, it is most intense in parent-offspring and sibling matings. Inbreeding tends to cause deviations from the Hardy-Weinberg equilibrium, in particular by increasing the

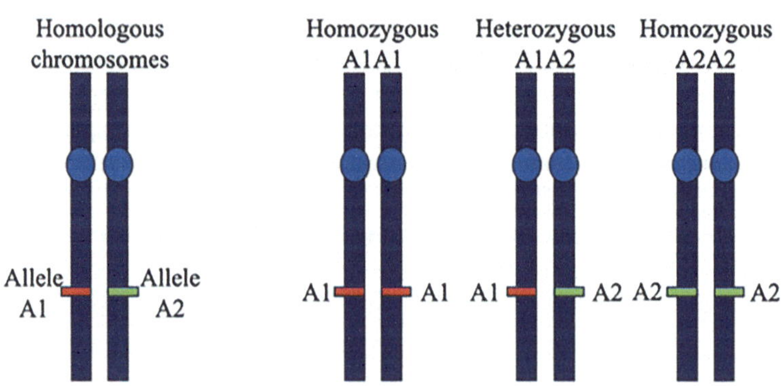

Fig. 7.1 A pair of homologous chromosomes (left) and the different possible combinations (homozygotes and heterozygotes) for a single pair of alleles, *A1* and *A2* (right), the basic concepts for the understanding of inbreeding

proportion of homozygotes and reducing that of heterozygotes, even though the allele frequencies are not altered (Klug et al. 2008). Ultimately, inbreeding tends to reduce the genetic diversity of a population. In other words, endogamy promotes genetic homogenization.

In nature, self-fertilization, which is common in certain groups of plants, and is also known as autogamy, results in high levels of homozygosity in the genomes of many species. Despite the increase in homozygosity, as an evolutionary strategy, autogamy avoids the need for pollinators and will fix desirable traits in the species, thus providing reproductive assurance. This reproductive strategy nevertheless produces organisms that are less genetically diverse and thus less adaptable to environmental change (Frankham et al. 2002), and may also be more prone to genetic diseases. The frequency of heterozygous loci will decrease progressively after each generation of endogamy (Fig. 7.2). When a heterozygous (*A1A2*) individual self-fertilizes, it produces offspring with equal proportions of heterozygous (*A1A2*) and homozygous individuals (*A1A1*, *A2A2*). Consecutive generations will undergo a progressive reduction (by 50% per generation) in the proportion of heterozygous individuals, so that, by the nth generation, the proportion of heterozygotes will be $(1/2)^n$. After seven successive generations of self-fertilization (Fig. 7.2), in fact, over

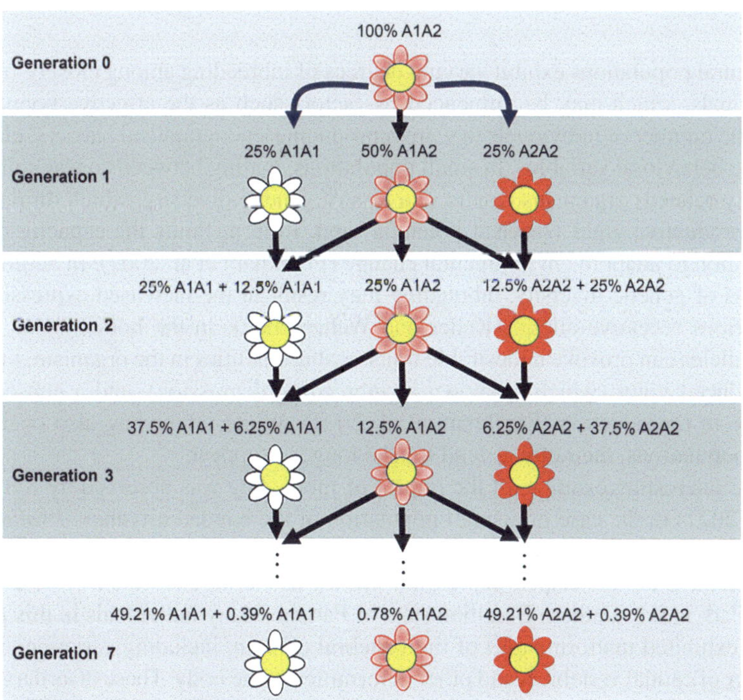

Fig. 7.2 Progressive reduction in the percentage of heterozygotes due to self-fertilization in a plant population through the segregation of the alleles *A1* and *A2*

99% of the individuals will be homozygous, half for one allele (*A1A1*), and half for the other allele (*A2A2*).

Box 7.1 What Is Self-Fertilization?
Self-fertilization is a reproductive process observed in some organisms, in which an individual is able to fertilize its own gametes (Wright 1922; Stebbins 1957). Obviously, this process is only possible in true hermaphrodites, in which the organs of both sexes are present in the same individual, permitting the production and fusion of male and female gametes in the same organism. Self-fertilization is common in hermaphroditic flowering plants, in which the male (stamens) and female (carpels) reproductive organs are present in the same flower (Jarne and Charlesworth 1993). Self-fertilization is not possible in dioecious plants, however, in which the individuals produce only female (ovules) or male (pollen) gametes. Self-fertilization is less common in animals, occurring only in hermaphroditic groups, such as the flatworms (*Taenia*) and some mollusks. In these organisms, the presence of the male and female reproductive organs in the same individual permits fertilization to occur (Jarne and Charlesworth 1993).

Natural populations exhibit varying degrees of inbreeding among closely-related individuals, which may be influenced by factors such as the effective population size (the number of individuals that are reproducing), geographical barriers, climate or even behavioral variables. In small populations, mating between consanguineous (closely-related) organisms results in a loss of genetic diversity, which diminishes the reproductive vigor (survival potential) and, in turn, limits the capacity of the population to adapt to environmental change (Frankham et al. 2002). In addition to the loss of genetic diversity, inbreeding may result in the increased expression of deleterious recessive alleles (Keller and Waller 2002). In the homozygous state, these alleles can provoke undesirable traits or abnormalities in the organism, including reduced vigor (which is known as inbreeding depression), and a consequent decline in reproduction (Frankham et al. 2002). While inbreeding also occurs in large populations, their effects tend to take longer to appear.

One interesting example of the impact of inbreeding was observed by Kerniske et al. (2021) in the case of a small population of the banded astyanax, *Psalidodon* aff. *fasciatus* (a small Neotropical fish) found in a completely isolated freshwater ecosystem, a doline-shaped depression known as Furna 2, located in Vila Velha State Park in the southern Brazilian state of Paraná. Many individuals in this population exhibited malformations of the vertebral column, including variations in the number of caudal vertebrae, and other deformities of the body. These disorders were attributed to the high level of inbreeding in the studied population, and the absence of gene flow, due to their isolation.

In addition to its occurrence in natural populations, intentional inbreeding is a practice used in various breeding programs of both animals and plants. In this case, genetically-similar individuals are bred together to preserve desirable genetic traits, which enhances the commercial value of the progeny. However, this practice can lead to increased homozygosity of both dominant and recessive alleles, which can eventually result in unforeseen losses for the breeder.

In this chapter, we will discuss the effects and consequences of inbreeding in natural populations, and the implications of this process in stocks of domesticated plants and animals, in which inbreeding is employed in breeding programs to select and reinforce the presence of desirable traits.

7.2 The Inbreeding Coefficient

The theoretical foundations for the study of inbreeding were first established in the early twentieth century. In 1922, Sewall Wright developed a statistical tool, the coefficient of inbreeding (*F*), to better evaluate the effects of endogamy in a population. This coefficient estimates the proportion by which heterozygosity is reduced in a population due to inbreeding (Wright 1922, 1951), that is, it calculates the probability of any individual having two identical alleles at the same locus (Chagas and Oliveira 2021). The higher the level of inbreeding, the lower the theoretical proportion of heterozygotes in the population. The value of *F* varies from 0, in the case of a totally outbred population, to 1, in a completely inbred one, such as that of an autogamous species (Klug et al. 2008).

When *F* is equal to 0, then, mating is totally random within the population (a scenario known as panmixia), with no inbreeding among closely-related individuals. When *F* reaches 1, on the other hand, mating occurs only between related individuals, which results in a high frequency of homozygotes in the population. A straightforward method to estimate *F* is to compare observed heterozygosity (*Ho*), that is, the proportion of heterozygous individuals in the population, with expected heterozygosity (*He*), that is, the proportion of heterozygous individuals expected in the population based on the Hardy-Weinberg equilibrium. The inbreeding coefficient is thus calculated by the following expression:

$$F = He - Ho / He$$

where:
F = the inbreeding coefficient
Ho = the observed heterozygosity
He = the expected heterozygosity

Another way to calculate the inbreeding coefficient is by considering the probability of the inheritance of a given allele during mating. An endogamous individual will result from successive generations of self-fertilization (see Fig. 7.2) or mating between closely-related individuals. If we consider a heterozygous plant with a pair of alleles (*A1*/*A2*), the descendants of a self-cross have a 50% chance of inheriting allele *A1* from the ovule and a 50% chance of inheriting this allele from the pollen.

Overall, then, 25% of the offspring of this self-fertilization will be homozygous for allele *A1* (*A1A1*) and 25% will be homozygous for allele *A2* (*A2A2*). To calculate the coefficient of endogamy, it is necessary to determine the probability that the descendants will inherit only one of the alleles (i.e., genotype *A1A1* or *A2A2*), that is, 0.25 + 0.25 = 0.50. The same calculation can be employed to determine the coefficient of endogamy in the mating of genetically similar individuals, which is a common practice in animal breeding programs. When full siblings are mated (Fig. 7.3), the steps represented by the arrows indicate that the probability of allele *A1* passing to the next generation is 1/2. In this example, the *A1* allele of one of the grandparents has a $(1/2)^4 = 1/16$ chance of reaching the grand offspring, given that there are four steps along both pathways that this gene would need to pass. The same calculation can be applied to each of the four alleles (*A1*, *A2*, *A3*, and *A4*), with the probability of the grand offspring being homozygous for any one of these alleles being 1/16.

Determining the coefficient of endogamy in natural populations or domesticated stocks provides important insights into the genetic structure of the organism. This coefficient also sheds light on the genetic diversity of the population, and permits the identification of potential risks that may threaten its long-term survival. The inbreeding coefficient is an important variable for decision-making on both *in situ* and *ex situ* conservation measures, given that it highlights the levels of homozygosity in the study population, which may have both positive or negative implications. The positive implications refer to the fixation or homogenization of desired traits in

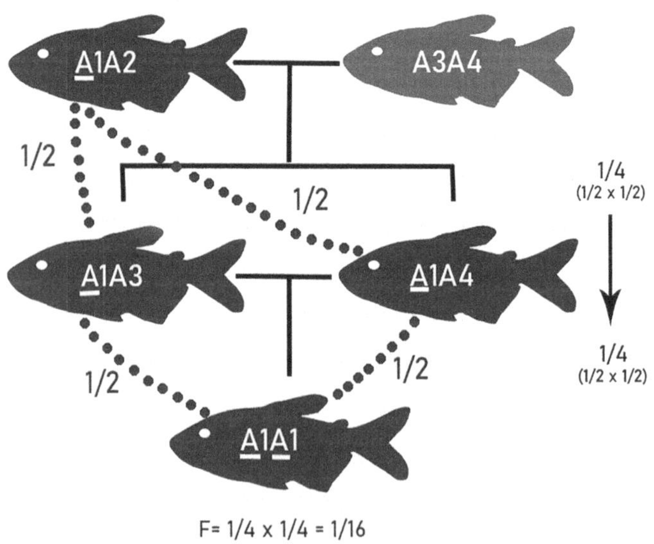

Fig. 7.3 Calculation of the inbreeding coefficient (*F*) in the offspring of a double sibling cross. The example shows the probability that an offspring of the second generation is homozygous for the *A1* allele. In each generation, ½ represents the probability of gamete formation with the allele *A1*, while ¼ is the frequency of genotypes carrying the *A1* allele in the next generation

domesticated species by genetic improvement programs, while the negative implications arise when homozygosity decreases the adaptive potential of a population.

7.3 The Consequences of Inbreeding

One potentially negative consequence of endogamy is a decline in the survival and reproductive capacity of a population, a phenomenon known as inbreeding depression. To understand this process, it is necessary to document the evolutionary history of the population. In the wild, deleterious alleles will be eliminated by natural selection, given that they reduce the evolutionary fitness of the carrier. If the deleterious allele is dominant, the probability of it being passed on to the next generation is virtually null, because any carrier will present the deleterious phenotype, and will thus be eliminated from the breeding pool (Keller and Waller 2002). When recessive, however, deleterious alleles are masked by their dominant, non-deleterious homologs, and are exempt from natural selection as long as they are present in the heterozygote condition. They can thus be passed down the generations and remain present in the population.

In large populations with panmictic mating, deleterious recessive alleles have a greatly reduced chance of being expressed because consanguineous encounters are minimal. In small populations, however, mating is more likely to occur between related individuals that carry deleterious alleles, with the chances of these alleles being expressed increasing in proportion to the reduction in population size (Wang et al. 1999). When a population remains small and isolated for many generations, the expression of undesirable traits becomes increasingly likely, leading to a decrease in adaptive success, reduced reproduction and survival, and eventually, an increased risk of extinction due to inbreeding depression (Lynch et al. 1995; Keller and Waller 2002).

Box 7.2 What Are Deleterious Recessive Alleles?

Deleterious recessive alleles harm their carriers when in the homozygous condition. In this context, the term deleterious refers to harmful genetic disorders, which may cause the early death of the embryo, disease or even a reduction in reproductive capacity. In the heterozygous condition, the expression of these alleles is masked by their dominant counterparts, and the presence of the deleterious recessive allele does not provoke any adverse effects (Dudash and Carr 1998). This permits these deleterious alleles to persist in large, panmictic populations, in which outbreeding prevails. When the size of the population is reduced, on the other hand, inbreeding tends to become more frequent, causing what is known as inbreeding depression, which is characterized by an increase in the frequency of homozygotes. Small, inbred populations thus tend to be more vulnerable to the expression of the recessive deleterious alleles that were previously masked by the dominant alleles through consanguineous matings.

Attias et al. (2023) describe an example of inbreeding in a small, natural population of giant anteaters (*Myrmecophaga tridactyla*) from Mato Grosso do Sul state, in central Brazil. This species is widely distributed in the Neotropics, although it is classified as vulnerable to extinction (IUCN 2023). The specimens sampled in this study included two albino individuals. Albinism is a genetic characteristic determined by recessive alleles, which impede the deposition of melanin in the skin and other body parts. The genetic analyses of this population revealed a relatively high inbreeding coefficient and a moderate level of genetic diversity, which indicate that the study population had undergone a recent process of decline, probably due to a population bottleneck (Barragán-Ruiz et al. 2021). The frequency of albinism tends to be associated with the size of the population, given that the chances of the manifestation of the recessive alleles, which grow as the relatedness of mates increases, tend to be greater in smaller populations. In a second study of giant anteaters, in the Emas National Park of Goiás state, in central Brazil, Collevatti et al. (2007) found evidence of a reduction in the size of the population, followed by the loss of genetic diversity and high rates of inbreeding, which had resulted from the isolation of the population in the national park by the expansion of farmland in the surrounding area. These two anteater populations exemplify the negative effects of population decline on genetic diversity and, in turn, the expression of undesirable traits.

A similar pattern has also been observed in plants. Tambarussi et al. (2016) studied three isolated populations of jequitibá (*Cariniana legalis*) trees, a neotropical species also classified as vulnerable (IUCN 2023) and found that population size influenced breeding rates (self-fertilization and reproducing with close relatives), with inbreeding increasing in the smaller populations. In the adult trees, allelic richness was higher in the larger populations, in comparison with the smaller ones. Due to their low genetic diversity and high levels of inbreeding, the fitness of the seeds produced by the trees in the smaller population may decline due to inbreeding depression, or these inbred descendants may not survive at all, increasing their risk of population decline or extinction. It is important to note, however, that some studies have shown that inbred plant populations can persist over the long term, possibly due to the purging of deleterious recessive alleles, which would reduce the "genetic costs" of inbreeding in these species (Ortiz-Barney and Ackerman 1999).

As inbreeding can potentially provoke the extinction of a species, it is hardly surprising that so many organisms have developed mechanisms to avoid breeding among closely-related individuals. Many plants have evolved recognition mechanisms that prevent pollen from fertilizing their own flowers or those of genetically similar individuals (Broz and Bedinger 2021). Similarly, most human societies have strict taboos against consanguineous marriage (Oniya et al. 2019).

From an evolutionary perspective, inbreeding reduces genetic variability and may limit the potential for a population to adapt to selective pressures. This has been stated before in this book, but it is always good to remind the reader that genetic variability provides populations with the raw material (alleles) needed to adapt more quickly to environmental change because genetically diverse populations tend to have a better chance of resisting diseases, tolerating shifts in the environment, and adjusting to new ecological niches. Inbreeding depression can thus be considered to be an obstacle to adaptive evolution, given that it limits the genetic diversity available for evolutionary adaptation.

One interesting example of the loss of genetic diversity in an endogamous population is provided by Vitorino et al. (2015, 2017), who analyzed the genetic diversity (microsatellite and ISSR markers) of the pirarucu (*Arapaima gigas*), a large freshwater fish found primarily in the Amazon basin and neighboring river systems. These researches focused out on five populations in the Araguaia-Tocantins basin, three in the Brazilian state of Mato Grosso—Araguaiana, Novo Santo Antônio, and São Félix do Araguaia—one (Luís Alves) in Goiás, and one from Itupiranga, in Pará state (Fig. 7.4). The analyses revealed the presence of two clusters within the Araguaia-Tocantins basin, one formed only by the population from Araguaiana and the other by the other four populations together. The population from Araguaiana, located further upstream on the Araguaia River than the other populations (Fig. 7.4), had considerably lower genetic diversity, with only 44.0% of the alleles identified in the two studies, as well as the lowest heterozygosity ($Ho = 0.133$, $He = 0.133$), and allelic richness ($Ar = 1.526$). By contrast, the Itupiranga population, from an area downstream from the confluence of the Tocantins and Araguaia rivers, had the highest genetic diversity, with 68.0% of the alleles identified in the two studies, as well as the highest heterozygosity ($Ho = 0.429$, $He = 0.450$) and allelic richness ($Ar = 2.429$). In general, however, all the populations from the Araguaia-Tocantins basin had lower genetic diversity in comparison with a population from the Amazon basin (Araripe et al. 2013). The low genetic variability found in the populations of the Araguaia-Tocantins basin is likely related primarily to predatory fishing and other anthropogenic impacts, which have greatly reduced the size of these

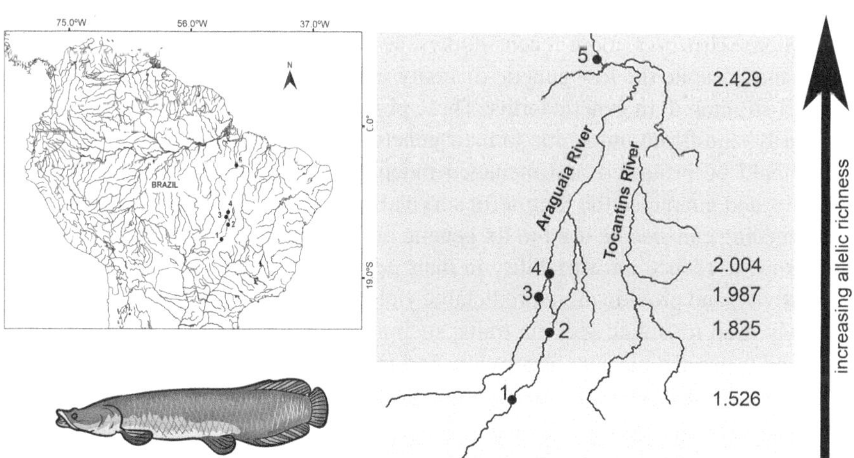

Fig. 7.4 Partial map of South America, showing the sites in the Araguaia-Tocantins basin in central Brazil, from which the *Arapaima gigas* (lower left) specimens were collected for analysis by Vitorino et al. (2015, 2017) and the allelic richness (Ar) recorded in the respective populations: (1) Araguaiana (Mato Grosso state), (2) Luís Alves (Goiás), (3) Novo Santo Antônio (Mato Grosso), (4) São Felix do Araguaia (Mato Grosso) and (5) Itupiranga (Pará)

populations. The reduced diversity of these may also be related, in part, to the sedentary behavior of the pirarucu, which engages only in short-range, lateral movements within the floodplain which, given the hydrological characteristics of the basin, limits the potential for longitudinal migrations and gene flow among the studied populations.

Despite its negative potential, inbreeding is not entirely harmful. In some circumstances, it can be used temporarily as a strategy to preserve populations in imminent danger of extinction. In fact, some breeding programs promote mating between close relatives to ensure the survival of a species (Klimova et al. 2022), although it is important to ensure the reintroduction of genetic diversity into the population as soon as possible to avoid negative long-term effects, in particular, inbreeding depression. Finding ways to preserve the genetic variability of a population, even in the face of threats to its survival, is essential, given that this variability ensures the capacity of a species to adapt to changes in the environment and, ultimately, survive.

The Neotropical hyacinth macaw (*Anodorhynchus hyacinthinus*) is one example of a bird with reduced genetic variability. This species is vulnerable to extinction (IUCN 2023), due primarily to the destruction of its natural habitat and the illegal trafficking of captive individuals. In their genetic and ecological study, Presti et al. (2015) showed that the hyacinth macaw populations have low genetic variability. There are a number of hypotheses that have attempted to account for this reduced variability. One hypothesis proposes that this is an intrinsic characteristic of *A. hyacinthinus*, possibly resulting from climatic shifts occurring during the Quaternary period, which contributed to the current disjunct geographic distribution of the species. A second hypothesis is that this low genetic variability is the result of specific behaviors, such as philopatry, which supports inbreeding, or genetic drift in isolated populations. However, more recent studies, such as de Almeida et al. (2019), have shown that, despite the low genetic diversity in *A. hyacinthinus*, some populations are well-structured, in genetic terms. These populations were considered to be evolutionarily significant units, due to their genetic differentiation, which implies that they should be protected and managed independently, in order to preserve their diversity, and guarantee the long-term survival of the species.

Inbreeding can also be used to fix genetic uniformity in economically-important organisms, to reduce the variability in their performance, that is, their growth and productivity, and provide more predictable yields. In plants, crossing close relatives is widely used to isolate specific traits, such as resistance to pests and pathogens, improved water and nutrient absorption, and increased productivity.

In commercial crop plants, the fixation of specific genetic traits through inbreeding is used to create new cultivars. One good example is the common bean (*Phaseolus vulgaris* L.), one of the most globally important legume crops for the production of human food, given its high protein, dietary fiber, vitamin, carbohydrate, and mineral content (Broughton et al. 2003). Many bean cultivars reproduce rapidly and have a short life cycle, although they are vulnerable to variation in the environment, that is, the genotype x environment interaction. To overcome this limitation, inbred lineages are selected for resistance to different growing environments, with the objective of identifying the best adapted cultivar. This selection process often involves the evaluation of the performance of alternative genotypes in different environments and conditions, such as annual fluctuations, locations, and sowing times, to obtain the most productive cultivars (Batista et al. 2017).

In domestic livestock, inbreeding can also be beneficial, in particular for the improvement of productivity in the dairy, meat, and wool sectors. A careful selection and mating of genetically similar individuals can be used to fix desirable traits in a herd. In the case of milk production, inbreeding has been used to boost yields in cattle breeds such as the Holstein, resulting in animals with enhanced feed conversion rates and milk production (Mastrangelo et al. 2018; Sieklicki et al. 2020). Similarly, inbreeding can also enhance traits that are beneficial for the production of meat, such as growth rates, carcass conformation, and the quality of the meat, increasing profitability for producers and providing consumers with higher quality products (Szmatoła et al. 2019). Inbreeding has also been used to improve wool yields in sheep, by increasing the quality of the wool, producing finer and more uniform fibers (Justinski et al. 2023). Despite this potential, inbreeding must be applied with caution, considering the potential risks, such as an increase in the prevalence of recessive genetic diseases. Given this, responsible genetic management and continuous monitoring are essential to ensure that the benefits of inbreeding outweigh its potential drawbacks.

7.4 Final Remarks

This chapter has revealed two fundamental facets of inbreeding. The first is the negative consequences that are inherent to the phenomenon. These consequences result primarily from the expression in individuals of negative traits encoded in recessive genes, as well as the loss of genetic diversity in the population. In small and isolated populations, these processes can lead to a decline in the population or even their extinction. The second facet is the strategic employment of endogamy for the creation of genetically uniform organisms. This practice can isolate or reinforce desired traits, primarily for agricultural productivity, but also for other objectives, such as ornamental uses.

In general, however, the terms endogamy and consanguinity are most often associated with the potential problems arising from inbreeding, both in natural populations and domestic stocks. In fact, knowledge of the potential for population decline derived from consanguineous mating has long been evident in the common practices of farmyard animal breeding. Most chicken farmers, for example, are well aware that it is essential to replace their breeding rooster periodically, because they know that, if a rooster is allowed to mate with its own daughters, the resulting offspring may be prone to a decline in quality or even the appearance of congenital disorders, due to undesirable crosses. It is interesting to note here that an in-depth understanding of genetics is not necessarily required for the selection of good breeding animals or maintaining a genetically healthy stock. An attentive breeder will know how to manage potential problems related to kinship, and will avoid or minimize potentially undesirable pairings, even in small-scale operations.

In addition to domestic stock and plant cultivars, relatively complex examples can be found in some human populations, in particular, among the numerous indigenous groups of Brazil, which are often represented by relatively small and isolated communities. An inevitable question arises here—how are the potential risks of endogamy and consanguinity avoided in these communities? One example here can be seen in the case of the Baniwa, an ethnic group of the Juivitera community found on the middle Içana River in São Gabriel da Cachoeira, Amazonas state, Brazil. Baniwa (2012) observed that the occupation of territory by the Baniwa is structured according to mythological precepts and rules that govern kinship relationships. The Baniwa are organized in fraternities, and individuals from each fraternity invariably marry individuals from other fraternities, thereby avoiding the possibility of consanguineous mating. This clearly reflects traditional knowledge, which has guided cultural norms to minimize the potential consequences of endogamous pairings that could have evolutionary disadvantages for the group.

Based on the available data, Luciano (2006) concluded that the entire social, cultural, and economic organization of an indigenous community is intricately woven into its worldview and way of life. In other words, there is a fundamental cosmology, which is expressed through myths and rituals that guide social interactions, marriages, and numerous other daily events. It is from these cosmological principles that exogamous or endogamous marriage arrangements will emerge. Following this same line of thought, the author highlights the problems that may arise from "disrespecting certain natural rules". In this way a social rule, for instance, refers to marriage. Each indigenous community has its own marriage customs based on kinship, affinity, and established social relationships that everyone must adhere to. Commonly, social groups potentially form marriage alliances. If, for any reason, one group denies a woman to a man from another group, the shaman is called upon to avenge this breach of rule or sociopolitical relations by invoking the original principles upon which these rules were established in human history. If the decision to violate the rule is collective, the retaliation affects the entire community, leading to group illnesses or epidemics (according to Western medicine).

This passage states clearly that disregarding a traditional rule can generate "group illnesses or epidemics". This raises an interesting question—could these be the effects of endogamy that have been imprinted in the collective consciousness of these peoples? Has this guided them to establish a comprehensive set of rules that, from their own perspective, represents a more spiritual response? After reading this chapter we leave you to your thoughts on the matter.

Key Points

1. Inbreeding occurs when organisms mate with individuals from the same genetic lineage or family, including self-fertilization and identity by descent.
2. Inbreeding can result in a reduction of genetic diversity within a population, which increases the risk of the expression of recessive genetic traits and may decrease the evolutionary fitness of the descendants.

3. Inbreeding can be used as a genetic management tool, when applied in a controlled and strategic manner. When conducted appropriately, inbreeding can fix desirable traits within a population, creating purebred lineages that are consistent in terms of both their phenotype and genotype. This can be especially useful in plant and animal breeding programs, in which specific traits are desired.
4. In conservation programs, inbreeding can be used to establish founder populations or breeding groups that maintain the existing genetic diversity found in an endangered population.

Questions

1. What is inbreeding, and how does it differ from other breeding patterns?
2. What are the potential negative effects of inbreeding on a population?
3. What is the potential role of inbreeding in conservation programs for endangered species?
4. How can inbreeding be used to fix desirable traits in a population?
5. How does inbreeding affect the adaptation and evolution of a population over time?

References

Araripe J, Rêgo PS, Queiroz H et al (2013) Dispersal capacity and genetic structure of Arapaima gigas on different geographic scales using microsatellite markers. PLoS One 8:e54470. https://doi.org/10.1371/journal.pone.0054470

Attias N, Yogui DR, Alves M et al (2023) Presence and monitoring of albino giant anteaters (Myrmecophaga tridactyla) in the Cerrado savanna. Bol Mus Para Emílio Goeldi, Ciências Naturais 18. https://doi.org/10.46357/bcnaturais.v18i2.866

Baniwa TT (2012) Saúde e alimentação em áreas rurais e urbanas: um testemunho indígena. In: Garnelo L, Pontes AL (eds) Saúde Indígena: uma introdução ao tema. MEC-SECADI, Brasília, 296 pp

Barragán-Ruiz CE, Silva-Santos R, Saranholi BH et al (2021) Moderate genetic diversity and demographic reduction in the threatened giant anteater, Myrmecophaga tridactyla. Front Genet 12:669350. https://doi.org/10.3389/fgene.2021.669350

Batista LG, dos Anjos RSR, Poersch NL et al (2017) Multigeneration index in the selection of common bean inbred families. Crop Sci 57:2354–2360. https://doi.org/10.2135/cropsci2016.08.0665

Broughton WJ, Hernández G, Blair M et al (2003) Beans (Phaseolus spp.): model food legumes. Plant Soil 252:55–128. https://doi.org/10.1023/A:1024146710611

Broz AK, Bedinger PA (2021) Pollen-pistil interactions as reproductive barriers. Annu Rev Plant Biol 17(72):615–639. https://doi.org/10.1146/annurev-arplant-080620-102159

Chagas FEO, Oliveira M (2021) Endogamia. In: Cruz CD, Oliveira M (eds) Genética de populações com o aplicativo GPOP, 1nd edn. Brazil Publishing, Curitiba, pp 196–230

Collevatti RG, Leite KCE, Miranda GHB et al (2007) Evidence of high inbreeding in a population of the endangered giant anteater, Myrmecophaga tridactyla (Myrmecophagidae), from Emas National Park, Brazil. Genet Mol Biol 30(1):112–120. https://doi.org/10.1590/S1415-47572007000100020

de Almeida TRA, Presti FT, Cruz VP et al (2019) Genetic analysis of the endangered Hyacinth Macaw (Anodorhynchus hyacinthinus) based on mitochondrial markers: different conservation

efforts are required for different populations. J Ornithol 160:711–720. https://doi.org/10.1007/s10336-019-01652-z
Dudash M, Carr D (1998) Genetics underlying inbreeding depression in Mimulus with contrasting mating systems. Nature 393:682–684. https://doi.org/10.1038/31468
Frankham R, Ballou JD, Briscoe DA (2002) Introduction to conservation genetics. Cambridge University Press, Cambridge. 545 pp
IUCN (2023) The IUCN Red list of threatened species. Version 2023-1. https://www.iucnredlist.org. Accessed on. [2024-04-24]
Jarne P, Charlesworth D (1993) The evolution of the selfing rate in functionally hermaphrodite plants and animals. Annu Rev Ecol Syst 24(1):441–466. https://doi.org/10.1146/annurev.es.24.110193.002301
Justinski C, Wilkens J, Distl O (2023) Effect of individual rate of inbreeding, recent and ancestral inbreeding on wool quality, muscling conformation and exterior in German Sheep Breeds. Animals 13(21):3329. https://doi.org/10.3390/ani13213329
Keller LF, Waller DM (2002) Inbreeding effects in wild populations. Trends Ecol Evol 17:230–241. https://doi.org/10.1016/S0169-5347(02)02489-8
Kerniske FF, Pena Castro J, De la Ossa-Guerra LE et al (2021) Spinal malformations in a naturally isolated Neotropical fish population. PeerJ 9:e12239. https://doi.org/10.7717/peerj.12239
Klimova A, Gutiérrez-Rivera J, Sánchez-Sotomayor V (2022) The genetic consequences of captive breeding, environmental change and human exploitation in the endangered peninsular pronghorn. Sci Rep 12:11253. https://doi.org/10.1038/s41598-022-14468-4
Klug WS, Cummings MR, Spencer CA et al (2008) Concepts of genetics, 9th edn. Pearson Benjamin Cummings, San Francisco. 876 pp
Luciano GS (2006) Quem são e quantos são os índios no Brasil. In: O Índio Brasileiro: o que você precisa saber sobre os povos indígenas no Brasil de hoje. Brasília, Ministério da Educação, Secretaria de Educação Continuada, Alfabetização e Diversidade, LACED Museu Nacional, 232 pp
Lynch M, Conery J, Burger R (1995) Mutation accumulation and the extinction of small populations. Am Nat 146:489–518. https://doi.org/10.1086/285812
Mastrangelo S, Sardina MT, Tolone M et al (2018) Genome-wide identification of runs of homozygosity islands and associated genes in local dairy cattle breeds. Animal 12(12):2480–2488. https://doi.org/10.1017/S1751731118000629
Oniya O, Neves K, Ahmed B (2019) A review of the reproductive consequences of consanguinity. Eur J Obstet Gynecol Reprod Biol 232:87–96. https://doi.org/10.1016/j.ejogrb.2018.10.042
Ortiz-Barney E, Ackerman JD (1999) The costs of selfing in Encyclia cochleata (Orchidaceae). Plant Syst Evol 219:55–64. https://doi.org/10.1007/bf01090299
Presti FT, Guedes NM, Antas PT et al (2015) Population genetic structure in hyacinth macaws (Anodorhynchus hyacinthinus) and identification of the probable origin of confiscated individuals. J Hered 106(1):491–502. https://doi.org/10.1093/jhered/esv038
Sieklicki M d F, Mulim HA, LFB P et al (2020) Population structure and inbreeding of Holstein cattle in southern Brazil. R Bras Zootec 49:e20190052. https://doi.org/10.37496/rbz4920190052
Snustad DP, Simmons MJ (2015) Principles of genetics. John Wiley & Sons, pp 41–47
Stebbins GL (1957) Self fertilization and population variability in the higher plants. Am Nat 91(861):337–354. https://doi.org/10.1086/281999
Szmatoła T, Gurgul A, Jasielczuk I et al (2019) Comprehensive analysis of runs of homozygosity of eleven cattle breeds representing different production types. Animals (Basel) 9(12):1024. https://doi.org/10.3390/ani9121024
Tambarussi EV, Boshier DH, Vencovsky R et al (2016) Several small: how inbreeding affects conservation of "Cariniana legalis" Mart. Kuntze (Lecythidaceae) the Brazilian Atlantic Forest's largest tree. Int For Rev 18(4):502–510. https://doi.org/10.1505/146554816820127550
Vitorino CA, Oliveira RCC, Margarido VP et al (2015) Genetic diversity of Arapaima gigas (Schinz, 1822) (Osteoglossiformes: Arapaimidae) in the Araguaia-Tocantins basin estimated by ISSR marker. Neotrop Ichthyol 13(3):557–568. https://doi.org/10.1590/1982-0224-20150037

Vitorino CA, Nogueira F, Souza IL et al (2017) Low genetic diversity and structuring of the arapaima (Osteoglossiformes, Arapaimidae) population of the Araguaia-Tocantins Basin. Front Genet 8:159. https://doi.org/10.3389/fgene.2017.00159

Wang J, Wg H, Charlesworth D et al (1999) Dynamics of inbreeding depression due to deleterious mutations in small populations: mutation parameters and inbreeding rate. Genet Res 74(2):165–178. https://doi.org/10.1017/S0016672399003900

Wright S (1922) Coefficients of inbreeding and relationship. Am Nat 56(645):330–338. https://doi.org/10.1086/279872

Wright S (1951) The genetical structure of populations. Ann Eugenics 15:323–354. https://doi.org/10.1111/j.1469-1809.1949.tb02451.x

Chapter 8
What Is the Genetic and Phenotypic Basis of Quantitative Variation in Populations?

Evandro Vagner Tambarussi

Contents

E. V. Tambarussi (✉)
Departamento de Produção e Melhoramento Vegetal, Faculdade de Ciências Agronômicas,
Universidade Estadual Paulista "Júlio de Mesquita Filho", Botucatu, SP, Brazil
e-mail: evandro.tambarussi@unesp.br

M. V. Bohrer Monteiro Siqueira et al. (eds.), *Population Genetics in the Neotropics*,
https://doi.org/10.1007/978-3-031-83685-5_8

Abstract After Fisher, Haldane and Wright's iconic contributions to population genetics, scientists have been trying to isolate genetic from non-genetic factors and thereby understand the basis of genetic variation in populations. Identifying the phenotypes of a population to differentiate their components present challenges that are inherent to the analysis itself. Often this decomposition is inefficient and significantly influenced by environmental factors, posing major challenges for scientists. However, at the end of the 1960s, with the emergence of protein electrophoresis methods and a substantial theoretical production in population genetics, researchers were able to access genetic variation more efficiently, independent of the influence of the environment. Currently, with the most advanced DNA sequencing approaches, along with powerful computational equipment, the genetic bases of quantitative variation have become clearer, and important scientific advances have been made. In this chapter we expect the reader to understand the main concepts and applications of additive and non-additive components of variation computed from quantitative traits, using neotropical organisms as examples.

Keywords Genetic variation · Molecular markers · Additive variation · Quantitative traits

Learning Objectives

After reading this chapter, readers should be able to understand:

- The definition and elucidation of the fundamental characteristics of quantitative variation, drawing comparisons with Mendelian characteristics and discrete variation.
- Examine the influence of environmental factors on quantitative variation, exploring the concepts of phenotypic plasticity and epigenetic mechanisms.
- Provide an overview of population genetics and how quantitative variation shapes genetic diversity within and among populations.
- Explore the statistical methods and approaches used to measure and analyze quantitative characteristics, while also addressing the challenges and limitations inherent in studying such characteristics.
- Featuring case studies that illustrate real-world examples of quantitative variation, such as plant height and crop yield.
- Discuss the evolutionary significance of quantitative variation and its role in adaptation and speciation.
- Future directions in quantitative genetics, considering advances in genomics and other technologies that will improve our understanding of quantitative traits and their implications for various fields of study.

By examining these aspects, this chapter seeks to provide a comprehensive and up-to-date overview of the genetic and phenotypic foundations of quantitative variation, elucidating its meaning and applications in diverse areas of biology and beyond.

8.1 Definition of Quantitative Variation

Quantitative variation, also known as continuous variation or polygenic variation, refers to the range of phenotypic differences observed among individuals in a population for a specific trait (Lynch and Walsh 1998). Unlike Mendelian traits, which are determined by a single gene with discrete phenotypic categories (e.g., dominant or recessive), quantitative traits are influenced by the combined effects of multiple genes (polygenes) and environmental factors. These characteristics often present a bell-shaped distribution (Gaussian distribution) in a population, with most individuals clustered around the average value and fewer individuals presenting extreme values at each end.

Quantitative traits in plants and animals are numerous and diverse, such as height, weight, productivity, etc. Complex tree traits, such as height or wood productivity, are classic examples of variables with quantitative variation because they are influenced by the interaction of numerous genetic factors and environmental constraints (Namkoong et al. 1988). In plants, quantitative traits often show a range of variation within a population due to the combined effects of multiple genetic and environmental factors (Fig. 8.1).

Discussing a few traits a little further, we can understand a little bit how complex are quantitative traits. For starters, let's take plant height, which is influenced by multiple genes and environmental factors, such as nutrient availability, spacing and water supply. Another important characteristic is flowering time, which is influenced by the action of multiple genes concurring for the phenotype alongside the environmental influence. For instance, seed size, in some plant species such as *Arabidopsis thaliana*, a widely studied model in plant biology, is controlled by several genes that have been identified and play important roles in their determination. It has been shown these genes are involved in regulating ovule growth, endosperm (seed nutritive tissue) development, and other metabolic pathways related to seed size. Furthermore, seed size can also be influenced by environmental factors, such as the availability of water, nutrients and light during plant development. Therefore, the exact number of genes involved in determining seed size can vary widely among different plant species and is an active area of research in plant genetics. Understanding the genetic and molecular mechanisms that control seed size is important for improving agricultural production and understanding plant biology in general. We can also mention the germination rate, which can be affected by factors such as temperature, soil moisture and seed genetics, the nutrient content of plant tissues, covering factors such as protein levels in grains or vitamin content in fruits, is influenced by the interaction of genetic factors and environmental conditions.

Fruit ripening time in plants is also controlled by many genes and its relevance extends to agricultural production and assists in effective harvest planning. A very important quantitative trait is productivity, which is the crop yield, such as number of grains, fruit weight, wood productivity (volume), this depends on a combination of genetic factors and environmental conditions.

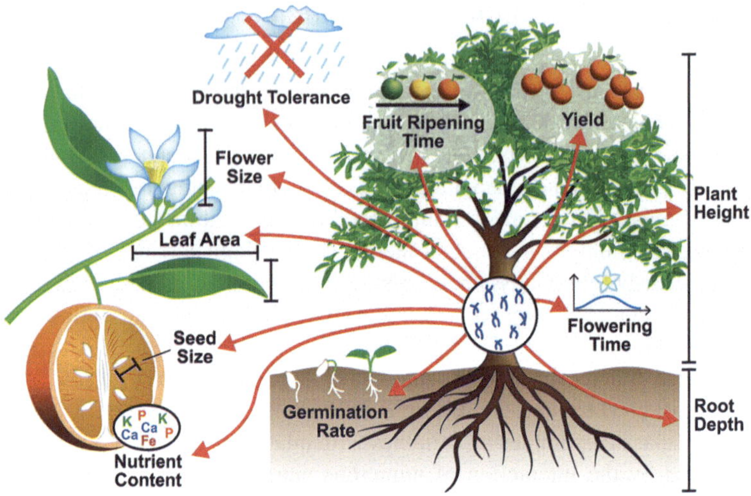

Fig. 8.1 Examples of quantitative traits in plants

Wood is composed of a variety of tissues, including xylem and wood, and its development is influenced by several genetic and environmental factors. Different tree species can also have different genetic profiles related to wood production. Currently, with climate change constraints such as drought, the tolerance to water deficit, which is the ability of a plant to surpass dry conditions, is a quantitative trait influenced by several genes and environmental factors.

Quantitative trait analysis is essential in plant and animal breeding and genetics to understand the genetic basis of these traits and to drive more efficient breeding programs, producing new plant varieties with improved traits (Lynch and Walsh 1998). Researchers use techniques such as quantitative trait locus (QTL) mapping and association studies to identify the specific genes and genomic regions responsible for variation in these traits (Ferrão et al. 2017).

8.2 Genotype and Phenotype

Genetics is the branch of biology that explores the inheritance of traits and characteristics in living organisms. It is a fundamental field that seeks to understand how the information encoded in an organism's DNA is transmitted through generations and how it ultimately shapes the physical and biochemical characteristics of individuals. One of the key concepts in genetics is the idea of phenotypes, which are observable traits or characteristics of an organism. To delve deeper into the realm of genetics and phenotypes, let's explore some fundamental concepts.

Genes are the fundamental units of heredity. They are segments of DNA (deoxyribonucleic acid) that contain the instructions for building and maintaining an organism. DNA appears as a double-stranded, twisted ladder-shaped structure, with each rung composed of pairs of nucleotide bases, namely adenine (A), thymine (T), cytosine (C) and guanine (G). These bases collectively encode the genetic information that governs an organism's characteristics. Genes exist in different forms known as alleles, representing variations of a gene that can result in different traits or characteristics. For example, resistance to corn leaf spot caused by *Helminthosporium carbonum* VII is a monogenic and qualitative resistance mechanism (Carson and Carson 1989). An individual inherits one allele for each gene from each parent, and the combination of these alleles determines the specific expression of the trait. An organism's genotype refers to its genetic makeup. It is the complete set of genes that an individual carries in their DNA, including dominant and recessive alleles. The genotype is crucial because it provides the basis for an organism's potential characteristics, but does not necessarily dictate the observable traits. The phenotype is the external appearance or observable characteristics of an organism, resulting from the interaction between genotype and environment.

In many cases, one allele can exert dominance over the other in determining a specific variant of the trait. A dominant allele is characterized by its ability to manifest itself in the phenotype even if only one copy is present. In contrast, a recessive allele is only expressed if an individual carries two copies of it. Codominance is another effect that occurs when two different alleles of a gene are both expressed in the phenotype of a heterozygous individual. In other words, neither allele is dominant over the other, and both are fully expressed. This results in a phenotype that shows traits of both alleles simultaneously. A classic example of codominance is seen in the ABO blood group system in humans. There are three alleles involved: A, B, and O. An individual with genotype AO or BO will express either the A or B antigen, but not both, along with the O antigen on their red blood cells. Consequently, their blood type will be AB or BB, respectively. This means that both alleles, A and O (or B and O), are expressed equally, resulting in a unique phenotype that reflects traits from both alleles. Codominance differs from incomplete dominance, where the phenotype of the heterozygous individual is an intermediate blend of the phenotypes associated with the two homozygous genotypes. In codominance, there's no blending; both alleles are expressed fully and independently. This phenomenon can lead to a wide range of trait variations within a population. Genetic characteristics can follow specific inheritance patterns. Some traits are controlled by a single gene with two or more alleles (Mendelian inheritance), while others are influenced by multiple genes (polygenic inheritance). The concepts of dominance and codominance are important to understand the nature of molecular markers, which are the subject of the next chapter (Chap. 9).

Genetics also plays a critical role in the theory of evolution. Over time, the interaction between genetic variation and natural selection can give rise to the development of new species (Kimura and Crow 1963). Genetic diversity within populations provides the raw material for evolution, allowing adaptation to changing environments (Walsh and Lynch 2018). Thus, genetics is a complex and fascinating field that investigates the intricate mechanisms that govern the inheritance and expression of traits in living organisms. Phenotypes, the visible results of genetic information, are a product of the interaction between an individual's genotype and the surrounding environment. This understanding of genetics not only enriches our knowledge of life on Earth, but also has great potential for applications in areas such as medicine, agriculture and conservation.

Phenotypic variation arises from the interaction of genetic and environmental factors ($P = G + E$). A comprehensive understanding of the origins of this variation is fundamental when investigating the study of quantitative traits and their patterns of inheritance (Lynch and Walsh 1998). Genetic variation is the result of differences in the alleles present in the genotypes of individuals in a population. These variations are inherited from parents and contribute significantly to the spectrum of observable phenotypic distinctions present in a population.

Environmental factors such as nutrition, temperature, and exposure to toxins can profoundly impact an organism's phenotype. Environmental influences can modify gene expression, leading to phenotypic plasticity—the ability of an organism to adjust its phenotype in response to environmental signals. Gene-environment interactions occur when the effect of a gene on the phenotype depends on the specific environmental conditions experienced by an individual referent. The same genetic variant can result in different phenotypic outcomes in different environments.

Epigenetic mechanisms involve changes in gene expression without altering the underlying DNA sequence. These modifications can be influenced by environmental factors and have the potential to be inherited by subsequent generations, contributing significantly to phenotypic variation (Al Aboud et al. 2023) (for more information on epigenetic mechanisms, see Chap. 3).

As an example, we have the study by Silva et al. (2023) who revealed that genomic and epigenomic variation in different *Psidium* species is not limited only to the difference in nuclear genome sizes but extends to the geographic distribution of polyploid *Psidium* species compared to diploids. These genomic discrepancies are particularly evident between *Psidium cattleyanum* and *Psidium guajava*, which suggest adaptive responses of individuals to the specific conditions of the environments where they develop. The authors found a positive correlation between nuclear 2C values and the percentage of GC bases, together with its negative correlation with certain essential oil compounds, which points to the influence of environmental factors on gene expression, indicating a possible epigenetic mechanism which can be passed on to the next generations.

Some sources of phenotypic variation are purely stochastic, resulting from chance events during development or cellular processes, and are called crossover events. These random events can contribute to subtle differences in phenotypes, even between genetically identical individuals. The genotype-phenotype relationship is a complex interaction of genetic and environmental factors that ultimately determine the observable characteristics of an organism (Lynch and Walsh 1998). Phenotypic variation arises from multiple sources, including genetic diversity, environmental influences, gene-environment interactions, epigenetic modifications, and random events. A comprehensive understanding of these fundamental concepts is essential for examining the genetic and phenotypic underpinnings of quantitative variation within populations.

8.3 Mendelian Versus Quantitative Inheritance

Mendelian inheritance and quantitative inheritance are two fundamental patterns of genetic inheritance that underlie the transmission of characteristics from one generation to another. They differ in their underlying genetic mechanisms and the nature of the phenotypic variation they produce. Mendelian inheritance, named after Gregor Mendel, follows the principles of classical genetics. It involves the transmission of traits controlled by single genes, each with distinct and easily recognizable phenotypic categories.

The main characteristics of Mendelian Inheritance are: (I) traits are controlled by a single gene with two distinct alleles; (II) phenotypic variation falls into discrete categories (e.g. tall vs. short seeds, round vs. wrinkled seeds); (III) the genotype-phenotype relationship is direct and predictable and (IV) the segregation and independent classification of alleles occur during meiosis.

Quantitative inheritance, on the other hand, involves the inheritance of traits that are influenced by multiple genes with little contribution of effect on the phenotype (Walsh and Lynch 2018). One example is plant height, for which it is estimated that hundreds of genes are associated. The main features of quantitative inheritance are: (I) characteristics are influenced by multiple genes (polygenes) and environmental factors; (II) phenotypic variation is continuous and falls within a spectrum (for example, tree shape, height, DBH, etc.); (III) The genotype-phenotype relationship is complex and involves interactions between multiple genes and the environment and (IV) segregation and independent distribution are not as easily discernible as in Mendelian inheritance.

As an example of characteristics influenced by multiple genes (polygenes) and environmental factors, we have the study by Collevatti et al. (2019) with the neotropical species *Handroanthus impetiginosus*. The authors examined 13 populations of the species, using 75,838 single nucleotide polymorphisms (SNP) in order to reconstruct demographic history and identify loci under adaptive selection. The results revealed 14 outlier loci for adaptive selection in specific groups of populations, indicating less genetic differentiation, and 32 loci showed evidence of local adaptation, although their correlations with edaphoclimatic variables were limited. Many of these genes, although conserved, show structural and functional variations, suggesting adaptation to environmental conditions. The analysis suggests that, despite high genetic differentiation between populations, differences in the distribution of allele frequencies between contrasting environments were minimal. This indicates the species' ability to deal with environmental changes, probably due to phenotypic plasticity and polygenic adaptation.

The influence of the environment on the manifestation of the phenotype is evident in the study by Schmitt et al. (2021), which explores how genetic adaptation in tropical trees such as *Symphonia* and *Eschweilera* directly reflects local environmental conditions. The article highlights that different species within these tree complexes demonstrate significant phenotypic variations according to topography and soil characteristics. These phenotypic variations reflect the presence of multiple genes involved in the adaptation of trees to the local environment, highlighting the polygenic nature of adaptation. For example, *Symphonia* exhibits three distinct species, each adapted to specific topographic niches, varying in response to water availability and terrain conditions. Likewise, *Eschweilera* shows a preference for different types of soils, reflecting genotypic variations that are in tune with local environmental variations.

8.4 Polygenic Inheritance and Complex Traits

Polygenic inheritance represents a form of quantitative inheritance in which traits are influenced by the combined effects of multiple genes, each contributing a small portion to the overall phenotype (Walsh and Lynch 2018). Such traits are also influenced by environmental factors, further increasing the complexity of their expression.

Polygenic traits involve the contribution of multiple genes, each with their own additive effect on the phenotype. The more genes that contribute, the greater the potential for continuous variation (Walsh and Lynch 2018). The effects of individual genes are cumulative, leading to a wide range of phenotypic outcomes. Each additional favorable allele contributes incrementally to the expression of the trait (Lynch and Walsh 1998). Furthermore, polygenic traits are often analyzed using QTL mapping to identify regions of the genome associated with the trait. These regions can contain several genes, each contributing to the observed variation (Ferrão et al. 2017). The expression of polygenic traits is influenced by environmental factors. Environmental conditions can modify trait expression and can even override genetic contributions in certain circumstances (Lynch and Walsh 1998).

In the study by Zhang et al. (2023), the genetic mapping of *Anthurium andraeanum* Linden allowed the identification of QTLs associated with 15 different morphological traits, such as distance from the left spathe (LED) and distance from the right spathe (RED), which presented coefficients of variation notably greater than other characteristics. Based on this mapping, the authors identified 59 QTLs associated with the 15 traits. These QTLs were distributed across several linkage groups and explained phenotypic variations ranging from 6.21% to 17.74%. Traits such as spathe length, spathe width, pedicel length, pedicel diameter, among others, multiple QTLs were identified in different LGs, highlighting the genetic complexity underlying these traits.

8.5 The Genetic Basis of Quantitative Variation

8.5.1 The Role of Multiple Genes in Phenotypic Expression in Trees

Trees, like many other organisms, exhibit a wide range of phenotypic variation, and this diversity is influenced by the combined effects of multiple genes. Different aspects of tree morphology, physiology, and reproduction are often controlled by a complex interplay of genetic factors, making them excellent examples of polygenic traits (Groover 2005).

Tree growth, for example, is a multifaceted trait influenced by the coordinated action of several genes involved in cell division, cell elongation and hormonal signaling pathways. Likewise, traits such as wood density, leaf size, flowering time and disease resistance are all controlled by the concerted influence of multiple genes. To provide an example, within the *Parkia platycephala* Benth tree, a neotropical species found in Northeast Brazil, the values calculated for individual heritability in the strict sense (h_a^2) and heritability within the progeny (h_w^2) ranged from low to moderate (0.01 to 0.39) in different populations. Notably, the most substantial values were recorded for diameter at ground level (0.39) (Silva et al. 2023).

In the study by Valderrama et al. (2022) on the neotropical genus *Costus* L. the authors found that floral traits, such as the presence of appendages on the bracts, which have the function of both attracting pollinators and defending against herbivory, are influenced by the action of multiple genes.

Environmental factors have a great impact on the growth and development of trees. Interactions between genetic factors and the environment further contribute to the observed phenotypic variation. As a result, trees growing in different habitats or climates can exhibit varying phenotypes, even when they share similar genetic origins.

As an example, we have the study by Brancalion et al. (2018) on *Euterpe edulis*, a palm tree widely distributed in the Brazilian Atlantic Forest, popularly known as heart-of-palm. The authors showed how environmental factors can shape the phenotypic expression of trees with similar genetic backgrounds. This research highlighted that phenotypic plasticity plays a significant role in the palm's ability to establish itself in different forest types, enabling its expansion into biogeographically related habitats. The authors observed that palms of this species presented distinct local adaptations for establishment in divergent environments, highlighting the influence of the interaction between genotype and environment on the expression of specific traits. In the Semideciduous Forest, for example, the authors found that local adaptations were associated with the emergence of seedlings, resulting in greater density, as well as a greater proportion of root-to-shoot dry mass, which could benefit their growth in lower soils. In the Dense Ombrophylous Forest, the seedlings had greater aerial mass and lower mortality, suggesting adaptations to a shadier and more humid environment. On the other hand, between the Dense Ombrophilous Forest and restingas, more closely related ecosystems, the authors found no significant differences in genetic differentiation, and there was no clear evidence of local adaptation.

8.5.2 Additive and Non-additive Genetic Effects

In the context of tree genetics, the expression of polygenic traits is influenced by two main types of genetic effects: additive and non-additive effects ($G = A + D + E$). Additive effects are associated with the independent contributions of individual alleles at different gene loci. In other words, an individual's phenotype is the sum of the effects of each allele it carries. For example, if two alleles at a specific locus each contribute one unit of phenotypic value to a trait, an individual carrying two of the same alleles will show a higher value for that trait than an individual carrying only one copy of the allele.

Non-additive effects arise when interactions between alleles at different loci lead to phenotypic outcomes that cannot be predicted solely based on the sum of their individual effects. These interactions can result in synergistic (positive epistasis) or antagonistic (negative epistasis) effects on the phenotype.

In trees, non-additive effects are often observed in traits with complex developmental pathways, such as fruit production, where the interaction between multiple genes can influence fruit size and quality, in addition to the additive contributions of each individual gene.

Fruit production by different plant species is influenced by various environmental factors, plant characteristics and evolutionary inheritance, resulting in phenological responses that are not merely the sum of isolated effects, but rather a complex interaction between multiple elements. This is evidenced in the study by Cortés-Flores et al. (2019) carried out in a neotropical dry forest in Mexico, over two years. The authors documented the fruiting phenology of 151 species belonging to different growth forms. The morphological characteristics of the fruits, such as size and dispersal syndrome, showed a significant association with the form of plant growth, but with a striking phylogenetic signal, pointing to the hereditary influence in this process. Environmental seasonality, represented by precipitation and day length, had a strong influence on fruit maturation time, but without a direct association with the phylogeny of plant species. In woody species, fruiting was directly related to different dispersal syndromes (mainly autochory and endozoochory). On the other hand, in herbaceous species, fruiting phenology was restricted by the length of the life cycle and the seasonality of rainfall, selecting specific fruit traits to maximize seed dispersal during the dry season.

8.5.3 The Concept of Heritability

Heritability is a key concept in quantitative genetics, as it measures the proportion of phenotypic variation in a trait that can be attributed to genetic variation within a population. It provides an estimate of how much of the observed differences in a trait between individuals are due to genetic factors.

Quantitative traits typically exhibit low heritability values. As we already know, quantitative traits are polygenic in nature, meaning they are controlled by multiple genes. As these characteristics involve the contribution of numerous genes to phenotypic expression, they are consequently more influenced by environmental factors.

For trees, heritability estimates help foresters and breeders to predict the success of certain traits in passing on to the next generation. High heritability indicates that a substantial portion of the phenotypic variation is due to genetic differences and suggests that selective breeding for this trait could be more effective (Visscher et al. 2008).

However, heritability is also influenced by environmental factors, age and characteristics. Traits with high heritability in one environment may have lower heritability in another. It is therefore crucial to consider the specific environmental conditions in which trees are grown when interpreting heritability estimates.

Heritability is determined by calculating the ratio between genetic variance and phenotypic variance, as obtained from experimental data:

$$H^2 = \frac{\sigma_G^2}{\sigma_F^2}$$

This value varies between 0 and 1, indicating that, when closer to zero, the trait is predominantly influenced by the environment, while values closer to 1 suggest that a smaller number of genes are involved in controlling the trait. The expression above accounts for the broad sense heritability (represented as H^2), that considers all the genetic variance (both additive and non-additive effects). A similar estimate, the narrow sense heritability (represented as h^2), considers only the additive component of the genetic variance, and it is very useful for predicting genotypic effects in next generations of breeding.

8.5.4 Mapping Quantitative Trait Loci (QTL) in Trees

QTL mapping is a powerful technique used to identify the genetic regions (loci) associated with quantitative traits. By studying patterns of trait inheritance across multiple individuals and generations, researchers can link specific genomic regions to the expression of specific traits.

The research by Lanaud et al. (2009) with *Theobroma cacao*, which is an economically very important tropical species, exemplifies the identification of genomic regions associated with specific traits, such as disease resistance. This study analyzed 76 QTL mapping experiments for resistance to different diseases in cocoa. The authors observed 'hot spots' in the genome, identifying key regions linked to disease resistance, allowing the accumulation of favorable alleles to improve cocoa resistance. The comparative analysis of multiple studies allowed us to reduce the number of QTLs, refining their location and reliability, highlighting the importance of broad investigations to link traits to specific regions of the genome.

QTL mapping involves genotyping individuals in a mapping population (e.g., a family of trees) and measuring their phenotypes for the trait of interest. Statistical analyses are then employed to identify regions of the genome that show a significant association with the trait. These regions are known as QTLs.

QTL mapping has provided valuable information about the genetic basis of several traits in trees, including growth rate, wood properties, disease resistance, and fruit quality (Lauer and Isik 2021). Information obtained through QTL mapping can guide breeding efforts aimed at improving desired traits in tree populations, leading to the development of more resilient, productive and sustainable tree varieties for diverse applications, including forestry and agroforestry.

8.6 Environmental Influence on Quantitative Traits

8.6.1 Gene-Environment Interactions

Gene-environment interactions refer to the phenomenon in which the expression of a gene or the effect of a genetic variant on a trait is influenced by the environment in which an organism lives. In other words, the outcome of a specific genotype may differ depending on the specific environmental conditions an individual experiences.

Gene-environment interactions contribute to the wide range of phenotypic variation observed within a population. Individuals with the same genetic origin can show different characteristics in response to different environmental signals. Some genes can only influence a trait under specific environmental conditions. For example, a gene involved in drought tolerance can only be activated and effective in arid environments.

Another interesting definition is that of reaction norm, which describes the range of phenotypic results that can be produced by a single genotype under different environmental conditions. Illustrates the flexibility of gene expression in response to varying environments. Gene-environment interactions play a crucial role in adaptation and fitness. Organisms that can adjust their phenotype in response to environmental changes have better chances of survival and reproduction.

Many other forms of G x E can be found in the literature. For example, some plants may have genes that are expressed differently at different temperatures, leading to altered growth patterns or flowering times depending on the climate. The availability of essential nutrients in the soil can influence the expression of genes related to nutrient absorption and utilization in plants. Plants may have genes sensitive to daylight length, affecting flowering time and other developmental processes.

8.6.2 Phenotypic Plasticity and Flexibility in Plants

Plasticity in plants refers to the ability of a single genotype to produce different phenotypes in response to different environmental conditions. This phenotypic flexibility allows plants to optimize their performance in different habitats and face constantly changing environmental challenges (Monroe 2023). Knowledge of this concept is fundamental in the daily life of a breeder, and widely used in gene-environment interactions.

During development, plants can change their growth patterns and resource allocation in response to environmental stimuli. For example, plants may grow taller in low-light conditions to compete for light or allocate more resources to root growth in nutrient-poor soils. Furthermore, they can acclimate to specific environmental factors, such as temperature or light intensity, adjusting their physiological and biochemical processes. This allows them to maintain optimal operation under varying conditions.

Phenotypic plasticity can directly influence plant fitness and survival, allowing an optimization of resource acquisition and reproductive success in their current environment (Monroe 2023). Plasticity allows plants to respond to stressors such as drought, high salinity or herbivory by adjusting their traits to minimize damage and conserve resources.

For example, plants can produce different leaf shapes and sizes in response to light availability, with leaves wider in low-light conditions to capture more light energy. Root systems can exhibit plasticity to exploit nutrient-rich patches or penetrate compacted soils. Plants can delay or accelerate flowering in response to variations in photoperiod and temperature.

8.6.3 Epigenetic Mechanisms and Their Role in Quantitative Variation

Epigenetic mechanisms involve heritable changes in gene expression that do not involve changes in the DNA sequence. Epigenetic modifications, such as DNA methylation, histone modifications, and regulation of small RNAs, can influence gene activity and have significant implications for quantitative variation in populations (Al Aboud et al. 2023).

Epigenetic modifications can be transmitted from one generation to the next, potentially contributing to heritable phenotypic variation that is not based solely on genetic differences. Epigenetic mechanisms can mediate the plant responses to environmental changes. Environmental stresses can induce epigenetic modifications that alter gene expression and potentially lead to adaptive responses (Sahu et al. 2013).

Epigenetic changes can contribute to phenotypic plasticity, influencing gene expression patterns that allow plants to adjust their phenotype to different environmental conditions. Such modifications can also influence gene expression by activating or silencing specific genes, affecting the phenotype expressed by an organism. They also may impact the expression of multiple genes involved in quantitative traits, adding complexity to the genetic basis of variation.

Examples of epigenetic mechanisms in quantitative variation:

- DNA methylation: changes in DNA methylation patterns can influence gene expression, leading to phenotypic differences in traits such as flowering time, stress tolerance and seed development;
- Histone modifications: changes in histone structure can affect chromatin accessibility, affecting gene expression and potentially leading to variation in complex traits;
- Small RNA regulation: Small RNAs, such as microRNAs, can target specific mRNAs for degradation or translational repression, playing a role in fine-tuning gene expression and phenotypic outcomes;

Understanding the role of epigenetic mechanisms in quantitative variation is an ongoing area of research and promises to provide valuable insights into the complex interplay between genetics, epigenetics, and the environment in shaping the phenotypic diversity observed in plant populations.

8.7 Population Genetics and Quantitative Variation

8.7.1 Gene Flow and Genetic Drift in Plants

Gene flow and genetic drift are two important processes that influence the genetic composition of plant populations over time. Gene flow is the movement of genetic material (alleles) between populations of the same species. It occurs through the migration of individuals, seeds or pollen from one population to another. It is a crucial mechanism that promotes genetic exchange and maintains genetic diversity within and between plant populations.

Gene flow facilitates the exchange of genetic variants between populations, reducing genetic differentiation and contributing to the homogenization of the gene pool. The introduction of new alleles can increase genetic diversity within a population, providing a source of potentially advantageous traits that can be subject to natural selection. This phenomenon can also counteract the effects of genetic drift, preventing the fixation or loss of alleles in small populations and promoting the persistence of genetic variation.

Human activities, such as the intentional or unintentional movement of plants, can lead to gene flow between geographically distant populations and impact the genetic structure of plant species.

Genetic drift refers to the random fluctuations in allele frequencies that occur in small populations due to chance events (go back to Chap. 4 for details on genetic drift). It is more pronounced in small, isolated populations and can lead to loss of genetic diversity over time.

Some key aspects of genetic drift are:

- Founder effect: genetic drift generally occurs when a small group of individuals colonizes a new, isolated habitat. The limited genetic diversity of the founding population can lead to the loss of certain alleles or the fixation of others over successive generations;
- Population bottleneck: population bottlenecks occur when a significant reduction in population size results in a subset of individuals contributing to the next generation. This reduction can lead to a loss of genetic variation and an increased influence of genetic drift on allele frequencies;
- Impact on rare alleles: genetic drift can lead to the loss of rare alleles in a population, reducing genetic diversity and potentially affecting the adaptability of the population to changing environments.

8.7.2 Natural Selection and Its Impact on Quantitative Traits in Plants

Natural selection is a fundamental mechanism of evolution, shaping the genetic composition of plant populations by favoring individuals with advantageous characteristics that increase their chances of survival and reproduction (go back to Chap. 5 for further details on natural selection).

Natural selection acts on the phenotypic variation of a population, favoring individuals with characteristics that provide greater fitness (reproductive success) in their specific environment, leading to the evolution of adaptations that improve the ability of a plant to survive and reproduce in its specific habitat. These adaptations can be morphological, physiological or behavioral. Another type is directional selection that occurs when environmental conditions favor individuals at one extreme of the phenotypic distribution. Over time, allele frequencies change in the direction of the favored trait.

Stabilizing selection occurs when intermediate phenotypes have greater fitness than extreme phenotypes. This type of selection tends to reduce phenotypic variation in a population and, ultimately, disruptive selection that occurs when extreme phenotypes have greater fitness than intermediate phenotypes, leading to the diversification of the population into two or more distinct forms.

8.7.3 Selection of Balance and Maintenance of Variation in Plants

Balancing selection is a type of natural selection that maintains genetic diversity within a population by favoring multiple alleles or phenotypes. It works when no single allele has a clear advantage, and the fitness of different genotypes is balanced over time. To understand balancing selection, some important aspects must be known. For example, balancing selection can occur through heterozygote advantage, where individuals carrying two different alleles (heterozygotes) have greater fitness than individuals with two identical alleles (homozygotes) at a given locus.

Frequency dependent selection occurs when the fitness of a genotype depends on its frequency in the population. Rare alleles may have greater fitness when they are infrequent, which maintains genetic diversity. Equilibrium selection can be maintained in environments that fluctuate or change over time. Different alleles may be favored under different environmental conditions, contributing to the maintenance of variation.

Overdominance can also be mentioned, which can be known as heterozygous superiority, occurring when the heterozygous genotype presents greater fitness than any of the homozygous genotypes, promoting genetic diversity.

Balancing selection plays a critical role in maintaining genetic variation in plant populations and can contribute to the persistence of diverse phenotypic traits within species. This diversity provides a reservoir of potential adaptive traits that may become advantageous under changing environmental conditions.

8.8 Measuring and Analyzing Quantitative Variation

8.8.1 Phenotypic Measurements and Data Collection in Plants

Phenotypic measurements and data collection in plants are essential for studying quantitative variation and unraveling the genetic and environmental factors that influence complex traits. Accurate and reliable data are crucial to conducting meaningful analyses and drawing valid conclusions.

To better illustrate the importance of phenotypic measurements and data collection in plants, it is essential to consider the challenges faced in making this information available, as well as establishing appropriate standards. As highlighted in the study by Ćwiek-Kupczyńska et al. (2016) on plant phenotyping, the wealth of phenotypic data is a valuable source of knowledge about the vital mechanisms of plants. However, the absence of unified standards for describing phenotypic data has hampered their exchange and reuse. To this end, the authors proposed a document entitled "Minimum Information About a Plant Phenotyping Experiment" (MIAPPE) in ISA-Tab format as guidelines for organizing, describing, and storing phenotypic data, aiming for better interpretation, replicability and interoperability of sets of data. These tools enable the inclusion of essential metadata in datasets, making them accessible and independent of the metadata policies of a specific repository, and allow compatibility with other experimental results, ensuring data interoperability.

The conclusion of the study highlights practical challenges, such as managing large volumes of data generated by high-speed phenotyping experiments, and the need to determine the level of detail and size of data to be published. These challenges reflect the complexity of standardizing and making phenotypic data available, but they highlight the importance of tools like ISA-Tab for the efficient construction and management of these datasets, aiming to promote open data policies and the implementation of FAIR data principles (Findable, Accessible, Interoperable, and Reusable) in the plant phenotypic studies community.

For good phenotypic measurements and data collection, some aspects need to be considered:

- Feature selection: researchers need to carefully choose which features to measure based on their relevance to the research objectives. Traits can include growth-related parameters, yield components, disease resistance, biochemical characteristics, and more.
- Standardized protocols: consistency in data collection is vital for comparing results between different studies. Researchers must follow standardized protocols for measurement, sampling, and data recording techniques.
- Sample size: a sufficient sample size is needed to capture the true range of phenotypic variation within a population. Larger sample sizes increase the statistical power of analyzes and improve the precision of estimates.
- Replication: replicating measurements across multiple individuals or experimental units helps account for natural variation and random errors, increasing the reliability of results.
- Environmental control: when studying the genetic basis of quantitative traits, researchers often grow plants in controlled environments to minimize the impact of environmental variation on measured traits;
- High-throughput phenotyping: advances in technology have led to the development of high-throughput phenotyping methods, allowing rapid and automated measurement of multiple traits in large plant populations.

8.8.2 *Statistical Methods for the Analysis of Quantitative Traits*

Quantitative trait analysis requires robust statistical methods to uncover the underlying genetic and environmental factors that influence phenotypic variation. We can use descriptive statistics to summarize and describe the distribution of phenotypic data. Measures such as mean, median, standard deviation, and variance provide basic descriptive statistics of traits in the population.

Another very important statistic in genetics is the correlation analysis, which evaluates the relationships between different characteristics or between phenotypic data and environmental variables. It assists in the identification of possible trait associations and gene-environment interactions.

We cannot forget the analysis of variance (ANOVA) which is used to determine whether there are significant differences in the means of characteristics between different groups or treatments, such as different genotypes or environmental conditions. Heritability estimation, for example, can be obtained through various methods, such as parent-child regression and through decomposition of the components of variances.

Currently, genome-wide association studies (GWAS) identify genetic variants associated with quantitative traits, testing the correlation between trait variation and genome-wide genetic markers. And finally, QTL mapping involves identifying genomic regions (quantitative trait loci) that are linked to the expression of specific traits. This method helps unravel the genetic architecture of complex traits.

8.8.3 *Challenges in Studying Quantitative Variation in Plants*

Analyzing quantitative variation in plants poses several challenges due to the complex nature of these traits and the interaction of genetic and environmental factors, so it is necessary to understand that each population you are working with will have specific challenges. We need to remember that quantitative traits are influenced by multiple genes and environmental factors, challenging the identification of specific genetic variants responsible for the variation in that trait.

As already mentioned, phenotypic plasticity can obscure genetic associations, as the same genotype can produce different phenotypes under varying environmental conditions. Quantitative traits are shaped by the collective influence of multiple genes and environmental factors, making it difficult to identify specific genetic variants responsible for the variability in that trait. Natural environmental variation can introduce noise into phenotypic data, making it difficult to distinguish between genetic and environmental effects.

A highly important factor is sample size. Adequate sample size and replication are essential to obtain reliable results. However, collecting large, well-replicated datasets can be resource-intensive and time-consuming.

Genetic interactions, such as epistasis (gene-gene interactions), can complicate the genetic basis of quantitative traits, requiring more sophisticated statistical models for analysis.

Advances in high-throughput phenotyping and genotyping have led to large datasets with high-dimensional data, posing computational and analytical challenges. Developing automated phenotyping methods that accurately capture complex traits across diverse plant species is an ongoing technological challenge. Nevertheless, advances in genomics, statistical methods, and phenotyping technologies continue to push the limits of our understanding of quantitative variation in plants. The combination of multidisciplinary approaches and the integration of genomic, phenotypic, and environmental data hold promise for unraveling the intricacies of complex traits and their genetic basis.

8.9 Evolutionary Significance of Quantitative Variation

8.9.1 Role of Quantitative Variation in Adaptation

The presence of quantitative variation provides genetic flexibility to populations. This means that, in a population, there are a range of possible phenotypes for a given trait. This genetic diversity is crucial for rapid adaptation to new environmental challenges, such as climate change or the introduction of new predators. Natural selection acts on quantitative variation, favoring characteristics that confer adaptive advantages. For example, in a colder environment, organisms with traits that increase cold resistance may be favored, leading to a gradual change in the frequency of these traits in the population.

Quantitative variation allows for the continuous evolution of traits over time. Rather than abrupt changes, evolution is often a gradual process, shaped by selective pressures acting on existing genetic diversity.

Polygenic traits and those influenced by gene interactions contribute to quantitative variation. This genetic complexity offers more possibilities for adaptation, as different combinations of alleles can lead to a variety of phenotypes.

In more stable environments, quantitative variation can be maintained at relatively constant levels. In contrast, in more variable environments, selective pressure may favor the maintenance of broader genetic diversity.

8.10 Future Directions in Quantitative Genetics

8.10.1 Advances in Genomic Technologies

Advances in genomic technologies are catalyzing a revolution in quantitative genetics, providing unprecedented insights into the complexity of polygenic traits. As next-generation sequencing techniques and other genomic tools become more accessible and efficient, a fascinating window is opening for demystifying the mechanisms underlying the multifactorial inheritance of complex traits. This introduction explores the rapidly evolving landscape of genomic technologies and their impact on quantitative genetics. From identifying genetic variants associated with specific traits to understanding the underlying molecular mechanisms, this convergence between genomics and quantitative genetics promises to revolutionize our ability to decipher and manipulate the genetic architecture of complex phenotypes.

8.10.2 Integration of Genetic and Environmental Data

The integration of genetic and environmental data represents an innovative and comprehensive approach to understanding the mechanisms that control plant performance and adaptation. As the need to boost agricultural production, optimize resources and face the challenges posed by climate change increases, synergy between genetic and environmental information emerges as an essential tool. This convergence allows for a deeper analysis of the complex interactions between a plant's genotype and the environmental conditions in which it grows. In this era of advanced sequencing technologies and high-precision environmental sensors, the integration of genetic and environmental data offers valuable insights to improve variety selection, understand plant adaptive responses, and ultimately drive sustainability and efficiency in agriculture. This introduction explores the importance, challenges and opportunities provided by the integration of these two data domains, highlighting their transformative potential for agricultural science and sustainable crop management.

8.10.3 Predictive Modeling and Synthetic Biology Applications

The convergence between predictive modeling and synthetic biology applications outlines an exciting frontier in the quest for precise understanding and manipulation of biological systems. Predictive modeling, fueled by computational advances and massive biological data acquisition, has become a key tool in simulating complex phenomena. At the same time, synthetic biology has emerged as an innovative discipline, enabling the reengineering of biological systems to perform specific functions. This intersection offers extraordinary promise, enabling the creation of organisms tailored for specific tasks and the anticipation of biological behaviors based on precise models. By uniting the power of computational prediction with the ingenious manipulation of biology, this convergence promises to open new horizons in creating innovative, personalized solutions to complex biological challenges.

8.11 Final Remarks

Quantitative variation is a fundamental aspect of biology that impacts the diversity and resilience of populations. This chapter explored the genetic and phenotypic basis of this variation, emphasizing the complex interplay between genes, environments, and natural selection. Advances in genomics and statistical methods have revolutionized our understanding of the genetic architecture underlying quantitative traits. As we venture into the future, greater exploration of quantitative variation promises to unlock new knowledge about the evolutionary processes that shape life on Earth. By deciphering the intricacies of quantitative genetics, we pave the way for tackling real-world challenges, from crop improvement to personalized medicine.

Key Points

- Quantitative variation, also known as continuous or polygenic variation, encompasses the range of phenotypic differences observed among individuals for a specific trait.
- Environmental factors play a significant role in quantitative variation, influencing traits through mechanisms such as phenotypic plasticity and epigenetic mechanisms, which can modify the expression of traits and contribute to adaptation.
- Population genetics offers insights into how quantitative variation shapes genetic diversity within and among populations, providing a framework for understanding the genetic basis of traits and their evolution.
- Quantitative variation holds evolutionary significance, influencing adaptation and speciation processes.
- Future directions in quantitative genetics involve leveraging advances in genomics and other technologies to further understand the genetic basis of quantitative traits, offering opportunities for interdisciplinary research with implications across various fields.

Questions

1. How do environmental factors interact with genetic mechanisms to shape quantitative variation in traits, and what are the implications of understanding these interactions for adaptation and species survival in changing environments?
2. What are the current challenges and limitations in accurately measuring and analyzing quantitative traits, particularly in plants and animals, and how can the advances in genomics and statistical methods contribute to overcoming these challenges for a more comprehensive understanding of polygenic inheritance?

References

Al Aboud NM, Tupper C, Jialal I (2023) Genetics, epigenetic mechanism. In: StatPearls. StatPearls Publishing

Brancalion PHS, Oliveira GCX, Zucchi MI et al (2018) Phenotypic plasticity and local adaptation favor range expansion of a Neotropical palm. Ecol Evol 8:7462–7475. https://doi.org/10.1002/ece3.4248

Carson SD, Carson MJ (1989) Breeding for resistance in forest trees—a quantitative genetic approach. Annu Rev Phytopathol 27:373–395. https://doi.org/10.1146/annurev.py.27.090189.002105

Collevatti RG, Novaes E, Silva-Junior OB et al (2019) A genome-wide scan shows evidence for local adaptation in a widespread keystone Neotropical Forest tree. Heredity 123:117–137. https://doi.org/10.1038/s41437-019-0188-0

Cortés-Flores J, Cornejo-Tenorio G, Urrea-Galeano LA et al (2019) Phylogeny, fruit traits, and ecological correlates of fruiting phenology in a Neotropical dry forest. Oecologia 189:159–169. https://doi.org/10.1007/s00442-018-4295-z

Ćwiek-Kupczyńska H, Altmann T, Arend D et al (2016) Measures for interoperability of phenotypic data: minimum information requirements and formatting. Plant Methods 12:44. https://doi.org/10.1186/s13007-016-0144-4

Ferrão LFV, Ortiz R, Garcia AAF. Genomic selection: state of the art. In: Campos G, Caligari, PDS Genetic improvement of tropical crops. 1ed.: Springer International Publishing, 2017, 19–54

Groover AT (2005) What genes make a tree a tree? Trends Plant Sci 10:210–214. https://doi.org/10.1016/j.tplants.2005.03.001

Kimura M, Crow JF (1963) The measurement of effective population number. Evolution 17:279–288. https://doi.org/10.1111/j.1558-5646.1963.tb03281.x

Lanaud C, Fouet O, Clément D, Boccara M et al (2009) A meta–QTL analysis of disease resistance traits of Theobroma cacao L. Mol Breed 24:361–374. https://doi.org/10.1007/s11032-009-9297-4

Lauer E, Isik F (2021) Major QTL confer race-nonspecific resistance in the co-evolved *Cronartium quercuum* f. Sp Fusiforme-*Pinus taeda* pathosystem. Heredity 127:288–299. https://doi.org/10.1038/s41437-021-00451-8

Lynch M, Walsh B (1998) Genetics and analysis of quantitative traits. Sinauer Associates Inc Publishers, Sunderland

Monroe JG (2023) Potential and limits of (mal)adaptive mutation rate plasticity in plants. New Phytol 237:2020–2026. https://doi.org/10.1111/nph.18640

Namkoong G, Kang HC, Brouard HC (1988) Tree breeding: principles and strategies. Springer-Verlag

Sahu PP, Pandey G, Sharma N et al (2013) Epigenetic mechanisms of plant stress responses and adaptation. Plant Cell Rep 32:1151–1159. https://doi.org/10.1007/s00299-013-1462-x

Schmitt S, Tysklind N, Hérault B et al (2021) Topography drives microgeographic adaptations of closely related species in two tropical tree species complexes. Mol Ecol 30:5080–5093. https://doi.org/10.1111/mec.16116

Silva MA, Soares FAF, Clarindo WR et al (2023) Genomic and epigenomic variation in *Psidium* species and their outcome under the yield and composition of essential oils. Sci Rep 13:1385. https://doi.org/10.1038/s41598-023-27912-w

Valderrama E, Landis JB, Skinner D et al (2022) The genetic mechanisms underlying the convergent evolution of pollination syndromes in the Neotropical radiation of *Costus L.* Front Plant Sci 13:874322. https://doi.org/10.3389/fpls.2022.874322

Visscher P, Hill W, Wray N (2008) Heritability in the genomics era — concepts and misconceptions. Nat Rev Genet 9:255–266. https://doi.org/10.1038/nrg2322

Walsh B, Lynch M (2018) Evolution and selection of quantitative traits. Oxford University Press

Zhang L, Chen Y, Leng Q et al (2023) A high-resolution linkage map construction and QTL analysis for morphological traits in Anthurium (*Anthurium andraeanum* Linden). Plan Theory 12:4185. https://doi.org/10.3390/plants12244185

Part II
Tools for Population Genetics

Chapter 9
How Have Biochemical and Molecular Markers Been Applied in Population Genetics?

Dulcinéia de Carvalho, Alison Gonçalves Nazareno, Adelson Lemes da Silva Junior, Lucimara Cruz de Souza, and Enéas Ricardo Konzen

Contents

Abstract During the last decades, the use of biochemical and PCR-based molecular markers has revealed genetic variation at both the protein and DNA levels for several neotropical organisms, playing an important role in population genetic studies in the Neotropics. Taking into account the crucial importance of standing genetic

D. de Carvalho (✉) · A. L. da Silva Junior · L. C. de Souza
Departamento de Ciências Florestais, Universidade Federal de Lavras, Lavras, MG, Brazil
e-mail: dulce@ufla.br

A. G. Nazareno
Instituto de Ciências Biológicas, Universidade Federal de Minas Gerais, Belo Horizonte, MG, Brazil

E. R. Konzen
Centro de Estudos Costeiros, Limnológicos e Marinhos, Universidade Federal do Rio Grande do Sul, Imbé, Brazil

M. V. Bohrer Monteiro Siqueira et al. (eds.), *Population Genetics in the Neotropics*,
https://doi.org/10.1007/978-3-031-83685-5_9

variation (i.e., genetic diversity) in conservation and management policies, here we describe the use of biochemical (i.e., allozyme) and PCR-based molecular markers (e.g., amplified fragment length polymorphism—AFLP, simple sequence repeats—SSR, inter simple sequence repeats—ISSR, and single nucleotide polymorphism—SNP) to assess intra- and inter-population genetic diversity in a plethora of tropical species. Although numerous advances on population genetics methods and computational tools for their analyses have been made, the development of modern molecular methods (eg, high-throughput DNA sequencing methods to identify SNPs) could be sometimes misunderstood, resulting in an inappropriate choice of the molecular marker that is sometimes more expensive and labor intense than others. Here, we emphasize that a straightforward experimental design encompassing eminent questions/hypotheses must be aligned with the adequate choice of the molecular marker technology, that in turn is sample size-dependent. We also proposed to discuss the use of bi- and polyallelic-based markers, dominantly or co-dominantly inherited, to investigate the genetic diversity patterns in tropical species with distinct life history traits and ecological requirements. We summarize the genetic diversity patterns in a few plants and forest trees as examples, following the timeline of the molecular markers evolution, and depicting fundamental population genetics questions related to conservation genetics from a fine spatial scale perspective to a landscape genetics scenario.

Keywords Conservation genetics · Experimental design · PCR-based markers · Plant population genetics · Standing genetic variation

Learning Objectives

After going through this chapter, readers should be able to understand the following:

- The main types of biochemical and molecular markers that have been used for population genetic studies;
- How consecutive advances in technologies addressing biochemical and molecular markers have enhanced population genetic and genomic studies;
- The main applications of biochemical and molecular markers into conservation and breeding of neotropical species.

9.1 Introduction

Biochemical and molecular markers provide insights into the biology of a wide diversity of species. The implementation of practical and optimal management strategies based on results from molecular markers is still limited for both animals and plants, such as those of neotropical regions, especially due to the high biodiversity of this realm. However, before applying molecular markers to their samples, students and researchers must first consider theoretical and practical foundations and issues that lie behind molecular genetic studies. These include whether molecular approaches are really the best way to address a specific question, if sufficient resources are available to do so effectively, and if the right kind of material is available for testing. With prior knowledge of the attributes of a molecular method (e.g. the dominant or codominant inheritance pattern) we can plan how to collect field samples and prepare them for the genetic analyses.

Addressing theoretical foundations and selecting adequate markers can improve the effectiveness of studies that employ biochemical and molecular markers. Population genetic studies using molecular markers enable assessing genetic variation and how it is structured within and among populations of a species. Such knowledge is crucial for conservation purposes, for instance. Studies on population genetic structure involve the quantification of allelic diversity and genotype frequencies in sampled populations. Studies involving molecular markers assume that alleles are inherited according to the Mendelian laws, i.e. for a given locus, alleles segregate during the gamete formation (check Box 9.1 for further clarification on basic concepts for this chapter). After the characterization of the genetic structure, it is possible to understand variation in the biological context and get further signals of the forces shaping diversity, such as migration, mutation, recombination, selection, and drift, described in the chapters of the first section of this book. This understanding provides better means of how to use and to devise conservation plans of the targeted species.

Several molecular markers are found in nuclear, mitochondrial and chloroplast DNA. Such genomes are inherited by different means, so that markers belonging to each one of them can reveal distinct mechanisms shaping populations. Mitochondrial and chloroplast genomes are frequently inherited through the maternal parent only, but that can vary among species (Birky Jr 1995). Molecular markers can be used to study natural, managed and cultivated tree stands, and can measure the extent that individuals and populations are connected to each other. They provide information for establishing the breeding systems of populations, determine phylogenetic relationships among different taxa, assess hybridization and other interactions among species, and can help to determine human impacts on animal and plant populations through habitat fragmentation, selection, and cultivation. In theory, knowledge in all these areas can be used to devise better management strategies for tree species within natural and human landscapes. At the simplest level, by assessing whether or not variation is geographically structured among stands, it should be possible to design appropriate strategies for sampling populations for domestication and genetic improvement programs. In addition, such testing should assist the identification of genetically distinct groups that may be targets of conservation (Konzen et al. 2021).

Box 9.1
Basic concepts for molecular population genetics

A few of the following concepts are described throughout this chapter along with their definitions, but this box is aimed at compiling the most important for learning.

Genetic marker: either a morphological trait or a molecular variation (protein, DNA or RNA) that segregates according to Mendelian expectations.

Biochemical marker: protein, or specifically an enzyme, that has variation in their amino acid sequence. The biochemical marker can be a molecular marker for population genetic studies when it fits Mendelian expectations.

Molecular marker: generally applied to variation directly at the level of DNA or even RNA, shown as a difference of at least one nucleotide of one sample to another along their sequences.

Locus: in a specific and biological sense, the location of a determined gene and their variants. For technical application, locus can be defined as the location of any genetic marker along the chromosomes of a given species.

Allele: in a biological consideration, any variant of the same gene that produces alternative proteins that could differ based on a single amino acid only, that is, a single non-synonymous mutation at the DNA level. For technical consideration, any variant of a genetic marker, whether morphological, biochemical or molecular. That is, a single nucleotide variation between two samples, e.g. A to T, can be considered an allele for a population genetic analysis.

Isozymes and allozymes: isozymes are amino acid polymorphisms of enzymes located at different loci, but with the same function. In turn, allozymes are amino acid polymorphisms (alleles) originated from the same locus.

Polymorphism: any variation of amino acid or nucleotide sequence that can be further considered in genetic analyses.

RFLP: restriction fragment length polymorphism, one of the first molecular marker methods widely employed for population genetic studies and linkage mapping. It is based on the principle that restriction enzymes cut the DNA at very specific locations, but mutations or other nucleotide variants cause enzymes to still cut or fail in doing so, or cutting at distinct sites, producing DNA fragments with different sizes, that is, variation in restriction fragments size. An original reference is available from Grodzicker et al. (1974).

PCR: polymerase chain reaction, a powerful method that enables the rapid multiplication, that for technical reasons is called amplification, of an initial amount of DNA, producing multiple copies of an initial template. It usually involves three main steps: DNA denaturation, primers annealing and DNA polymerase extension for ending fragment duplication. Theoretically, 30 to 40 cycles of amplification result in millions or billions of copies of the original template. For an original reference, check Mullis et al. (1986).

RAPD: random amplified polymorphic DNA. Random 10-nucleotide-sized initiators (primers) amplify a few fragments that match their sequence along the genome, using PCR. Any mutation that halts fragments from being amplified results in a lack of band in an agarose gel after electrophoresis, in comparison to their presence in individuals where the amplification of that fragment was successful. Therefore, RAPD markers are usually dominant (presence or absence of bands). For an original reference, check Williams et al. (1990).

AFLP: amplified fragment length polymorphism, a method that combines PCR and restriction enzymes, to produce multiple copies of several restriction targets. Since restriction enzymes cut DNA at several locations along the genome, primers composed by adaptors that anneal to restriction sites aided by one, two or three random nucleotides selectively amplify among all markers available. While RAPD markers usually do not surpass 10 bands per primer, AFLP markers can overcome the limit of 100 or more markers. AFLP markers are also dominant. For an original reference, check Vos et al. (1995).

SSR or microsatellites: simple sequence repeats. Microsatellites are repetitive sequences that occur virtually in all genomes. Consist of small tandemly repeated motifs of two to six nucleotides in specific loci of the genome, e. g., ATC, a three-nucleotide motif that can be repeated multiples times such as: ATCATCATCATCATCATCATC. By counting the number of repeats, you can simplify the sequence as $(ATC)_7$. Microsatellites are hypervariable markers as high mutation rates affect the number of repeats per motif, that is, allele callings are based on repeat size variation among individual samples. That variation can occur even in the same individual, when the same locus varies in the number of repeats after comparing the two homologous chromosomes. For an original reference, check Litt and Luty (1989).

SNP: single nucleotide polymorphism, that is, a variation that occurs in a single nucleotide when comparing at least two sequences at a specific nucleotide position. They can be detected mainly by direct sequence comparisons, although allele-specific PCR approaches are possible and enable detection through gel electrophoresis for the purposes of genotyping larger samples with lower cost.

INDEL markers: insertions or deletions that commonly occur in genomes. Those markers vary in size, not rarely occurring a deletion or an insertion of a single nucleotide. Without prior phylogenetic inference of the origin of either deletions or insertions, we collectively call them insertions or deletions, the INDEL markers.

Gene markers: nucleotide variation either directly on DNA or RNA of specific genes under analysis.

Taking forest tree species (which will be frequently aimed at exemplifying molecular markers in this chapter) to set an example, since most of them are outcrossing, inbreeding depression will most likely occur if genetic variation is low (Frankham et al. 2010). Inbreeding depression includes limited fruit set, reduced overall seed yield, lower seed germination rates, reduced seedling survival, vigor loss and poor growth (see Chap. 7 to remind you of the fundamentals of inbreeding). Molecular markers can therefore be used to identify highly diverse natural stands that are suitable for cultivation, and that information can be used to plan management strategies to avoid inbreeding, that is, making crosses among genetically distinct individuals as prior detected from molecular screening. Detecting the levels of genetic diversity through molecular markers also assists to further comprehend the adaptive capacity of trees to respond to pressures such as changes in pest and disease prevalence, climate change, industrial quality and preferences, among others (Govindaraj et al. 2015). More subtly, through working out the reproductive biology of tree species, it should be possible to use molecular markers to better take into account the behavior of pollinators and seed dispersers when devising management strategies. Moreover, through a combination of field and laboratory work, it should be possible to determine appropriate densities and configurations for tree planting on farms, identify relevant niches for cultivation, find ways to manage pollinators that maximize productivity (e. g., to maximize fruit yields), and design suitable corridors for linking forest fragments.

Assessing patterns of genetic variation also provides information on the long-term changes that have occurred in the landscape in which a tree species is found. Any unexpected 'disjunctions' in genetic structure may relate to past fluctuations in climate that have led to repeated expansions and contractions of different vegetation types. Molecular markers may assist in the identification of forest refugia—particularly diverse regions that have survived past contractions in forest cover—that are especially worthy of conservation. Such information is also relevant for predicting how tree species will respond to current changes in climate (Aitken and Bemmels 2016) through human-induced global warming, allowing management strategies to be developed that minimize impacts on natural forests and agroforestry ecosystems (Bishaw et al. 2013).

Although we addressed the application of molecular markers specifically to the case of trees here, the reader must be aware that they can be applied to any type or organism (other plants, animals, fungi, and microorganisms) for the most diverse set of applications. Neotropical species certainly have been given focus by multiple researchers aimed at investigating population genetic mechanisms and breeding strategies with them, but much more has to be done for better acknowledging the immense biodiversity that such dominion harbors.

9.2 A Few Types of Molecular Markers: Definition, Genomic Location, Inheritance and Function

Several genetic markers have been used for genetic characterization of animals, plants, fungi and microorganisms. Initially, most studies were conducted with isozymes or allozymes, starting from the mid 1960s. In the 1980s, with more advanced protocols for DNA extraction, digestion with restriction enzymes and even DNA sequencing, novel methods were implemented, such as restriction fragment length polymorphisms (RFLP). In the mid 1980s, the polymerase chain reaction (PCR) was developed, paving the way for several new methods for genotyping populations in the next years and decades: random amplified polymorphic DNA (RAPD); amplified fragment length polymorphism (AFLP); and simple sequence repeats (SSR) or microsatellites. The rapid advances in sequencing technologies have further allowed the study of numerous single nucleotide polymorphisms (SNP) in a diverse range of species. Let's point out a few characteristics of four of the aforementioned methods.

9.2.1 Isozymes

Isozymes are variants of the same enzyme with identical or similar functions, and present in the same individual (Markert and Møller 1959). Enzyme markers represent electrophoretically detectable forms of enzymatic proteins visualized by substrate-specific staining. In the genetic studies, isozymes are defined as the products of loci; an isozyme is coded by one structural gene locus.

In turn, allozymes are proteins coded by different alleles of a locus, therefore an isozyme can consist of a set of allozymes. Isozymes are codominant markers, that is, they allow distinguishing homozygotes from heterozygotes. The codominant nature, low cost and technical simplicity of the analyses are among the main reasons for the use of this biochemical marker. The allelic variation at isozyme loci has had great impact on population genetics, in studies of gene flow, mating systems and drift effects, as further referenced. A disadvantage of isozymes is their low number of loci detected and low genomic coverage. Currently, these markers are scarcely used yet, as much more advanced techniques based on DNA, especially through sequencing, have been developed.

The typical workflow for an isozyme assay is shown in Fig. 9.1, for the case of plants, specifically trees. First, we need to conduct protein/enzyme extraction protocol, usually with 100–200 mg leaf tissue that is grinded in an extraction solution (e.g. in Alfenas 1998). After centrifuging, the extracts are placed in the gel wells to proceed to the electrophoretic runs. Discontinuous vertical electrophoresis in a polyacrylamide gel is normally used (Alfenas 1998). Staining protocols and the genetic basis of allozyme banding patterns are inferred from segregation patterns with reference to typical subunit structure and conceptual methods described in the literature (Alfenas 1998).

9.2.2 ISSR

Inter Simple Sequence Repeat (ISSR) markers are generated from repeated sequences of microsatellites located at the 5′ and 3′ ends of DNA. These markers consist of a single primer of 10 to 25 base pairs and are amplified by polymerase chain reaction (PCR). The development and first applications of ISSRs were carried out by various research groups in the early 1990s (Meyer et al. 1993; Gupta et al.

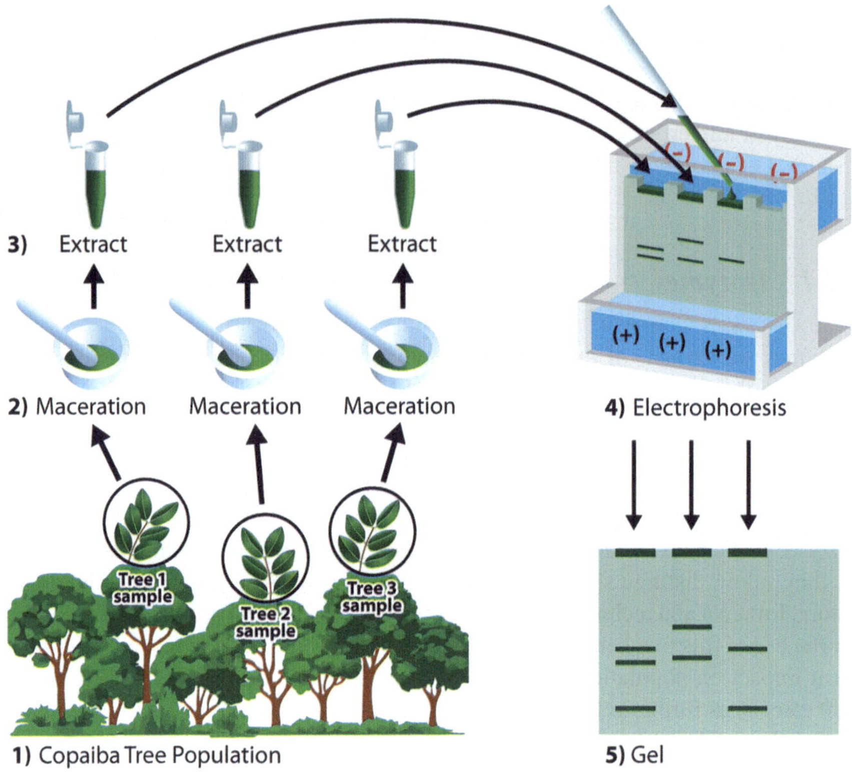

Fig. 9.1 A simple workflow for the analysis of isozymes: (1) Tissue sampling; (2) and (3): protein/enzyme extraction; (4) Polyacrylamide gel electrophoresis; and (5) Gel staining and genetic interpretation

1994; Zietkiewicz et al. 1994). They are classified as dominant, meaning that they do not distinguish between homozygous dominant (*AA*) and heterozygous (*Aa*) genotypes, which may limit their utility under a few circumstances (Adhikari et al. 2017). However, they are considered universal because they do not require prior knowledge of DNA sequences and provide a wide range of information by amplifying multiple loci. In addition, they are advantageous due to their low cost in both development and application compared to other types of molecular markers (Ng and Tan 2015). Due to such specificities, ISSR markers have been widely used in a variety of animal and plant species, especially those with little or no studies.

In an ISSR workflow, several steps should be performed, as shown in Fig. 9.2, starting with the DNA extraction step, where activities such as collection and preparation of the material of interest are performed, as well as the use of effective extraction methods to ensure DNA purity and integrity. PCR reactions are then performed to select primers, i. e., to select those with the best amplification profile. With the selected primers, PCR reactions are conducted on the entire set of individuals

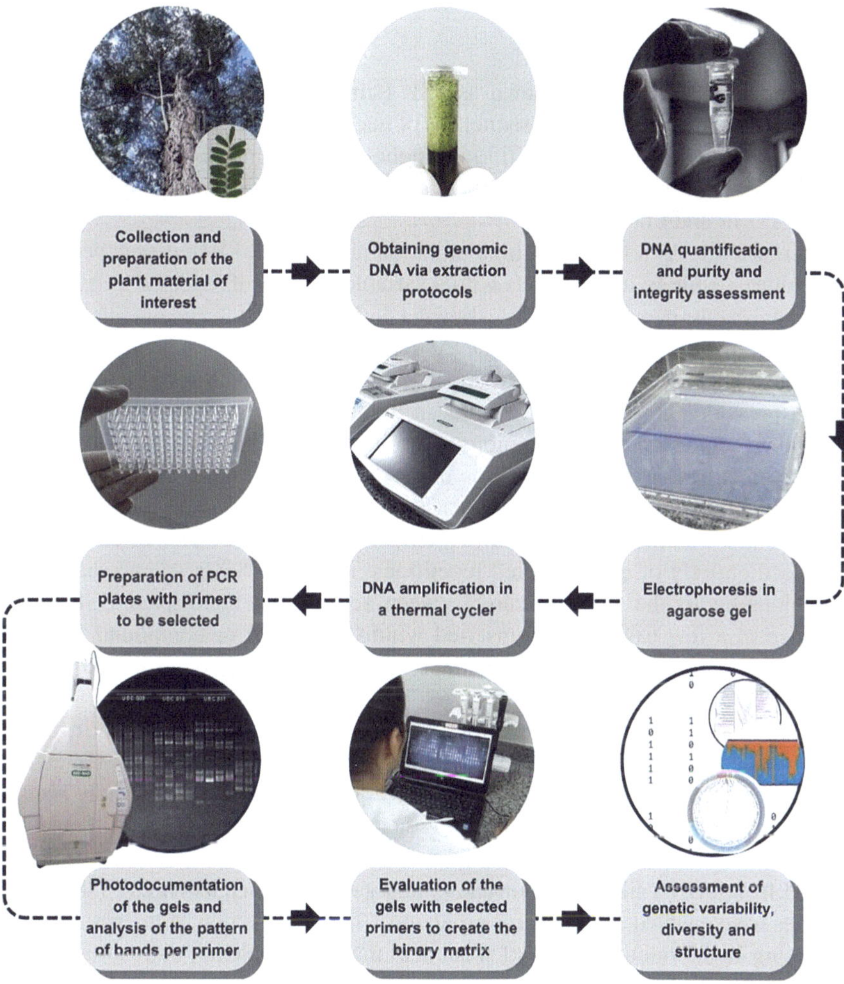

Fig. 9.2 Flowchart of the application of ISSR molecular markers in tree species

sampled, involving amplification cycles that include DNA denaturation, primer annealing, and DNA chain extension. Electrophoresis, usually in agarose gels, is then performed to separate the amplified fragments. The gel is then photodocumented and the results are analyzed by observing band patterns and making comparisons between different samples. Each band line on the gel is considered a locus, creating a binary matrix in which the presence of a band is considered 1 and its absence is marked 0. Finally, results are interpreted based on the identification of DNA polymorphisms among samples, analysis of genetic variability within and between populations, and the use of statistical methods to assess the genetic structure of populations (Fig. 9.2).

9.2.3 *Microsatellites*

Microsatellite markers, also known as SSR (simple sequence repeats), are DNA segments characterized by short sequences of nucleotides (1 to 6 base pairs) that are repeated in tandem (sequence motifs) throughout the genome. They are flanked by highly conserved sequences and can be located in coding or non-coding regions of the DNA (Litt and Luty 1989).

Microsatellite regions exhibit high mutation rates (10^{-6} to 10^{-2} events per locus/generation), resulting in fragments with a variable number of repeated units. These mutational events occur due to DNA slippage during replication and unequal recombination or incorrect alignment of DNA strands. Due to this high mutation rate, microsatellites have a lower occurrence in gene regions, which could compromise gene expression. This limitation results in them being more frequently found in non-coding regions, preserving genomic stability (Vieira et al. 2016).

Microsatellites can be classified as di-, tri-, tetra-, penta-, or hexanucleotides, depending on whether the repetitive unit consists of 2 to 6 bases, respectively. In terms of their composition, they can be classified as perfect SSRs when they are composed entirely of uninterrupted repetitions; imperfect SSRs when a base pair not belonging to the motif occurs between repetitions; interrupted SSRs if a sequence of a few base pairs is inserted within the motif; or compound SSRs if formed by two repetitions of distinct adjacent sequences (Turchetto-Zolet et al. 2017).

In the identification and acquisition of microsatellites, the initial step involves the prior identification of nucleotide sequences where the SSRs are present. This process can be accomplished through the detection of sequences in publicly accessible internet databases or by creating and selecting DNA libraries. The latter can be constructed from entire genomic DNA or from ESTs (expressed sequence tags) derived from cDNA libraries. Currently, next-generation sequencing (NGS) technologies have enabled advances for rapid and efficient detection of microsatellites, generating an enormous amount of sequences made available in public databases.

Once identified, primers (18 to 25 base pairs in length) complementary to the regions flanking the SSRs should be synthesized. The primers need validation through PCR by individual genotyping or multiplexing, and the amplification products separated by agarose gel electrophoresis, polyacrylamide gel, or capillary electrophoresis. This latter technique exhibits higher efficiency and precision (distinguishing fragments with only 1 bp) when compared to gels (which differentiate fragments with only 2 bp) (Turchetto-Zolet et al. 2017). The detection of microsatellites can also be conducted from a direct search of microsatellites within reads provided through high-throughput sequencing methods (Pimentel et al. 2018). After that, polymorphisms among individuals should be detected through the construction of a data matrix based on the size of the amplified/sequenced alleles. Population genetic statistics computations follow the data obtained.

Microsatellites have numerous advantages when compared to other classes of molecular markers. SSR sequences show remarkable variation among genotypes, giving this category of markers a high polymorphism rate and the presence of multiple alleles, characterizing them as codominant markers. They are easily reproducible, quickly detected, and in comparison with other technologies, they can be more cost-effective, especially in small-scale studies. Furthermore, they are transferable between related species, with relative abundance and broad coverage across eukaryotic genomes (Hodel et al. 2016; Vieira et al. 2016). The main limitation of these markers lies in the requirement for prior knowledge of DNA sequences for the development of species-specific primers, if no NGS method is affordable for direct sequencing. Additionally, it is important to note that the application of these markers in polyploid organisms may be restricted due to the difficulty in estimating the number of copies of each allele, therefore implicating in significant challenges for their genotyping (Mason 2015).

SSRs have various applications in plant genetics and improvement. Notably, they are used in the construction of linkage maps and mapping of loci involved in quantitative traits (QTLs), allowing for the identification and study of genes related to complex traits. Studies on diversity and genetic structure within and among populations, with estimates of gene flow and crossing rates, provide insights for conservation and management strategies of genetic resources. The applications also extend to marker assisted selection (MAS), a process that allows breeders to efficiently and accurately select individuals with desirable traits, accelerating the plant breeding process. Additionally, SSRs play a crucial role in defining DNA fingerprints, which can be used to identify and distinguish varieties or cultivars (Vieira et al. 2016).

9.2.4 SNP

Single nucleotide polymorphisms (SNP) are the most abundant type of molecular markers throughout genomes (Mammadov et al. 2012). A SNP is a variation in a single base at a specific genomic position when two or more sequences are compared. For instance, consider partial sequences of a hypothetical genomic region in three distinct individuals:

Individual 1: ATCTGCTTGCGCTGCTA
Individual 2: AT**T**TGCTTGCG**G**TGCTA
Individual 3: ATCTGCTTGCG**G**TGCTA

By comparing the three individuals, you can notice a difference in the third nucleotide of individual 2 (T) in relation to the others (C). Another difference can be seen in 12th position, where individual 1 has C, while the others have G. Both sites along the sequences represent SNP markers.

With the advances of sequencing technologies, numberless studies have enabled the detection of SNP markers across genomes of multiple organisms. SNP markers are codominantly inherited, that is, you can identify heterozygotes based on sequence comparison. The current sequencing strategies allow the identification of multiple reads of both homologous chromosomes, enabling the detection of heterozygotes. A heterozygote, in this case would be such as in the following for a diploid species:

Maternal chromosome: AT**C**TGCTTGCGCTGCTA
Paternal chromosome: AT**T**TGCTTGCGCTGCTA

The third position in the maternal chromosome is C, while the same position in the paternal chromosome is T. A CT at a specific locus in homologous chromosomes in a single individual confers a heterozygote.

A workflow to identify SNP markers involves the preparation of DNA for sequencing and posterior sequence analyses. The simplest strategy today is Sanger sequencing, originally developed in the 1970s (Sanger et al. 1977), but that has been automated in the following decades, using fluorescent dyes and capillary electrophoresis. This method typically sequences DNA fragments up to ~700 bp in length (an example using Sanger sequencing is shown later in this chapter with common bean).

A second approach for detecting SNP markers is the use of SNP arrays or SNP chips, small chips that are marked with synthesized oligos (immobile probes) that match specific positions (allele-specific markers) along genomes with known references. Several plant and animal models have SNP-chips developed for genotyping thousands of SNP markers using the principle of fluorescent hybridization of genomic DNA with the known markers that are developed for them (see an example at Song et al. 2015).

In the 2010s and 2020s, large-scale SNP calling from more advanced methods using high-throughput sequencing technologies have been also widely used, as they provide large amounts of markers. In a method called genotyping-by-sequencing (Elshire et al. 2011), the DNA is usually fragmented using restriction enzymes that cut it into pieces between ~100 and ~1000 bp size. The fragmented DNA is then submitted to standard protocols for DNA libraries according to the company and equipment to be used (e.g. Illumina, Nanopore, etc). The libraries are frequently prepared with amplification with PCR, to augment the representation of the same fragments. Libraries are then submitted to sequencing. Millions of sequences, hereafter called reads, are produced. After sequencing quality and filtering, the final reads are ready for alignment and comparisons among individuals. Great part of the effort taken for SNP identification requires a quite extensive bioinformatics pipeline in high performance computers. Genotyping-by-sequencing can be conducted for species with a genome reference sequence, but were also adapted for non-model species, using adapted pipelines for their analyses (Poland and Rife 2012). More on genotyping-by-sequencing and other methods for the detection of a high number of markers are described in Chap. 10.

SNP have become the most popular markers among the scientific community, as they have been used for several applications in all organisms, including populations of neotropical species: population genetic diversity and structure, demographic modeling, molecular phylogenomics and phylogeography, linkage and QTL and association mapping for agricultural traits such as drought tolerance, pest and disease resistance, yield and growth, among others. SNP markers have also provided opportunities for in-depth analyses of threatened species and strategies for their conservation in the Neotropics (Mammadov et al. 2012).

9.3 The Application of Molecular Markers to Forest Trees and a Few Case Studies with Distinct Genetic Markers

As previously stated, this chapter will mainly cover examples of the applications of molecular markers in forest tree species as examples. At the end, for the case of SNP markers, an agricultural crop will be placed into context. Nonetheless, several of the applications that are described can frequently be extended to animal models (including humans) and other organisms.

With a few exceptions—most of which are important timber species—the genetic structure of tropical trees has received limited research attention. For most species, therefore, the partition (within and among populations) of the genetic diversity across all the range of distribution of the species is yet unknown. And besides molecular markers, biological information on the species is of critical importance for depicting population genetic processes, as well as coordinating breeding programs. The distribution of genetic variation relies on the system of mating of each species under consideration, aided by their floral evolution, and also how these traits have been shaped by climate change, forest management and cultivation, among other processes. Molecular markers can aid information on a comprehensive analysis of all these processes. Particular features of trees, such as longevity, long generation interval and large size, impact in unique ways on their genetic structure.

With all that described, molecular markers provide unique opportunities to quantify the genetic diversity in tropical trees. But for species not having any information on genetic diversity, which is the type of markers that should be used? Maybe, for beginners, classical molecular markers could still be the choice for researchers, especially if that information could sufficiently address the question that is at stake:

- Is there genetic diversity in each population?
- Are the populations genetically structured? Are there populations that are reproductively isolated?
- How much selfing and outcrossing rates do the populations have?
- Is there gene flow among populations?
- Are there different levels of genetic diversity among populations?
- Is there inbreeding within populations?
- How does environmental variation relate to genetic diversity and structure?
- Is there enough genetic variation for breeding for traits of interest?
- Do desirable genotypes for human food and industrial applications have genetic differentiation to others?

9.3.1 Examples with Isozymes

Studies on allozyme variation have revealed variable levels of genetic variation across multiple species in the Neotropics. Table 9.1 provides a description of several tree species belonging to different families. In general, the tree species showed high genetic diversity within populations. Genetic diversity values are frequently higher than others described for tropical tree species using biochemical markers (for a brief description of the main population genetic statistics, read Box 9.2). That is, even old-fashioned biochemical markers have enabled the detection of high levels of diversity in the Neotropics, showing the outstanding genetic reservoir that this biogeographic dominion holds.

Table 9.1 Genetic diversity statistics in neotropical trees, based on allozyme markers N: number of individuals, $\hat{A}$: mean number of alleles per locus, $\hat{H}_o$: Observed heterozigosity, $\hat{H}_e$: expected heterozygosity, $\hat{f}$: fixation index, $\hat{\theta}_p$: population differentiation and $\hat{N}_m$ gene flow

Species	N	*Loci*	$\hat{A}$	%P	$\hat{H}_o$	$\hat{H}_e$	$\hat{f}$	$\hat{\theta}_p$	Reference
Annonaceae									
Xylopia emarginata Mart	89	11	2.0	0.576	0.403	−0.439	0.083	13.8	Jaeger et al. (2007)
Xylopia brasiliensis Sprengel	20	16	2.1	0.457	0.313	−0.419	–	–	Pinto and Carvalho (2004)
Arecaceae									
Geonoma schottiana Mart.	60	10	2.0	0.570	0.428	−0.342	0.010	6.4	Silva et al. (2011)
Burseraceae									
Protium spruceanum (Benth.) Engler	150	10	2.0	0.510	0.445	−0.098	0.037	4.2	Vieira et al. (2010)
Caryocaraceae									
Caryocar brasiliense Camb.	240	10	2.8	0.721	0.498	−0.444	0.020	12.3	Melo Junior et al. (2004)
Clusiaceae									
Calophyllum brasiliense Camb	120	14	2.0	0.409	0.484	0.155	0.086	0.296	Botrel et al. (2006)
Fabaceae									
Copaifera langsdorffii Desf.	60	12	2.4	0.408	0.605	0.033	0.130	0.4	Oliveira et al. (2002)
Dimorphandra mollis Benth.	80	10	2.0	0.472	0.464	−0.018	0.025	4.0	Gonçalves et al. (2010)
Machaerium villosum Vog.	95	10	2.4	0.592	0.487	−0.213	0.061	1.7	Botrel and Carvalho (2004)
Moraceae									
Ficus arpazusa Casaretto			2.0		0.409	−0.125	0.068		Nazareno and Carvalho (2009)
Myrtaceae									
Podocarpaceae									
Podocarpus sellowii Klotzsch	232	10	2.0	0.629	0.490	−0.292	0.021	8.8	Gonçalves et al. (2016)

Box 9.2

The fundamental population genetic statistics

Throughout this book you studied several concepts applied in population genetics. This chapter also includes various population genetic statistics that can be computed from allele frequencies (see Chap. 4 for calculating allele frequencies). Let's go through a few of them.

Mean number of alleles (A, frequently an estimate, represented by $\hat{A}$): the average number of alleles per locus in a given group of loci under analysis. Similarly, we can also compute the *effective number of alleles*, that shows the number of alleles that would be expected in the population.

Observed heterozygosity (Ho, estimated as $\hat{H}_o$): the proportion of heterozygotes within a sample of individuals.

Expected heterozygosity (He, estimated as $\hat{H}_e$): it is also known as Nei's genetic diversity, the likelihood of an individual being heterozygous in a locus. This is generally referred to as the genetic diversity estimate in a population. Molecular markers such as isozymes, microsatellites and SNP usually use their estimation. RAPD, AFLP and ISSR markers also allow the calculations of that statistic, however, considering the dominant inheritance of such markers, other estimates of diversity such as the Shannon diversity index from ecology are also calculated. The higher the estimates, the higher is the diversity of the population.

Inbreeding coefficient or fixation index (F or $\hat{f}$): the deviation of the observed heterozygosity from the expected heterozygosity (see Chap. 7).

Genetic differentiation coefficient (F_{ST} or Θ_p *in* this book): the proportion of the genetic diversity that is due to the differences among populations.

These are just a few statistics that are usually computed in population genetic studies. A good review on the matter, and that provides more details as well as other statistics is available from Kanaka et al. (2023).

The high genetic diversity observed in populations of the tree species shown here indicates their potential use in genetic conservation programs. Mitigation processes, such as the introduction of immigrants or vegetation corridors (Brandão et al. 2015), may be carried out in areas that have suffered genetic bottlenecks, or when gene flow is not sufficient to counteract the effects of drift. The implementation of in situ conservation strategies by introducing divergent genotypes may be utilized to increase effective population size and minimize inbreeding depression (Vieira et al. 2010). Management and compliance with laws that protect these areas are required so that the high genetic diversity can be maintained.

9.3.2 Inter Simple Sequence Repeats (ISSR)

Using ISSR markers, the genetic diversity statistics used to evaluate the different species and their respective populations show values ranging from low to high (Table 9.2). These results provide valuable insights into the history of fragmentation

Table 9.2 Genetic diversity statistics for neotropical trees, using ISSR markers *N*: number of individuals; *LP*: polymorphic loci; *%P*: percentage of polymorphism; *H**: Nei's genetic diversity index; *I**: Shannon's diversity index; *Na*: mean observed allele number; *Ne*: mean expected allele number

Species	*N*	*Loci*	*LP*	*%P*	*H**	*I**	*Na*	*Ne*	**Reference**
Anacardiaceae									
Astronium concinnum Schott ex Spreng.	48	121	89	73.55	0.31	0.47	–	–	Vieira et al. (2022)
Schinus terebinthifolius Raddi	162	181	157	86.74	0.20	–	–	–	Álvares-Carvalho et al. (2015)
Asteraceae									
Eremanthus erythropappus (DC.) MacLeish	60	123	116	94.31	0.36	0.53	1.94	1.62	Arriel et al. (2023)
Eremanthus incanus (Less.) Less.	60	119	102	85.71	0.34	0.50	1.86	1.58	Arriel et al. (2023)
Eremanthus glomerulatus Less.	60	122	116	95.08	0.40	0.57	1.95	1.71	Arriel et al. (2023)
Bignoniaceae									
Paratecoma peroba (Record) Kuhlm.	93	112	98	87.50	0.26	0.40	1.88	1.41	França et al. (2022)
Euphorbiaceae									
Senefeldera verticillata (Vell.) Croizat	39	179	136	75.97	0.32	0.50	–	1.53	Vieira et al. (2018)
Fabaceae									
Dalbergia nigra (Vell.) Allemão ex Benth.	180	162	122	75.30	0.35	0.53	2.00	1.62	Silva Júnior et al. (2022)
Melanoxylon brauna Schott	59	183	117	63.93	0.36	0.54	2.00	1.63	
Plathymenia reticulata Benth.	149	156	101	64.74	0.40	0.59	–	–	Souza et al. (2017)
Schizolobium parahyba var. *amazonicum* (Huber ex. Ducke) Barneby	171	136	79	58.08	0.37	0.55	2.00	1.65	Silva Júnior et al. (2017)
Myrtaceae									
Myrcia splendens (Sw.) DC.	168	70	70	100.00	0.37	0.53	2.00	1.70	Brandão et al. (2015)
Moraceae									
Ficus bonijesulapensis R.M. Castro	189	75	75	100.00	0.30	0.55	–	–	Duarte et al. (2015)
Pentaphylacaceae									
Freziera atlantica Zorzan. & Amorim	44	179	86	46.6	0.25	0.39	2.00	1.40	Zorzanelli et al. (2022)

and exploitation of the vegetation in the study areas. Over time, the extent and intensity of anthropogenic activities have played a crucial role in shaping the genetic structure of populations, which is reflected in the different patterns of genetic variability that we observe today. This scenario calls for effective conservation and management strategies to safeguard and, where possible, restore the genetic richness of these species in face of environmental and anthropogenic challenges. Moreover, it is imperative to emphasize the importance of studies aimed at the genetic improvement of certain species in order to enhance desirable traits and increase their resilience to environmental pressures and socio-economic demands.

9.3.3 Simple Sequence Repeats: SSR

Table 9.3 compiles studies conducted with SSR markers for population genetic studies in neotropical species. It is frequent that high levels of diversity are encountered within populations of forest tree species with these codominant markers. However, insights into the effects of fragmentation on genetic diversity and increasing levels of inbreeding or other processes that negatively affect populations have also been given by such studies, once more driving the attention of conservationists into the urgency of taking effective and practical measures for conservation and sustainable management.

9.3.4 SNP Markers

Unlike with the previous molecular tools, for SNP markers we will follow with an example of agricultural crop, the common bean (*Phaseolus vulgaris* L.), to demonstrate their application. *P. vulgaris* is a widespread neotropical plant that is naturally

Table 9.3 Genetic diversity index in tropical tree species, based on microsatellite markers N: number of individuals, $\hat{A}$: mean number of alleles per locus, $\hat{H}_o$: observed heterozigosity, $\hat{H}_e$: expected heterozygosity, $\hat{f}$: fixation index, and $\hat{\theta}_p$: population differentiation

Species	N	Locos	$\hat{A}$	$\hat{H}_o$	$\hat{H}_e$	$\hat{f}$	$\hat{\theta}_p$	Reference
Araucariaceae								
Araucaria angustifolia (Bert) O Ktze	450	11	6.7	0.82	0.76	−0.09	0.019	de Pádua (2015)
Asteraceae								
Eremanthus erythropappus (DC) MacLeish	275	8	2.5	0.66	0.42	−0.53	–	Rocha et al. (2021)
Fabaceae								
Dalbergia nigra (Vell) Allemão ex Benth	180	7	7.4	0.56	0.57	0.01	0.1483	Silva Júnior et al. (2022)
Anadenanthera peregrine (L) Speg	166	6	7.0	074	076	0.001	–	Cortelete et al. (2021)

distributed from northern Mexico to northern Argentina, mostly across the highlands along that spanning range. It produces the edible common bean, a nutritious grain widely consumed in the world, especially in Latin America. Allozyme, RAPD, SSR and AFLP markers have shown two main gene pools of common bean, the Andean (from northern Argentina to part of Colombia) and the Mesoamerican (from Colombia to Mexico) (Kwak and Gepts 2009). Frequently a third pool has also been revealed by those markers within a smaller region between Ecuador and Peru (Ariani et al. 2018). SNP markers have frequently confirmed the same gene pools of the previous tools, however with much deeper insights into their origin, diversification and domestication (Ariani et al., 2018). Moreover, SNP markers have enabled the detection of specific traits associated with nucleotide polymorphisms, that is, through QTL and association mapping.

A set of dehydration responsive element-binding genes (*DREB*), genes involved in abiotic stress responses in plants, have specifically been addressed for detecting nucleotide polymorphisms among Mesoamerican and Andean genotypes, whether those were wild or domesticated (Konzen et al. 2019). Using Sanger sequencing among a set of 17 common bean genotypes, 18 SNP markers were encountered for *PvDREB6B*, a homolog of the *RAP24* gene from *Arabidopsis thaliana*. That is, even in a single gene, and being a transcription factor, several polymorphisms were detected in a region spanning 957 bp only. Most polymorphic sites enable the differentiation between Andean and Mesoamerican genotypes within *PvDREB6B*.

Another study with the same species allowed the identification of SNP markers associated with pod indehiscence, the ability to hold pods from shattering and releasing their seeds. Using hundreds of genotypes and sequencing their DNA with a predefined set of SNP markers developed as an array (SNP-chip, with more than 5000 SNP probes available for genotyping in common bean—Song et al. 2015), polymorphisms were mapped flanking *PvIND*, with high association with that trait with high precision (Parker et al. 2020). Their study can be considered a great model of how a phenotype can be linked straight to a gene. It is to be highlighted, however, that the authors also found significant peaks in other regions than the location of *PvIND*, that is, that gene partly explains pod indehiscence, but other genes are probably involved, each explaining part of the phenotypic variation. This makes SNP markers powerful tools for unraveling the loci involved in the manifestation of a target trait, and association mapping has been widely used to that end thanks to such markers.

9.4 Final Remarks

Molecular markers are essential tools for population genetics, as they allow the detection of allelic variation for computing statistics of genetic diversity and structure across populations of a diverse range of species. Although allozyme markers have been frequently used in the past, they were rapidly replaced by DNA-based markers, especially those based on PCR. Among the most important tools, we highlight microsatellites as valuable markers for genetic diversity and structure discovery. Their codominant nature allows precise estimates of gene flow among populations. Another class of high importance are SNP markers, which have become the most popular markers for deciphering population genetic processes, whether recent or ancient, as well as the correlation or association between nucleotide polymorphisms and phenotypic variation. In the genomic era, SNP and microsatellites can be accessed directly from sequencing, enabling accurate estimates of population genetic statistics. Neotropical organisms need to be given much more focus to gather a deeper genetic diversity panorama of each species to better plan for their long-term conservation in such a biodiverse region, and molecular markers remain essential tools in such noble endeavors.

Keypoints

- Genetic markers are phenotypic traits or molecular polymorphisms, either from isozymes or DNA, which variation fits Mendelian expectations;
- Molecular markers are differences in size and/or the sequence of DNA that allow the differentiation among samples;
- The most used molecular markers in population genetics are currently microsatelites (or simple sequence repeats—SSR markers) and single nucleotide polymorphisms (SNP markers);
- Molecular markers have important applications for conservation and breeding of species, as they allow to discover the levels of genetic diversity and structure, as well as population genetic mechanisms underlying species evolution;
- Neotropical species have been the focus of several population genetic studies using molecular markers. However, compared to other realms, molecular studies are scarcely available with the native species. In this chapter, we showed examples of tree species that were evaluated with allozymes, ISSR, SSR and SNP markers.

Questions

1. How can you distinguish a genetic marker from a molecular marker? Is there actually a difference? On what terms?
2. Despite being codominant, why were allozyme markers rapidly replaced by microsatellites in population-level studies? What are the advantages and disadvantages of both markers?
3. What types of molecular markers are dominant? What are their limitations in comparison to codominant markers?
4. Why are molecular markers important for species conservation in the Neotropics?
5. Make a summary of the molecular tools described in this chapter, describing: mode of inheritance, genomic location, function, and main applications to conservation and breeding.

References

Adhikari S, Saha S, Biswas A et al (2017) Application of molecular markers in plant genome analysis: a review. Nucleus 60:283–297. https://doi.org/10.1007/s13237-017-0214-7

Aitken SN, Bemmels JB (2016) Time to get moving: assisted gene flow of forest trees. Evol Appl 9(1):271–290. https://doi.org/10.5683/SP2/BQBI7S

Alfenas AC (1998) Eletroforese de proteinas e isoenzimas afins: fundamentos e aplicações em plantas e microorganismos Viçosa, UFV, p 574

Álvares-Carvalho SV, Duarte JF, Carvalho D et al (2015) *Schinus terebinthifolius*: population structure and implications for its conservation. Biochem Syst Ecol 58:120–125. https://doi.org/10.1016/j.bse.2014.10.002

Ariani A, Berny Mier y Teran JC, Gepts P (2018). Spatial and temporal scales of range expansion in wild Phaseolus vulgaris. Molecular biology and evolution, 35(1):119–131

Arriel DAA, Fajardo CD, Vieira FDA et al (2023) Spatial genetic structure of sympatric populations of *Eremanthus* species in Brazil. J Trop For Sci 35(1):56–65

Birky CW Jr (1995) Uniparental inheritance of mitochondrial and chloroplast genes: mechanisms and evolution. Proc Natl Acad Sci 92(25):11331–11338. https://doi.org/10.1073/pnas.92.25.11331

Bishaw B, Neufeldt H, Mowo J, Abdelkadir A, Muriuki J, Dalle G, Assefa T, Guillozet K, Kassa H, Dawson IK, Luedeling E, Mbow C (2013) Farmers' strategies for adapting to and mitigating climate variability and change through agroforestry in Ethiopia and Kenya

Botrel MCG, Carvalho D (2004) Variabilidade isoenzimática em populações naturais de jacarandá paulista (*Machaerium villosum* Vog). Rev Bras Bot 27(4):621–628. https://doi.org/10.1590/S0100-84042004000400002

Botrel MCG, Souza AM, Pinto SIC et al (2006) Caracterização genética de *Calophyllum brasiliense* Cambem populações de Mata Ciliar. Revista Árvore 30(5):821–827. https://doi.org/10.1590/S0100-67622006000500016

Brandão MM, Vieira FA, Nazareno AG et al (2015) Genetic diversity of neotropical tree *Myrcia splendens* (Myrtaceae) in a fragment–corridor system in the Atlantic rainforest. Flora 216:35–41. https://doi.org/10.1016/j.flora.2015.07.006

Cortelete MA, Silva Júnior AL, Pereira MLS et al (2021) Molecular characterization as strategy for ex situ conservation of *Anadenanthera peregrina* (L) Speg. Scientia Forestalis 49:e3443. https://doi.org/10.18671/scifor.v49n129.23

de Pádua JAR (2015) Genetic diversity of natural populations of Araucaria angustifolia (Araucariaceae) at Minas Gerais state, Brazil.

Duarte JF, Carvalho D, Vieira FA (2015) Genetic conservation of *Ficus bonijesulapensis* R M Castro in a dry forest on limestone outcrops. Biochem Syst Ecol 59:54–62. https://doi.org/10.1016/j.bse.2015.01.008

Elshire RJ, Glaubitz JC, Sun Q et al (2011) A robust, simple genotyping-by-sequencing (GBS) approach for high diversity species. PLoS One 6(5):e19379. https://doi.org/10.1371/journal.pone.0019379

França TCC, Tavares LR, Silva Júnior AL et al (2022) Genetic characterization of remaining populations of *Paratecoma peroba*, an endangered and endemic species of the Atlantic Forest. Cerne 28:1–11. https://doi.org/10.1590/01047760202228013055

Frankham R, Briscoe DA, Ballou JD (2010) Introduction to conservation genetics. Cambridge University Press

Gonçalves AC, Vieira FA, Reis CAF et al (2010) Estrutura genética espacial em populações naturais de *Dimorphandra mollis* (Fabaceae) na região norte de Minas Gerais, Brasil. Rev Bras Bot 33(2):325–332. https://doi.org/10.1590/S0100-84042010000200013

Gonçalves FR, Vieira FA, Carvalho D (2016) Naturally fragmented but not genetically isolated populations of Podocarpus sellowii Klotzsch (Podocarpaceae) in Southeast Brazil. Genet Mol Res 15(4):1–17. https://doi.org/10.4238/gmr.15048871

Govindaraj M, Vetriventhan M, Srinivasan M (2015) Importance of genetic diversity assessment in crop plants and its recent advances: an overview of its analytical perspectives. Genet Res Int 2015(1):431487. https://doi.org/10.1155/2015/431487

Grodzicker T, Williams J, Sharp P, Sambrook J (1974) Physical mapping of temperature-sensitive mutations of adenoviruses. In: Cold Spring Harbor symposia on quantitative biology, vol 39. Cold Spring Harbor Laboratory Press, pp 439–446

Gupta M, Chyi YS, Romero-Severson J et al (1994) Amplification of DNA markers from evolutionarily diverse genomes using single primers of simple-sequence repeats. Theor Appl Genet 89:998–1006. https://doi.org/10.1007/BF00224530

Hodel RG, Segovia-Salcedo MC, Landis JB, Crowl AA, Sun M, Liu X et al (2016) The report of my death was an exaggeration: a review for researchers using microsatellites in the 21st century. Appl Plant Sci 4(6):1600025. https://doi.org/10.3732/apps.160002

Jaeger P, Moura MCO, Van Den Berg E et al (2007) Caracterização genética de populações naturais de *Xylopia emarginata* Mart (Annonaceae). Sci For 73:91–99

Kanaka KK, Sukhija N, Goli RC, Singh S, Ganguly I, Dixit SP et al (2023) On the concepts and measures of diversity in the genomics era. Curr Plant Biol 33:100278

Konzen ER, Recchia GH, Cassieri F et al (2019) DREB genes from common bean (*Phaseolus vulgaris* L.) show broad to specific abiotic stress responses and distinct levels of nucleotide diversity. Int J Genom 2019(1):9520642. https://doi.org/10.1155/2019/9520642. eCollection 2019

Konzen ER, Pozzobon LC, Souza DMSC, Fernandes SB, Campos WF, Brondani GE et al (2021) Molecular markers in bamboos: understanding reproductive biology, genetic structure, interspecies diversity, and clonal fidelity for conservation and breeding. In: Biotechnological Advances in Bamboo: The "Green Gold" on the Earth, pp 33–62. https://doi.org/10.1007/978-981-16-1310-4_2

Kwak M, Gepts P (2009) Structure of genetic diversity in the two major gene pools of common bean (*Phaseolus vulgaris* L., Fabaceae). Theor Appl Genet 118:979–992. https://doi.org/10.1007/s00122-008-0955-4

Litt M, Luty JA (1989) A hypervariable microsatellite revealed by in vitro amplification of a dinucleotide repeat within the cardiac muscle actin gene. Am J Hum Genet 44(3):397–401

Mammadov J, Aggarwal R, Buyyarapu R et al (2012) SNP markers and their impact on plant breeding. Int J Plant Genom 2012(1):728398. https://doi.org/10.1155/2012/728398

Mason AS (2015) Challenges of genotyping polyploid species. In: Plant genotyping: methods and protocols. Springer New York, pp 161–168. https://doi.org/10.1007/978-1-4939-1966-6_12

Markert CL, Møller F (1959) Multiple forms of enzymes: tissue, ontogenetic, and species specific patterns. Proceedings of the National Academy of Sciences, 45(5):753–763.

Melo Júnior AF, Carvalho D, Póvoa JRS et al (2004) Estrutura genética de populações naturais de pequizeiro (*Caryocar brasiliense* Camb). Cerne 66:56–65

Meyer W, Mitchell TG, Freedman EZ et al (1993) Hybridization probes for conventional DNA fingerprinting used as single primers in the polymerase chain reaction to distinguish strains of *Cryptococcus neoformes*. J Clin Microbiol 31:2274–2280. https://doi.org/10.1128/jcm.31.9.2274-2280.1993

Mullis K, Faloona F, Scharf S, Saiki R, Horn G, Erlich H (1986) Specific enzymatic amplification of DNA in vitro: the polymerase chain reaction. In: Cold Spring Harbor symposia on quantitative biology, vol 51. Cold Spring Harbor Laboratory Press, pp 263–273. https://doi.org/10.1101/SQB.1986.051.01.032

Nazareno AG, Carvalho D (2009) What the reasons for no inbreeding and high genetic diversity of the neotropical fig tree *Ficus arpazusa*? Conserv Genet 10:1789–1793. https://doi.org/10.1007/s10592-008-9776-x

Ng WL, Tan SG (2015) Inter-simple sequence repeat (ISSR) markers. ASM Sci J 9(1):30–39

Oliveira AF, Carvalho D de, Rosado S (2002) Taxa de cruzamento e sistema reprodutivo de uma população natural de Copaifera langsdorffii Desf. na região de Lavras (MG) por meio de isoenzimas. Brazilian Journal of Botany, 25, 331–338.

Parker TA, Berny Mier y Teran JC, Palkovic A, Jernstedt J, Gepts P (2020) Pod indehiscence is a domestication and aridity resilience trait in common bean. New Phytol 225(1):558–570. https://doi.org/10.1111/nph.16164

Pimentel JSM, Carmo AO, Rosse IC et al (2018) High-throughput sequencing strategy for microsatellite genotyping using Neotropical fish as a model. Front Genet 9:73. https://doi.org/10.3389/fgene.2018.00073

Pinto SIDC, Carvalho DD (2004) Genetic structure in populations of pindaíba (Xylopia brasiliensis Sprengel) by isozymes. Rev Bras Bot 27:597–605. https://doi.org/10.1590/S0100-84042004000300019

Poland JA, Rife TW (2012) Genotyping-by-sequencing for plant breeding and genetics. Plant Genome 5(3). https://doi.org/10.3835/plantgenome2012.05.0005

Rocha LF, Paula NRD, De Carvalho D (2021) Fine-scale analysis reveals a potential influence of forest management on the spatial genetic structure of *Eremanthus erythropappus*. J For Res 32(4):1567–1578

Sanger F, Nicklen S, Coulson AR (1977) DNA sequencing with chain-terminating inhibitors. Proc Natl Acad Sci 74(12):5463–5467. https://doi.org/10.1073/pnas.74.12.5463

Silva Júnior AL, Souza LC, Pereira AG et al (2017) Genetic diversity of *Schizolobium parahyba* var *amazonicum* (Huber ex Ducke) Barneby, in a forest area in Brazil. Genet Mol Res 16(3):1–10. https://doi.org/10.4238/gmr16039774

Silva Júnior AL, Cabral RLR, Sartori L et al (2022) Molecular markers applied to the genetic characterization of *Dalbergia nigra*: implications for conservation and management. Trees 36(5):1539–1557. https://doi.org/10.1007/s00468-022-02309-w

Silva MS, Vieira FA, Carvalho D (2011) Diversity and genetic structure in natural populations of *Geonoma schottiana* Mart (Arecaceae): implications for conservation. Cerne 17:01–07. https://doi.org/10.1590/S0104-77602011000200006

Song Q, Jia G, Hyten DL et al (2015) SNP assay development for linkage map construction, anchoring whole-genome sequence, and other genetic and genomic applications in common bean. G3: genes, genomes. Genetics 5(11):2285–2290. https://doi.org/10.1534/g3.115.020594

Souza LC, Silva Júnior AL, Souza MC et al (2017) Genetic diversity of *Plathymenia reticulata* Benth in fragments of Atlantic Forest in southeastern Brazil. Genet Mol Res 16:1–13. https://doi.org/10.4238/gmr16039775

Turchetto-Zolet AC, Turchetto C, Zanella AM et al (2017) Marcadores Moleculares na Era genômica: Metodologias e Aplicações Sociedade Brasileira de Genética, Ribeirão Preto: São Paulo, 181 p

Vieira FA, Fajardo CG, de Souza AM, de Carvalho D (2010) Landscape-level and fine-scale genetic structure of the neotropical tree Protium spruceanum (Burseraceae). Int J For Res 2010(1):120979. https://doi.org/10.1155/2010/120979

Vieira MLC, Santini L, Diniz AL et al (2016) Microsatellite markers: what they mean and why they are so useful. Genet Mol Biol 39:312–328. https://doi.org/10.1590/1678-4685-GMB-2016-0027

Vieira AAR, Silva Júnior AL, Souza LC et al (2018) Genetic structure and diversity of *Senefeldera verticillata* (Euphorbiaceae) in semideciduous seasonal forest fragments. Genet Mol Res 17(3):1–11. https://doi.org/10.4238/gmr18059

Vieira AAR, Souza AC, Silva Júnior AL (2022) Diversity and genetic structure of *Astronium concinnum* Schott ex Spreng in conservation units. Plant Genet Resour 19:530–537. https://doi.org/10.1017/S1479262121000654

Vos P, Hogers R, Bleeker M, Reijans M, Lee TVD, Hornes M et al (1995) AFLP: a new technique for DNA fingerprinting. Nucleic Acids Res 23(21):4407–4414. https://doi.org/10.1093/nar/23.21.4407

Williams JG, Kubelik AR, Livak KJ, Rafalski JA, Tingey SV (1990) DNA polymorphisms amplified by arbitrary primers are useful as genetic markers. Nucleic Acids Res 18(22):6531–6535. https://doi.org/10.1093/nar/18.22.6531

Zietkiewicz E, Rafalski A, Labuda D (1994) Genome fingerprinting by simple sequence repeat (SSR) – anchored polymerase chain-reaction amplification. Genomics 20:176–183. https://doi.org/10.1006/geno.1994.1151

Zorzanelli JPF, Kunz SH, Carrijo TT et al (2022) Bases for genetic conservation of *Freziera atlantica*, an endangered wood species and endemic to the Atlantic Forest hotspot. Trees 36:1005–1015. https://doi.org/10.1007/s00468-022-02266-4

Chapter 10
How Can We Generate Genomic Resources for Neotropical Species?

Leonardo Carlos Jeronimo Corvalán, Larissa Resende Carvalho, Héctor Antônio Assunção Romão, Rhewter Nunes, and Renata de Oliveira Dias

Contents

Abstract Genomics has revolutionized our understanding of genetic variability within and among species. The vast amount of information available from genome sequencing is a source for developing new strategies for understanding population dynamics, among other things. However, these resources are generally scarce for non-model species, even those of high economic interest. This chapter describes the

L. C. J. Corvalán · L. R. Carvalho · H. A. A. Romão · R. de Oliveira Dias (✉)
Instituto de Ciências Biológicas, Universidade Federal de Goiás, Goiânia, GO, Brazil
e-mail: renata_dias@ufg.br

R. Nunes
Universidade Estadual de Goiás, Iporá, GO, Brazil

M. V. Bohrer Monteiro Siqueira et al. (eds.), *Population Genetics in the Neotropics*,
https://doi.org/10.1007/978-3-031-83685-5_10

most commonly used strategies for sequencing, assembling, and annotating new genomes and how complete or draft genomes have been used to discover new polymorphic sites for studying population genetics in the Neotropics. In this context, this chapter addresses issues related to genome sequencing, including the selection of target species, the best sequencing methods to use considering genome size and financial cost, and the most commonly employed strategies for nuclear genome assembly and annotation. In addition, given the still high cost of sequencing the genomes of new species, this chapter includes examples of how even low-coverage genomic data can be used to develop powerful resources for population genomics studies. Finally, this chapter discusses the multiple uses of newly assembled genomes for unraveling genetic variability within and between species, including their use in molecular marker development. The discussions presented here will provide the reader with the main challenges and strategies to overcome them in studies of neotropical species, especially those still poorly understood.

Keywords Bioinformatics · DNA sequencing · Genotyping · Populational genomics · Genome assembly

Learning Objectives

This chapter should provide a concise guide through the most recent applications of genomics for neotropical species by describing:

- The diverse applications of genomic resources and the methods used to their production;
- The methods for genome sequencing, including the strategic evaluation of sequencing, assembly, and annotation approaches and the trade-offs associated with each one of them;
- The fundamental genomic resources that are used in population genomic studies, including analysis of genetic diversity, population structure, and evolutionary dynamics;
- Examples of the application of population genomics methods to our understanding of neotropical species.

10.1 How Does Population Genomics Work?

With the ever-decreasing cost and time of DNA sequencing, the availability of genomic-scale data is revolutionizing classical population genetics studies. The term "population genomics" began to appear in the literature around the 1990s, mainly in analyzing large-scale human polymorphisms. From this period onwards, some biologists began to predict that large-scale population genetics approaches would be critical for studying natural populations.

On a large scale, obtaining molecular markers (and genotyping data) is associated with two main approaches. The first is related to the increasing availability of genome sequences of target species. Once a complete or partial genome sequence is available, researchers can generate resequencing data on individuals from populations of interest, obtain genotyping data from them, or even develop genotyping chips for specific loci. Resequencing data can be obtained using high-coverage strategies, such as whole genome (WGS) and whole exome (WES) sequencing, or using low-coverage strategies, such as the sequencing of pools of individuals (Pool-seq) and low-coverage genome sequencing (lcWGS) (Fig. 10.1). A second approach considers the absence of a reference genome for the species or the lack of need to sample the genome as a whole in a resequencing analysis. The set of methods

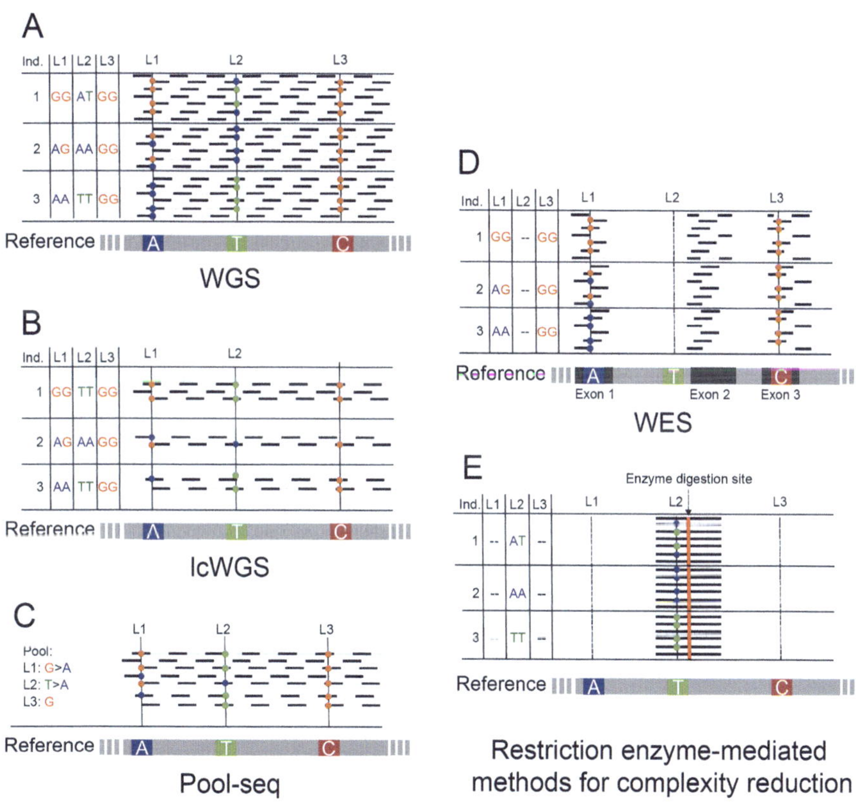

Fig. 10.1 Schematic review of the most popular population genomics approaches. (**a**) In the Whole Genome Sequencing (WGS) strategy, sequencing reads are produced for all portions of the genome and each individual separately. To reduce sequencing costs, researchers can also opt for (**b**) low-coverage Whole Genome Sequencing (lcWGS), which also sequences the entire genome but with lower coverage for each position, or (**c**) Pool-seq, where the individuals are analyzed as a pool of DNA. Other strategies for sampling populations with less breadth include (**d**) Whole Exome Sequencing, in which only the exons are sequenced, and (**e**) strategies that reduce the complexity of the genome by sequencing only the regions around restriction enzyme cleavage sites, such as RAD-seq and GBS-Cornell. The Figure represents a hypothetical population with three individuals and three single nucleotide polymorphic loci, represented by L1, L2, and L3. The hypothetical individuals sampled in the figure have the genotypes for loci L1, L2, and L3, respectively: 1 = GG, AT, and GG; 2 = AG, AA, and GG; and 3 = AA, TT, and GG. However, not all strategies can detect all of these variants due to their technical limitations described in this chapter

related to this practice is called genotyping-by-sequencing (check Chap. 9, that also described a little on the method). It uses methods that reduce genome complexity to sample a subset of regions from which the markers and genotypes will be detected. The techniques can achieve this reduction in complexity through restriction enzymes, primers, or probes, where the targeting of the sequencing effort to particular genomic regions may or may not be specific. Cornell genotyping-by-sequencing (GBS) and restriction site-associated DNA sequencing (RAD-Seq) are the best-known reduced representation sequencing methods. Besides their technical differences, both methods are similar since both use restriction enzymes to fragment the DNA and NGS platforms to sequence the obtained fragments (Fig. 10.1). GBS and RAD-seq were milestones in population genetics studies that vastly increased the number of detected markers sampled for any population of organisms at reasonable costs.

A typical pipeline for a population genomic study has a few critical steps: (i) design of sequencing strategy; (ii) generation of sequence data; (iii) mapping of sequence reads to assembly; (iv) variant calling (genotyping); and (v) downstream population genetic or molecular evolutionary analyses. The sequencing strategy includes aspects such as the depth of coverage, the choice of sequence for individual samples or groups of individuals, and issues common to any population genetic study (e.g., number of individuals per population and number of populations).

With the decreasing cost of DNA sequencing, researchers are increasingly replacing sequencing strategies of reduced complexity with strategies of greater genomic breadth. These advances make it urgent to provide suitable reference genomes for neotropical species. This chapter explains how to obtain reference genomes to enable large-scale population genetics studies that require this resource, and describes how even draft genomes can be a powerful tool for these and other studies aimed at conserving Neotropical species.

10.2 How Do Researchers Sequence the DNA?

Obtaining complete genomes remains challenging just over four decades after the emergence of the first DNA sequencing methods and three generations of DNA sequencing methods (Box 10.1). However, with the significant advances in the development of sequencing technologies since the DNA sequencing performed by Maxam and Gilbert (1977) and Sanger et al. (1977), and the culminating reduction in the cost of sequencing and the increase in the volume of data generated, the significant challenges today are related to the assembly, annotation, and further analysis of these sequences.

The DNA sequencing strategies used in the genome assembly projects can be divided into two categories based on the length of the reads generated: short (second-generation) and long (third-generation). Short reads have a lower cost per

Box 10.1: The evolution of DNA sequencing strategies

First-generation sequencing methods began with Maxam and Gilbert (1977), who used radioactive deoxynucleotides printed on radiographic plates to identify each nucleotide added to the DNA strand. However, it was popularized by Sanger et al. (1977) sequencing method, which uses radiolabeled deoxynucleotides that prevent the synthesis of the DNA strand and electrophoresis to separate fragments of different sizes and identify each of the bases sequentially. The principles of Sanger sequencing are still used today, but in an automated way using modified deoxynucleotides and capillary electrophoresis, which allows each nucleotide to be read by a computer. This sequencing method is limited to DNA fragments up to 700 base pairs (bp) but with high confidence in base assignment.

Second-generation sequencing is also known as next-generation sequencing (NGS). Through this technology, the DNA is fragmented and then amplified using emulsion (Roche 454) or bridge PCR (Illumina), which allows the formation of clusters of cloned DNA fragments. Both techniques include different reading and base identification mechanisms and are characterized by massively parallel sequencing (thousands or millions of sequence clusters being analyzed simultaneously). Currently, Illumina technology dominates the market. This technology can generate billions of reads per run, but sequence lengths usually do not exceed 600 bp of length (300 bp forward and 300 bp reverse).

Third-generation sequencing, characterized by generating long sequencing reads, has Pacific biosciences (PacBio) and Oxford Nanopore technology (ONT) as its most popular technologies. Both technologies read nucleotide sequences directly without amplification, avoiding noise that can occur during amplification. As well as the second-generation technologies, they generate large volumes of data ranging from tens to hundreds of giga base pairs (gb). However, the accuracy of base calling in third-generation sequencing is lower than in second-generation sequencing. Given the recent gains in the accuracy of third-generation sequencing, third-generation technologies have shown a tendency to improve accuracy in the coming years.

sequencing base. However, during the assembly process, where the read overlap is critical to producing contigs, short reads can potentially create more fragmented genome assemblies due to difficult-to-resolve repetitive regions and their potential to produce chimeric assemblies, as short reads from the different regions of the genome may overlap during assembly. Long-read sequencing can meet this need for more significant overlapping regions, resulting in more continuous genomes. Combining the two methods has resulted in less fragmentation, greater completeness, and higher primary sequence coverage (i.e., a higher average number of times each base in a genome is sequenced).

Although there is no consensus on how much of the genome should be recovered to be considered a complete genome, most genomes, especially for non-model and neotropical species, are draft genomes (incomplete genomes with a high degree of fragmentation). Genomic drafts can clarify various questions regarding evolution and the conservation of multiple species, although genomes at the chromosomal level are required for structural analysis. However, obtaining a complete genome from telomere to telomere is challenging, as exemplified by the human genome, which was only recently come to a completion, in 2022 (Nurk et al. 2022).

10.3 How Do We Assemble a Reference Genome?

Sequencing and assembling the complete genome of a eukaryotic organism can be challenging. This is because not all organisms are equally amenable to sequencing. Genome size, the amount of repetitive DNA, and the level of heterozygosity are the main factors determining the project's feasibility. Essentially, a genome project will involve the following steps: (i) species selection and collection; (ii) library preparation and sequencing; (iii) genome assembly; (iv) genome annotation; and, potentially, (v) revisions and version updates.

The algorithms used for the genome assembly stage face specific challenges, such as dealing with repetitions, correcting reading errors, and filling gaps. Many modern assemblers address these challenges through sophisticated strategies, including read error correction, reliable overlap detection, and resolution of problematic regions.

Hundreds of genome assemblers are available; many have specificities in their genome sequence recovery strategy. However, some algorithms are widely used in some stages of these assembler pipelines (Fig. 10.2). One class of algorithms commonly used in this context is the De Bruijn Graph-Based. These algorithms divide sequence reads into smaller fragments, known as k-mers, and build a De Bruijn Graph to represent the overlaps between these k-mers. In this case, overlaps between adjacent k-mers are represented by directed edges between nodes. If two k-mers have an overlap of (k-1) bases, an edge connects the corresponding nodes. Subsequently, these algorithms attempt error correction by removing low-coverage paths and resolving ambiguous bifurcations. This way, contigs are constructed from the longest and most consistent paths in the De Bruijn Graph that connect all k-mers. Examples of assemblers that follow this approach include Velvet, SOAPdenovo, and SPAdes.

Another commonly used algorithm is the Overlapping Layout Consensus (OLC), an alternative approach to genome assembly, especially suitable for third-generation sequencing data, such as those generated by PacBio or Oxford Nanopore platforms. The basic idea behind OLC is to identify overlaps between long sequence reads to build contigs and, eventually, the complete genome (Fig. 10.2). The process begins

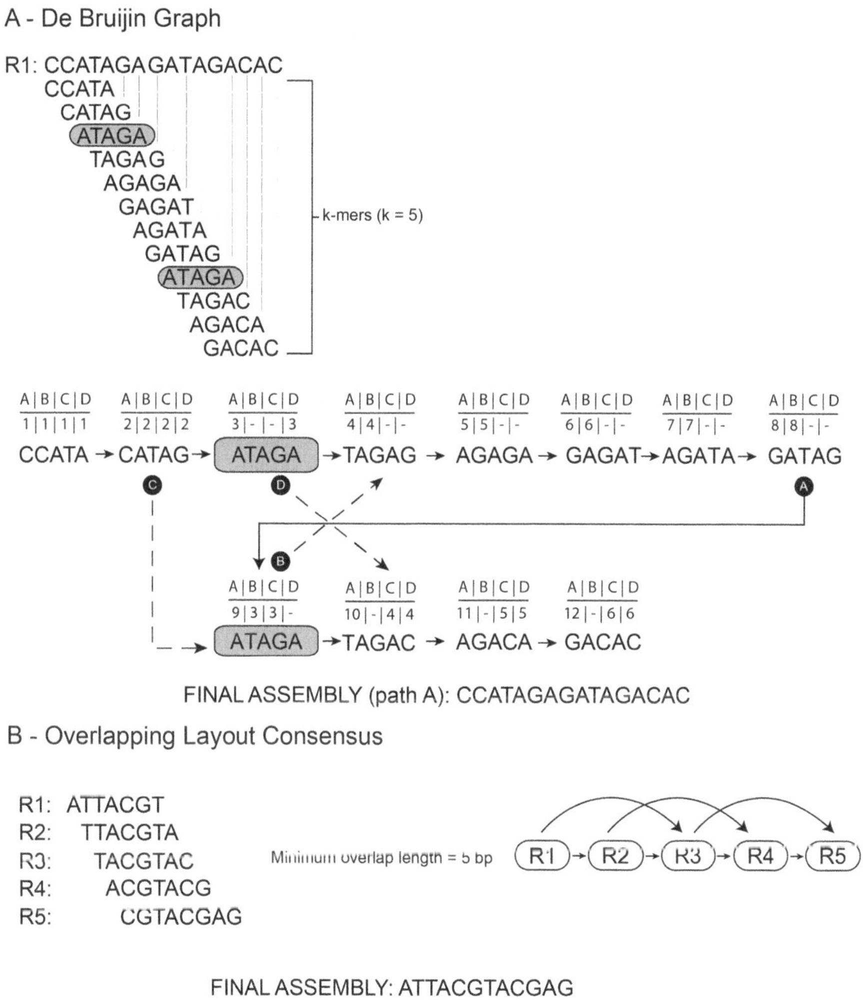

Fig. 10.2 Schematic representation of two standard algorithms used in short-read genome assembly. (**A**) De Bruijn Graph-Based Assembly – The input sequence is fragmented into k-mers (k = 5). Nodes represent k-mers, and directed edges indicate overlaps of k-1 bases between consecutive k-mers. Four possible paths (A to D) are highlighted in the figure, each following a different route to k-mer connection. These paths can be traced using the numbers in the table, which correspond to each k-mer. The final assembled sequence is reconstructed by following path A, which covers all k-mers while resolving branches and redundancies. (**B**) Overlapping Layout Consensus (OLC) Assembly – Reads are aligned based on their sequence overlaps (minimum overlap length = 5 bp) to construct a directed graph, where nodes represent reads and edges indicate significant overlaps. The final consensus sequence is determined by identifying the most consistent and contiguous path through the graph

with identifying overlaps between long sequence reads. Each read is compared to all others to find regions where they overlap. The identified overlaps are used to construct an overlap graph. In this graph, nodes represent reads and edges indicate where the reads overlap. The next step is to build a consensus layout. This involves identifying contiguous paths in the overlap graph representing consensus sequences of overlapping reads. Examples of assemblers that use the OLC approach include Canu and Flye. These assemblers are particularly useful for handling long sequence reads and overcoming challenges associated with repetitive or complex regions of the genome. The OLC approach is a valuable alternative to De Bruijn Graph-based algorithms, especially when dealing with third-generation sequencing data.

Additionally, some assemblers adopt a hybrid approach to leverage the advantages of data from different technologies (such as more prolonged or more accurate sequences). This approach aims to improve the accuracy of genome assembly, particularly in challenging regions like complex repetitive sequences. Genome assemblers like SPAdes and MaSuRCA are examples of programs that can employ this strategy.

The genome may remain incomplete even after sequencing with reads from different technologies. After the initial assembly, contigs may represent sequences from different genomic regions, and their relative positions and orientations are unknown. In this context, the Hi-C methodology has been widely employed in finalizing these genomes. Hi-C is a genomics technique used to study chromatin's three-dimensional (3D) organization in a genome. The methodology begins by capturing and fixing the chromatin conformation *in vivo*, which helps to identify cross-linked regions and the overall structure. Next, chromatin is fragmented by digestion using endonucleases. Proximity ligation is then used to generate chimeric DNA fragments. These fragments are subjected to paired-end sequencing. The resulting sequencing data is used to construct a DNA library, which facilitates a draft genome assembly mapping. Hi-C data provides information about the proximity of different DNA fragments in the 3D space of the nucleus, helping to link and order contigs into larger structures known as scaffolds.

10.4 How Can Scientists Identify the Regions Coding for Genes and Repetitive Regions in Genomes?

Once genomic data are obtained, the task is to associate this information with observable traits or specific phenotypes through annotation. Annotation involves identifying genes and repetitive sequences, elucidating their genomic loci, and proposing their putative functions. This endeavor poses significant challenges, particularly for genomes characterized by low contiguity or for species genetically distant from those whose genetic makeup has been extensively studied.

First, researchers must decide whether the genome assembly is ready for annotation or whether further sequencing efforts are needed to annotate a genome, since highly fragmented genomes may have contigs smaller than the median gene length for the species. Today, several tools are available to assess the completeness of the genome, such as estimating the N50 scaffold length (i.e., the length of the shortest scaffold containing 50% of the total assembly sequence length) and checking if the number of single-copy genes predicted for the genome is close to that found in closely related species. Secondly, researchers must decide which of the several available annotation pipelines to use.

Yandell and Ence (2012) provided an excellent beginner's guide to eukaryotic genome annotation, containing the common genome annotation steps in these pipelines described below (Fig. 10.3).

1. Repeat identification and masking. Genomic repeats include low-complexity sequences (tandem repeats) and transposable elements, which must be masked before gene annotation to avoid influencing gene identification processes. Genomic repeats are typically not conserved even within species, making their identification difficult and requiring personalized pipelines and careful post-identification curation.
2. Computational prediction of the gene location. Different strategies can be used to identify the gene sequences along the masked genome sequence. (a) *Evidence alignment.* Transcripts and protein sequences previously identified for the target species are often used to improve gene annotation, as they provide evidence of the portions of the genome that are transcribed to RNA molecules. With the same goal, protein sequences from closely related species can also be added to the pipeline. All these sequences are aligned to the masked genome to identify exon boundaries. As the most popular tool for identifying transcripts today, reads obtained from RNA sequencing (RNA-seq) strategies are frequently used. These reads can be aligned directly to the genome or assembled *de novo* into transcriptome contigs and then aligned to the genome. (b) *Ab initio* gene prediction. Genes can be identified without external evidence, using mathematical models based on organism-specific genomic features, that rely on signal sensor elements such as splice sites, branch points, start codons and stop codons, and content sensor features, related to patterns such as intron-exon lengths and codon frequency distribution, to predict the most likely gene-coding sequences. However, these features can be highly variable among organisms, even closely related species, making it easier to train the predictors with evidence. For this reason, some pipelines use pre-calculated parameters from closely related model species or from pipelines that can estimate the parameters from the available evidence for the target species. (c) *Evidence-driven gene prediction.* Some software can merge *ab initio* to external evidence to improve gene prediction accuracy.
3. Gene annotation. Once the genetic features are identified, their biological function must be elucidated. Automated strategies have entirely replaced manual genome annotation. These strategies can use only a single ab initio predictor, a consensus between several ab initio predictors, or even a full-scale annotation pipeline considering the best gene models according to external evidence.

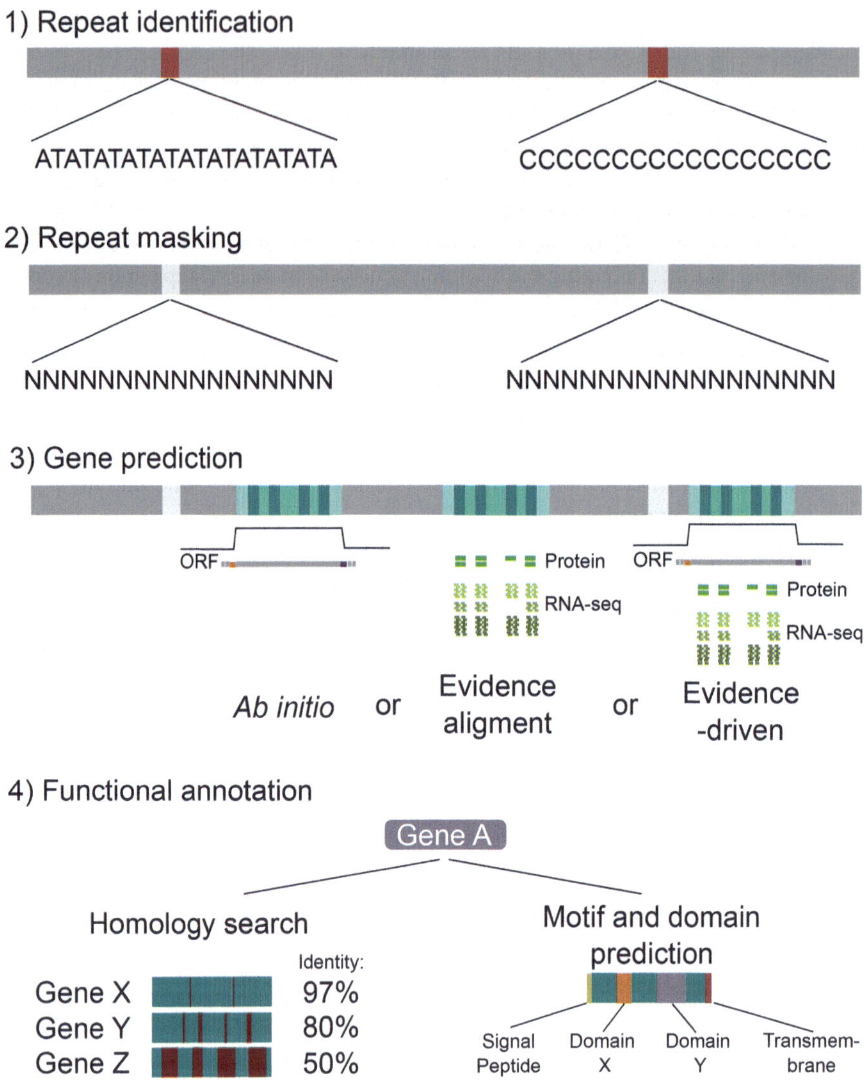

Fig. 10.3 The main steps involved in a genome annotation pipeline. (1) Repetitive region identification, which involves identifying sequences composed of consecutive identical nucleotides to perform the (2) repeat masking step, in which the identified repetitive region is masked, necessary to avoid errors and unnecessary computational effort. (3) Gene prediction can be performed by *ab initio* methods that identify Open Reading Frames (ORFs), branch points, and other genetic features indicative of putative genes; alternatively, evidence alignment methods use previously identified genes from related species to identify similar regions; and evidence-driven prediction combines these two strategies for gene prediction. (4) Functional annotation is achieved through homology or domain identification, allowing functional characterization of identified genes

10.5 Do All the Species Offer the Same Challenges to Genome Sequencing Projects?

Achieving high levels of completeness for some species is even more challenging. Factors such as large genome size, recent genome duplication (such as polyploidy), and high heterozygosity make genome assembly more complex (Fig. 10.4). The size of the genomes of target organisms is one of the primary characteristics that should be evaluated for species selection. This is because the larger the genome, the greater the sequencing effort required. To sequence a genome, the amount of raw sequencing data that have to be generated is several times larger than the genome size, a measure known as sequencing coverage (the relationship between the genome size and the amount of raw sequencing data). This makes it even more costly to sequence genomes of species with large genomes, increasing the computational demands for assembly.

When selecting new organisms for *de novo* sequencing, three approaches are recommended to estimate the genome size of the species. First, to avoid unnecessary effort to quantify this value, searching databases such as the Plant DNA C-values Database (https://cvalues.science.kew.org/search), the Animal Genome Size Database (http://www.genomesize.com/index.php), and the Fungal Genome Size Database (http://www.zbi.ee/fungal-genomesize/?q#bobx), as well as the literature is recommended. Second, if the genome size has never been estimated for the target species, experimental analyses such as flow cytometry can be used to calculate it. Finally, post-sequencing analyses can be used to estimate genome size, such as k-mer distribution analysis using the popular GenomeScope software, which can estimate heterozygosity and repeat content in addition to genome size estimation.

Another critical factor to assess when selecting a new organism for a genome sequencing project is to estimate the level of heterozygosity. Ideally, the two haplotypes are combined in a single consensus sequence in the genome assembly process (Fig. 10.2). The presence of divergent haplotypes can lead to problems in the assembly algorithms, resulting in larger genomes with possibly duplicated regions. For some organisms, such as plants, it is possible to make inferences about the level of heterozygosity since high levels of inbreeding can result in highly homozygous organisms. Selecting these organisms with higher levels of homozygosity can facilitate the algorithm for assembling a reference genome for the species.

Another bottleneck associated with the sequencing of non-model species is polyploidy, generally associated with plants but also present in the animal kingdom. The duplication of complete genomes results in larger genomes with more homologous regions. It can affect the level of heterozygosity of organisms, increasing the complexity of assembling and sequencing these genomes. It is advisable to study the cytogenetics of the sequenced organism, selecting haploid tissues or individuals, if possible. An example is the genomes of social bees, where the haploid male individuals are usually sequenced.

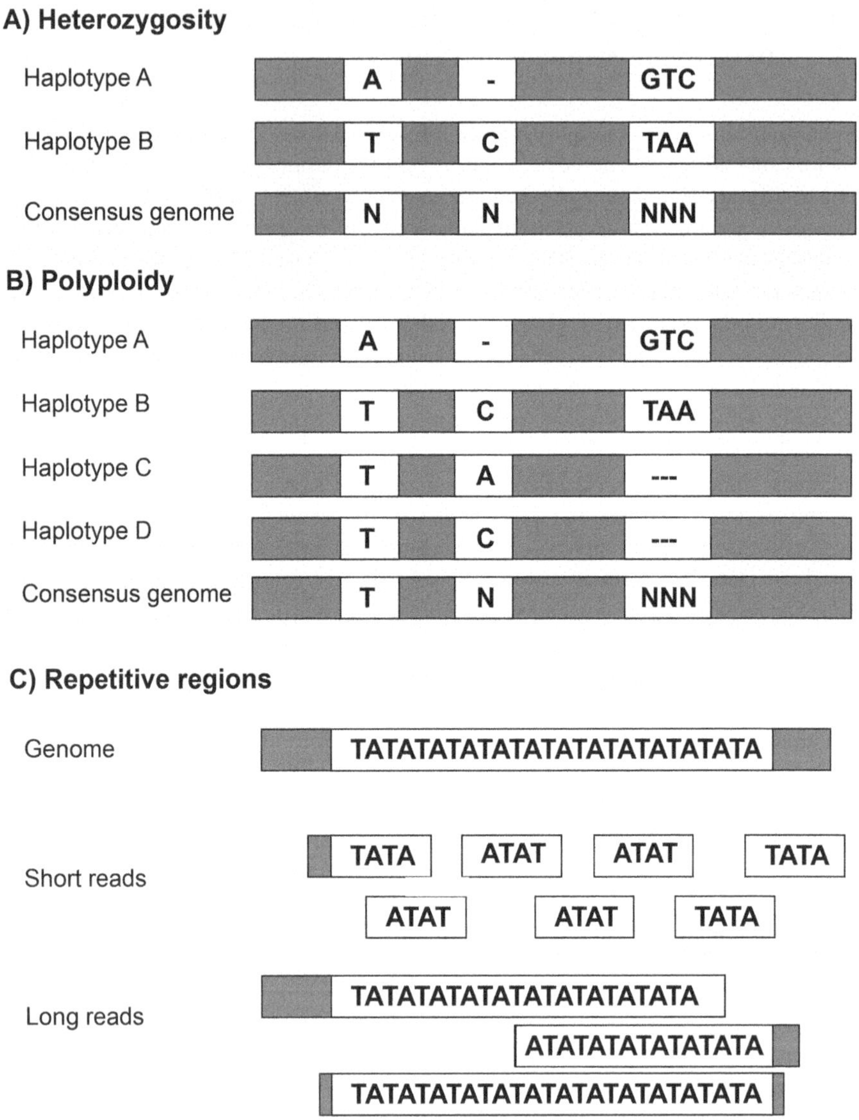

Fig. 10.4 Schematic representation of factors that make genome assembly difficult. (**a**) Genetic variation between the two alleles complicates assembly in a heterozygosity scenario because no precise consensus sequence can be established. Similarly, (**b**) polyploid species present challenges to genome assembly, as multiple homologous sequences may be present in different regions of their genome, further complicating the establishment of a consensus sequence. For genomes with high levels of (**c**) repetitive sequences, the assembly process is complicated by ambiguous overlaps and read boundaries within the sequence. The letter N in the consensus genome represents positions with base ambiguity. Different algorithms may address this ambiguity in various ways, such as using the IUPAC base coding system, where symbols like W denote uncertainty between A and T

10.6 Do We Need the Whole Genome Sequenced to Do Populational Genomics Studies?

When carrying out sequencing, one can think of four strategies: (1) sequencing the whole genome with high coverage (Whole Genome Sequencing—WGS); (2) sequencing the whole genome with low coverage per position an individual (low-coverage WGS—lcWGS) or, a pool of individuals from a population (Pool-seq); (3) sequencing the whole exome (Whole Exome Sequencing-WES); or (4) sequencing reduced representation regions, using strategies such GBS and RAD-seq (Fig. 10.1).

The preparation of the material subjected to sequencing differs between the technologies. However, they follow some basic principles: fragmentation of the initial material, linking the barcode sequences for individual identification of the samples, insertion of the sequencing adapters, and amplification of the material obtained. In the case of WES, GBS, or panels for specific regions, probes or primers enrich these sequences, thus obtaining sequences only for the areas of interest. The characteristics of each type of sequencing are described below.

10.6.1 Whole Genome Sequence

Whole genome sequencing (WGS) aims to sequence an individual's entire genome, including coding regions (exons) and non-coding regions (introns and intergenic spacer regions). Through WGS, point variations, indels, and some structural rearrangements can be identified. This type of sequencing helps to analyze and identify genetic variations in populations, contributing to population and evolutionary studies and comparative genomic studies between species.

Despite their many advantages, WGS presents significant challenges. The cost of whole genome sequencing is still high, and large genomes may require several rounds of sequencing using different strategies (short and long reads) to obtain a complete sequence. Additionally, high computational resources are needed to analyze whole genome data.

Another challenge involving large genomes is their complexity. Large genomes have many repetitive regions (such as microsatellite regions) and transposable elements. These regions do not have well-defined biological functions but are in high numbers in genomes. Even in genomes of lower complexity, large insertions and deletions with lengths greater than the short reads may be missed. During the genome assembly process, these regions can cause ambiguity, leading to assembly errors.

10.6.2 Pool-Seq and Low-Coverage Whole Genome Sequencing

In the Pool-seq strategy, the DNA from a pool of individuals is sequenced together in a unique NGS run. This strategy has proven efficient when using an optimum number of individuals per sample. However, this strategy does not allow individual-level analyses. Moreover, it also presents a challenge in distinguishing rare polymorphisms from sequencing errors since each individual contributes only to a small fraction of the DNA sequenced in the run, and polymorphisms identified in only a few individuals can easily be confounded with sequencing errors.

An option to obtain information for individual samples at a relatively low cost is to use the low-coverage WGS (lcWGS) strategy. This strategy can obtain polymorphisms from different genomic regions, even if the sequencing coverage is insufficient to cover each polymorphic position satisfactorily. In this strategy, barcodes are added separately to each DNA sample before the sequencing. In this way, it is possible to use the obtained reads to perform analysis at both the populational and individual levels.

10.6.3 Whole Exome Sequence

Whole exome sequencing (WES) consists of sequencing all the coding regions of the genome, known as the exome. These regions contain the information to make proteins, and variations in these regions can cause significant functional alterations. WES is widely used in diagnostics and is one of the main strategies used in medicine today, enabling a better cost-benefit ratio for diagnosing genetic conditions and in large-scale comparative genomic and association studies.

Exome sequencing allows the exploration of genes and alleles, regardless of the tissue, without being biased by the abundance of transcripts, as is the case with RNA-Seq. WES is an essential method for studying target traits. Compared to WGS, WES has some advantages, such as (a) the possibility of analyzing a more significant number of samples per round of sequencing, (b) a reduction in the complexity of the data to be analyzed since smaller regions will be sequenced, (c) data collected through WES can be used in population and/or evolutionary studies, and (d) it provides high coverage for identifying sequence variants.

10.6.4 GBS—Genotyping-by-Sequencing

Genotyping-by-sequencing method (GBS) involves sequencing specific loci to determine an individual's genotype. This method employs a reduction of the complexity of the genome by using restriction enzymes to fragment desired portions of the genome. After fragmentation, barcode adapters are ligated, and the reduced library is sequenced. The GBS method allows genotyping on a large scale, making it more cost-effective and enabling high-resolution genotyping. GBS can identify point mutations, small insertions, and deletions in the analyzed loci.

The GBS approach has been widely used to map SNPs in commercially essential plants to increase understanding of complex quantitative traits. Through restriction enzymes and advances in analysis pipelines, GBS has become possible even for species without a reference genome. SNPs identified through GBS are widely used for genetic diversity analysis, genome-wide association studies (GWAS), Quantitative trait loci (QTL) mapping, and genomic prediction.

10.7 Fundamental Applications of Population Genomics

As the population genomics approach has numerous applications, two are discussed below: one using neutral loci in kinship estimates and the other with outlier loci to study local adaptation.

One of the most direct contributions of population genomics is to increase the accuracy and precision of parameter estimates that require neutral loci (e.g., effective population size (*Ne*) and migration rate (*m*)) by genotyping hundreds to thousands of neutral loci in numerous individuals. The accuracy of parameter estimates will be improved because examining several loci makes it easier to identify and exclude loci under selection (outlier loci) that cause biased parameter estimates. For example, a small proportion (1 5%) of non-neutral loci can alter estimates of the average F_{ST} by 30–50% and alter evolutionary tree topology and branch lengths (Allendorf and Seeb 2000; Luikart et al. 2003; Storz 2005; Giger et al. 2006). Similarly, estimating demographic variables, such as population bottlenecks or growth rates, requires many loci to identify outliers and reliably infer the change in population size. Selection can decrease (by bottlenecks) or expand genealogies at a locus. Therefore, inferences about population growth should be more robust if outlier loci are removed, for example, using a hierarchical Bayesian model to evaluate the parameters of each locus separately.

Furthermore, increasing the number of these loci can improve the accuracy and precision of kinship estimates. The study by Santure et al. (2010) observed that just over 700 SNP markers could estimate kinship relationships equivalent to pedigree data (with genealogy already known) in zebra finches. This suggests that evaluations of correlations between phenotypes and genetic kinship, and therefore estimates of heritability, are feasible in natural populations as long as most individuals are sampled. Pedigree reconstruction enables scientists to estimate the effects of inbreeding and outbreeding on fitness, for example. Simulation data indicates that ~500,000 is enough to identify the kinship of distant categories, such as cousins (Skare et al. 2009).

One of the most promising applications of population genomics, identifying neutral and candidate loci under selection (outliers), is in studying local adaptation (adaptation to local conditions that vary in a metapopulation) (go back to Chap. 5 for more details on the matter). Detecting genomic regions responsible for local adaptation in a species depends on comparisons between related populations that may or may not be linked by ongoing gene flow. In this case, the most appropriate analyses generally focus on the differentiation in allele frequencies between populations (F_{ST}).

There are two main approaches to identifying outlier loci. The first is the relationship between population differentiation (F_{ST}) and expected heterozygosity. This approach can be applied using markers in genes that are already known and are candidates for being responsive to local adaptation, for example. In a second approach, called "genome scan", the large number of markers identified throughout the genome makes it possible to generate continuous distributions of F_{ST} values per locus along the chromosomes. Once a significance threshold has been established, loci with F_{ST} values above this threshold are considered outliers. As there is information on the position of the locus under selection on a chromosome or scaffold, it is possible to carry out a gene annotation of the region and investigate which genes or genomic components are present in the region, allowing evidence to be generated for the cause of the discrepant behavior of variation compared to the rest of the genome.

Despite the appeal of these methods, especially for non-model organisms, they need to improve from three glaring weaknesses from the point of view of ecological and evolutionary functional genomics when they are applied isolatedly. First, and perhaps most importantly, in cases where anonymous genetic markers are used to scan the genome, it is highly likely that any anonymous locus showing "outlier" behavior is not the causal locus itself but is physically linked or in linkage disequilibrium (LD) with the selected site(s). The extent of LD between the marker locus and the functionally relevant mutation can vary dramatically across the genome.

10.8 How Have Population Genomics Studies Contributed to Our Understanding of Neotropical Species?

In recent years, population genomics has experienced significant growth, providing valuable insights essential for the conservation and scientific appreciation of native species. Public databases now contain preliminary and complete genomes for several dozens of species, serving as invaluable resources for advancing population genomics analyses. Although the Neotropics are one of the most biodiverse regions in the world, population genomics studies in the region are scarce, and the number of genomes available in public databases does not commensurate with this vast diversity. However, the following examples of the use of genomic information from neotropical species through genome sequencing, whole exome sequencing, and reduced representation sequencing, show a little of the promising achievements of this field so far.

10.8.1 Whole Genome Sequencing

The genome sequences of neotropical species represent only a fraction of those available in public databases and are predominantly available as drafts. Nevertheless, draft genomes have proven to be instrumental in the development of molecular tools for species conservation. For example, the draft genomes of *Caryocar brasiliense* and *Eugenia klotzschiana*, assembled using Illumina technology (second-generation sequencing), have facilitated the construction of molecular tools for population studies, including the prediction of microsatellite primers (Nunes et al. 2019; Corvalán et al. 2023). Using a combination of Illumina paired-end and mate-pair approaches, the researchers also assembled two genomes with more than 90% coverage, for the plant *Handroanthus impetiginosus* and the stingless bee *Frieseomelitta varia* (Silva-Junior et al. 2018; de Paula Freitas et al. 2020).

Only a few genomes have been assembled using only long-read sequencing. The genome of *Ateles fusciceps* (brown-headed spider monkey) was reconstructed exclusively using Oxford Nanopore Technologies (third-generation sequencing), yielding a genome of 2.6 gigabases with more than 20× sequencing coverage (Pozo et al. 2024). The MinION sequencer used in the *A. fusciceps* genome project is characterized by a minimal initial investment, making it an attractive option for carrying out genome assembly efforts in neotropical species.

A newer method involves integrating short reads from Illumina with long reads from PacBio to create assemblies that significantly improve coverage and contiguity. This fusion of long and short reads is referred to as hybrid strategies. When comparing the use of short reads and hybrid assembly for the assembly of the *Tetragonisca angustula* (Jatai bee) genome, hybrid strategies resulted in a more contiguous genome (Ferrari et al. 2024). This approach was also used in the assembly of *Melipona bicolor*, shedding light on the evolutionary mechanisms of genes involved in venom evolution and providing insights into genome organization (Araujo et al. 2024). The enhanced continuity of genomes provided by this strategy allows for more comprehensive insights into genome organization and its impact on specific traits.

10.8.2 Whole Exome Sequencing

Exome sequencing offers a valuable opportunity to understand the interactions between parasites and their hosts, explore adaptive issues, and analyze the dynamics of genetic variation in populations. Additionally, it is a crucial tool for conserving and managing endangered species. An example is the case of the Harlequin Frogs, belonging to the genus *Atelopus*. In recent years, there has been a substantial decline in the population of these species. This decline is strongly related to infection by the fungus *Batrachochytrium dendrobatidis* (Bd), which is lethal. Some isolated populations in Costa Rica, Venezuela, Ecuador, and Panama have shown resistance to the fungus, but these populations are still small and have low genetic flow, making them vulnerable to extinction.

A study conducted by Byrne et al. (2020) analyzed the exomes of two species of amphibians of the genus *Atelopus* (*A. varius* and *A. zeteki*) in Panama, seeking to understand the genetic structure of populations and identify coding regions associated with disease resistance. Samples were collected throughout the distribution of the species in Panama, including captive colonies in the Panamá Amphibian Rescue and Conservation Project (PARC). Due to the lack of a reference genome, the researchers first assembled the transcriptome from different tissues and samples of infected animals at various stages of infection, using the Illumina HiSeq4000 platform in paired-end mode. Then, capture probes were synthesized from the assembled transcriptome to capture the coding regions, generating a reference exome for the species on the Illumina platform.

The results revealed a sharp decline in genetic diversity and effective population size of *A. varius* and *A. zeteki* populations related to *Batrachochytrium dendrobatidis* infection. Additionally, genes associated with the immune system that may be involved in resistance to infection by pathogens have been identified, providing valuable information for managing and conserving these species.

10.8.3 Reduced Representation Sequencing

Reduced representation sequencing methods, such as GBS, provide accessible approaches to sample desired genomic regions at lower cost. Using GBS, researchers assessed the diversity and genetic structure of *Copernicia prunifera* (Carnaúba palm), a palm tree that is a vital economic resource for Brazil's local people and is threatened and overexploited (Costa et al. 2022). This study analyzed 160 individuals from 14 populations scattered throughout the Caatinga biome. Genome complexity reduction was achieved using two restriction enzymes (*NsiI* and *MspI*) to digest the DNA samples, further sequenced on an Illumina HiSeq 2500 instrument. This study allowed the identification of the structure and diversity of the population of *C. prunifera*, allowing the detection of several polymorphisms, some of which bore loci putatively under selection. The identification of this molecular diversity is essential for the development of conservation strategies to preserve the genetic wealth of the population. As a high genetic diversity was observed, it was indicated that large conservation areas must be dedicated to carrying out in situ conservation strategies to gather all the existing diversity within this population.

10.9 Final Remarks

Population genomic studies have become reliable means to propose several inferences in the environmental and biological landscape. The advances in genomic resource gathering, the detailed information within the genetic composition of species can be explored and translated into practical data that enables decision-making regarding several applications (Rajora Om 2019).

As this genomic data collection makes progress, opportunities for further exploration emerge, facilitating the adoption of more precise, faster, and newer methodologies for analyzing variations and diversity. Analytical tasks such as genotyping, identifying genetically informative regions, neutral loci, and other variations have become increasingly feasible due to advancements in sequencing technology and the availability of extensively reviewed public data. Therefore, promoting state-of-the-art advancements related to the current generation of genomic resources is crucial. This promotion facilitates reliable data communication by making libraries and genomic data available in public databases.

The continuous evolution of genomic data also propels the development of bioinformatics and statistical analytical tools associated with its processing. These tools ensure high data quality and enable error-free analysis. Consequently, new methodologies can emerge, transforming other genetic features into practical inferences with biological significance.

The trend in population genomics research is now toward developing an interdisciplinary field that integrates data from the omics science spectrum with spatial and environmental variables. Therefore, it is essential for researchers working with neotropical species to become more familiar with bioinformatics tools and population theory to navigate this interdisciplinary landscape effectively.

Key Points

- Population genomic analyses are concerned with large-scale genetic variation and diversity within and among organisms;
- The general workflow for developing new reference genomic resources consists of sample collection, DNA extraction and sequencing, genome assembly, and annotation;
- Population genomic studies require genetic data that can be obtained by sequencing strategies sensitive to research goals and cost trade-offs;
- Results to date have demonstrated the power of population genomics to improve understanding and conservation efforts in the Neotropics.

Questions

1. Using Fig. 10.1 as a reference, make a list of the variants detected by each genomic population approach and argue on the advantages and disadvantages of each strategy.
2. Why is the choice of a target species critical to the design of genomic research? What are the challenges associated with the genomes of some species?
3. How may population genomics revolutionize the study of neotropical species?

References

Allendorf FW, Seeb LW (2000) Concordance of genetic divergence among sockeye salmon populations at allozyme, nuclear DNA, and mitochondrial DNA markers. Evolution 54:640–651. https://doi.org/10.1111/j.0014-3820.2000.tb00065.x

Araujo NS, Ogihara F, Martins PM, Arias MC (2024) Insights from Melipona bicolor hybrid genome assembly: a stingless bee genome with chromosome-level scaffold. *BMC Genomics 25*(1):1–13. https://doi.org/10.1186/s12864-024-10075-x

Byrne AQ, Richards-Zawacki CL, Voyles J et al (2020) Whole exome sequencing identifies the potential for genetic rescue in iconic and critically endangered Panamanian harlequin frogs. Glob Change Biol 27:50–70. https://doi.org/10.1111/gcb.15405

Corvalán LCJ, Carvalho LR, Melo-Ximenes AA et al (2023) Data of SSRs primers for high-throughput genotyping-by-sequencing (SSR-Seq) based on the partial genome assembly of *Eugenia klotzschiana* (Myrtaceae). Data Brief 47:108917. https://doi.org/10.1016/j.dib.2023.108917

Costa MF, Morales-Marroquín JA, de Araújo Batista CE et al (2022) Population genomics of the neotropical palm Copernicia prunifera (Miller) H. E. Moore: implications for conservation. PLoS One 17(11):e0276408. https://doi.org/10.1371/journal.pone.0276408

de Paula Freitas FC, Lourenço AP, Nunes FMF et al (2020) The nuclear and mitochondrial genomes of Frieseomelitta varia—a highly eusocial stingless bee (Meliponini) with a permanently sterile worker caste. BMC Genomics 21(1):386. https://doi.org/10.1186/s12864-020-06784-8

Ferrari RR, Ricardo PC, Dias FC et al (2024) The complete genome of *Tetragonisca angustula* (Apidae: Meliponini), the most commonly reared stingless bee in Brazil. PREPRINT (Version 1) available at Research Square. https://doi.org/10.21203/rs.3.rs-3833242/v1

Giger T, Excoffier L, Philip JRD et al (2006) Life history shapes gene expression in salmonids. Curr Biol 16:R281–R282. https://doi.org/10.1016/j.cub.2006.03.053

Luikart GH, England P, Tallmon DA et al (2003) The power and promise of population genomics: from genotyping to genome-typing. Nat Rev Genet 4:981–994. https://doi.org/10.1038/nrg1226

Maxam AM, Gilbert W (1977) A new method for sequencing DNA. Proc Natl Acad Sci USA 74(2):560–564. https://doi.org/10.1073/pnas.74.2.560

Nunes R, Gonçalves AR, Telles MPC (2019) Data on the draft genome sequence of *Caryocar brasiliense* Camb. (Caryocaraceae): an important genetic resource from Brazilian savannas. Data Brief 26:104543. https://doi.org/10.1016/j.dib.2019.104543

Nurk S, Koren S, Rhie A et al (2022) The complete sequence of a human genome. Science 376(6588):44–53. https://doi.org/10.1126/science.abj6987

Pozo G, Albuja-Quintana M, Larreátegui L, Gutiérrez B, Fuentes N et al (2024) First whole-genome sequence and assembly of the Ecuadorian brown-headed spider monkey (*Ateles fusciceps* fusciceps), a critically endangered species, using Oxford Nanopore Technologies. G3 Genes Genom Genet 14(3):1. https://doi.org/10.1093/g3journal/jkae014

Rajora Om P (2019) Population genomics: advancing understanding of nature. In: Luikart G, Kardos, Hand BK, Rajora Omp P, Aitken SN, Hohenlohe PA (eds) Population Genomics - Concepts, Approaches and Applications. Springer, pp 3–82

Sanger F, Nicklen S, Coulson AR (1977) DNA sequencing with chain-terminating inhibitors. Proc Natl Acad Sci USA 74(12):5463–5467. https://doi.org/10.1073/pnas.74.12.5463

Storz JF (2005) Using genome scans of DNA polymorphism to infer adaptive population divergence. Mol Ecol 14:671–688. https://doi.org/10.1111/j.1365-294X.2005.02437.x

Santure AW, Stapley J, Ball AD et al (2010) On the use of large marker panels to estimate inbreeding and relatedness: empirical and simulation studies of a pedigreed zebra finch population typed at 771 SNPs. Mol Ecol 19:1439–1451. https://doi.org/10.1111/j.1365-294X.2010.04554.x

Silva-Junior OB, Grattapaglia D, Novaes E, Collevatt RG (2018) Genome assembly of the pink Ipê (*Handroanthus impetiginosus,* Bignoniaceae), a highly valued, ecologically keystone Neotropical timber forest tree. GigaScience 1(7):1–16. https://doi.org/10.1093/gigascience/gix125

Skare O, Sheehan N, Egeland T (2009) Identification of distant family relationships. Bioinformatics 25:2376–2382. https://doi.org/10.1093/bioinformatics/btp418

Yandell M, Ence D (2012) A beginner's guide to eukaryotic genome annotation. Nat Rev Genet 13(5):329–342. https://doi.org/10.1038/nrg3174

Chapter 11
How to Integrate Population Genetics with Ecological Modeling?

Gustavo Reis de Brito

Contents

Abstract The Neotropics realm is a significant biogeographic region in tropical and subtropical areas of the Americas, with high biodiversity and endemism. This region serves as a prime example of how genetics and ecology intersect, especially when considering that genetic diversity in this region plays a crucial role in shaping ecological processes and species interactions. The genetic makeup of species within the Neotropics also influences their ecological roles and functions: genetic variation

G. R. de Brito (✉)
Departamento de Ciências Biológicas, Universidade Estadual Paulista "Júlio de Mesquita Filho" - UNESP, Faculdade de Ciências e Letras, Campus de Assis, Assis, SP, Brazil
e-mail: reis.brito@unesp.br

M. V. Bohrer Monteiro Siqueira et al. (eds.), *Population Genetics in the Neotropics*,
https://doi.org/10.1007/978-3-031-83685-5_11

can affect traits such as reproductive strategies and physiological adaptations, which in turn influence the ecological relationships and species interactions. The intricate relation between genetics and ecology highlights the importance of preserving both the genetic diversity and the ecological integrity of the landscape. Fortunately, with all the data available on the Neotropics, computational approaches such as ecological niche modeling (ENM) can be used to uncover insights into how environmental factors and historical events have influenced the distribution of populations and their genetic diversity, especially under climate change scenarios. ENM provides means for identifying barriers to gene flow, areas of historical isolation, and test biogeographical hypotheses on how environmental factors have affected the genetic diversity of populations. This chapter is aimed at introducing the core concepts of ecological niche models, their intersection with population genetics studies, and applications of ENMs in conservation efforts and biodiversity management studies in the Neotropics to ensure the long-term viability and sustainability of the neotropical ecosystems in the face of human interference and environmental challenges.

Keywords Ecology · Ecological niche models · Conservation genetics · Neotropical ecology · Population genetics

Learning Objectives
1. To understand the interplay between population genetics and ecological processes within the Neotropics;
2. To investigate how genetic diversity shapes ecological interactions and functions, impacting the roles of species in the neotropical region;
3. To learn about Ecological Niche Modeling (ENM) and its applications in predicting species distribution based on environmental factors;
4. To recognize the significance of integrating ENMs with population genetics studies for uncovering patterns in the distribution of genetic diversity;
5. To understand the broader implications of the integrated approach for conservation efforts, biodiversity management, and the long-term viability of neotropical ecosystems.

11.1 Introduction

The Neotropics, an extensive biogeographic region spanning the tropical and subtropical zones of the Americas, serves as a compelling testament to the remarkable diversity of life on our planet, standing as a biodiversity hotspot of global significance (Myers et al. 2000). At its core lies a mosaic of ecosystems, ranging from the Amazon rainforest to the arid landscapes of the Andes, hosting a diverse array of plant and animal species, many of them exclusive to each geographic domain. Its unique geological history and climatic conditions have given rise to diverse habitats, each fostering distinct biological communities (Myers et al. 2000; Mittermeier et al. 2004; Antonelli et al. 2018).

Beneath the biological abundance of the Neotropics lies a complex puzzle. Questions regarding how genetic patterns within populations influence broader ecological dynamics and the role genetic diversity plays in shaping interactions between species and their environments emerge and, to address such questions, the convergence of population genetics and ecological modeling offers a compelling approach to study the Neotropics. Basically, this chapter acknowledges that genetics and ecology are inherently linked. The genetic composition of populations not only influences their survival but also defines their roles within the intricate web of ecological relationships. To navigate through the intersection of population genetics and ecological modeling, this chapter starts with an exploration of the fundamental connection between genetics and ecology, setting the stage for the integration of population genetics with ecological modeling, focusing on the potential applications of Ecological Niche Models to demonstrate how this integrated approach can contribute to the conservation and sustainable management of the Neotropics amid ongoing environmental challenges.

11.2 Genetics and Ecology: The Influence of Genetic Diversity in Ecological Processes and Interactions

The Neotropics, with their expansive biogeographic reach and extraordinary biodiversity, serve as an ideal canvas for exploring the profound impact of genetic diversity on ecological processes and interactions. Genetic diversity, the variability within and among populations in terms of their genetic makeup, plays a pivotal role in shaping the dynamics of ecosystems (Hughes et al. 2008).

Understanding the link between genetic diversity and ecological processes begins with recognizing the intricate connections that exist at the genetic level. The genetic makeup of populations influences a myriad of traits, from reproductive strategies to physiological adaptations. These traits, in turn, dictate how species interact with their environment and with each other (e.g., Leimu et al. 2006; Whitham et al. 2006; Franks et al. 2007).

Take, for instance, the diverse array of plant species found in the Neotropics. The genetic diversity within these plant populations can determine factors such as growth patterns, reproductive strategies, and resistance to diseases. Such genetic differences, subtle as they may seem, have cascading effects on the ecological interactions within the ecosystem (e.g., the types of herbivores that feed on the plants). Similarly, in animal populations, genetic diversity can determine traits related to foraging behaviors, migration patterns, and even communication methods.

As the intricate relationship between genetic diversity and ecological processes is revealed, it becomes evident that preserving the rich genetic background of the Neotropics is not only a matter of biological conservation but a key strategy for maintaining the health and vigor of their ecosystems. The genetic nuances within populations contribute to the resilience of species in the face of environmental challenges, ensuring their ability to adapt and thrive (Kardos et al. 2021).

Genetic makeup, the pool of genes within populations, serves as the blueprint for the traits exhibited by individuals. These traits, ranging from morphological features to behavioral patterns, are not merely expressions of genetic information but are integral components of ecological roles played by species within their ecosystems (e.g., Li et al. 2021). Consider the diverse array of bird species that populate the Neotropics: the genetic makeup of these avian populations dictates traits such as beak shape, coloration, and mating behaviors, traits that affect the ecological roles these birds play. For instance, birds with different beak shapes may specialize in feeding on distinct types of food sources, influencing the distribution of plant seeds and the composition of plant communities (e.g., Melo et al. 2020).

Examples of the interplay between population genetics and ecology in biodiversity show that preservation of genetic diversity is synonymous with safeguarding ecological roles (Kardos et al. 2021). Each genetic variation represents a thread in the ecological tapestry, contributing to the resilience and adaptability of populations. Understanding the impact of the genetic makeup on traits and ecological roles is not just an academic pursuit: it is a roadmap for conservation strategies that seek to maintain the delicate balance of species interactions within a rich and vast region known as the Neotropics.

11.3 Ecological Niche Models

11.3.1 Historical Background and Definitions

Ecological Niche Models (ENMs) emerged as powerful tools in ecology and conservation biology around the year 2000 (Guisan and Zimmermann 2000), providing a means to understand and predict environmental suitability and species distribution patterns in response to environmental variables. Since the 2000s, ENMs have become a valuable tool in ecological and conservation studies (Guisan and Thuiller 2005; Elith and Leathwick 2009; Peterson 2011; Franklin 2010).

At its core, an ENM is a mathematical approximation (i.e., a model) that tries to correlate the occurrence of a species with environmental variables in the locations where it occurs. This process quantifies the ecological niche of a species—the realized niche, mapping the environmental suitability for a species (Sillero et al. 2021). With the rise of computational power, much software and many routines have been developed, allowing ENMs to be widespread in various fields, including biodiversity conservation, climate change impact assessment, and species-environment relationships.

Developing an ENM is a simple, yet complex process. There are many possibilities of choice when looking for computer-assisted tools (e.g., programming languages such as R and Python); the literature surrounding ENMs is quite rich, too.

But almost limitless access to information can lead to confusion (Peterson and Soberon 2012; Sillero and Barbosa 2021). An ENM routine is based on a simple framework:

- Modeling conceptualization is based on the target species and research questions;
- Collection of occurrence data (e.g., fieldwork, online databases);
- Defining which modeling algorithms will be used (e.g., Maxent, GLM, Random Forests);
- Model fitting;
- Statistical validation of the models based on statistical metrics (e.g., AUC, TSS, Kappa);
- Model projection.

Although the framework presented here looks basic, it is important to know that each step of the process has its own challenges, and every model is tailored to address specific questions. Always review the literature to avoid "modeling pitfalls" (e.g., Peterson and Soberon 2012; Sillero and Barbosa 2021; Sillero et al. 2021).

11.3.2 Key Components of ENMs

Environmental Variables
ENMs use a range of environmental variables, such as temperature, precipitation, elevation, and land cover, to characterize the ecological conditions associated with the presence or absence of a species. Given that there are many sources for environmental variables, these can appear in different formats, sizes, and resolutions. Therefore, the selection and standardization of these variables are crucial, as they directly influence the accuracy and reliability of the model predictions (Sillero et al. 2021).

Species Occurrence Data
Input data for ENMs include records of species occurrence, compiled from field surveys, museum collections, online databases, research papers, or even citizen science initiatives. These data points serve as the foundation for the model, allowing it to identify patterns and associations (i.e., correlations) between species and environmental conditions (Sillero et al. 2021). As with environmental variables, occurrence data can influence the modeling process and must be checked for errors and biases (Sillero et al. 2021).

Model Algorithms
Various algorithms are employed to develop ENMs, each with their strengths and limitations. Common algorithms include Maxent (Maximum Entropy) (Phillips et al. 2006), Random Forest (Breiman 2001), and Generalized Linear Models (McCullagh and Nelder 1998; Venables and Ripley 1999). These algorithms use statistical relationships between environmental variables and species occurrences to

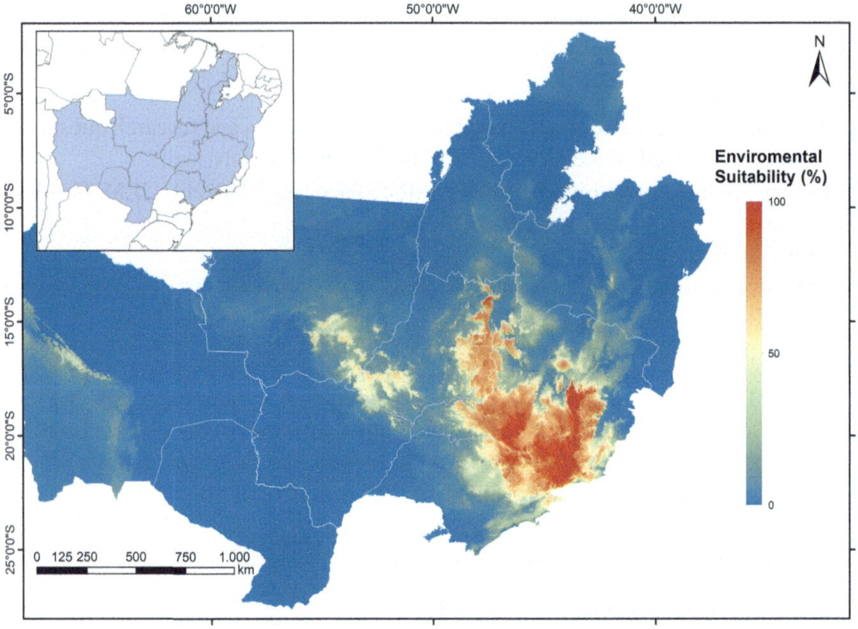

Fig. 11.1 Example of a suitability map as seen in Meireles et al. (2023)

generate predictive maps (e.g., probability of occurrence, habitat suitability) (Fig. 11.1).

It is important to note that models can be trained using a single algorithm or a combination of two or more (i.e., model ensembling), with the selection depending on what is the target of the model: presence-absence methods (e.g., Generalized Linear Models) are used to predict potential occurrences; presence-only and presence-background (e.g., Bioclim, Maxent) are used to model habitat suitability, climate relations, and also potential distributions. Choosing an algorithm is a crucial task and the literature on ENMs has been well developed since the advent of Bioclim (Busby 1991), with several applications being described over the years.

11.3.3 Applications of ENMs

Conservation Planning

ENMs play a crucial role in conservation planning by identifying suitable habitats for endangered or vulnerable species. These models assist in the design of protected areas and restoration efforts, aiming to ensure the persistence of biodiversity in the face of habitat loss and climate change (e.g., Marini et al. 2010; Gonçalves et al. 2021).

Climate Change Impact Assessment
With the ongoing changes in global climate patterns, ENMs are valuable tools for assessing the potential impacts on species distributions. By projecting how species' ranges may shift in response to climate change, researchers and conservationists can prioritize areas for intervention and management (e.g., Kolanowska et al. 2020; De Oliveira et al. 2024).

Invasive Species Management
ENMs are instrumental in predicting the potential spread of invasive species. Understanding the ecological conditions that facilitate invasions enables proactive management strategies to mitigate the impact of invasive species on native ecosystems (e.g., Santos and Fabricante 2018; Lopes et al. 2023).

Understanding Ecological Interactions
ENMs contribute to our understanding of ecological interactions by revealing the environmental factors influencing species coexistence and competition. These insights can inform ecological studies and guide field research (e.g., Mercado Gómez et al. 2020; Preuss and Padial 2021).

In conclusion, ecological niche models represent a sophisticated approach to studying species-environment relationships. Their versatility and applicability make them indispensable tools for researchers and practitioners working towards the conservation and sustainable management of biodiversity in a rapidly changing world.

11.4 Integration of Population Genetics and ENMs

The Neotropics, with their vast array of species and intricate ecological dynamics, beckons researchers to explore the juncture where genetic insights and ecological realities converge. This intersection is particularly vibrant when considering the synergies between population genetics studies and the computational prowess of ENMs. Together, these approaches illuminate the complex relationships between genetic diversity, environmental factors, and the distribution of populations.

At the heart of population genetics studies lies the exploration of genetic variation within and among populations. Studies on this subject seek to understand the evolutionary forces shaping the gene pools of species, including factors such as gene flow, genetic drift, and natural selection. Given the complexity of these studies, the intersection between population genetics and ENMs enhances our comprehension of how genetics and ecology intersect.

These models can offer a complementary perspective by predicting the spatial distribution of species (i.e., the realized niche) based on its correlation with environmental variables. When integrated with population genetics data, ENMs become powerful tools for uncovering the drivers behind the observed genetic patterns. The synergy lies in the ability of ENMs to unravel the environmental factors that have influenced the distribution of populations, acting as a bridge between genetic variations and ecological contexts.

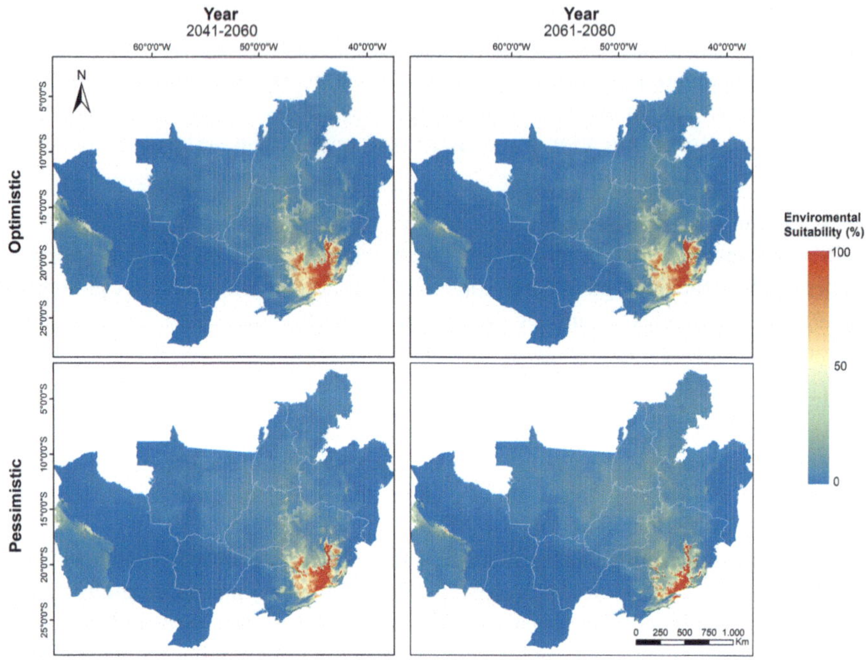

Fig. 11.2 Examples of future suitability maps project under two climate scenarios (optimistic and pessimistic) as seen in Meireles et al. (2023)

For instance, consider a neotropical plant species scattered across diverse habitats. Population genetics studies may reveal patterns of genetic diversity within these populations, highlighting regions of historical isolation or signatures of adaptation to specific environmental conditions. Concurrently, ENMs can provide predictive maps displaying the suitability of habitats for this species based on current, future, and past environmental parameters (Fig. 11.2).

The intersection becomes particularly profound when ENMs are used to model the potential distribution of populations under different climate change scenarios. By overlaying the predictions of changing habitats with genetic data, researchers gain insights into how populations may respond to environmental shifts. This proactive approach allows for the identification of vulnerable populations, potential areas of genetic bottleneck, and the formulation of conservation strategies tailored to ensure the resilience of species in the face of ongoing environmental changes. Moreover, the integration of population genetics and ENMs contributes to our understanding of how historical events have shaped the genetic diversity of Neotropical populations. Whether it is the influence of past climatic fluctuations, the formation of geographical barriers, or the interactions between species, the combined insights offer a holistic perspective on the evolutionary journey of biodiversity in the Neotropics.

The intersection of population genetics studies and ENMs in the Neotropics creates a dynamic synergy, where genetic insights are contextualized within the ecological theater (e.g., De Alencar et al. 2018). This approach not only advances our scientific understanding of the region's biodiversity but also holds practical implications for conservation (e.g., Valencia-Rodríguez et al. 2021). By harnessing the collective power of genetic studies and predictive modeling, researchers can unlock a deeper comprehension of how species respond to environmental challenges, guiding conservation efforts in preserving the genetic diversity and ecological integrity of the Neotropics.

11.4.1 Identifying Barriers to Gene Flow

One of the key challenges for populations is the movement of genes across landscapes. ENMs, by integrating environmental variables, aid in identifying barriers to gene flow—geographical features or environmental conditions that limit the exchange of genetic material between populations. For example, a mountain range or a river might act as a barrier, influencing the distribution of genetic diversity (e.g., Muell and Brown 2024). By pinpointing these barriers, researchers gain insights into the factors shaping the genetic structure of populations, fostering a deeper understanding of how landscapes influence gene flow.

11.4.2 Detecting Areas of Historical Isolation

ENMs shine as temporal travelers, delving into the historical isolation of populations. By modeling the distribution of species across time (e.g., Myers et al. 2015), these models can unveil regions that served as refugia during climatic upheavals. Such refugia, whether in isolated mountain valleys or persistent patches of suitable habitat, become reservoirs of genetic diversity. ENMs, when coupled with genetic data, can help discerning the imprints of historical isolation, guiding researchers that study neotropical regions through the evolutionary journey of populations (e.g., Raposo do Amaral et al. 2021; Cooper 2024).

11.4.3 Testing Biogeographical Hypotheses

The Neotropics, with their complex geological and climatic history, offer a fascinating opportunity for testing biogeographical hypotheses (e.g., historical migration routes, colonization patterns, impact of past environmental changes on the distribution of genetic diversity). Given that the focus of ENMs is modeling environmental suitability, which directly affects species distribution, biogeographical and population genetics studies can intersect with the help of the models (e.g., Leal et al. 2019; López-Caamal et al. 2019).

11.5 Challenges and Future Directions

One key point to remember is that, when integrating insights from ENMs and population genetics, it is crucial to navigate the nuanced terrain of challenges and limitations inherent in this comprehensive approach. Acknowledging these complexities is pivotal in refining methodologies, fostering a more robust scientific foundation, and charting a course forward that respects the intricacies of a biogeographically significant region such as the Neotropics.

One prominent challenge lies in the data requirements for both ENMs and population genetics studies. Accurate predictions from ENMs demand extensive and high-quality environmental data (Sillero et al. 2021), which may be lacking in certain regions of the Neotropics. Similarly, population genetics studies require comprehensive genetic datasets, posing logistical challenges in collecting representative samples across vast and often remote landscapes. Addressing these data limitations requires collaborative efforts, technological advancements, and a commitment to expanding data repositories to ensure the reliability of findings.

Modeling future scenarios, especially in the context of climate change, introduces another layer of complexity. ENMs predict potential shifts in species distributions based on current environmental conditions, assuming species will track their preferred climatic niches (i.e., niche conservatism). However, these models may not capture the full range of adaptive responses or the complexities of species interactions (Lissovsky et al. 2021). Incorporating dynamic ecological processes and improving the predictive capacity of models under changing environmental conditions is an ongoing challenge that requires continuous refinement.

The integration of ENMs and population genetics also encounters challenges in the interpretation of results. Correlating patterns identified by ENMs with genetic data requires careful consideration of ecological and evolutionary processes. For instance, a predicted barrier to gene flow may be a consequence of historical events or ecological factors rather than a true genetic barrier.

Furthermore, the potential for oversimplification in both modeling approaches poses a limitation. ENMs often assume species respond uniformly to environmental variables, neglecting individual variations within populations. Similarly, population genetics studies may oversimplify the genetic structure by focusing on neutral markers, potentially overlooking adaptations that contribute to species resilience. Striking a balance between model simplicity and ecological/genetic realism is an ongoing challenge.

Ethical considerations regarding the impact of data collection on species and ecosystems present yet another challenge. Population genetics studies may involve invasive sampling, while intensive data collection for ENMs can have ecological implications. Balancing the benefits of scientific knowledge with ethical concerns requires a thoughtful and responsible approach, incorporating principles of conservation ethics and sustainability.

In navigating these challenges and limitations, it is crucial to view them not as roadblocks but as opportunities for refinement and growth. Collaborative efforts between scientists, conservationists, and stakeholders, along with advancements in technology and data accessibility, can address many of these challenges. The journey to unravel the complexities of the Neotropics demands a commitment to continuous improvement, transparency, and a holistic understanding of the intricate web of ecological and genetic interactions that shape this remarkable region.

11.5.1 Fine-Scale Environmental Data Integration

Future research could focus on refining the resolution of environmental data integrated into ENMs. High-resolution data would provide a more accurate representation of the complex and heterogeneous landscapes within the Neotropics. Advanced remote sensing technologies, satellite imagery, and on-the-ground data collection methods could be harnessed to enhance the precision of environmental predictors. This would enable more detailed insights into species distributions, facilitating the identification of microhabitat preferences critical for certain species.

11.5.2 Incorporating Genomic Data and Functional Traits

Advances in genetic technologies offer an exciting avenue for future research. Integrating genomic data into population genetics studies would provide a more comprehensive understanding of the adaptive potential of species. Exploring functional traits related to ecological roles and responses to environmental changes can augment ENMs, facilitating mechanistic approaches and capturing a more nuanced picture of species' ecological niches.

11.5.3 Dynamic Modeling Under Climate Change Scenarios

As the impacts of climate change intensify, future research could enhance the dynamic modeling capabilities of ENMs. Incorporating climate change scenarios with a focus on temporal dynamics would provide more realistic predictions of species distributions under changing conditions. Dynamic models that consider evolutionary processes, such as adaptation and dispersal, can offer a more comprehensive understanding of how species respond to environmental shifts over time. This is especially crucial for informing conservation strategies aimed at preserving biodiversity in the face of ongoing climate change.

11.5.4 Integrating Socio-Ecological Factors

An interdisciplinary approach that incorporates socio-ecological factors would enrich the integrated framework. Understanding how human activities, land-use changes, and socio-economic factors influence both ecological processes and genetic diversity is essential. Research could explore the reciprocal interactions between human communities and neotropical ecosystems, considering the implications for biodiversity conservation and sustainable resource management.

11.5.5 Ethical Considerations and Community Engagement

Future research should place a heightened emphasis on ethical considerations in data collection and research practices. Collaborative efforts with local communities and indigenous peoples' groups can foster mutual understanding and respect. Engaging different ethnic groups in the research process not only contributes to the ethical conduct of studies but also enriches the knowledge base by incorporating traditional ecological knowledge and cultural perspectives.

11.5.6 Cross-Species and Ecosystem-Level Interactions

Expanding the scope of research to explore cross-species interactions and ecosystem-level dynamics would provide a holistic understanding of the Neotropics' biodiversity. Investigating how distinct species within an ecosystem interact, including mutualistic relationships, trophic cascades, and competition dynamics, enhances our comprehension of the intricate ecological web. Such investigations can contribute to the development of more inclusive conservation strategies that consider the broader ecosystem context.

In charting the future of neotropical research, collaboration among scientists, policymakers, local communities, and stakeholders is paramount. Establishing research networks that span geographical boundaries and disciplinary domains fosters a holistic and inclusive approach. As science continues to unravel the complexities of the Neotropics, the dedication to innovation, ethical conduct, and collaborative exploration will propel researchers toward a more profound understanding of this biogeographic treasure, helping navigate the challenges of conservation and sustainable biodiversity management in the decades to come.

11.6 Final Remarks

Presenting itself as an expansive biogeographic region graced by lush rainforests, diverse ecosystems, and unparalleled biodiversity, the Neotropics are, without a question, an important hotspot for global biodiversity. Through the lens of population genetics studies, it is possible to study and better understand the genetic intricacies that dictate the survival and evolution of species within this dynamic realm. The integration of ENMs adds a layer of predictive power, offering insights into the spatial distribution of populations based on environmental factors. Together, these approaches can help uncover the hidden patterns of gene flow, historical isolation, and the nuanced interplay between genetics and ecology.

This chapter emphasized the fundamental connection between genetics and ecology, portraying them as two sides of the same coin shaping the destiny of species in the Neotropics. Genetic makeup, from reproductive strategies to physiological adaptations, emerged as a key influencer of ecological roles and functions. The application of ENMs illuminated the potential impacts of environmental factors and historical events on the distribution of genetic diversity, providing a comprehensive understanding of how the Neotropics' biodiversity has been sculpted over time.

The significance of the integrated approaches is evident: the interplay of population genetics studies and ENMs not only enriches the scientific understanding of the Neotropics (and other regions) but emerges as a pragmatic tool for conservation and biodiversity management. This approach encourages scientists, policymakers, and conservationists to collaborate by identifying conservation priorities, informing targeted strategies, and predicting species responses to environmental changes, furthering the efforts to better understand and preserve this natural treasure called the Neotropics.

Key Points

- The complexity of the Neotropics requires holistic approaches that transcend individual disciplines: combining genetics and ecology is a strong combination for effective conservation and management.
- ENMs have become significant tools in ecology and conservation biology since the early 2000s, aiding in the prediction of species distributions based on environmental variables. Functioning as mathematical models that correlate species occurrences with environmental factors, they allow for the mapping of ecological niches and environmental suitability, and are widely used in biodiversity conservation, climate change impact assessments, and understanding species-environment relationships, making them versatile tools in ecological research.
- Understanding how genetic diversity influences ecological processes, species interactions and distributions is crucial, particularly in the context of the Neotropics.
- Each genetic variation within a population contributes to the ecological tapestry, enhancing the resilience and adaptability of species in changing environments.

The integration of ENMs in studies of population genetics can help identify conservation priorities, inform targeted strategies, and predict species responses to environmental changes.

- The combination of ENMs and population genetics studies is a strong framework for complex and rich regions such as the Neotropics, allowing not only for the comprehension of how genetics and ecology intersect but also facilitating conservation planning and policy making.

Questions

1. What are the historical developments that led to the emergence of Ecological Niche Models (ENMs) as tools in ecology and conservation biology?
2. How does genetic diversity contribute to the resilience and adaptability of populations within the Neotropics?
3. In what ways can the integration of population genetics and ENMs inform conservation strategies and biodiversity management?
4. What challenges are associated with integrating fine-scale environmental data and genomic information into ecological modeling?
5. How can understanding the interplay between genetics and ecology enhance our ability to predict species responses to climate change and other environmental pressures?

References

Amaral RP, Thom G, Lima-Ribeiro MS et al (2021) Rugged relief and climate promote isolation and divergence between two neotropical cold-associated birds. Evolution 75(10):2371–2387. https://doi.org/10.1111/evo.14318

Antonelli A, Zizka A, Carvalho FA et al (2018) Amazonia is the primary source of Neotropical biodiversity. Proc Natl Acad Sci 115:6034–6039. https://doi.org/10.1073/pnas.1713819115

Breiman L (2001) Random forests. Mach Learn 45:5–32. https://doi.org/10.1023/A:1010933404324

Busby, J.R. (1991) BIOCLIM – a bioclimate analysis and prediction system. *Plant Protection Quarterly, 6*, 8–9

Cooper JC (2024) Ecological niche divergence or ecological niche partitioning in a widespread neotropical bird lineage. PeerJ 12:e17345. https://doi.org/10.7717/peerj.17345

De Alencar JLM, Kaltchuk-Santos E, Fachinetto J et al (2018) Genetic and ecological niche modeling of Calydorea crocoides (Iridaceae): an endemic species of Subtropical Highland Grasslands. Genet Mol Biol 41:327–340. https://doi.org/10.1590/1678-4685-gmb-2017-0078

De Oliveira K, Novaes RLM, Weber MM, Moratelli R (2024) Forecasting climate change impacts on neotropical Myotis : insights from ecological niche models for conservation strategies. Ecol Evol 14:e11419. https://doi.org/10.1002/ece3.11419

Elith J, Leathwick JR (2009) Species distribution models: ecological explanation and prediction across space and time. Annu Rev Ecol Evol Syst 40:677–697. https://doi.org/10.1146/annurev.ecolsys.110308.120159

Franklin J (2010) Mapping species distributions: spatial inference and prediction, 1st edn. Cambridge University Press

Franks SJ, Sim S, Weis AE (2007) Rapid evolution of flowering time by an annual plant in response to a climate fluctuation. Proc Natl Acad Sci USA 104:1278–1282. https://doi.org/10.1073/pnas.0608379104

Gonçalves F, Sales LP, Galetti M, Pires MM (2021) Combined impacts of climate and land use change and the future restructuring of Neotropical bat biodiversity. Perspect Ecol Conserv 19:454–463. https://doi.org/10.1016/j.pecon.2021.07.005

Guisan A, Thuiller W (2005) Predicting species distribution: offering more than simple habitat models. Ecol Lett 8:993–1009. https://doi.org/10.1111/j.1461-0248.2005.00792.x

Guisan A, Zimmermann NE (2000) Predictive habitat distribution models in ecology. Ecol Model 135:147–186. https://doi.org/10.1016/S0304-3800(00)00354-9

Hughes AR, Inouye BD, Johnson MTJ et al (2008) Ecological consequences of genetic diversity. Ecol Lett 11:609–623. https://doi.org/10.1111/j.1461-0248.2008.01179.x

Kardos M, Armstrong EE, Fitzpatrick SW et al (2021) The crucial role of genome-wide genetic variation in conservation. Proc Natl Acad Sci 118:e2104642118. https://doi.org/10.1073/pnas.2104642118

Kolanowska M, Rewicz A, Baranow P (2020) Ecological niche modeling of the pantropical orchid Polystachya concreta (Orchidaceae) and its response to climate change. Sci Rep 10:14801. https://doi.org/10.1038/s41598-020-71732-1

Leal BSS, Graciano VA, Chaves CJN et al (2019) Dispersal and local persistence shape the genetic structure of a widespread Neotropical plant species with a patchy distribution. Ann Bot 124:499–512. https://doi.org/10.1093/aob/mcz105

Leimu R, Mutikainen P, Koricheva J, Fischer M (2006) How general are positive relationships between plant population size, fitness and genetic variation? J Ecol 94:942–952. https://doi.org/10.1111/j.1365-2745.2006.01150.x

Li Q-M, Cai C-N, Xu W-M et al (2021) Adaptive genetic diversity of dominant species contributes to species co-existence and community assembly. Plant Divers 44:271–278. https://doi.org/10.1016/j.pld.2021.11.002

Lissovsky AA, Dudov SV, Obolenskaya EV (2021) Species-distribution modeling: advantages and limitations of its application. 1. General Approaches. Biol Bull Rev 11:254–264. https://doi.org/10.1134/S2079086421030075

Lopes A, Demarchi LO, Piedade MTF et al (2023) Predicting the range expansion of invasive alien grasses under climate change in the Neotropics. Perspect Ecol Conserv 21:128–135. https://doi.org/10.1016/j.pecon.2023.02.005

López-Caamal A, Ferrufino-Acosta LF, Díaz-Maradiaga RF et al (2019) Species distribution modelling and cp SSR reveal population history of the Neotropical annual herb *Tithonia rotundifolia* (Asteraceae). Plant Biol 21:248–258. https://doi.org/10.1111/plb.12925

Marini MÂ, Barbet-Massin M, Martinez J et al (2010) Applying ecological niche modelling to plan conservation actions for the Red-spectacled Amazon (Amazona pretrei). Biol Conserv 143:102–112. https://doi.org/10.1016/j.biocon.2009.09.009

McCullagh P, Nelder JA (1998) Generalized linear models, 2nd edn. Chapman & Hall/CRC, Boca Raton

Meireles RC, Lopes LE, Brito GR, Solar R (2023) The future of suitable habitats of an endangered Neotropical grassland bird: a path to extinction? Ecol Evol 13:e9802. https://doi.org/10.1002/ece3.9802

Melo MA, Silva MAGD, Piratelli AJ (2020) Improvement of vegetation structure enhances bird functional traits and habitat resilience in an area of ongoing restoration in the Atlantic Forest. An Acad Bras Ciênc 92:e20191241. https://doi.org/10.1590/0001-3765202020191241

Mercado Gómez JD, Prieto-Torres DA, Gonzalez MA et al (2020) Climatic affinities of Neotropical species of Capparaceae: an approach from ecological niche modelling and numerical ecology. Bot J Linn Soc 193:263–275. https://doi.org/10.1093/botlinnean/boz092

Mittermeier RA, Mittermeier RA, Cemex SA, de CV (eds) (2004) Hotspots revisited, First English Edition. Cemex, Mexico City, Mexico

Muell MR, Brown JL (2024) Integrating ecological niche modeling and rates of evolution to model geographic regions of mimetic color pattern selection. Evol Ecol 38:599–619. https://doi.org/10.1007/s10682-024-10290-8

Myers N, Mittermeier RA, Mittermeier CG et al (2000) Biodiversity hotspots for conservation priorities. Nature 403:853–858. https://doi.org/10.1038/35002501
Myers CE, Stigall AL, Lieberman BS (2015) PaleoENM: applying ecological niche modeling to the fossil record. Paleobiology 41:226–244. https://doi.org/10.1017/pab.2014.19
Peterson AT (ed) (2011) Ecological niches and geographic distributions. Princeton University Press, Princeton
Peterson AT, Soberon J (2012) Species distribution modeling and ecological niche modeling: getting the concepts right. Nat Conserv 10:102–107. https://doi.org/10.4322/natcon.2012.019
Phillips S, Anderson R, Schapire R (2006) Maximum entropy modeling of species geographic distributions. Ecol Model 190:231–259. https://doi.org/10.1016/j.ecolmodel.2005.03.026
Preuss G, Padial AA (2021) Increasing reality of species distribution models of consumers by including its food resources. NBC 16:411–425. https://doi.org/10.3897/neotropical.16.e64892
Santos JPB, Fabricante JR (2018) Biological invasion by Thespesia populnea in sites under fluviomarine influence. Neotrop Biol Conserv 13:356–360. https://doi.org/10.4013/nbc.2018.134.11
Sillero N, Barbosa AM (2021) Common mistakes in ecological niche models. Int J Geogr Inf Sci 35:213–226. https://doi.org/10.1080/13658816.2020.1798968
Sillero N, Arenas-Castro S, Enriquez-Urzelai U et al (2021) Want to model a species niche? A step-by-step guideline on correlative ecological niche modelling. Ecol Model 456:109671. https://doi.org/10.1016/j.ecolmodel.2021.109671
Valencia-Rodríguez D, Jiménez-Segura L, Rogéliz CA, Parra JL (2021) Ecological niche modeling as an effective tool to predict the distribution of freshwater organisms: the case of the Sabaleta Brycon henni (Eigenmann, 1913). PLoS One 16:e0247876. https://doi.org/10.1371/journal.pone.0247876
Venables WN, Ripley BD (1999) Modern applied statistics with S-PLUS. Springer New York, New York
Whitham TG, Bailey JK, Schweitzer JA et al (2006) A framework for community and ecosystem genetics: from genes to ecosystems. Nat Rev Genet 7:510–523. https://doi.org/10.1038/nrg1877

Chapter 12
How Have Genetics and Genomics Made it Possible to Understand the Structure of Populations in the Neotropical Landscape?

Patricia Amavet

Contents

Abstract In population genetic studies, the structure of populations is an extremely important attribute for understanding the microevolutionary processes that shape their characteristics. Two decades ago, the terminology landscape genetics was proposed to refer to a discipline that integrates population genetics studies developed using molecular methods and landscape ecology, which analyzes the interactions between spatial patterns and ecological processes. The two key aspects of landscape genetics are the detection of genetic discontinuities and their correlation with landscape features and environmental conditions. In the current context where climate change is a reality and anthropogenic activities are a threat to many species' conservation due to the disruption of gene flow, landscape genetics is an essential discipline to gather information about the genetic status of populations, especially in the Neotropics. In addition, the data collected through these studies are necessary to evaluate the existence of evolutionarily significant units, management units, or

P. Amavet (✉)
Laboratorio de Genética, Facultad de Humanidades y Ciencias, Universidad Nacional del Litoral-Consejo Nacional de Investigaciones Científicas y Técnicas, Santa Fe, Argentina
e-mail: pamavet@fhuc.unl.edu.ar

M. V. Bohrer Monteiro Siqueira et al. (eds.), *Population Genetics in the Neotropics*,
https://doi.org/10.1007/978-3-031-83685-5_12

conservation units in the range of the species analyzed, which informs different management decisions. This chapter includes a brief review of the history of this discipline, the main methods currently used in landscape genetics and genomics studies, as well as examples of their application in studies on neotropical species.

Keywords Population genetics · Landscape ecology · Gene flow · Molecular markers · Conservation actions

Learning Objectives
This chapter will help readers to understand the following:

1. The importance of the knowledge of the genetic structure of a population.
2. The concept of landscape genetics and methods for analyses.
3. Current genetic technologies that facilitate landscape genetic studies.
4. Studies developed in the area of landscape genetics and their future applications.

12.1 Population Structure

The structure of a population is one of its most important characteristics since it determines the dynamics of gene flow, gain or loss of variability, and the direction in which it will evolve. A population is structured when it has local populations (or demes) within which the reproduction of individuals is more likely.

The population structure is defined as the amount of genetic variability and its distribution within and among local populations and individuals within a species (Templeton 2021). This definition emphasizes the spatial patterns of genetic variation that occur through the exchange of gametes. Thus, genetic population structure has three major components:

- System of mating;
- Genetic drift;
- Gene flow.

In this system, the type of population structure that exists is important. Usually, the continent-island or island models are analyzed, but the population structure is usually more complex than these models. In these analyses, it is important to determine if there is isolation by distance (IBD), that is if the genetic distances between subpopulations are related to the geographical distances that separate them (Fig. 12.1). The population size (N) of each local population is crucial, since genetic drift can have a notable effect.

The most commonly used index to know if a population is structured is Wright's (1951) F_{ST} index, which expresses the average reduction in heterozygosity of a subpopulation due to genetic drift. Therefore:

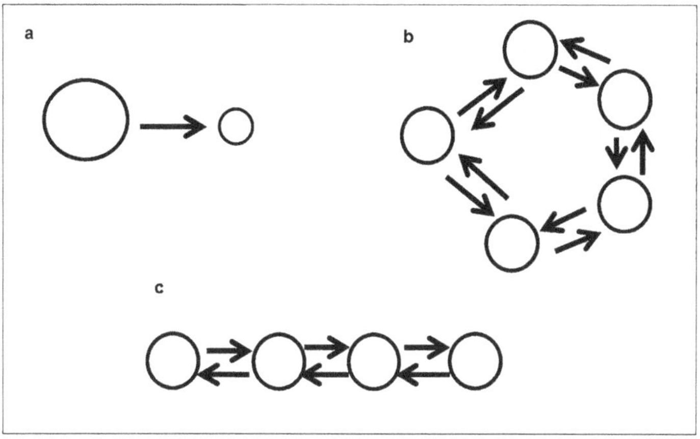

Fig. 12.1 Population structure models. The circles represent subpopulations and their area represents population size. Arrows show the gene flow. (**a**) Continent-island model. (**b**) Island model. (**c**) Island model with an isolation by distance structure

$$F_{ST} = \frac{H_T - H_S}{H_T}$$

where:

H_S = represents the level of heterozygosity that would be found within a subpopulation if random crossings occurred, therefore, is the expected heterozygosity (*He*) for a subpopulation.

H_T = represents the heterozygosity that would be found if all the subpopulations were grouped, and random crossings occurred in the total population. Then, if the average allele frequency for all subpopulations is p_0 the H_T value will be $2p_0q_0$.

The deviation from Hardy–Weinberg genotype frequencies in one species as a whole, caused by population subdivision, was defined as the Wahlund *effect*, after the researcher who first identified this phenomenon (Wahlund 1928). The F_{ST} value is always greater than or equal to zero, due to the Wahlund effect which ensures that $H_T > H_S$ average.

If all subpopulations are in Hardy-Weinberg equilibrium with the same allele frequencies, then $F_{ST} = 0$, which means that there is no structure in the population, and mating occurs between all individuals in the metapopulation with the same probability. Besides, F_{ST} values of 0–0.05 are generally considered to indicate little population genetic differentiation, values of 0.05–0.25 indicate moderate genetic differentiation, and values over 0.25 represent pronounced levels of genetic differentiation (Freeland et al. 2011).

The study of the structure of populations in the environment of a given landscape is the objective of landscape genetics, the primary subject that this chapter covers.

12.2 A Brief Historical Perspective on Landscape Genetics

The beginnings of landscape genetics date back to the studies of the botanist and biogeographer Augustin-Pyramus de Candolle (1778–1841) and Alfred Russel Wallace (1823–1913). De Candolle observed that there are differences in the distribution pattern of organisms (taxa) throughout the landscape, which depend on both "physical causes" that are acting in the present, and also on other forces that no longer exist (Crisci 2001). These forces (environmental factors, climate, etc.) operate on different time scales, currently included in disciplines such as ecological and historical biogeography (Brown and Lomolino 1998; Crisp 2001).

Another initial contribution to the discipline was made by Alfred Russell Wallace, who while traveling through the Malay Archipelago, observed a break ("boundary") in the fauna between the Australian region and the eastern region (Wallace 1860).

In recent years, the discipline of landscape genetics has been growing thanks to the rapid development of genetic-molecular tools to examine organisms on a large scale, in addition to having effective spatial statistics, and geographic information systems.

12.3 Landscape Genetics Approach

Landscape genetics combines population genetics studies applying molecular markers, with landscape ecology (Manel et al. 2003). This discipline aims to obtain information about the interaction between landscape characteristics and microevolutionary processes, such as gene flow, genetic drift, and natural selection, that occur in populations. It also allows identifying cryptic boundaries, which are interruptions in gene flow between populations without any apparent cause, or the secondary contact between previously isolated populations.

This approach is useful for understanding how environmental and geographic characteristics structure genetic variation at both individual and population levels, and does not require the identification of discrete populations.

The analysis involves the detection of genetic discontinuities and their correlation with landscape variables. The "landscape" in landscape genetic studies can be a basin, one or several valleys, hundreds of square kilometers of forest area, a section a highway and its surroundings, or an area of urban expansion around a city center. The landscape feature or barrier could be: waterfalls, forest, mountains, humidity gradients, etc.).

The most usual spatial genetic patterns described by Manel et al. (2003) are:

(a) *clines*: continuous variation in a character through a series of contiguous or adjacent populations;
(b) *isolation by distance* (IBD): when genetic differentiation between individuals (or populations) increases with their geographical distance (because gene flow declines at larger distances);

(c) *genetic boundaries (discontinuities) to gene flow*: geographic zone of sharp genetic change;
(d) *metapopulations*: a group of subpopulations that exchange occasional migrants and that might be subject to local extinction and recolonization;
(e) *random patterns*: spatial distribution pattern in which the presence of one individual has no influence on the distribution of other individuals.

Landscape ecology and population genetics converge in order to analyze how habitat loss and spatial isolation or habitat fragmentation affect the movement of species in different landscapes, which is why these are the main topics of study in this discipline.

The concept of landscape also encompasses seascapes. Marine ecosystems have been scarcely studied in landscape genetics, because of their environmental differences with terrestrial ecosystems. In marine ecosystems, seawater currents, water temperatures, or resource gradients are of particular importance. Consequently, *seascape genetics* has emerged as an independent but expanding branch of landscape genetics (Selkoe et al. 2008).

Another key concept in this topic is the study of *dispersal*, which is the process by which individuals are expelled from their birth habitat (or population) to reproduce in another habitat (or population). This process is a key factor for the survival of the population, especially in fragmented landscapes, and it is related to *connectivity*, which has two components: functional and structural. The *functional connectivity* is "the degree to which the landscape facilitates or impedes movement among resource patches" (Taylor et al. 1993), while *structural connectivity* is related to existence of habitat patches (and the populations inhabiting) which are linked by spatial structures in a landscape (e.g., hedgerows, field edges, stepping stones), irrespective of whether they provide functional connectivity. In the framework of landscape genetics, it is important to measure genetic connectivity by gene flow (at historical and contemporary level) from genetic distances or assignment and parentage tests (Manel and Holderegger 2013).

Additionally, landscape genetic studies allow us to determine what constitutes a natural rupture within or between populations to delineate *Evolutionary Significant Units* (ESU), *Management Units* (MU), or *Conservation Units* (CU). These terms have different concepts and applications: an ESU is a set of individuals or populations of a given species with a unique evolutionary history and a genetic pool that makes them worth protecting. Uniqueness is often inferred from phylogenies combined with ecological data. On the other hand, MU has a broader concept: it comprises "demographically distinct populations that should be managed to ensure the viability of the larger evolutionary significant unit, subspecies, or regional populations". The latter is similar to CU, which "refers to either evolutionary significant unit or management unit, or any geographical units that managers feel important to conserve" (Manel et al. 2003). These are very useful concepts to managers of wildlife and conservation biologist to design management and use plans.

12.4 Genomics Advances Have Improved Landscape Genetics Studies

Technological advances have allowed researchers to use markers with different resolution levels (for example, mitochondrial DNA, microsatellites, amplified fragment length polymorphisms- AFLP-, and Y chromosomes) to perform temporal or spatial studies (see Chaps. 9 and 10 to reviewing that). Numerous tools are available for different taxa and current studies involve the analysis of a larger number of markers, including single nucleotide polymorphisms (SNPs), mapped loci and candidate genes, combined in the framework of *population genomics*.

Recently, the availability of large single nucleotide polymorphism (SNP) datasets (with millions of molecular markers genotyped per individual) based on next-generation sequencing (NGS) techniques has allowed the development of new methods for genome-wide association studies (GWASs) (Segura et al. 2012). GWAS looks for associations between SNPs and particular phenotypic traits that are stronger than would be expected by neutral evolution. Such SNPs indicate genomic areas that are potentially under natural selection for a given trait. The GWAS approach is conceptually close to landscape genomics approaches that assess the association of SNPs with environmental factors rather than their association with phenotypic traits.

Population genomics involves sampling many individuals from each population and many loci within the genome, usually using functional gene loci whose distribution is already known on the map. The importance of studying many loci is to increase the power to identify "outlier" loci that might be under selective effects (Vitalis et al. 2001). It is important to identify these selected loci because: those loci are not suitable for studying population parameters as: demographic history, migration rates, genetic distance, or structure, because natural selection biases their estimation; comparing distributions of neutral markers and markers under natural selection, and/or quantitative traits can explain how were generated the patterns were generated; and the identification of outlier loci can allow to understand the genetic basis of local adaptation, adaptive differentiation and speciation (Manel et al. 2003).

12.5 Methods in Landscape Genetics and Genomics

Landscape genetics uses two approaches to study gene flow between populations: (a) an individual-based approach to fine-scale genetic structure; (b) the direct study of recent or current gene flow (Holderegger and Wagner 2008) (Fig. 12.2).

Current genomic methods such as QTL analysis, gene expression profiling (genes expressed under certain environmental conditions from mRNA), and candidate genes mainly provide molecular markers for developing these studies. However, an important limitation of these methods is that it is difficult to associate them with environmental variation in natural landscapes and apply them to multiple populations, as is necessary in landscape genetics.

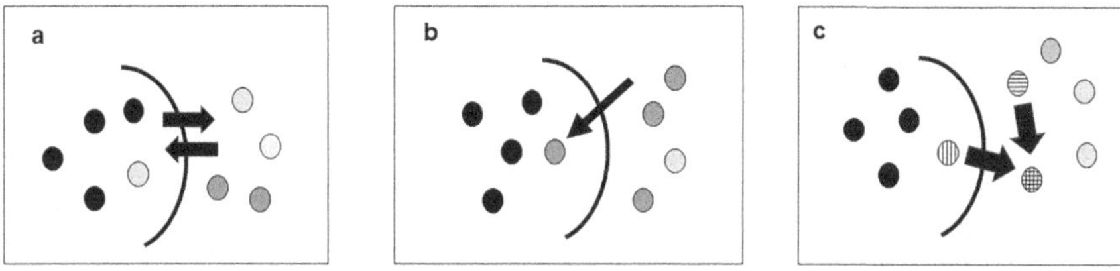

Fig. 12.2 The black curve line represents a landscape feature; the circles represent individuals and the different colors of these circles refer to different genotypes. (**a**) approach based on individual genetic distances: genetic distances between individuals are correlated with landscape structure. The largest genetic distances occur across the landscape feature. (**b**) Recent gene flow assessed by assignment testing. Individuals are grouped into two populations on each side of the landscape element. An assignment test identifies an individual as a recent immigrant from the other side of the landscape feature (arrow). (**c**) Current gene flow is assessed by parentage analysis of offspring. The marked offspring has one parent on the other side of the landscape feature (arrow). (Modified from Holderegger and Wagner (2008))

An alternative approach uses genome scans and genetic sampling along ecological gradients or in different types of habitats where the environmental variation can be characterized using Geographic Information Systems (GIS). This approach uses massive screening of molecular markers in the genome to infer genes or loci that show signs of adaptive selection and correlate them with data on environmental heterogeneity. To do this, large numbers of individuals and populations are sampled in different types of habitats along environmental gradients and are analyzed with neutral genetic markers (dozens of microsatellites, hundreds of AFLPs, and hundreds or thousands of SNPs). Loci showing greater genetic differentiation between populations or habitat types compared to the vast majority of neutral markers are then analyzed using statistical approaches. The differentiation of these "atypical loci" (outliers) compared to a neutrality model suggest that natural selection is acting on them or their genetic linkage (Manel et al. 2003).

Correlating outlier loci with environmental data from sample locations reveals which local environmental conditions or landscape features are related to which outlier loci. Thus, we could identify and genetically characterize loci that are probably implicated in adaptive genetic variation related to certain characteristics within landscapes much more quickly. The advantage of this approach is that it provides clear clues to those environmental factors that act as selective forces.

Most of landscape genetic studies have used Mantel tests, which relate a matrix of genetic distances between individuals or populations to geographic (Euclidean) distances or landscape matrices distances to investigate landscape effects on gene flow. These models are based on single or multiple *least-cost paths*, i.e. a path that minimizes the costs of movement through a landscape. Different *resistance values* to movement are assigned to several landscape elements in a cost surface. Sartor et al. (2022) evaluated the accuracy of two commonly used approaches, habitat suitability modeling and landscape genetics, in estimating landscape resistance to genetic connectivity of two species of Neotropical cats (*Leopardus guttulus* and *Leopardus geoffroyi*) across their ranges. For the studied species, the transformation of habitat suitability models into resistance surfaces is a poor proxy for permeability to dispersal, and the use of genetic data is more reliable in modeling connectivity for species conservation.

However, Mantel tests and other related methods such as multiple regression on distance matrices have been criticized. Because of non-independence in the response and predictor variables, the actual degrees of freedom are unknown and common methods to select the best model from a set of alternative models such as R2 or the Akaike information criterion (AIC) are invalid. One alternative to deal with the problem of non-independence is using mixed effect models incorporating the covariance structure of allele frequencies. Landscape genetics needs to find methods to measure gene flow directly from raw data (e.g., allele frequencies) and then use multiple regression (in the case of a single response variable) or canonical redundancy analysis (for multiple response variables) when investigating gene flow in response to landscape configuration.

Geo-referenced genetic data using a geographic information system is a prerequisite for spatial-based analysis of spatial ecological processes. One important application of GISs in landscape genetics is the overlay of land use, land cover, or topographical maps with genetic groups or genetic boundaries (Manel and Holderegger 2013).

12.6 Study Cases in the Neotropics

Among vertebrates, most landscape genetic studies carried out in the Neotropics area refer to mammals. Flores-Manzanero and Vázquez-Domínguez (2019) surveyed landscape genetics work on American mammals. Their results showed that the order most represented in American ecosystems was Rodentia (n = 20); in deserts it was Artiodactyla (n = 4). Of the total number of works, studies focused on estimating "connectivity" (n = 14) and "structure" (n = 12) were the majority, and the most used analyses were based on Mantel tests (simple and partial). Dispersal capacity and the presence of vegetation were key variables. Although the evaluation of connectivity through correlation analysis between distance matrices and Mantel tests predominated, the use of novel methodologies was identified, such as the construction of genetic and demographic networks from resistance matrices. It should be noted that the ecological variables of the species, particularly rodents, and the vegetation cover were key in deserts. North America is the most studied region, while there are few genetic landscape works with mammals in Mexico and South America and in desert ecosystems, which is why the authors raised the urgency of carrying out studies in these regions and in deserts.

In Argentina, Gallo et al. (2022) explored spatial segregation of pumas (*Puma concolor*) by testing for hierarchical structure within previously identified clusters, genetic differentiation among regions, to investigate whether isolation-by-distance (IBD) or/and isolation-by-resistance (IBR) could explain the previously reported putative correlation between gene flow and hunting pressure. The dataset comprised

83 pumas from three Argentinian provinces, which were genotyped at 25 species-specific microsatellite loci. Based on two previously identified genetic populations the authors tested for hierarchical genetic structure for each population analyzing population differentiation using pairwise F_{ST} and different clustering methods. The results show a significant correlation between genetic and geographic distance suggesting the presence of an isolation-by-distance process among individuals. The authors established that genetic structure and gene flow among south-central Argentine pumas can be partially related to the landscape connectivity pattern observed in the area.

Later, Silva et al. (2020) wanted to understand how Amazonian anthropogenic landscapes (areas comprising rainforest remnants, more or less fragmented embedded in distinct matrices of human-altered land covers and other land uses), might affect connectivity between populations, analyzing the distribution of genetic diversity and population structure in four phyllostomid bat species. The four species studied were two understory frugivores: *Carollia perspicillata* (n = 91) and *Rhinophylla pumilio* (n = 57); and two canopy frugivores: *Artibeus obscurus* (n = 34) and *Dermanura gnoma* (n = 37). Genomic DNA from all tissue samples was amplified using three mitochondrial DNA loci: *cytochrome oxidase subunit I* (*COI*), *cytochrome b* (*CYTB*) and *control region* (*D-LOOP*).

According to expectations, smaller and less vagile species (*R. pumilio* and *D. gnoma*) exhibited lower nucleotide diversity than the bigger-sized species, and larger forested patches seem to sustain higher genetic diversity than longer-impacted landscapes. The landscape structure analyses showed responses that are mainly species-specific. The results found a certain structure for all species, identifying geographical congruence and a certain tendency towards a north-south structuring. This latitudinal trend of differentiation was also found with F_{ST} pairwise comparisons. The authors proposed the forest dependency of some of the bat-studied species and highlighted how the loss of habitat connectivity due to deforestation seems to be already shaping genetic diversity and population structure in phyllostomid bats in eastern Amazonia.

Then, Latorre-Cardenas et al. (2021) assessed the spatial genetic structure of the neotropical otter (*Lontra longicaudis*) and to evaluate the landscape-riverscape attributes that shape its genetic structure and gene flow at local sites (habitat patches) and between sites (landscape matrix). The authors used a landscape genetics individual-based approach to test the effect stream hierarchical structure, IBR and IBR on genetic structure and gene flow. The molecular markers used were 11 microsatellite loci. A few of the variables (slope, elevation, Normalized Difference Vegetation Index—NDVI, and ecosystem integrity index) were included in the landscape-riverscape analysis to create the environmental suitability model because they wanted to analyze their individual contribution to the genetic structure of the otter. They found genetic structure that corresponded to the latitudinal and altitudinal heterogeneity of the landscape and riverscape, as well as to the hierarchical structure of the streams. The authors proposed it is likely that differences of the environmental suitability between the upper and lower basins such as the availability of water along the year and the degree of human perturbation, may be reducing gene flow of *L. longicaudis*. Then, several landscape and riverscape attributes influenced the genetic structure and gene flow of *L. longicaudis*, supporting the hypothesis that IBR explains better the genetic patterns of the species than IBD. Open areas and steep slopes were the variables affecting genetic structure at local sites, whereas areas with suitable habitat conditions, higher ecosystem integrity, and larger streams enhanced gene flow between sites.

Regarding landscape genetic studies in plants, Rodriguez et al. (2015) analyzed the organization of genetic variation of the common bean (*Phaseolus vulgaris*) in centers of domestication. They used 131 single nucleotide polymorphisms to investigate 417 wild common bean accessions and 160 domesticated genotypes, including Mesoamerican and Andean genotypes, being a total of 577 accessions. By analyzing the genetic spatial patterns of the wild common bean, they found several genetic groups and the occurrence of variable levels of diversity in Mesoamerica and the Andes. Furthermore, using a landscape genetics approach, they demonstrated that both demographic processes and adaptation were responsible for the observed genetic structure.

The authors showed that the study of correlations between markers and ecological variables at a continental scale can help in identifying local adaptation genes. Also, they located putative areas of common bean domestication in Mesoamerica, in the Oaxaca Valley, and the Andes, in southern Bolivia-northern Argentina. These findings are important for the conservation and exploitation of the genetic diversity preserved within this species and other plant genetic resources.

More recently, Silva-Arias et al. (2021) analyzed the genetic diversity and population structure of the perennial nightshade *Calibrachoa heterophylla* to infer the influence of abiotic landscape features on the level of gene flow in the South Atlantic Coastal Plain using ten microsatellites. The *C. heterophylla* populations located on early-deposited coastal plain regions showed higher genetic diversity than those closer to the sea. The genetic differentiation followed a pattern of IBD. Landscape features, such as water bodies and wind corridors, and geographical distances explained the observed genetic differentiation, whereas the precipitation seasonality exhibited a strong signal for isolation-by-environment in marginal populations. The estimated levels of gene flow suggested that marginal populations had restricted immigration rates enhancing differentiation. The authors concluded that topographical features related to coastal plain deposition history influence population differentiation in *C. heterophylla*, and the gene flow was mainly restricted to nearby populations and facilitated by wind fields, although large water bodies did not appear to influence gene flow. Furthermore, differential rainfall regimes in marginal populations seem to promote genetic differentiation.

An example of the most modern molecular methods used in landscape genetics analysis is the study performed by Varas-Myrik et al. (2022). Using restriction-site associated DNA sequencing (RAD-Seq) and more than 49,000 SNPs screened from 12 locations of *Araucaria araucana* in Chile, the authors assessed the genetic structure and predicted the genetic offset of this tree species under two future climate scenarios.

Using Generalized Dissimilarity Modeling (GDM) they found that the temperature annual range was the most important variable shaping the observed patterns of adaptive divergence. The results showed that populations living in the piedmont of the southern Andes Mountain range are at the greatest risk of maladaptation, while populations living in the high-elevation zones in the Andes Mountain range are at the lowest risk.

We can also cite a work on human populations. Barbieri et al. (2019) explored the genetic structure of 176 individuals from the Pacific Coast, the Andes and Amazonia, genotyped with the Affymetrix Human Origins array. Analyzing 633,994 SNPs the authors inferred multiple sources of ancestry within the Native American ancestry component; one with predominance on the Coast and in the Andes, and at least two distinct substrates in neighboring Amazonia, including a previously undetected ancestry characteristic of northern Ecuador and Colombia. Amazonian populations are also involved in recent gene flow with each other and across ecogeographic domains, which does not accord with the traditional view of small, isolated groups. Long-distance genetic connections between speakers of the same language family suggest that indigenous languages here were spread not by cultural contact alone. The authors concluded that native American populations admixed with post-Columbian European and African sources at different times, with few cases of prolonged isolation.

12.7 Future Directions

There is still a long way to go in the area of landscape genetics. Following Manel and Holderegger (2013), the most important topics are:

- *Prediction and conservation management studies*: multispecies landscape genetic or landscape-genomic studies are promising approaches to achieving successful conservation strategies because it is particularly challenging to infer general landscape effects on gene flow or local adaptation from single-species studies. Landscape-genetic or landscape-genomic studies could also compare species belonging to diverse taxonomic groups or different habitat types in the same landscape. High-throughput next-generation sequencing and the reduction in processing costs per sample open perspectives to study large genomic datasets for a large number of organisms;
- *Landscape genetics analysis in urban areas:* urban areas are highly fragmented landscapes. Landscape genetics can be extremely useful in assessing connectivity in urban species and it is important to develop studies in these areas because urban areas covered 0.5% of the planet's land area in 2000 and are predicted to expand 12-fold between 2000 and 2050;
- *Global climate change effects studies using landscape genetics*: it is a general agreement that anthropogenic alterations, including climate change, produce biodiversity loss. But long before a species is finally lost, its genetic diversity

will be substantially affected. The disappearance of populations is a process directly related to the loss of intraspecific genetic diversity. The main objective of current landscape genetics is to improve our understanding of the effect of global change on genetic patterns to address these two questions: (a) How has recent global change affected patterns of neutral and adaptive genetic variation?; (b) Are species likely to adapt to ongoing global change on an ecological time scale?

Functional connectivity is of crucial importance for tracking changing niches under climate change. Some studies, developed in species that grow in cold environments, analyzed how the genetic structure will be affected under various climate change scenarios due to the modification of functional connectivity. The first step is to conduct studies of the current distribution of environmental factors within a landscape and determine how they influence the distribution of adaptive genetic variants. Several studies have recently found that annual temperature regimes influence the distribution of adaptive genetic variation in model and non-model organisms. A second step is to study current gene flow using neutral genetic variation to investigate the potential for adaptive genes to spread across landscapes. If this potential is high, gene flow will allow populations and species to adapt to changing environmental conditions, not only due to current genetic variation but also due to gene migration. The third step combines spatial patterns of adaptive genetic variation with information on the pace of gene flow and allows for predicting the future spatial distribution of adaptively relevant genetic variation in different species under various global change scenarios. Although such applications have not yet been carried out on ecological time scales, they now appear feasible thanks to the increasing sophistication of landscape genetic and genomic tools.

12.8 Final Remarks

The importance of landscape genetics studies is that they have increased our understanding of the large effects that spatial patterns of landscape elements and environmental factors exert on neutral and adaptive genetic variation. In other words, landscape genetics has brought population genetics and evolutionary research closer to the empirical world as organisms perceive and interact with it. This effort must continue and be expanded.

Key Points

- Study methods of population genetics structure: main indexes and Mantel test.
- Most usual spatial genetic patterns and the landscape features related to these patterns.
- Current and future applications of landscape genetics studies.

Questions

1. Genetic corridors: imagine two fragments of habitat separated by a road. How would the highway affect genetic connectivity between populations on either side? Provide solutions to mitigate the negative impact of the road on genetic diversity.
2. Effect of geographic barriers: suppose there is a mountain range separating two valleys. How do you think this geographic barrier would affect the genetic structure of the populations in each valley? Describe how genetic differences might arise between valley populations.
3. Metapopulations and patch extinction: consider a series of habitats dispersed across a landscape. How would the extinction of a patch affect the genetic diversity of the metapopulation? Describe strategies to conserve genetic diversity in a scenario of habitat loss.
4. Genetic flow modeling uses a map of a landscape to model gene flow between different habitat patches. How would genetic structure be affected if some patches have more gene flow than others? Explain how you could model this scenario.
5. Effect of climate changes on landscape genetics: consider that climate change alters conditions in a landscape. How would this affect the distribution of species and their genetic diversity? Provide specific examples and how populations might adapt.
6. Landscape fragmentation assessment: examines a landscape fragmented by human activities. How does fragmentation affect the genetic diversity of populations? It provides examples of how fragmentation can be detrimental and how ecological corridors could be implemented.

References

Barbieri C, Barquera R, Arias L et al (2019) The current genomic landscape of Western South America: Andes, Amazonia, and Pacific Coast. Mol Biol Evol 36:2698–2713. https://doi.org/10.1093/molbev/msz174

Brown JH, Lomolino MV (1998) Biogeography, 2nd edn. Sinauer Associates Inc. Publishers, Sunderland., 691 pp. https://doi.org/10.2307/1383194

Crisci JV (2001) The voice of historical biogeography. J Biogeogr 28:157–168. https://doi.org/10.1046/j.1365-2699.2001.00523.x

Crisp M (2001) Historical biogeography and patterns of diversity in plants, algae and fungi: introduction. J Biogeogr 28:153–156. https://doi.org/10.1046/j.1365-2699.2001.00522.x

Flores-Manzanero A, Vázquez-Domínguez E (2019) Landscape genetics of mammals in American ecosystems. Therya 10:381–393. https://doi.org/10.12933/therya-19-833

Freeland J, Kirk H, Petersen S (2011) Molecular ecology, 2nd edn. Wiley. 464 pp. https://doi.org/10.1002/9780470979365

Gallo O, Castillo DF, Godinho R et al (2022) Assessing landscape connectivity for South-Central Argentine pumas dispersing under genetic source-sink dynamics. Landsc Ecol 38:999–1012. https://doi.org/10.1007/s10980-022-01585-8

Holderegger R, Wagner H (2008) Landscape genetics. Bioscience 58(3):199–207. https://doi.org/10.1641/B580306

Latorre-Cardenas MC, Gutiérrez-Rodríguez C, Rico Y et al (2021) Do landscape and riverscape shape genetic patterns of the Neotropical otter, *Lontra longicaudis*, in eastern Mexico? Landsc Ecol 36:69–87. https://doi.org/10.1007/s10980-020-01114-5

Manel S, Schwartz K, Luikart G et al (2003) Landscape genetics: combining landscape ecology and population genetics. Trends Ecol Evol 18:189–197. https://doi.org/10.1016/S0169-5347(03)00008-9

Manel S, Holderegger R (2013) Ten years of landscape genetics. Trends Ecol Evol 28:614–621. https://doi.org/10.1016/j.tree.2013.05.012

Rodriguez M, Rau D, Bitocchi E et al (2015) Landscape genetics, adaptive diversity and population structure in Phaseolus vulgaris. New Phytol 209:1781–1794. https://doi.org/10.1111/nph.13713

Sartor CC, Wan HY, Pereira JA et al (2022) Landscape genetics outperforms habitat suitability in predicting landscape resistance for congeneric cat species. J Biogeogr 49:2206–2217. https://doi.org/10.1111/jbi.14498

Segura V, Vilhjálmsson BJ, Platt A et al (2012) An efficient multi-locus mixed-model approach for genome-wide association studies in structured populations. Nat Genet 44:825–830. https://doi.org/10.1038/ng.2314

Selkoe KA, Henzler CM, Gaines SD (2008) Seascape genetics and the spatial ecology of marine populations. Fish Fish 9:363–377. https://doi.org/10.1111/j.1467-2979.2008.00300.x

Silva SM, Ferreira G, Pamplona H et al (2020) Effects of landscape heterogeneity on population genetic structure and demography of Amazonian phyllostomid bats. Mammal Res 66:217–225. https://doi.org/10.1007/s13364-020-00546-3

Silva-Arias GA, Caballero-Villalobos L, Giudicelli GC et al (2021) Landscape and climatic features drive genetic differentiation processes in a South American coastal plant. BMC Ecol Evol 21:1–13. https://doi.org/10.1186/s12862-021-01916-4

Taylor PD, Fahrig L, Henein K et al (1993) Connectivity is a vital element of landscape structure. Oikos 68:571–573. https://doi.org/10.2307/3544927

Templeton AR (2021) Population genetics and microevolutionary theory, 2nd edn. Wiley-Blackwell, Hoboken. 768 pp

Varas-Myrik A, Sepúlveda-Espinoza F, Fajardo A et al (2022) Predicting climate change-related genetic offset for the endangered southern South American conifer *Araucaria araucana*. For Ecol Manag 504:119856. https://doi.org/10.1016/j.foreco.2021.119856

Vitalis R, Dawson K, Boursot P (2001) Interpretation of variation across marker loci as evidence of selection. Genetics 158:1811–1823. https://doi.org/10.1093/genetics/158.4.1811

Wahlund S (1928) Zuzammensetzung von Poulationen und Korrelationserscheinungen vom Standpunkt der Vererbungslehre aus betrachtet. Hereditas 11:65–106. https://doi.org/10.1111/j.1601-5223.1928.tb02483.x

Wallace AR (1860) On the zoological geography of the Malay archipelago. J Proc Linn Soc Zool 4:172–184

Wright S (1951) The genetical structure of populations. Ann Eugenics 15:323–354. https://doi.org/10.1111/j.1469-1809.1949.tb02451.x

Chapter 13
What Does Phylogeography Teach Us About Population Genetics in the Neotropics?

Ossman Barrientos-Díaz, Caroline Turchetto, Ana Lucia Anversa Segatto, and Andreia Carina Turchetto-Zolet

Contents

Abstract The neotropical region is a highly significant and biodiverse area, spanning through a large portion of Central and South America and encompassing diverse landscapes. The region holds immense global importance for biodiversity, as it offers essential ecological services crucial to the planet's health. Ensuring the protection of the region and its unique biodiversity is of utmost importance for the well-being of current and future generations. To this end, phylogeography, a relatively recent field, provides a historical and evolutionary context for biodiversity studies since it investigates the historical processes that have influenced genetic diversity. Phylogeography enables the formulation of biogeographic hypotheses,

O. Barrientos-Díaz · C. Turchetto
Instituto de Biociências, Universidade Federal do Rio Grande do Sul, Porto Alegre, RS, Brazil

A. L. A. Segatto
Universidade Federal de Santa Maria, Centro de Ciências Naturais e Exatas, Departamento de Bioquímica e Biologia Molecular, Santa Maria, RS, Brazil

A. C. Turchetto-Zolet (✉)
Universidade Federal do Rio Grande do Sul, Porto Alegre, RS, Brazil
e-mail: carina.turchetto@ufrgs.br

M. V. Bohrer Monteiro Siqueira et al. (eds.), *Population Genetics in the Neotropics*,
https://doi.org/10.1007/978-3-031-83685-5_13

the description of the evolution of reproductively isolated population units, the inference of processes relating to the origin, distribution, and maintenance of biodiversity, and the creation of hypotheses concerning the influence of past environmental changes, physical and biotic characteristics of populations, population structure, and genetic diversity. This chapter presents an overview of how phylogeography can be instrumental in identifying divergent lineages, understanding species' responses to climatic changes, identifying refuge areas, dating geographical events, and maintaining the neotropical biota. Phylogeography has shed light on high levels of genetic diversity, genetic structure, hybridization, and various historical events, thereby serving as a pillar in population genetics research. Furthermore, advancements in molecular techniques and more efficient machine learning have facilitated the availability of larger datasets for the region, leading to fundamental insights into the evolutionary processes that have given rise to the extensive biodiversity we observe in the Neotropics today.

Keywords Biodiversity · Environmental changes · Lineages · Genetic diversity · Phylogeography

Learning Objectives
- Learn about the principles of phylogeography.
- Develop an understanding of the importance of phylogeographic analysis in population genetic studies.
- Appreciate the rapidly evolving molecular and statistical methods of phylogeography.
- Synthesize the existing phylogeographic patterns in Neotropical regions.

13.1 The Origin of Phylogeography and Its Importance in Population Genetic Studies

It has been over three decades since the word phylogeography was first mentioned when Avise et al. (1987) reported the utility of mitochondrial DNA (mtDNA) in genealogical inference at the intraspecific level. This study drove the creation of *phylogeography*, a discipline that combines evolutionary biology and geography to investigate the evolutionary processes responsible for the geographic distribution of genetic lineages, allowing links between phylogenetics and population genetics (Avise 2009). The phylogeographic approach seeks to understand how evolutionary processes have shaped the genetic diversity of species in different regions worldwide. Phylogeographic studies often use genetic information such as DNA sequences, to trace the evolutionary history of a species and understand patterns of population connectivity, such as migration, isolation, and population expansion over time. These analyses contribute significantly to the understanding of the current distribution of genetic diversity of species, providing crucial information to conservation strategies, as well as preserving biological diversity.

Phylogeography integrates phylogenetic and population genetics approaches in a single analysis, unveiling the connection between microevolution and macroevolution. The phylogeographic approach allows us to further understand how historical events contributed to form the current geographic distribution of genes, populations, and species (Avise 2000, 2009). In this context, phylogeographic studies can reveal which microevolutionary processes operate within species, and that can be extrapolated to explain macroevolutionary differences among species and higher taxa (Hickerson et al. 2010). In this line, we understand that both population genetics and phylogenetics have a focus on ancestral relationships.

Phylogeography has a temporal and spatial genealogical dimension (Avise 2009), including methods from population genetics, coalescent theory (Nielsen 2006), and phylogenetics. Although phylogeography uses historical information inherent to the gene tree, it is not a mere extension of phylogenetic principles at an intraspecific level. More than that, phylogeography characterizes the population subdivision by recognizing the geographic patterns of the genealogical structure throughout the distribution of the species. Phylogeography can overcome the limitations inherent in classical population genetics and systematics by synthesizing the influence of both contemporary and historical genetic exchanges, contributing to a better understanding of the evolutionary relationships between populations and closely related species.

Phylogeography has improved significantly following novel analytical tools (analysis methods and datasets) and continuous updates in theoretical foundations (revealing different patterns worldwide). Since its origins, phylogeography has rapidly advanced from descriptive to statistical methods based on coalescence (Statistical Phylogeography) (Carstens et al. 2005; Fagundes et al. 2007; Kuhner 2009). Over time, phylogeography has become increasingly integrative (Avise 2009).

The inferred genealogical relationships can be used to trace the biogeographical history of intraspecific populations, and address evolutionary questions such as gene flow, fragmentation, range expansion, and colonization. By managing these questions, phylogeographical studies can better understand biodiversity, dispersal modes, diversification times, extinctions, refuge areas, and other species/population-level processes (Turchetto-Zolet et al. 2013). Therefore, phylogeography can be instrumental in identifying divergent lineages, understanding responses of species to climate change, identifying refuges, date geographical events, and maintaining the neotropical biota (Fig. 13.1).

13.2 Molecular Data and Statistical Methods in Phylogeographic Studies

Phylogeographic research aims to uncover the geographic distribution of evolutionary lineages over time. Advanced methods in DNA sequencing, molecular phylogeny, molecular clocks, migration analyses, and niche modeling play a crucial role in

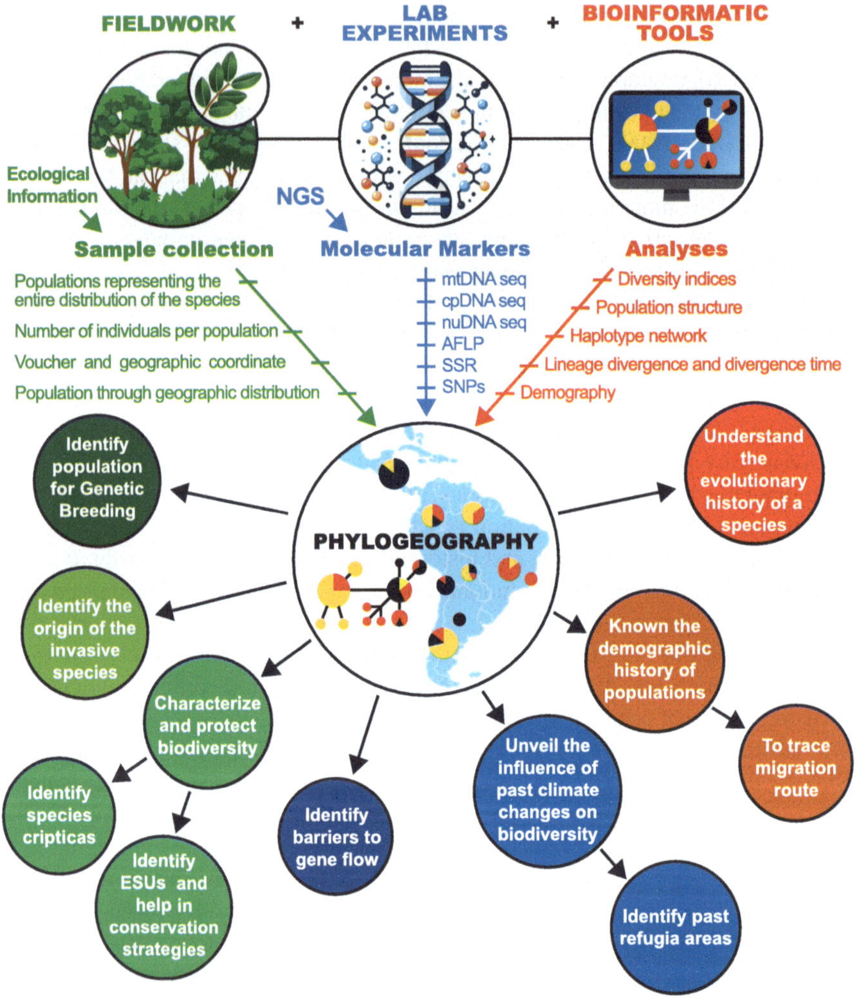

Fig. 13.1 Workflow of a phylogeographic study showing the main steps to design the study from obtaining samples to results and their applications. NGS (Next generation sequencing); ESUs (Evolutionary Significant Units); cpDNA (chloroplast DNA); mtDNA (mitochondrial DNA); nuDNA (nuclear DNA); AFLP (Amplified fragment length polymorphism); SSR (simple sequence repeated); SNP (single nucleotide polymorphisms)

this investigation. Through these methods, researchers can obtain valuable insights into the evolution and dispersal of organisms. By combining molecular data and statistical analysis, this holistic approach allows scientists to reconstruct a species evolutionary history and better understand how geographical and environmental factors have shaped genetic diversity and population distribution over time. These findings give a deeper insight into the complex interplay between biological and environmental factors.

13.2.1 Molecular Data

The choice of molecular markers in a phylogeographic study is critical since the main objective of phylogeography is to reveal the history behind the geographic distribution of lineages in the present (Table 13.1). One reason that facilitated the advance of phylogeography was the use of mitochondrial molecular markers, which overcame limitations such as the ability to resolve phylogenetic relationships and understand processes of population dispersal and diversification. In the 1980s and 1990s, mitochondrial DNA (mtDNA) emerged as a pivotal marker, leveraging its high mutation rate and maternal inheritance (Avise et al. 1987). Since the beginning, mitochondrial DNA (mtDNA) and chloroplast DNA (cpDNA) have been widely used molecular markers in phylogeographic studies in animals and plants, respectively, due to their rapid evolutionary rate and maternal inheritance (Avise 2000; Garrick et al. 2015). In the next decades, attention shifted to nuclear DNA (nuDNA), with microsatellites and single nucleotide polymorphisms (SNPs) offering increased resolution and versatility (Brumfield et al. 2003; Petit and Excoffier 2009). After that, multimarker approaches emerged, recognizing that the combination of various molecular markers, such as mitochondrial and nuclear DNA (or plastidial and nuDNA in plants), offers a more comprehensive perspective in phylogeography, enabling the assessment of different aspects of the evolutionary history of species (Edwards and Beerli 2000).

Improvements in DNA sequencing technology have benefited the composition of genetic data sets in phylogeography (Garrick et al. 2015), which has been instrumental in understanding evolutionary relationships and species distribution patterns over time. The advent of next-generation sequencing (NGS) drove the prominence of multi-locus approaches such as RAD sequencing, enabling comprehensive genomic analyses (Davey et al. 2011; Catchen et al. 2013; Taberlet et al. 2012). As we progress into the present and future, whole-genome sequencing (WGS) transforms phylogeography, providing an in-depth view of the genome (Lamichhaney et al. 2015). Furthermore, the emerging use of environmental DNA (eDNA) presents new perspectives for non-invasive monitoring and large-scale study of phylogeographic patterns (Thomsen and Willerslev 2015). The ongoing progress in phylogeography reflects the ever-increasing sophistication of molecular tools and their capacity to unravel intricate evolutionary processes shaping biodiversity.

In phylogeography, mitochondrial (or plastidial in plants) and nuclear DNA sequences are employed to study the evolutionary history and relationships between populations on broader temporal scales, while molecular markers in population genetics, such as microsatellites and SNPs, focus on intrapopulation genetic variability and are crucial for analyzing genetic structure and relationships within a specific population (Avise 2000). While population genetic markers aid in understanding population dynamics and genetic diversity within close groups, markers in phylogeography reveal evolutionary patterns and geographic distribution over time, highlighting dispersal and adaptation events. The choice of markers depends on the temporal scale and specific objectives of the study (Avise 2000; Beerli and

Table 13.1 A summary of the different attributes of various marker types used in phylogeography studies and suitability ranking for applications: -, not suitable; /, to be avoided if possible; +, suitable;++, well suited;+++, very well suited. Based to Ouborg et al. (1999)

	MARKER TYPES					
	cpDNA and mtDNA	nuDNA	Allozymes	RAPD/AFLP	SNPs	Microsatellites
Polymorphism	/	+++	+++	++	+++	+++
Dominance	Haploid	Codominant	Codominant	Dominant	Codominant	Codominant
Mode of inheritance	Uniparental	Biparental	Biparental	Biparental	Biparental	Biparental
Recombination	No	Yes	Yes	Yes	No	Yes
Costs (development and application)	Low	Low	Low	Intermediate	High	High
Applications						
Indirect, within populations	++	++	++	+	++	+++
Indirect, between populations	++	++	+++	++	++	+++
Direct	++	+	+	++	+++	+++
Phylogeography	+++	+++	+	++	+++	/

Felsenstein 2001; Knowles and Maddison 2002). Microsatellites (SSRs) and single nucleotide polymorphisms (SNPs) are frequently used in population genetics. In contrast, mitochondrial, plastidial, and nuclear DNA sequences and SNPs are prioritized in phylogeography to trace evolutionary events on broader temporal scales and comprehend geographic distribution over time.

Therefore, to succinctly represent the most commonly used molecular markers in research according to recent publications (~417) in plant phylogeography (Morris and Shaw 2018) and review of current South American phylogeographical (Turchetto-Zolet et al. 2013), we created the following a comparative table, which is based on the classification proposed by Beheregaray (2008) and Morris and Shaw (2018). Classes primarily differ in the division of dominant and codominant markers, identifying markers across five classes (Table 13.2). It is imperative to highlight that although phylogeography researchers can use different classes of molecular markers, it is always necessary to choose at least one class of markers that meets the principles of phylogeography.

Table 13.2 Distinct molecular markers used in Phylogeography

			CLASS		
	I	II	III	IV	V
Beheregaray (2008)	**Organellar sequences, chloroplast [cpDNA] and mitochondrial [mtDNA]**	**Nuclear DNA (nuDNA)** sequences, amplified fragment length polymorphisms (**AFLPs**), single nucleotide polymorphisms (**SNPs**), or randomly amplified polymorphic DNA (**RAPD**)	**Microsatellites**	**Allozymes**	**Combined marker classes**
Morris and Shaw (2018)	**Organellar sequences, cpDNA and mtDNA.**	**nuDNA**, **ITS** (nuclear ribosomal internal transcribed spacer; nrDNA ITS), external transcribed spacer (**ETS**), expressed sequence tags (**EST**), low-copy nuclear genes and **SNPs**	Dominant markers: **AFLP, RAPD and ISSR.**	Codominant markers including **Microsatellites, both nuclear [nSSR]** and **chloroplast [cpSSR]**	**Allozymes**

13.2.2 Statistical Methods

In the early phase of phylogeography coinciding with the rise of the polymerase chain reaction (PCR), interpretations were often based on qualitative assessments of single locus gene genealogies, employing geographically contextualized interpretations for demographic inferences. Descriptive phylogeography entails delineating patterns of geographic distribution and evolutionary relationships of populations without necessarily incorporating detailed statistical analyses (Avise 2000). Methods like nested clade phylogeographic analysis (NCPA) associated with gene genealogies and spatially explicit summary statistics, applying permutation tests for demographic patterns (Slatkin and Maddison 1989; Templeton 1998; Knowles 2009) were considered the best practices. However, coalescent simulation studies questioned the reliability of single locus implementations of NCPA (Knowles and Maddison 2002), highlighting the importance of statistical rigor in the field (Nielsen and Beaumont 2009). In the current wave of studies, there is a shift towards simulation-based statistical methods, using explicit coalescent models to estimate parameters and test historical hypotheses, acknowledging the limitations of direct interpretation from gene genealogies (Hey and Machado 2003).

The field of statistical phylogeography, employing coalescent models, is rooted in Chamberlin's approach (1890, reprinted 1965) of considering competing hypotheses for unobservable historical events (Knowles and Maddison 2002). This approach aligns with Bayesian/likelihood-based frameworks, evaluating hypotheses by fitting data to models and employing decisions based on theoretical methods. External data, such as packrat middens or ancient DNA, inform plausible models, strengthening statistical phylogeography. Simulation-based approaches are widely used, like calculating summary statistics from simulated data sets (Knowles 2001). Full likelihood/Bayesian approaches face challenges with complex models, but the composite likelihood and approximate Bayesian computation (ABC) offer solutions, simultaneously allowing estimation and hypothesis testing (Beaumont et al. 2002; Nielsen and Beaumont 2009; Chan et al. 2006). Statistical phylogeographic studies unveil novel demographic insights, emphasizing the impact of model-based multilocus approaches on historical inference, including coalescent models and hypothesis testing, to analyze phylogeographic data, thereby evaluating evolutionary and migratory patterns (Beerli and Felsenstein 2001). Moreover, they establish a foundation for exploring a range of evolutionary and ecological inquiries by analyzing population genetic parameters and geographic genetic patterns (Knowles 2009).

In this context, comparative phylogeography emerged aiming to test hypotheses on abiotic and ecological evolution drivers within communities, and employing innovative methods beyond single taxon analyses (Arbogast and Kenagy 2001; Avise 2000). One approach estimates topological congruence across taxa, creating a supertree depicting geographic linkages tested against drivers of lineage divergence (Lapointe and Rissler 2005) and it shares a common conceptual framework with historical biogeography (Zink 1996). Another method, a coalescent model-based ABC with a hierarchical Bayesian model, allows for variability in demographic parameters across taxa, offering insights into congruence in divergence times and historical demographic models (Hickerson and Meyer 2008). Comparative phylogeographic ABC methods have successfully tested simultaneous divergence times and biogeographic scenarios across taxa in various settings (Hickerson et al. 2006; Leaché et al. 2007; Carnaval et al. 2009). We highlight the argument of Bonatelli et al. (2022) advocating for the promotion of comparative studies investigating trait-mediated deterministic processes in regions characterized by exceptional diversity and various hypotheses to explain species diversification, as occurred in the Neotropics.

Finally, as Knowles and Maddison (2002) describe, comparative phylogeography involves integrating molecular, genetic, and geographic data with ecological or morphological information, enhancing our understanding of the evolutionary history and distribution of species. Method selection depends on the research question and available data sources (Alvarado-Serrano and Knowles 2014).

13.3 Large Scale Sequencing in Phylogeography

Numerous approaches have been developed to study changes and population dispersion over time. For instance, a variety of methods have been used in the Neotropics, from next generation sequencing (NGS) (Mardis 2011; Goodwin et al. 2016) to spatial modeling, landscape ecology, and multidisciplinary data integration (Peterson and Soberón 2012; Rissler 2016; Ehrendorfer and Guo 2006). These methods provide valuable insights into the distribution and evolution of populations over time, contributing to a better understanding of the life history of organisms in such a heterogeneous and contrasting area as the neotropical region.

Applying NGS in phylogeography studies in the Neotropics has revolutionized genetic research. This method enables efficient sampling of populations across extensive geographic areas by using specific molecular markers such as SNPs and microsatellites. Platforms such as Illumina and PacBio offer diverse approaches to obtaining comprehensive genomic data, facilitating detailed analyses of genetic structure and historical events influencing genetic diversity (Kulski 2016). This demographic approach contributes to estimating population growth rates and divergence times (Hoffman and Blouin 2004). NGS provides a more comprehensive insight into species evolution, providing detailed analyses of dispersal patterns and connections among populations in the Neotropics (Turchetto-Zolet et al. 2013; Hughes et al. 2013).

As highlighted in the preceding discussion, the progress in phylogeography is propelled by the evolution of diverse molecular markers, statistical methods, and applicable approaches. Phylogeography encompasses four main approaches: descriptive, statistical, comparative, and integrative. Each of these approaches has played a role in advancing our comprehension of patterns and evolutionary relationships among populations based on the information they furnish.

There is a pressing need to increase the number of species with comprehensive phylogeographical data across their entire range. The application of next-generation sequencing to acquire higher-resolution datasets is crucial for incorporating evolutionary responses and often overlooked factors such as soil type and species distribution models (Cavers and Dick 2013). A vital connection exists between data on barriers to gene flow and provenance trials, aiming to identify drivers of evolutionary divergence and establish adaptive units for more effective management strategies (Funk et al. 2012). These methodological advances coincide with increased opportunities to integrate comparative phylogeographic data with other rapidly advancing fields. This includes ecological niche models (Waltari et al. 2007), spatial analysis of genomic signatures in natural selection (Joost et al. 2007), and studies of ecological speciation (Schluter 2009). With this interdisciplinary synthesis, integrating comparative phylogeography is poised to tackle fundamental questions concerning the role of climate, geography, and ecological interactions in shaping the composition and evolution of communities (Hickerson et al. 2010).

Given this premise, numerous challenges arise to improve our understanding of the evolutionary history of tropical species, particularly those less studied through phylogeography. This encompasses using analytical approaches derived from systematics and population genetics, considering technological advances and collaborative international research efforts.

13.4 Phylogeography in Neotropical Regions

The neotropical region has a complex geological history, including the formation of the Andes and the emergence of the Panama land bridge. These events have significantly impacted the evolution and diversification of species in the region. Phylogeographic studies provide valuable insights into how historical processes unfolded across temporal and spatial scales (Santos Leal et al. 2016) and have shaped the distribution and genetic diversity of species in the neotropical region. In this section, we will show some examples of phylogeographic studies in the neotropical regions and how they contribute to the understanding and protection of biodiversity.

THE ANDES: several phylogeographic studies conducted in the Andes have demonstrated that wide-ranging species exhibit phylogeographic breaks in their lowland populations, which are separated by the significant elevation of the Andes (Cavers et al. 2003; Eberhard and Bermingham 2004; Turchetto-Zolet et al. 2013). Here are some specific examples from the Andes: *Quercus humboldtii*—Zorrilla-Azcué et al. 2021; ant-plants (*Azteca* spp./*Tococa guianensis*)—Torres et al. 2021; *Abrothrix olivacea*, *Phyllotis darwini* (small mammals)—Palma et al. 2017). These studies aimed to assess the effects of climatic fluctuations, especially during the Last Glacial Maximum (LGM), on Andean populations from various parts of the mountain range. These studies sought to establish whether the timing of diversification events aligns with the primary uplift activity in the Cordillera.

In the Neotropics, high elevations have played a prominent role in the development and distribution of the biota in the American continent (Antonelli et al. 2018a). For instance, the uplift of the Andes between the Miocene and Oligocene is a significant driver of neotropical biodiversity (Antonelli et al. 2009; Lagomarsino et al. 2016). That stated, the Andes can be regarded as a stage for rapid and extensive evolutionary radiations, standing out as one of the most biodiverse regions on the planet, both in terms of species numbers and levels of endemism (Antonelli et al. 2009, 2018a, b; Ulloa Ulloa et al. 2017; Castello et al. 2020; Barboza et al. 2019).

AMAZON: the Amazon rainforest is one of the biomes/ecoregions in the Neotropics, with the highest number of phylogeographic studies compared to others. It is frequently reported that the Amazon River and its tributaries constitute a historical barrier to gene flow for various organisms (Turchetto-Zolet et al. 2013). Studies have identified distinct phylogeographic patterns ascribed to multiple factors such as marine introgressions, tectonic activity, and Pleistocene refugia. The examples from the Amazon region: trees, Myristicaceae and Moraceae family (Honorio Coronado et al. 2019); tetrapods (Antonelli et al. 2010); birds (Dornas et al. 2022), were characterized by studies that specifically compared and contrasted the evolutionary differences among lineages and the distinctions and similarities within the various groups under investigation. These studies elucidated the genetic diversity, genetic structure, and phylogeographic relationships among populations.

This biome showcases heterogeneity in both floristic and faunistic composition (Silva et al. 2005; Araújo-Silva et al. 2017). The Amazon Basin hosts a disproportionate share of global biodiversity; a probable influence on the evolution of the Amazonian landscape has been a combination of tectonic events shaping river characteristics and climatic changes induced by orbital factors (Haffer and Prance 2001; de Fátima et al. 2005). Despite challenges posed by vast diversity, understanding the systematic relationships of Amazonian species is pivotal for assessing hypotheses explaining the high diversity levels of the region (da Silva and Patton 1998; Leite and Rogers 2013). Significant progress has been achieved for various Amazonian taxa through systematic molecular investigations, shedding light on phylogenetic and biogeographic relationships, palynology, paleontology, and climatology (Symula et al. 2003; Fernandes et al. 2013).

ATLANTIC FOREST: phylogeographic studies focused on organisms inhabiting the Atlantic Forest are still limited (Carnaval et al. 2009; Thomé et al. 2010; Cabanne et al. 2011; d'Horta et al. 2011; Turchetto-Zolet et al. 2013) in relation to the high species diversity of the biome. Therefore, only a few estimates of divergence dates for populations and/or lineages are available (Batalha-Filho et al. 2012).

In the selected examples from the Atlantic Forest, such as *Eugenia uniflora* (Turchetto-Zolet et al. 2016); *Xiphorhynchus fuscus* (Cabanne et al. 2008); *Bothrops jararaca* (Grazziotin et al. 2006), the primary objectives of these investigations were exploring the phylogeographic and demographic history along their natural distributions through plastid DNA, mitochondrial DNA, nuclear DNA, or the cytochrome b gene. The Atlantic Forest is influenced by various factors and biotic and abiotic gradients (Martins 2011; Turchetto-Zolet et al. 2016), and has been previously acknowledged in the literature as a center for the diversification and richness of numerous species (Murray-Smith et al. 2009). Various theories, such as the forest refuge theory, the riparian theory as gene flow barriers, and the ecological gradient theory further support the rationale behind this phenomenon (Brown and Ab'sáber 1979; Ribas et al. 2012; Smith et al. 2001; Cabanne et al. 2011). These theories consider the ecological pressures within the Atlantic Forest, considering the interplay between continental and oceanic dynamics (Barrientos-Díaz et al. 2024).

WEST INDIES: over the past decade, numerous articles focused on phylogeographic studies within the Caribbean region (e.g., Malone et al. 2000; Seal et al. 2011; Camargo et al. 2009; Jowers et al. 2019). Seal et al. (2011) reports that recent advances in molecular methods have significantly facilitated this research (Hedges 2006), in particular, approaches that utilize non-recombining molecular markers to deduce founder effects, population expansion, and bottlenecks (Freeland 2006; Avise 2009). However, addressing fine-scale patterns for the region requires a remarkable multidisciplinary effort, as the primary challenge stems from the vastly different geological histories of the islands, leading to variation in their histories of dispersal and vicariance (de Queiroz 2005; Crisci and Katinas 2009; Crews and Esposito 2020).

Various studies have explored potential dispersal routes for terrestrial organisms to colonize distant islands, including over-water transports via natural rafts and potential land connections (Cornee et al. 2021). For the Antilles or West Indies, the examples (genus *Cyclura*—Malone et al. 2000; *Zenaida aurita*—Monceau et al. 2013; *Zamia pumila*—Salas-Leiva et al. 2017) are focused on identifying historical processes of colonization and expansion across the Caribbean islands by examining the potential roles that vicariance and/or dispersal mechanisms have played in species in terms of genetic divergence, gene flow, and population structure within the Antilles. West Indies, comprising the Greater and Lesser Antilles on the northeastern edge of the Caribbean Plate, are a crucial hub of island biodiversity (Myers et al. 2000; Mittermeier et al. 2011). Despite extensive research spanning decades, debates persist over the phylogenetic origin of this remarkable biodiversity and their historical biogeography. From ecological and evolutionary standpoints, research on island biotas offer crucial insights into speciation, adaptive radiation, community assembly, and biogeography (Losos and Ricklefs 2009).

SOUTH AMERICAN DIAGONAL OF OPEN FORMATIONS (DOF): despite the rich biodiversity of 19,114 plant species in the dry diagonal, only 0.15% have phylogeographic information (Collevatti et al. 2020). With an alarming annual extinction rate of 0.01%, two plant species are expected to vanish yearly. These data highlight the urgent need for increased funding in research due to the stark contrast between scientific exploration and species extinction. The global neglect of dry biomes exacerbates threats to Seasonally Dry Forests in South America, particularly in Brazil, that face challenges such as agricultural expansion and increasing fire episodes (Collevatti et al. 2013; Sunderland et al. 2015; DryFlor et al. 2016).

The examples cited here (*Phyllopezus pollicaris* (geckos)—Werneck et al. 2012; *Lepidocolaptes angustirostris*—Rocha et al. 2020; *Mandirola hirsuta*—Fiorini et al. 2020) were focused on describing the phylogeographic patterns and structure of these species, taking into account the underlying processes that shaped the evolutionary history of the dry diagonal. These results were achieved through various approaches, such as environmental distribution models to explore climatic fluctuations and divergence under biogeographic models and genetic diversity gradients to infer signs of expansion and colonization. Three domains occur in this region: the Caatinga, the primary nucleus of seasonally dry tropical forests (SDTFs) in northeastern Brazil; the Cerrado, a mosaic of savanna physiognomies in central Brazil; and the Chaco, composed of subtropical seasonally dry forests in northern Argentina, Paraguay and Bolivia (Werneck 2011; Bonatelli et al. 2022).

13.5 Case Studies

Phylogeographical studies of neotropical species have contributed to unveiling the evolutionary history of several taxa. Here, we show some examples of studies performed with neotropical plant species, which highlight the importance of phylogeography in conservation efforts for plant biodiversity.

***Eugenia uniflora* L.**

Eugenia uniflora (pitanga or Brazilian cherry) belongs to the Myrtaceae family and is widely distributed in the Atlantic Forest. It can grow in various habitats, from the sandy coastal plain, known as restinga, in northeastern Brazil to the riparian vegetation in the southern regions extending up to 400 km inland in Brazil, Argentina, and Uruguay. Throughout its distribution, *E. uniflora* shows remarkable phenotypic variability. It grows as a woody tree, averaging 10 meters in height in the riparian forests and up to shoulder-height bushes in the dunes of Restinga. Some population genetics and phylogeographical studies based on molecular data were performed for this species. Population genetics studies can assess genetic diversity, population structure, and differentiation (Margis et al. 2002; Salgueiro et al. 2004; Ferreira-Ramos et al. 2008, 2014). A phylogeographical study revealed the demographic history and lineage divergence (Turchetto-Zolet et al. 2016), suggesting that the evolutionary history of the species was shaped by distinct factors in the southern and northern ranges of its distribution.

The study of Turchetto-Zolet et al. (2016) examined plastid DNA sequence markers of 313 individuals from 46 sites spanning the distribution of *E. uniflora*. The authors were aimed at investigating phylogeographic patterns and the demographic history of the species, as well as to understand how past climatic oscillations have affected its current genetic variation. The analyses of this dataset allowed this study to identify two divergent lineages with an estimated divergence time of c. 4.9 Mya. Furthermore, the study revealed a phylogeographic break between groups of populations located south of Torres, in the State of Rio Grande do Sul (c. 29°-30°S), which is recognized as a critical phytogeographical boundary. This division separated the two lineages, one distributed in the south of Brazil and Argentina (riparian forest and restinga) and the other in the southeast and northeast of Brazil (restinga). This pattern of lineage divergence was also found for the orchid

Epidendrum fulgens (Pinheiro et al. 2011), a neotropical coastal species occurring on coastal dunes and granitic outcrops in the Brazilian Atlantic Forest. Low number of haplotypes were shared between *E. uniflora* populations from northern and southern distributions. The maternal inheritance of the markers used, and zoochoric dispersal syndrome of the species may explain this differentiation. The authors also observed low migration rates between the genetic units in the north and south, implying the existence of seed dispersal barriers.

Eugenia uniflora populations from southeastern and northeastern Brazil associated with restinga have low genetic diversity and structure, while southern populations have high genetic diversity and are also highly structured. The southern populations grow in more diverse ecosystems, such as sandy coastal plains, riparian forests, and open vegetation in the Pampa Biome. Pinheiro et al. (2011) also found higher genetic diversity in southern areas of distribution of *E. fulgens*.

Demographic analyses based on cpDNA revealed distinct demographic histories for populations in the northern and southern parts of the spanning range of the species. Populations from northern regions experienced moderate changes in their effective population sizes, exhibiting signatures of recent demographic expansion. In contrast, populations from the southern regions experienced smaller or no changes in effective sizes, thus presenting demographic stability. This result suggests that *E. uniflora* likely maintained stable populations in its southern distribution throughout the late Pleistocene, even during peaks of glaciations when forests became fragmented, and the climate turned temperate.

The findings of Turchetto-Zolet et al. (2016) underscore the role of phylogeographic studies in elucidating the evolutionary history of this widely distributed species. Despite its abundance within the Atlantic Forest, this species exhibits divergent lineages and is associated with different ecosystems, reinforcing attention to the conservation of their populations.

The results of this study contributed to better acknowledge the evolutionary history of a tree species widely distributed in neotropical humid forests, which has taxonomic problems. The two varieties classified within the species require a taxonomic review since the phylogeographic study showed they are quite divergent. Thus, this study contributes to the systematics of the group and provides important information for conservation efforts in this species.

Schizolobium parahyba (Vell.) Blake

Schizolobium parahyba is a member of the Fabaceae family and is widely distributed in neotropical rainforests, such as the Atlantic Forest, Amazonian, Andean, and Central American forests. This species has two varieties: *S. parahyba* var. *parahyba* and *S. parahyba* var. *amazonicum* (Ducke) Barneby. These varieties have distinct distributions: *S. parahyba* var. *parahyba* is prevalent in the Brazilian Atlantic Forest (ranging from Bahia 13°S to Santa Catarina 25°S), as well as in Paraguay and Argentina. On the other hand, *S. parahyba* var. *amazonicum* is found from the Amazon basin in Brazil and Bolivia to southeastern Mexico. Notably, *S. parahyba* shows no adaptation to savannas or the typical xeromorphic Caatinga, nor does it grow at elevations above 700 meters above sea level.

Turchetto-Zolet et al. (2012) analyzed sequence variation of chloroplast (cpDNA) and nuclear (nrDNA) markers, which exhibit different modes of inheritance, to elucidate the evolutionary history of this species. They aimed to examine the pattern of genetic diversity, population structure, and the role of the dry corridor formed by Cerrado–Caatinga–Chaco and the Andean uplift as a barrier to the gene flow of the species. The authors sampled 198 trees from 20 populations of *S. parahyba* spanning most of its distribution range. Of the total, 118 trees were classified as *S. parahyba* var. *parahyba* and 80 as *S. parahyba* var. *amazonicum*. A high level of genetic diversity was found in a few sites of *S. parahyba*, with two primary genetic diversity centers: one located within the southeastern Atlantic Forest, predominantly represented by *S. parahyba* var. *parahyba*, and the other located within the Amazonian basin, primarily characterized by *S. parahyba* var. *amazonicum*. These regions also exhibit the presence of fixed haplotypes, which may have been the result of a founder effect.

The authors observed a notable trend of southward expansion within the Atlantic Forest upon independently analyzing the *S. parahyba* var. *parahyba* populations. This finding may be related to the more recent establishment of these forests and is consistent with an expansion of distribution and recolonization of the southern portion of the forest by migrants from the Atlantic Forest of southeastern Brazil. Phylogeographic studies of *Passiflora actinia* and *Passiflora elegans* (Lorenz-Lemke et al. 2005) and *Vriesea gigantean* (Palma-Silva et al. 2009) also report expansion towards the south of the Atlantic Forest. These findings corroborate fossil pollen records, which unveiled a replacement of forests by grasslands during the Last Glacial Maximum in the southern Atlantic Forest (Behling and Negrelle 2001; Carnaval and Moritz 2008).

Populations of *S. parahyba* var. *parahyba* showed high genetic differentiation compared to populations of *S. parahyba* var. *amazonicum*, with almost no shared haplotypes. This result indicates that Atlantic and Amazonian Forest populations of *S. parahyba* were once fragmented by dry corridors formed by the Cerrado and Caatinga ecoregions, which form potential barriers to *S. parahyba* gene flow between these two wet forests. Considering only the populations of S. *parahyba* var. *amazonicum*, which are distributed in the Amazon and Central American forests, the work of Turchetto-Zolet et al. (2012) found divergent lineages among populations on both sides of the Andes Mountain range. This result suggests that the Andean uplift may have been a potential barrier to gene flow between Central American/Colombian and Amazonian Forest populations (cis–trans-Andean population separation).

Petunia integrifolia (Hook.) Schinz and Thell. complex

Traditionally, morphological traits have been used to describe species; however, with the advances of molecular tools, studies have recognized that morphological divergence often has little correlation with genetic divergence. In this context, morphologically very similar organisms may constitute different evolutionary lineages, be reproductively isolated, and be different species. Conversely, morphologically distinct individuals may belong to the same evolutionary lineage and be one species

(Orr 2001). In this context, cryptic species are a group of taxa that are morphologically identical or very similar, reproductively isolated, and frequently sympatric. Cryptic species often form what is known as species complexes, a group of organisms that are known not to be the same species but for which the biological classification is not precisely known. Cryptic species and species complexes lead to erroneous samplings that limit our understanding of biodiversity (Heath et al. 2008). In plants, species within a complex can often cross and produce fertile offspring; however, even if hybrids and introgression occur in nature, the parental species often remain separate entities (Pinheiro et al. 2018). A phylogeographical approach is frequently used to understand the evolutionary history of cryptic species or species complexes once the organisms are usually evolutionary very close and the gene lineages do not coalesce (Kuhner 2009).

The genus *Petunia* Juss (Solanaceae) is native to southern South America and is predominantly found in southern Brazil. The genus is known worldwide, since *Petunia hybrida* is a model plant for science and is also widely cultivated as an ornamental plant (Vandenbussche et al. 2016). Species of the genus diverged very recently in geological time scale (around two million years) and are genetically and morphologically similar (Särkinen et al. 2013). Currently, 17 species are recognized (Greppi et al. 2019), but many taxonomic questions remain regarding the classification of species and subspecies of the genus. One of these taxonomic issues concerns the species complex of *P. integrifolia (*Hook.) Schinz and Thell. The similarities among the species are purple and infundibuliform corolla, violaceous pollen, and a stigma that is located between the anthers of the didynamous stamens. According to Stehmann et al. (2009), the complex comprised five recognized taxa: *Petunia bajeensis* T. Ando and Hashim, *P. inflata* R. E. Fr., *P. integrifolia* subsp. *depauperata* R. E. Fr. (hereafter, *Petunia depauperata*), *P. integrifolia* subsp. *integrifolia* (hereafter *P. integrifolia*), and *Petunia interior* T. Ando and Hashim. However, some authors consider *Petunia littoralis* L. B. Sm. and Downs and *Petunia riograndensis* T. Ando and Hashim to be valid species in this group.

Phylogeographical approaches have been employed to unravel the number of evolutionary lineages in this complex. The first study, developed by Longo et al. (2014), used a descriptive phylogeographical approach to understand the evolutionary history of representatives of the *P. integrifolia* group (*P. integrifolia* subsp. *integrifolia*; *P. riograndensis*; *P. integrifolia* var. *depauperata* Smith and Downs; *P. littoralis*; and *P. inflata)* through the analysis of nuclear and plastid molecular markers. The network analysis of plastid markers divided the *P. integrifolia* group into three distinct and allopatric lineages: *P. integrifolia* (including *P. integrifolia* subsp. *integrifolia*, *P. integrifolia* var. depauperata, *P. littoralis* and *P. riograndensis*), *P. inflata* and *P. interior*. The phylogenetic tree for the combined nuclear (ITS) and plastidial markers showed two main clades: one consisting of sequences from individuals of the *P. inflata*, *integrifolia,* and *riograndensis* morphological groups (called mainland clade); and another with sequences of the *integrifolia*, *depauperata* and *littoralis* morphotypes (called coastal clade). This study suggests that *P. inflata* and *P. interior* are valid taxonomic entities and that the other taxa (*P. integrifolia* subsp. *integrifolia*, *P. integrifolia* var. depauperata, *P. littoralis* and *P.*

riograndensis) belong to two genetic divergently groups, one encompassing organisms found in the Atlantic Coastal Plain and another encompassing the rest of the organisms (mailand). Following this, Ramos-Fregonezi et al. (2015) evaluated the degree of genetic variation and structure among and within populations of the coastal plain and mainland organisms. They used plastidial markers and statistical phylogeography methods and concluded that the climate changes during the Pleistocene in the Atlantic Coastal Plain promoted speciation and divided the taxa into four lineages: mainland and coastal north, center, and south. In a phylogeography integrative study combined with genetic population analyses, Segatto et al. (2017) evaluated the genetic variability between the different taxa of the *P. integrifolia* complex (*P. bajeensis*, *P. inflata*, *P. depauperata*, *P. integrifolia* and *P. interior*. This study combined statistical phylogeography with ecological niche modeling and bioregions analysis. The authors concluded that four evolutionary lineages are associated with different environmental constraints (*P. integrifolia*, *P. depauperata*, *P. bajeensis*, and *P. inflata* + *P. interior*). The last work published evaluating taxons from the complex (Guzmán et al. 2022) uses SNPs and extensive genome coverage. Using this genomic approach, it is possible to investigate the association between neutral and potentially under-selection polymorphisms throughout the geographic distribution of the species. The approaches using massive SNPs are currently the golden standard of statistical phylogeography. The authors analyzed the same taxa that Ramos-Fregonezi et al. (2015) and concluded that the species were under differential environmental conditions during their diversification. Each taxon constitutes an independent genetic cluster, confirming the previous hypotheses that the coastal organisms are divided into three evolutionary lineages. Thus, the current taxonomic classification of *P. integrifolia* complex species does not reflect their evolutionary lineages, and studies that combine genetic and morphological variables are necessary to propose a new circumscription. Figure 13.2 summarizes the results within the *P. integrifolia* complex.

***Petunia axilaris* (Lam.) Britton, Sterns and Poggenb complex**

A common approach to phylogeographical studies is to identify genetic lineages across one or more species and then look for historical factors underlying the current distribution of these lineages. A lack of gene flow, drift, and adaptation to local environmental conditions among geographically separated populations can eventually result in genetically distinct lineages. The rise of genetically distinct lineages could be a by-product of population isolation, eventually leading to the evolution of reproduction barriers and speciation. Thus, the phylogeographic and population genetics approaches could be fundamental to identifying new or incipient species. One example was a study of an herbaceous plant, *P. axillaris*, widespread in southeastern South America (Turchetto et al. 2014a). This species belongs to the Solanaceae family and occurs in Pampa and Chaco open vegetation. Three geographical subspecies are recognized in the distribution, presenting distinct morphological traits (Turchetto et al. 2014b). This subspecies distribution is consistent with distinct genetic lineages, as found in Turchetto et al. (2014a), as identified through a phylogeographic and population genetic tool. That study analyzed plastid DNA

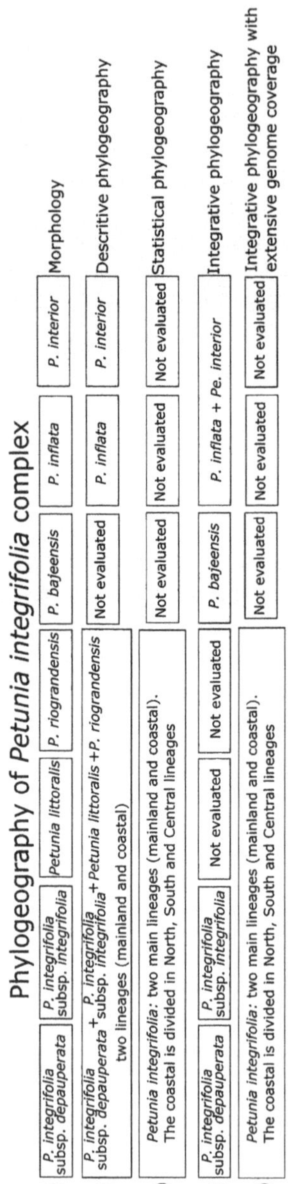

Fig. 13.2 Taxa of *Petunia integrifolia* complex and its evolutionary relationship disentangled by phylogeography approaches

sequence markers of 614 individuals from 102 collection sites covering a large part of the distribution of *P. axillaris*. The study also analyzed nuclear CAPS markers for a subset of samples and morphological and climatic data. The distinct genetic lineages also presented a high degree of climatic niche divergence and high population size with a recent divergence time estimate (~58 kya). The population dynamics over time was different for each genetic lineage (subspecies), suggesting a historical population growth for *P. axillaris* subsp. *parodii*, population decline for *P. axillaris* subsp. *subandina* and some of them moderate bottleneck (~ 20 kya) and later expansion since 10 kya for *P. axillaris* subsp. *axillaris*. The study also found a secondary contact region between two subspecies, but no extensive gene flow was found. These findings broadly agree with the different impact of Pleistocene climatic oscillations in the population dynamics of *P. axillaris* subspecies, according to the species distribution modeling results shown in Giudicelli et al. (2019). The population genetic and phylogeographical analyses suggest that differences between *Petunia axillaris* subspecies could represent an incipient speciation stage. A continued study with this model plant species analyzed 11,348 neutral SNPs from 96 individuals from 20 collection sites of *P. axillaris* subsp. *parodii* (Giudicelli et al. 2022). This study found three genetic lineages in this subspecies and identified the Pleistocene refuges and rivers acting as a barrier to gene flow between populations of this subspecies.

13.6 Final Remarks

Researchers have explored the interplay of historical, geographical, and environmental factors to comprehend species diversity in the Neotropics. Advanced genetic methods and spatial models have been used to investigate plant and animal dispersal, evolution, and speciation patterns. These phylogeographic studies have unveiled the significance of historical events, such as climatic and geological changes, in shaping genetic diversity and species distribution in this region. Moreover, they have underscored that it is imperative to conserve areas of high genetic diversity and uphold habitat connectivity to ensure the long-term survival of populations in this highly biodiverse zone. It has shed light on high levels of genetic diversity, genetic structure, hybridization, and various historical events, thereby serving as a pillar in population genetics research. Furthermore, advances in molecular methods and more efficient machine learning have facilitated the availability of larger datasets for the region, leading to fundamental insights into the evolutionary processes that have given rise to the extensive biodiversity we observe in the Neotropics today.

Key Points

- This chapter described the origin and importance of the field of phylogeography for neotropical species, by combining population genetics, phylogenetics and geographic datasets to unravel evolutionary factors that have implicated in the distribution of species across different environments;

- Molecular markers and statistical methods are crucial in unveiling the phylogeographic history of a species;
- Phylogeographic studies allow us to characterize and protect biodiversity, understand dispersal modes, identify refuge areas, get information about lineage diversification times, predict extinctions, and understand other species-/population-level processes.

Questions

1. Phylogeography has a very close relationship with population genetics. What are the main differences between these two disciplines?
2. What are the essential criteria for molecular markers to be considered suitable for a phylogeographic investigation?
3. What are the differences between descriptive, statistical, comparative, and integrative phylogeography?
4. How can phylogeography contribute to the knowledge and protection of neotropical biodiversity?

Acknowledgments We thank Conselho Nacional de Desenvolvimento Científico e Tecnológico (CNPq grant 313949/2023-9 to ACTZ), Fundação de Amparo à pesquisa do Estado do RS (FAPERGS) and Fundação Coordenação de Aperfeiçoamento de Pessoal de Nível Superior (CAPES) for financial support.

References

Alvarado-Serrano DF, Knowles LL (2014) Ecological niche models in phylogeographic studies: applications, advances and precautions. Mol Ecol Resour 14:233–248. https://doi.org/10.1111/1755-0998.12184

Antonelli A, Ariza M, Albert J et al (2018a) Conceptual and empirical advances in Neotropical biodiversity research. PeerJ 6:e5644. https://doi.org/10.7717/peerj.5644

Antonelli A, Kissling WD, Flantua SGA et al (2018b) Geological and climatic influences on mountain biodiversity. Nat Geosci 11:718–725. https://doi.org/10.1038/s41561-018-0236-z

Antonelli A, Nylander JAA, Persson C et al (2009) Tracing the impact of the Andean uplift on neotropical plant evolu-tion. Proc Natl Acad Sci 106(24):9749–9754. https://doi.org/10.1073/pnas.0811421106

Antonelli A, Quijada-Mascareñas A, Crawford AJ et al (2010) Molecular studies and phylogeography of Amazonian tetrapods and their relation to geological and climatic models. In: Amazonia, landscape and species evolution: a look into the past, vol 386, p 404

Araújo-Silva LE, Miranda LS, Carneiro L et al (2017) Phylogeography and diversification of an Amazonian understorey hummingbird: paraphyly and evidence for widespread cryptic speciation in the Plio-Pleistocene. Ibis 159:778–791. https://doi.org/10.1111/ibi.12500

Arbogast BS, Kenagy GJ (2001) Comparative phylogeography as an integrative approach to historical biogeography. J Biogeogr 28:819–825. https://doi.org/10.1046/j.1365-2699.2001.00594.x

Avise JC (2000) Phylogeography: the history and formation of species. Harvard University Press

Avise JC, Arnold J, Ball RM et al (1987) Intraspecific phylogeography: the mitochondrial DNA bridge between population genetics and systematics. Annu Rev Ecol Syst 18(1):489–522. https://doi.org/10.1146/annurev.es.18.110187.002421

Avise JC (2009) Phylogeography: retrospect and prospect. J Biogeogr 36:3–15. https://doi.org/10.1111/j.1365-2699.2008.02032.x

Barboza GE, Carrizo García C, Leiva González S et al (2019) Four new species of *Capsicum* (Solanaceae) from the tropical Andes and an update on the phylogeny of the genus. PLoS One 14(1):e0209792. https://doi.org/10.1371/journal.pone.0209792

Barrientos-Díaz O, Báez-Lizarazo MR, Enderle F et al (2024) The Atlantic forest is a potentially climatic suitable habitat for four Neotropical Myrtaceae species through time. Eco Inform 80:102490. https://doi.org/10.1016/j.ecoinf.2024.102490

Batalha-Filho H, Cabanne GS, Miyaki CY (2012) Phylogeography of an Atlantic Forest passerine reveals demographic stability through the last glacial maximum. Mol Phylogenet Evol 65(3):892–902. https://doi.org/10.1016/j.ympev.2012.08.010

Beaumont MA, Zhang W, Balding DJ (2002) Approximate Bayesian computation in population genetics. Genetics 162(4):2025–2035. https://doi.org/10.1093/genetics/162.4.2025

Beerli P, Felsenstein J (2001) Maximum likelihood estimation of a migration matrix and effective population sizes in n subpopulations by using a coalescent approach. Proc Natl Acad Sci 98(8):4563–4568. https://doi.org/10.1073/pnas.08106809

Beheregaray LB (2008) Twenty years of phylogeography: the state of the field and the challenges for the Southern Hemisphere. Mol Ecol 17:3754–3774. https://doi.org/10.1111/j.1365-294X.2008.03857.x

Behling H, Negrelle RR (2001) Tropical rain forest and climate dynamics of the Atlantic lowland, Southern Brazil, during the Late Quaternary. Quat Res 56(3):383–389. https://doi.org/10.1006/qres.2001.2264

Bonatelli IA, Gehara M, Carstens BC et al (2022) Comparative and predictive phylogeography in the South American diagonal of open formations: unravelling the biological and environmental influences on multitaxon demography. Mol Ecol 31(1):331–342. https://doi.org/10.1111/mec.16210

Brown JR, Ab'sáber AN (1979) Ice-age Forest refuges and evolution in the Neotropics: correlation of palaeoclimatological, geomorphological and pedological data with modern biological endemism. Paleoclimas (São Paulo) 5:1–30. Available from: biblio.fflch.usp.br/AbSaber AN_1351410_IceAgeForestRefugesAndEvolution.pdf

Brumfield RT, Beerli P, Nickerson DA et al (2003) The utility of single nucleotide polymorphisms in inferences of population history. Trends Ecol Evol 18(5):249–256. https://doi.org/10.1016/S0169-5347(03)00018-1

Cabanne GS, d'Horta FM, Sari EH et al (2008) Nuclear and mitochondrial phylogeography of the Atlantic forest endemic Xiphorhynchus fuscus (Aves: Dendrocolaptidae): biogeography and systematics implications. Mol Phylogenet Evol 49(3):760–773. https://doi.org/10.1016/j.ympev.2008.09.013

Cabanne GS, D'Horta FM, Meyer D et al (2011) Evolution of Dendrocolaptes platyrostris (Aves: Furnariidae) between the south American open vegetation corridor and the Atlantic Forest. Biol J Linn Soc 103(4):801–820. https://doi.org/10.1111/j.1095-8312.2011.01678.x

Camargo A, Heyer WR, de Sá RO (2009) Phylogeography of the frog Leptodactylus validus (Amphibia: Anura): patterns and timing of colonization events in the Lesser Antilles. Mol Phylogenet Evol 53(2):571–579. https://doi.org/10.1016/j.ympev.2009.07.004

Carnaval AC, Moritz C (2008) Historical climate modelling predicts patterns of current biodiversity in the Brazilian Atlantic forest. J Biogeogr 35(7):1187–1201. https://doi.org/10.1111/j.1365-2699.2007.01870.x

Carnaval AC, Hickerson MJ, Haddad CF et al (2009) Stability predicts genetic diversity in the Brazilian Atlantic forest hotspot. Science 323(5915):785–789. https://doi.org/10.1126/science.1166955

Carstens BC, Degenhardt JD, Stevenson AL et al (2005) Accounting for coalescent stochasticity in testing phylogeographical hypotheses: modelling Pleistocene population structure in the Idaho giant salamander Dicamptodon aterrimus. Mol Ecol 14(1):255–265. https://doi.org/10.1111/j.1365-294X.2004.02404.x

Castello LV, Chiapella JO, Barfuss MH et al (2020) Comparative phylogeography, morphological boundaries and climate envelopes of two sympatric widespread Bromeliaceae from the southern Andes. Bot J Linn Soc 192(4):726–743. https://doi.org/10.1093/botlinnean/boz108

Catchen J, Hohenlohe PA, Bassham S et al (2013) Stacks: an analysis tool set for population genomics. Mol Ecol 22(11):3124–3140. https://doi.org/10.1111/mec.12354

Cavers S, Dick CW (2013) Phylogeography of neotropical trees. J Biogeogr 40:615–617. http://wileyonlinelibrary.com/journal/jbi/doi:10.1111/jbi.1209

Cavers S, Navarro C, Lowe AJ (2003) Chloroplast DNA phylogeography reveals colonization history of a Neotropical tree, Cedrela odorata L., in Mesoamerica. Mol Ecol 12(6):1451–1460. https://doi.org/10.1046/j.1365-294X.2003.01810.x

Chamberlin TC (1965) The method of multiple working hypotheses: with this method the dangers of parental affection for a favorite theory can be circumvented. Science 148(3671):754–759. https://doi.org/10.1126/science.148.3671.754

Chan YL, Anderson CNK, Hadly EA (2006) Bayesian estimation of the timing and severity of a population bottleneck from ancient DNA. PLoS Genet 2(4):e59. https://doi.org/10.1371/journal.pgen.0020059

Collevatti RG, Lima-Ribeiro MS, Diniz-Filho JAF et al (2013) Stability of Brazilian seasonally dry forests under climate change: inferences for long-term conservation. Am J Plant Sci 04(04). https://doi.org/10.4236/ajps.2013.44098

Collevatti RG, Lima NE, Vitorino LC (2020) The diversification of extant angiosperms in the South America dry diagonal. In: Rull V, Carnaval A (eds) Neotropical diversification: patterns and processes. Fascinating life sciences. Springer, Cham. https://doi.org/10.1007/978-3-030-31167-4_21

Cornee JJ, Münch P, Philippon M et al (2021) Lost islands in the northern Lesser Antilles: possible milestones in the Cenozoic dispersal of terrestrial organisms between South-America and the Greater Antilles. Earth Sci Rev 217:103617. https://doi.org/10.1016/j.earscirev.2021.103617

Crews SC, Esposito LA (2020) Towards a synthesis of the Caribbean biogeography of terrestrial arthropods. BMC Evol Biol 20:12. https://doi.org/10.1186/s12862-019-1576-z

Crisci JV, Katinas L (2009) Darwin, historical biogeography, and the importance of overcoming binary opposites. J Biogeogr 36(6):1027–1032. https://doi.org/10.1111/j.1365-2699.2009.02111.x

D'Horta FM, Cabanne GS, Meyer D et al (2011) The genetic effects of Late Quaternary climatic changes over a tropical latitudinal gradient: diversification of an Atlantic Forest passerine. Mol Ecol 20:1923–1935. https://doi.org/10.1111/j.1365-294X.2011.05063.x

da Silva MNF, Patton JL (1998) Molecular phylogeography and the evolution and conservation of Amazonian mammals. Mol Ecol 7:475–486. https://doi.org/10.1046/j.1365-294x.1998.00276.x

Davey JW, Hohenlohe PA, Etter PD et al (2011) Genome-wide genetic marker discovery and genotyping using next-generation sequencing. Nat Rev Genet 12:499–510. https://doi.org/10.1038/nrg3012

de Fátima RD, Mann de Toledo P, Góes AM (2005) New geological framework for Western Amazonia (Brazil) and implications for biogeography and evolution. Quat Res 63(1):78–89. https://doi.org/10.1016/j.yqres.2004.10.001

De Queiroz A (2005) The resurrection of oceanic dispersal in historical biogeography. Trends Ecol Evol 20(2):68–73. https://doi.org/10.1016/j.tree.2004.11.006

Dornas T, de Carvalho CB, dos Anjos JS et al (2022) Avifauna Preliminar da Serra Geral, Região Central do Brasil: Conservação de Aves Campestres no Bioma Cerrado e a contribuição das Brigadas de Incêndio no conhecimento da Biodiversidade: Birds of The Serra Geral, Central Region of Brazil, Cerrado Biome. Revista Acta Ambiental Catarinense 19(1):01–31. https://doi.org/10.24021/raac.v19i1.5959

DryFlor B-RK, Delgado-Salinas A, Dexter KG et al (2016) Plant diversity patterns in neotropical dry forests and their conservation implications. Science 353(6306):1383–1387. https://doi.org/10.1126/science.aaf5080

Eberhard JR, Bermingham E (2004) Phylogeny and Biogeography of the *Amazona Ochrocephala* (Aves: Psittacidae) Complex. Auk 121(2):318–332. https://doi.org/10.1093/auk/121.2.318

Edwards S, Beerli P (2000) Perspective: gene divergence, population divergence, and the variance in coalescence time in phylogeographic studies. Evolution 54(6):1839–1854. https://doi.org/10.1111/j.0014-3820.2000.tb01231.x

Ehrendorfer F, Guo YP (2006) Multidisciplinary studies on Achillea sensu lato (Compositae-anthemideae): new data on systematics and phylogeography. Willdenowia:69–87. http://www.jstor.org/stable/3997683

Fagundes NJ, Ray N, Beaumont M et al (2007) Statistical evaluation of alternative models of human evolution. Proc Natl Acad Sci 104(45):17614–17619. https://doi.org/10.1073/pnas.0708280104

Fernandes AM, Gonzalez J, Wink M et al (2013) Multilocus phylogeography of the wedge-billed Woodcreeper Glyphorynchus spirurus (Aves, Furnariidae) in lowland Amazonia: widespread cryptic diversity and paraphyly reveal a complex diversification pattern. Mol Phylogenet Evol 66(1):270–282. https://doi.org/10.1016/j.ympev.2012.09.033

Ferreira-Ramos R, Accoroni KAG, Rossi A et al (2014) (2014). Genetic diversity assessment for *Eugenia uniflora* L., *E. pyriformis* Cambess., *E. brasiliensis* Lam. and *E. francavilleana* O. Berg neotropical tree species (Myrtaceae) with heterologous SSR markers. Genet Resour Crop Evol 61:267–272. https://doi.org/10.1007/s10722-013-0028-7

Ferreira-Ramos R, Laborda PR, Santos MO et al (2008) Genetic analyses of three populations of *Eugenia uniflora* L. through of newly developed SSR markers for *Eugenia uniflora* L. Conserv Genet 9:1281–1285. https://doi.org/10.1007/s10592-007-9458-0

Fiorini CF, Peres EA, da Silva MJ et al (2020) Phylogeography of the specialist plant Mandirola hirsuta (Gesneriaceae) suggests ancient habitat fragmentation due to savanna expansion. Flora 262:151522. https://doi.org/10.1016/j.flora.2019.151522

Freeland JR (2006) Molecular ecology. Wiley

Funk WC, McKay JK, Hohenlohe PA et al (2012) Harnessing genomics for delineating conservation units. Trends Ecol Evol 27(9):489–496. https://doi.org/10.1016/j.tree.2012.05.012

Garrick RC, Bonatelli IAS, Hyseni C et al (2015) The evolution of phylogeographic data sets. Mol Ecol 24:1164–1171. https://doi.org/10.1111/mec.13108

Giudicelli GC, Turchetto C, Guzmán-Rodriguez S et al (2022) Population genomics indicates micro-refuges and riverine barriers for a southern South American grassland nightshade. J Biogeogr 49:51–65. https://doi.org/10.1111/jbi.14277

Giudicelli GC, Turchetto C, Silva- Arias GA et al (2019) Influence of climate changes on the potential distribution of a widespread grassland species in South America. Perspect Plant Ecol Evol Syst 41:125496. https://doi.org/10.1016/j.ppees.2019.125496

Goodwin S, McPherson J, McCombie W (2016) Coming of age: ten years of next-generation sequencing technologies. Nat Rev Genet 17:333–351. https://doi.org/10.1038/nrg.2016.49

Grazziotin FG, Monzel M, Echeverrigaray S et al (2006) Phylogeography of the *Bothrops jararaca* complex (Serpentes: Viperidae): past fragmentation and Island colonization in the Brazilian Atlantic Forest. Mol Ecol 15:3969–3982. https://doi.org/10.1111/j.1365-294X.2006.03057.x

Greppi JA, Hagiwara JC, Stehmann JR (2019) A new species of Petunia (Solanaceae) from Corrientes, Argentina. Phytotaxa 414:289–295. https://doi.org/10.11646/phytotaxa.414.6.3

Guzmán S, Giudicelli GC, Turchetto C et al (2022) Neutral and outlier single nucleotide polymorphisms disentangle the evolutionary history of a coastal Solanaceae species. Mol Ecol 31:2847–2864. https://doi.org/10.1111/mec.16441

Haffer J, Prance GT (2001) Climatic forcing of evolution in Amazonia during the Cenozoic: on the refuge theory of biotic differentiation. Amazoniana 16(3/4):579–607. https://core.ac.uk/download/pdf/231715823.pdf

Heath TA, Hedtke SM, Hillis DM (2008) Taxon sampling and the accuracy of phylogenetic analyses. J Syst Evol 46:239–257. https://www.jse.ac.cn/CN/10.3724/SP.J.1002.2008.08016
Hedges SB (2006) Paleogeography of the Antilles and origin of West Indian terrestrial vertebrates. Ann Mo Bot Gard 93(2):231–244. https://doi.org/10.3417/0026-6493(2006)93[231:POTAAO]2.0.CO;2
Hey J, Machado C (2003) The study of structured populations — new hope for a difficult and divided science. Nat Rev Genet 4:535–543. https://doi.org/10.1038/nrg1112
Hickerson MJ, Meyer CP (2008) Testing comparative phylogeographic models of marine vicariance and dispersal using a hierarchical Bayesian approach. BMC Evol Biol 8:322. https://doi.org/10.1186/1471-2148-8-322
Hickerson MJ, Carstens BC, Cavender-Bares J et al (2010) Phylogeography's past, Catchenpresent, and future: 10 years after. Mol Phylogenet Evol 54(1):291–301. https://doi.org/10.1016/j.ympev.2009.09.016
Hickerson MJ, Stahl EA, Lessios HA (2006) Test for simultaneous divergence using approximate Bayesian computation. Evolution 60:2435–2453. https://doi.org/10.1111/j.0014-3820.2006.tb01880.x
Hoffman EA, Blouin MS (2004) Evolutionary history of the northern leopard frog: reconstruction of phylogeny, phylogeography, and historical changes in population demography from mitochondrial DNA. Evolution 58:145–159. https://doi.org/10.1111/j.0014-3820.2004.tb01581.x
Honorio Coronado EN, Dexter KG, Hart ML et al (2019) Comparative phylogeography of five widespread tree species: insights into the history of western Amazonia. Ecol Evol 9:7333–7345. https://doi.org/10.1002/ece3.5306
Hughes CE, Pennington RT, Antonelli A (2013) Neotropical plant evolution: assembling the big picture. Bot J Linn Soc 171(1):1–18. https://doi.org/10.1111/boj.12006
Joost S, Bonin A, Bruford MW et al (2007) A spatial analysis method (SAM) to detect candidate loci for selection: towards a landscape genomics approach to adaptation. Mol Ecol 16(18):3955–3969. https://doi.org/10.1111/j.1365-294X.2007.03442.x
Jowers MJ, Garcia Mudarra JL, Charles SP et al (2019) Phylogeography of West Indies Coral snakes (*Micrurus*): Island colonisation and banding patterns. Zool Scr 48:263–276. https://doi.org/10.1111/zsc.12346
Knowles LL, Maddison WP (2002) Statistical phylogeography. Mol Ecol 11(12):2623–2635. https://citeseerx.ist.psu.edu/document?repid=rep1andtype=pdfanddoi=2ee6ebcbb3664c6cf27ee3a5e7382caef3ad16e6
Knowles LL (2009) Statistical phylogeography. Annu Rev Ecol Evol Syst 40:593–612. https://doi.org/10.1146/annurev.ecolsys.38.091206.095702
Knowles LL (2001) Did the Pleistocene glaciations promote divergence? Tests of explicit refugial models in montane grasshopprers. Mol Ecol 10:691–701. https://doi.org/10.1046/j.1365-294x.2001.01206.x
Kuhner MK (2009) Coalescent genealogy samplers: windows into population history. Trends Ecol Evol 24:86–93. https://doi.org/10.1016/j.tree.2008.09.007
Kulski JK (2016) Next-generation sequencing—an overview of the history, tools, and "Omic" applications. In: Next generation sequencing-advances, applications and challenges, vol 10, p 61964. https://doi.org/10.5772/61964
Lagomarsino LP, Condamine FL, Antonelli A et al (2016) The abiotic and biotic drivers of rapid diversification in Andean bellflowers (Campanulaceae). New Phytol 210(4):1430–1442. https://doi.org/10.1111/nph.13920
Lamichhaney S, Berglund J, Almén M et al (2015) Evolution of Darwin's finches and their beaks revealed by genome sequencing. Nature 518:371–375. https://doi.org/10.1038/nature14181
Lapointe FJ, Rissler LJ (2005) Congruence, consensus, and the comparative phylogeography of codistributed species in California. Am Nat 166(2):290–299. https://doi.org/10.1086/431283
Leaché AD, Crews SC, Hickerson MJ (2007) Two waves of diversification in mammals and reptiles of Baja California revealed by hierarchical Bayesian analysis. Biol Lett 3(6):646–650. https://doi.org/10.1098/rsbl.2007.0368

Leite RN, Rogers DS (2013) Revisiting Amazonian phylogeography: insights into diversification hypotheses and novel perspectives. Org Divers Evol 13:639–664. https://doi.org/10.1007/s13127-013-0140-8

Longo D, Lorenz-Lemke AP, Mäder G et al (2014) Phylogeography of the Petunia integrifolia complex in southern Brazil. Bot J Linn Soc 174:199–213. https://doi.org/10.1111/boj.12115

Lorenz-Lemke AP, Muschner VC, Bonatto SL et al (2005) Phylogeographic inferences concerning evolution of Brazilian Passiflora actinia and P. elegans (Passifloraceae) based on ITS (nrDNA) variation. Ann Bot 95(5):799–806. https://doi.org/10.1093/aob/mci079

Losos JB, Ricklefs RE (2009) Adaptation and diversification on islands. Nature 457(7231):830–836. https://doi.org/10.1038/nature07893

Malone CL, Wheeler T, Taylor JF et al (2000) Phylogeography of the Caribbean rock iguana (Cyclura): implications for conservation and insights on the biogeographic history of the West Indies. Mol Phylogenet Evol 17(2):269–279. https://doi.org/10.1006/mpev.2000.0836

Mardis E (2011) A decade's perspective on DNA sequencing technology. Nature 470:198–203. https://doi.org/10.1038/nature09796

Margis R, Felix D, Caldas JF et al (2002) Genetic differentiation among three neighboring Brazil-cherry (*Eugenia uniflora* L.) populations within the Brazilian Atlantic rain forest. Biodivers Conserv 11:149–163. https://doi.org/10.1023/A:1014028026273

Martins FM (2011) Historical biogeography of the Brazilian Atlantic forest and the Carnaval–Moritz model of Pleistocene refugia: what do phylogeographical studies tell us? Biol J Linn Soc 104(3):499–509. https://doi.org/10.1111/j.1095-8312.2011.01745.x

Mittermeier RA, Turner WR, Larsen FW et al (2011) Global biodiversity conservation: the critical role of hotspots. In: Biodiversity hotspots: distribution and protection of conservation priority areas. Springer Berlin Heidelberg, Berlin, Heidelberg, pp 3–22. https://doi.org/10.1007/978-3-642-20992-5_1

Monceau K, Cezilly F, Moreau J et al (2013) Colonisation and diversification of the Zenaida Dove (Zenaida aurita) in the Antilles: Phylogeography, contemporary gene flow and morphological divergence. PLoS One 8(12):e82189. https://doi.org/10.1371/journal.pone.0082189

Morris AB, Shaw J (2018) Markers in time and space: a review of the last decade of plant phylogeographic approaches. Mol Ecol 27:2317–2333. https://doi.org/10.1111/mec.14695

Murray-Smith C, Brummitt NA, Oliveira-Filho AT et al (2009) Plant diversity hotspots in the Atlantic coastal forests of Brazil. Conserv Biol 23:151–163. https://doi.org/10.1111/j.1523-1739.2008.01075.x

Myers N, Mittermeier RA, Mittermeier CG et al (2000) Biodiversity hotspots for conservation priorities. Nature 403(6772):853–858. https://doi.org/10.1038/35002501

Nielsen R (2006) Reducing an intellectual deficit: coalescent theory for (almost) everyone. Evolution 60:208–209. https://doi.org/10.1111/j.0014-3820.2006.tb01097.x

Nielsen R, Beaumont MA (2009) Statistical inferences in phylogeography. Mol Ecol 18:1034–1047. https://doi.org/10.1111/j.1365-294X.2008.04059.x

Orr HA (2001) The genetics of species differences. Trends Ecol Evol 16:343–350. https://doi.org/10.1016/S0169-5347(01)02167-X

Ouborg NJ, Piquot Y, Van Groenendael JM (1999) Population genetics, molecular markers and the study of dispersal in plants. J Ecol 87:551–568. https://doi.org/10.1046/j.1365-2745.1999.00389.x

Palma RE, Gutiérrez-Tapia P, González JF et al (2017) Mountaintops phylogeography: a case study using small mammals from the Andes and the coast of Central Chile. PLoS One 12(7):e0180231. https://doi.org/10.1371/journal.pone.0180231

Palma-Silva C, Lexer C, Paggi GM et al (2009) Range-wide patterns of nuclear and chloroplast DNA diversity in *Vriesea gigantea* (Bromeliaceae), a neotropical forest species. Heredity 103:503–512. https://doi.org/10.1038/hdy.2009.116

Peterson AT, Soberón J (2012) Species distribution modeling and ecological niche modeling: getting the concepts right. Nat Conservação 10(2):102–107. https://doi.org/10.4322/natcon.2012.019

Petit RJ, Excoffier L (2009) Gene flow and species delimitation. Trends Ecol Evol 24(7):386–393. https://doi.org/10.1016/j.tree.2009.02.011

Pinheiro F, Dantas-Queiroz MV, Palma-Silva C (2018) Plant species complexes as models to understand speciation and evolution: a review of south American studies. Crit Rev Plant Sci 37:54–80. https://doi.org/10.1080/07352689.2018.1471565

Pinheiro F, de Barros F, Palma-Silva C et al (2011) Phylogeography and genetic differentiation along the distributional range of the orchid Epidendrum fulgens: a Neotropical coastal species not restricted to glacial refugia. J Biogeogr 38(10):1923–1935. https://doi.org/10.1111/j.1365-2699.2011.02539.x

Ramos-Fregonezi AM, Fregonezi JN, Cybis GB et al (2015) Were sea level changes during the Pleistocene in the South Atlantic Coastal Plain a driver of speciation in Petunia (Solanaceae)? BMC Evol Biol 15:92. https://doi.org/10.1186/s12862-015-0363-8

Ribas CC, Aleixo A, Nogueira AC et al (2012) A palaeobiogeographic model for biotic diversification within Amazonia over the past three million years. Proc Biol Sci 279:681–689. https://doi.org/10.1098/rspb.2011.1120

Rissler LJ (2016) Union of phylogeography and landscape genetics. Proc Natl Acad Sci 113(29):8079–8086. https://doi.org/10.1073/pnas.1601073113

Rocha AV, Cabanne GS, Aleixo A et al (2020) Pleistocene climatic oscillations associated with landscape heterogeneity of the South American dry diagonal explains the phylogeographic structure of the narrow-billed woodcreeper (Lepidocolaptes angustirostris, Dendrocolaptidae). J Avian Biol 51(9). https://doi.org/10.1111/jav.02537

Salas-Leiva DE, Meerow AW, Calonje M et al (2017) Shifting quaternary migration patterns in the Bahamian archipelago: evidence from the Zamia pumila complex at the northern limits of the Caribbean Island biodiversity hotspot. Am J Bot 104(5):757–771. https://doi.org/10.3732/ajb.1700054

Salgueiro F, Felix D, Caldas JF et al (2004) Even population differentiation for maternal and biparental gene markers in *Eugenia uniflora*, a widely distributed species from the Brazilian coastal Atlantic rain forest. Divers Distrib 10:201–210. https://doi.org/10.1111/j.1366-9516.2004.00078.x

Santos Leal BS, Silva CP, Pinheiro F (2016) Phylogeographic studies depict the role of space and time scales of plant speciation in a highly diverse Neotropical Region. Crit Rev Plant Sci 35(4):215–230. https://doi.org/10.1080/07352689.2016.1254494

Särkinen T, Bohs L, Olmstead RG et al (2013) A phylogenetic framework for evolutionary study of the nightshades (Solanaceae): a dated 1000-tip tree. BMC Evol Biol 13:214. https://doi.org/10.1186/1471-2148-13-214

Schluter D (2009) Evidence for ecological speciation and its alternative. Science 323(5915):737–741. https://doi.org/10.1126/science.1160006

Seal JN, Kellner K, Trindl A et al (2011) Phylogeography of the parthenogenic ant Platythyrea punctata: highly successful colonization of the West Indies by a poor disperser. J Biogeogr 38(5):868–882. https://doi.org/10.1111/j.1365-2699.2010.02447.x

Segatto ALA, Reck-Kortmann M, Turchetto C et al (2017) Multiple markers, niche modelling, and bioregions analyses to evaluate the genetic diversity of a plant species complex. BMC Evol Biol 17:234. https://doi.org/10.1186/s12862-017-1084-y

Silva JD, Rylands AB, Fonseca GD (2005) O destino das áreas de endemismo da Amazônia. Megadiversidade 1(1):124–131

Slatkin M, Maddison WP (1989) A cladistic measure of gene flow inferred from the phylogenies of alleles. Genetics 123(3):603–613. https://doi.org/10.1093/genetics/123.3.603

Smith TB, Kark S, Schneider CJ et al (2001) Biodiversity hotspots and beyond: the need for preserving environmental transitions. Trends Ecol Evol 16:431. https://doi.org/10.1016/S0169-5347(01)02201-7

Stehmann JR, Lorenz-Lemke AP, Freitas LB et al (2009) The genus petunia. In: Gerats T, Strommer J (eds) Petunia: evolutionary, developmental and physiological genetics. Springer, New York, pp 1–28. https://doi.org/10.1007/978-0-387-84796-2

Sunderland T, Apgaua D, Baldauf C et al (2015) Global dry forests: a prologue. Int For Rev 17(2):1–9. https://doi.org/10.1505/146554815815834813

Symula R, Schulte R, Summers K (2003) Molecular systematics and phylogeography of Amazonian poison frogs of the genus Dendrobates. Mol Phylogenet Evol 26(3):452–475. https://doi.org/10.1016/S1055-7903(02)00367-6

Taberlet P, Zimmermann NE, Englisch T et al (2012) Genetic diversity in widespread species is not congruent with species richness in alpine plant communities. Ecol Lett 15(12):1439–1448. https://doi.org/10.1111/ele.12004

Templeton AR (1998) Nested clade analyses of phylogeographic data: testing hypotheses about gene flow and population history. Mol Ecol 7:381–397. https://doi.org/10.1046/j.1365-294x.1998.00308.x

Thomé MTC, Zamudio KR, Giovanelli JG et al (2010) Phylogeography of endemic toads and post-Pliocene persistence of the Brazilian Atlantic Forest. Mol Phylogenet Evol 55(3):1018–1031. https://doi.org/10.1016/j.ympev.2010.02.003

Thomsen PF, Willerslev E (2015) Environmental DNA–an emerging tool in conservation for monitoring past and present biodiversity. Biol Conserv 183:4–18. https://doi.org/10.1016/j.biocon.2014.11.019

Torres MFJ, Stone GN, Sanchez A et al (2021) Comparative phylogeography of an ant-plant mutualism: an encounter in the Andes. Glob Planet Chang 205:103598. https://doi.org/10.1016/j.gloplacha.2021.103598

Turchetto C, Fagundes NJR, Segatto ALA et al (2014a) Diversification in the South American Pampas: the genetic and morphological variation of the widespread Petunia axillaris complex (Solanaceae). Mol Ecol 23:374–389. https://doi.org/10.1111/mec.12632

Turchetto C, Segatto ALA, Telles MPC et al (2014b) Infraspecific classification reflects genetic differentiation in the widespread Petunia axillaris complex: a comparison among morphological, ecological, and genetic patterns of geographic variation. Perspect Plant Ecol Evol Syst 16:75–82. https://doi.org/10.1016/j.ppees.2014.01.002

Turchetto-Zolet AC, Cruz F, Vendramin GG et al (2012) Large-scale phylogeography of the disjunct Neotropical tree species Schizolobium parahyba (Fabaceae-Caesalpinioideae). Mol Phylogenet Evol 65(1):174–182. https://doi.org/10.1016/j.ympev.2012.06.012

Turchetto-Zolet AC, Pinheiro F, Salgueiro F et al (2013) Phylogeographical patterns shed light on evolutionary process in South America. Mol Ecol 22(5):1193–1213. https://doi.org/10.1111/mec.12164

Turchetto-Zolet AC, Salgueiro F, Turchetto C et al (2016) Phylogeography and ecological niche modelling in Eugenia uniflora (Myrtaceae) suggest distinct vegetational responses to climate change between the southern and the northern Atlantic Forest. Bot J Linn Soc 182(3):670–688. https://doi.org/10.1111/boj.12473

Ulloa Ulloa C, Acevedo-Rodríguez P, Beck S et al (2017) An integrated assessment of the vascular plant species of the Americas. Science 358(6370):1614–1617. https://doi.org/10.1126/science.aao039

Vandenbussche M, Chambrier P, Rodrigues-Bento S et al (2016) ≃Petunia, your next supermodel? Front Plant Sci 7:72. https://doi.org/10.3389/fpls.2016.00072

Waltari E, Hijmans RJ, Peterson AT et al (2007) Locating Pleistocene refugia: comparing phylogeographic and ecological niche model predictions. PLoS One 2(7):e563. https://doi.org/10.1371/journal.pone.0000563

Werneck FP (2011) The diversification of eastern South American open vegetation biomes: historical biogeography and perspectives. Quat Sci Rev 30(13–14):1630–1648. https://doi.org/10.1016/j.quascirev.2011.03.009

Werneck FP, Gamble T, Colli GR et al (2012) Deep diversification and long-term persistence in the South American 'dry diagonal': integrating continent-wide phylogeography and distribution modeling of geckos. Evolution 66(10):3014–3034. https://doi.org/10.1111/j.1558-5646.2012.01682.x

Zink RM (1996) Comparative phylogeography in North American birds. Evolution 50(1):308–317. https://doi.org/10.1111/j.1558-5646.1996.tb04494.x

Zorrilla-Azcué S, González-Rodríguez A, Oyama K et al (2021) The DNA history of a lonely oak: *Quercus humboldtii* phylogeography in the Colombian Andes. Ecol Evol 11:6814–6828. https://doi.org/10.1002/ece3.7529

Part III
Population Genetics in Practice

Chapter 14
What Are the Strategies for Genetic Conservation in Plant Populations in the Neotropics?

Alessandro Alves-Pereira

Contents

Abstract The neotropics harbor an enormous biological diversity, including the largest tropical rainforest in the world and some of the major hotspots for biodiversity conservation. Such environments have a long history of interaction with various human communities that have intensively used their genetic resources. Therefore, it must not be surprising that some of the most important centers of plant domestication are in the neotropics. There are many crops of local and global importance among the plants domesticated in the neotropics, and the diversity of their wild relatives is also high. This highlights the importance of the conservation of the native biomes along with their associated people and useful plant species. *In situ* preservation (or conservation) consists of maintaining the species in their habitats, enabling them to keep evolving within their environments. Alternatively, *ex situ* conservation consists of maintaining species outside their native environment under human care, which in principle, enables that their genetic resources are readily available for use. Both of them may be implemented by different means, such as *in vitro* conservation, an *ex situ* strategy that consists of keeping biological parts of species within test tubes in the laboratory. Population genetic studies are important sources of

A. Alves-Pereira (✉)
Departamento de Genética, Escola Superior de Agricultura "Luiz de Queiroz", Universidade de São Paulo, Piracicaba, SP, Brazil
e-mail: alves-pereira@usp.br

M. V. Bohrer Monteiro Siqueira et al. (eds.), *Population Genetics in the Neotropics*,
https://doi.org/10.1007/978-3-031-83685-5_14

information for many applications, as discussed throughout this book. The aim of this chapter is to discuss how population genetics can be useful to assist the various conservation strategies aimed at plants of the neotropics.

Keywords Biological diversity · Genetic diversity · Genetic resources · Molecular markers · Population genetics

Learning Objectives
After comprehensively reading this chapter, the reader should be able to understand the following:

1. The importance of conserving biodiversity and their elements.
2. The key aspects and applications of *in situ* and *ex situ* conservation.
3. The link between basic genetic concepts and the practices of conservation.
4. How population genetics supports and informs conservation strategies.

14.1 Definitions for Conservation Strategies

This chapter describes some examples of how population genetics may be informative in the context of protecting and/or conserving Neotropical plant species, as well as highlighting the importance of these approaches to the different conservation strategies that are commonly applied. In this way, the reader is first introduced to the definitions and the principles underlying *in situ* and *ex situ* conservation. After that, connections between the main theoretical and methodological foundations of population genetics previously discussed in this book are established with the conservation approaches. Practical examples are also given to enclose the discussion.

The biological diversity on Earth is rapidly being depleted, mainly due to human activities (Frankham et al. 2010). Ever since humans have started to diversify and disperse through the planet, they have actively modified the natural landscapes, and the more human populations have grown the biggest has been the impact. However, only recently the negative impacts of human activity on nature started to drive more attention. In the end of the twentieth century, a multidisciplinary academic effort gave rise to the discipline of *Biological Conservation*. The major objectives of this discipline are: i) to understand the effects of human activities on species, communities, and ecosystems; and ii) developing practical approaches to prevent the extinction of species (Primack and Rodrigues 2001). For the latter, conservation approaches are further discussed.

Protecting biological diversity is not a trivial task. It demands a coordinated and cooperative plan of action of stakeholders and people in general. One of the most famous international instrument that drove the attention for the urgent need in taking action to protecting nature was the Convention on Biological Diversity (CBD), started in 1988 and finalized in 1992, being opened for signatures at the United Nations Conference on Environment and Development (the Rio "Earth

Summit"—ECO92) (https://www.cbd.int/history/). The CBD was very influential, serving to orient policies of signatory countries and inspiring other international treatings to reach three main goals: i) the conservation of biological diversity; ii) the sustainable use of their components; and iii) the fair and equitable sharing of benefits arising from the use of genetic resources (https://www.cbd.int/history/).

The CBD document (Secretariat of the CDB, 2011) describes fundamental definitions needed here. According to the CBD, biological diversity means the "*variability among living organisms from all sources (including inter alia, terrestrial, marine and other aquatic ecosystems) and the ecological complexes of which they are part; this includes diversity within species, between species and of ecosystems*". Note how complex this concept is when we consider that within the biological diversity there are, at least, three major components. If we are aimed at protecting nature, we should assure that all ecosystems, their living species, and their genetic diversity are also protected.

Within biological diversity we find what the CBD defined as biological resources, which "*include genetic resources, organisms or parts thereof, populations, or any other biotic component of ecosystems with actual or potential use or value for humanity*". This means that in nature we can find many species that are somehow useful to us, or that may eventually become useful. Nested in the concept of biological resources, there are other two of major importance: genetic resources and genetic material. Genetic resources were defined as "the genetic material of actual or potential value", and genetic material was defined as "*any material of plant, animal, microbial, or other origin containing functional units of heredity*". Note that the difference between biological resources and genetic resources is not very clear, in part by the wide range of interpretation of the concepts, but it leads us to an interesting reflection point. Theoretically, any species in nature is a biological resource that has or might have a value. Then, any species within the whole biological diversity may be regarded as a genetic resource as well. That is, every species has a value (either economic, cultural, ecological, yet-to-be-discovered, etc.), and may therefore be considered as a subject for biological conservation. The biological and genetic resources are targets of conservation strategies.

Okay, but how can we actually do conservation? What are the fundamental approaches? The CBD also defines the two major conservation strategies: *in situ* and *ex situ*. According to CBD, *in situ* conservation means "*the conservation of ecosystems and natural habitats and the maintenance and recovery of viable populations of species in their natural surroundings and, in the case of domesticated or cultivated species, in the surroundings where they have developed their distinctive properties*". Otherwise, *ex situ* conservation means "the conservation of components of biological diversity outside their natural habitats". Note how the definitions are wide, bringing terms like "natural habitats" and "components of biological diversity", sustaining the interpretation that we should conserve not only the species for what we find an interest, but any species whatsoever.

Now is the time to briefly account for the different strategies that may be adopted to take *in situ* or *ex situ* actions toward conservation. Remember that *in situ* is related to the natural habitat or the environment of origin, while *ex situ* means to being transported or transplanted outside the natural habitat. While it is not directly implied, there are plenty of *in situ* conservation strategies, in the sense that it does not mean simply delimiting an interesting area and protecting its biological diversity. As an example, we can cite the Brazilian law that created the National System for the Conservation of Nature (Table 14.1). By means of the law 9.985 (July 18th, 2000), Brazil created two major groups of *in situ* conservation units, which are based either on the full protection of nature, or in the sustainable use of their genetic resources (Brasil 2000). "*Full protection conservation units*" are aimed at maintaining the ecosystems without any human interference, admitting only the indirect use of their natural resources. The indirect use does not allow consumption, collection, damage, or destruction of the natural resources. Alternatively, in "*sustainable use conservation units*" the environment might be explored, but in a way that grants the persistence of natural resources and their associated ecological processes, maintaining the biodiversity and their ecological attributes. In sustainable use conservation units there might be the indirect use, as well as the direct use of the natural resources, i.e. the existence of collection and use (commercial or not) of natural resources. Additionally, note that these types of *in situ* strategies rely on two complementary concepts: protection and conservation. Both concepts are aimed at the long-term persistence of biological resources. However, protection presumes keeping an environment free of any human intervention, while conservation is associated with the human interest in using the biological diversity but in a way that little negative impacts should arise.

Table 14.1 contains a brief description of the different types of "full protection" and "sustainable use" conservation units that can be implemented in Brazil. The creation of these *in situ* conservation units are regulated by public administration, and among other conditions, must be preceded by technical studies and public inquiries, to identify their locations, dimensions, and the most adequate boundaries.

The other major strategy is *ex situ* conservation. Like *in situ* strategies, *ex situ* conservation is a long-term and continuous work, and it also demands much investment of financial resources, time, staff, equipment, and facilities (Jaramillo and Baena 2000). Of course, when we rely on economic investments, *ex situ* approaches may be justified by our needs, rather than the naïve intention of simply preserving some species (Jaramillo and Baena 2000). *Ex situ* conservation is very expensive, for many reasons, some of which are briefly discussed.

People involved in *ex situ* conservation need to know about the taxonomy and ecology of the species, the proper methods to adequately preserve samples, as well as how to appropriately represent the genetic variability of the targeted taxa. In a long-term project, it is important to gather as much information as possible, making that data organized and readily accessible. The facilities to keep samples need to ensure isolation from external environmental factors, especially pests and diseases. One of the ultimate goals of *ex situ* conservation is to protect and safeguard the biological resources against the catastrophic losses that might occur in their natural habitats. The management of *ex situ* collections require overqualified staff that are specialists in different fields. Therefore, it is convenient, although not always feasible, that the people involved in such conservation measures have stability in their

Table 14.1 Summary of the main categories for *in situ* protection or conservation areas according to the Brazilian law 9.985 (July 18th, 2000)

Conservation units	Name	Objectives
Full protection	Ecological station	Nature preservation and research
	Biological reserve	Full preservation of biota and other natural attributes, without direct human intervention or environmental modifications (except for restoration purposes and for management practices)
	National park	Preservation of the ecosystems for their ecological or scenic significance. Scientific research, environmental education and tourism are allowed
	Natural monument	Preservation of rare, unique, or scenic natural sites
	Refuge of wildlife	Protection of environments to ensure conditions for species reproduction of communities of local flora/fauna and of the resident or migratory fauna
Sustainable use	Area for environmental protection	Generally large areas with occasional human occupation. They have biotic, abiotic, cultural, or aesthetic attributes that are important for the well-being of their human population. The basic objectives include the protection of biological diversity; to discipline the occupation process; and ensuring the sustainable use of their natural resources
	Area of relevant ecological interest	Generally small areas with few or no human occupation. They have extraordinary natural features, or harbor regional/rare biota. Their objectives are the maintenance of regional important natural ecosystems, or to regulate the use of these areas to promote conservation
	National forest	Areas covered by forest of predominantly native species. The basic objective is the multiple sustainable use of forest resources, emphasizing the methods for the sustainable exploitation
	Reserve for extractivism	Areas used by traditional populations that rely on extractivism and small-farming practices and small-scale rearing. Such reserves are aimed at protecting the existence and cultural features of traditional populations, at the same time ensuring the sustainable use of natural resources
	Animal reserve	Natural areas with native populations of terrestrial or aquatic, resident or migratory animal populations. These areas are adequate for scientific studies and for the economic management of animal genetic resources
	Reserve for sustainable development	Natural areas occupied by traditional human populations that rely on sustainable systems of exploration of the natural resources, which have been developed for a long time and that are adapted to the local ecological conditions. These populations have a fundamental role in protecting nature and maintaining biological diversity
	Private reserve for natural heritage	Private areas that may be perpetual, with the objective of conserving biological diversity

jobs so that they can properly accomplish their work (Jaramillo and Baena 2000). Nonetheless, it is important to remember that simply creating an *ex situ* collection does not guarantee that the genetic resources of a species will be safe and conserved. This is because this kind of action requires continuous multi-institutional (and financial) support (Jaramillo and Baena 2000).

Probably the most well-known techniques of *ex situ* conservation are captive breeding of animals and germplasm banks for plants. Captive breeding is probably the unique solution for conserving species that require large areas to live, such as some large vertebrates, which are commonly reared under human care in zoos and aquariums (Primack and Rodrigues 2001). Additionally, botanical gardens and germplasm banks can also be considered a strategy to conserve plants in captivity. Germplasm is the component of the genetic resources that may be used to manage the genetic variability, and that is used for general or applied biotechnological research (Goedert 2006). For crops of economic significance, genetic variability should be available to develop new cultivars and to widen the genetic base of crops (Borém et al. 2012). A germplasm bank is the repository where the genetic variability of one or many species is stored for *ex situ* conservation. Although germplasm banks are commonly associated with their physical installation, that is, a fruit orchard, or a cold room where seeds are maintained, it is important to remember that it goes far beyond this. A germplasm bank cannot exist without the associated information of each sample stored in it, as well as without trained personnel for their maintenance.

A *germplasm bank* ideally contains a representative sample of the genetic variability of a species. However, this task is not simple, since there is not a standard number that defines a sample that could represent the genetic diversity of a species. This number varies according to the species and the source of germplasm that is being stored. In a broader sense, germplasm banks can also be implemented as "gene banks" or "genome banks". They may consist of collections of sperm, ova, embryos, seeds, tissues, or DNA, so there is no need to manage the whole individual (Allendorf et al. 2013). Genome banks may serve as an insurance against the extinction of populations or species; may increase the efficiency of captive breeding by reducing the number of individuals held in captivity; and may serve as a source of tissue and DNA for basic and applied research (Allendorf et al. 2013).

Germplasm or gene banks commonly represent the genetic diversity of a species by maximizing the number of samples, which are called accessions, of different nature and from different sources. These include (Jaramillo and Baena 2000): (i) wild species—these are generally conserved because they can be used as sources of new genetic variation for the target species; (ii) traditional varieties—samples that were under cultivation by traditional means, with low mechanization and few inputs. These represent the genetic variability selected according to different human preferences and that are employed for different uses, under distinct geographical and cultural contexts; (iii) varieties from breeding programs—modern and elite cultivars displaying higher yields or greater performance for other desirable traits; and (iv) genetically modified organisms—representing transgenic organisms. Table 14.2 presents a summary of the different types of germplasm banks.

Table 14.2 Summary of the types and objectives of *ex situ* germplasm banks

Type of germplasm (gene) banks	Objectives	Examples
Base collection	Aimed at encompassing the highest genetic variability as possible for a given species, also including their wild relatives, cultivars, traditional varieties, and elite germplasm. It is established for long-term germplasm conservation, as a safeguard for eventual catastrophic losses. It is not intended for germplasm distribution or sharing. Ideally it should be duplicated somewhere else	Perhaps, the Svalbard Global Seed Vault in Norway is the most well-known initiative for the long-term conservation of global genetic resources. It currently conserves more than 1.2 million seed samples of more than 6 thousand species. For further information: https://www.seedvault.no/
Active collection	It may consist of a base collection duplication. Therefore, it contains the largest genetic variability and as many accessions as possible. The main difference from a base collection is that the germplasm may be conserved alive	The International Center for Tropical Agriculture (CIAT, Colombia) maintains about 6500 manioc (*Manihot esculenta*) accessions conserved *in vitro*. https://www.genebanks.org/genebanks/ciat/
Core collection	It aims at conserving most of genetic variability with the minimum number of accessions as possible. It is established to facilitate the management and the use of their conserved genetic resources. In addition, it may also be implemented when financial resources to maintain larger collections are scarce	Cristo-Araújo et al. (2015) developed a core collection for Amazonian peach palm (*Bactris gasipaes*). The original germplasm collection evaluated by the authors consists of 371 accessions. Using molecular markers, the authors selected 40 accessions that represented more than 90% of the genetic diversity maintained at the original collection. This will facilitate the long-term conservation of the peach palm genetic resources because the costs of maintenance will be much lower for the reduced number of plants
Working collection	It is also called "breeding collection", and it is aimed at providing germplasm for researchers or institutions, and does not necessarily contain many accessions which represent the genetic variability of the species	A working collection is generally used when a plant breeder conducts experiments and genetic crossings aiming at, for example, developing pest-resistant or more productive plants. A working collection is generally used, because at this point the breeder has the knowledge of what are the most promising accessions that may be used to reach the goals of a breeding program

Field cultivated plants make germplasms banks very expensive and vulnerable to disasters, pathogens, and diseases. Another widely used strategy for *ex situ* conservation is based on *in vitro* culture, in which small propagules are induced by means of culture media to originate the whole plant. *In vitro* conservation is used when it is more advantageous than the other strategies or when there are no other suitable appropriate methods (Ramos 2015). This technique relies on the totipotency of plant cells that under specific stimuli may multiply and regenerate an entire plant from the initial propagule (Mansur et al. 2012). Generally, plant cells, organs or tissues are maintained at specific culture media in a sterilized and controlled room. Treatments for slow growth are used by inducing a reduction in metabolism and growth rates of the cultures, which can be obtained by reductions in the temperature, luminosity, and adjustments of nutrients of the growth media (Mansur et al. 2012). *In vitro* conservation is widely used when the propagation of plant clones is explored economically, but is also important for germplasm banks, mainly to share germplasm between institutions by facilitating shipment of accessions to institutions in different countries (Ramos 2015).

Another technique for *in vitro* conservation is cryopreservation, which consists of freezing plant cells, tissues or organs at ultra-low temperatures, in liquid nitrogen (at −196 °C) or in its gas phase (at about −150 °C) (Mansur et al. 2012). It may be the unique option for conserving some species. For example, plant species are classified according to the ability of their seeds to survive desiccation as either orthodox or recalcitrant (Roberts 1973). While orthodox seeds survive even when dried to less than 12% of water in the internal seed, and may be stored at freezing temperatures, recalcitrant seeds may tolerate some drying but cannot survive after drying or in conventional freezers (−20 °C). Additionally, some plants have intermediate seeds that can tolerate drying but remain viable for much less time than orthodox seeds (Ellis et al. 1990, 1991). Cryopreservation is the only available technique for long-term germplasm conservation of non-orthodox seed species and may be adapted to intermediate-seed species (Eira et al. 2007). Cryopreservation is successfully used not only in plants, but also in bacteria, fungi, bacteriophage, viruses, algae, yeasts, and animal cells (Stanbury et al. 2017). Collections conserved at slow growth require periodic renovation, while cryopreservation of seeds or plant parts virtually conserves the material for an unlimited period (Mansur et al. 2012). Although *in vitro* technologies are not adequate for all species such approaches have been contributing to the conservation and management of germplasm, and more advances should be expected in that matter (Ramos 2015).

Last, but not least, it is important to emphasize that, although very different in their rationale, means, and implementation, *ex situ* and *in situ* conservation strategies must not be seen as two separate and independent strategies (Jaramillo and Baena 2000). The ideal scenario includes the combination of both means for every species. This is because the species are constantly evolving in their natural habitat, in response to their interaction with other species, their own environment and other abiotic factors. Although studying and deciding the more adequate areas for the establishment of conservation units is not a simple task, the greatest difficulty of *in situ* conservation seems to be our own inability to preserve nature. Otherwise, bringing a certain species to *ex situ* conditions theoretically keeps them safer, in a controlled environment, but at the cost of limiting their evolution. For every *ex situ* collection we have a "snapshot" of a certain moment in the ecological history of a species. Because the species in nature tend to change, it would be interesting if *ex situ* collections could be periodically incremented with novel accessions brought from *in situ* areas. Conversely, if an in *situ* conservation unit undergoes a catastrophic event, *ex situ* collections might be used as the reservoir for ecological restoration of the affected areas. In practice, *ex situ* conservation appears more to be an accessory strategy to *in situ* since it requires much more funding (Jaramillo and Baena 2000).

14.2 Connecting Important Genetic Concepts

Let's now bring back a few of the concepts and theoretical foundations that were presented in the previous chapters to connect them with conservation practices. Along this book, population genetics has been defined as the study of the genetic diversity within and among populations, by means of extending the principles of Mendelian inheritance to the entire groups of individuals of a given species (Hartl and Clark 2007). For a given population, a key aspect is the reproduction of their individuals, as they intercross and pass their genetic information to their offspring.

The basic principles of Mendelian inheritance took almost four decades to be recognized, and after that, it took us about half a century to fully understand their molecular basis. After hard work of many prominent scientists, it became clear the DNA (desoxyribonucleic acid) was the molecule of inheritance, and that the equal and independent segregation of Mendelian factors could be associated to the behavior of chromosomes in the meiosis (Griffiths et al. 2015). After the elucidation of the DNA structure, it became clear that the genetic information can be stored as a given sequence of nucleotides (which we can generically call *gene*), transcribed in an intermediate mRNA (messenger ribonucleic acid) and translated into proteins. Each protein can be associated with a trait that may be influenced by the environment (which relates to the study of quantitative traits). Since both parents contribute to the gametes with one copy of each of its chromosomes, each individual has two copies of a given gene (what we call *genotype*) that is associated with a characteristic (what we call *phenotype*) (Chap. 3). From a simple observation, it is obvious that individuals vary in their phenotypes, which is ultimately explained by the variation in their genotypes (Allendorf et al. 2013). Therefore, the observable variation of phenotypes relates to the existence of variation within the same gene, i.e. for a given sequence of nucleotides associated with a certain protein, there may exist genes that differ slightly in their sequences. In genetics, the variable forms of the same gene are called *alleles*. They consist of the most important object of population genetic studies, that is to determine the diversity of alleles that exist in a population—which we refer to as *genetic diversity*.

We can also extend other genetic theories, such as the influence of evolutionary factors (natural selection, genetic drift, gene flow and mutation) to understand the genetic variation in populations. The evolutionary factors act modeling (that is, changing the patterns) the *genetic diversity within populations*. In its simplest form, this change in genetic diversity is what we call in biology as "evolution". However, if there were many separated groups of individuals of a given species that reproduce independently, there will be populations in which the Mendelian principles, and their association with the environment and evolutionary factors act independently. Not necessarily separated populations will evolve differently, but when this happens, we can also characterize the *genetic diversity among populations*. When the genetic diversity among populations is high, it means that they are divergent in the sense that they contain distinct genetic constitutions (either different sets of alleles, or the same set of alleles but with distinct frequencies). One of the most frequent biological factors that reduces the genetic diversity within populations is what we call *inbreeding* (Chap. 7)—the crossing between relatives. Inbreeding contributes to reducing the genetic diversity because relatives are more likely to share the same or similar alleles, so when they cross their offspring will have greater chances of having the same alleles (Hartl and Clarck 2007; Allendorf et al. 2013). The consequence is that inbred populations are more likely to have more homozygous and less heterozygous individuals, and the less heterozygotes in a population the less its genetic diversity. When studied among populations, inbreeding also contributes to an overall reduction of genetic diversity, because in each population (that may be genetic distinct from each other) their individuals will tend to cross only among themselves, decreasing the chances of *gene flow* (Chap. 6)—when individuals from different populations reproduce after migrating from one population to the other.

The section two of this book discussed the methodological advances that fostered the application of population genetics approaches to a wide panel of species so far. One of the most important advances was the advent of *genetic and molecular markers* (Chaps. 9 and 10). In a broad sense, a *genetic marker* is any kind of allelic variation that adheres to Mendelian segregation principles, and a molecular marker is an allele that is directly observed in the DNA (but in this context, a molecular marker may not be a gene, but any DNA sequence in which we can detect variation may be used as a molecular marker) (Ferreira and Grattapaglia 1998). Currently, we use molecular markers to obtain information about the diversity of alleles that exist in a group of populations. So, why is this important in the context of conservation strategies?

We can associate the use of the genetics theory and the methods of population genetics with the knowledge about the species that have been threatened by human activity to reduce their extinction risk. This is what we call *conservation genetics* (Frankham et al. 2010). One may correctly infer that some of the applications of conservation genetics are the following:

(i) The most appropriate strategy for protection/conservation;
(ii) Actions to support the management and the effective use of conserved resources;
(iii) The existence of populations that are evolving independently, and that thus deserve differential attention;
(iv) The need of supplementing the genetic diversity in *ex situ* collections;
(v) The adequate representation of genetic diversity of species within a protected area.

14.3 Applying the Concepts

14.3.1 Genetic Diversity and the Conservation of Native Species

Example 1 Population structure and genetic diversity of two *Prosopis* species in Brazilian Gran Chaco (Alves et al. 2018).

The Gran Chaco consists of the largest South American dry forest, spanning about 800,000 km^2 across Brazil, Paraguay, Bolivia and Argentina (Hueck 1972). In Brazil, the Gran Chaco borders with the south of Pantanal, and has been subject to anthropic disturbances (Silva and Abdon 1998). Pantanal has a recent history of anthropic degradation that has replaced the native flora with livestock agriculture (Abdon et al. 2007).

Alves et al. (2018) evaluated the genetic diversity of 11 populations of two *Prosopis* species, *P. rubriflora* and *P. ruscifolia* using 11 nuclear microsatellite (SSR) markers. Domestic and wild animals feed on *Prosopis* fruits, promoting their seed dispersal. *P. rubriflora* has ornamental potential, and its wood can be used to produce charcoal, while *P. ruscifolia* is mainly used for its wood, and is recommended for the restoration of degraded environments. The populations were under different degrees of disturbance; however, the genetic analysis found an overall high level of genetic diversity for both species. *P. rubriflora* populations had high number of alleles (ranging from 41 to 71), intermediate levels of genetic diversity (expected heterozygosity ranging from 0.49 to 0.64), but most importantly, the populations generally had low levels of inbreeding, with estimates ranging from −0.08 to 0.17 (only this latter value was statistically significant). The estimates of genetic diversity may vary from 0 to 1, and the higher the value, the higher the genetic diversity, while estimates of inbreeding close to 0 suggest that few crossings between parents are occurring. Similar results were observed for *P. ruscifolia* populations: the total number of alleles ranged from 50 to 87, the estimates of genetic diversity ranged from 0.54 to 0.67, and the estimates of inbreeding ranged from 0.01 to 0.28. These results suggested that, although variable levels of human disturbance are in course, these populations apparently had not significant changes in their genetic diversity. This also suggests that, at least at a short-term these populations are less prone to suffer negative consequences of inbreeding depression. In other words, they may still contain sufficient genetic diversity to keep evolving in response to human disturbance and climate change.

In the same study, most of the genetic variation was found within populations (92% for *P. rubriflora*, and 94% for *P. ruscifolia*) rather than among populations. This means that within each species, the populations are not very much genetically different from each other. The biological explanation is that there might be high levels of gene flow among populations within each species. Only one population of each species was a bit more genetically divergent: in both cases these ones were the geographically farthest from the others. This suggests that the geographic distance can limit the gene flow among populations of these species, and that future genetic studies should include more distant populations to compare the patterns of genetic diversity of these species in chaquenian areas. For conservation purposes, protecting each studied area (or, at least, most of them) should assist the maintenance of the genetic diversity of these species.

In practice, this study suggests that establishing conservation units that include the studied areas may be effective for *in situ* conservation of *P. rubriflora* and *P. ruscifolia*. Nonetheless, the high levels of genetic diversity observed in this study should not relax further conservation concerns: for both species, the authors detected populations with evidence of genetic bottleneck, which means that their genetic diversity has been reduced. Many of the evaluated populations were in isolated and small fragmented areas. It is possible that the negative consequences of this fragmentation plus human intervention are still to be observed. We may expect that the continuous fragmentation of chaquenian areas could lead to increased losses of alleles and genetic divergence of populations.

Example 2 Genetic diversity of wild annatto (*Bixa orellana* var. *urucurana*) in Brazilian Amazonia (Dequigiovanni et al. 2018).

Conservation of wild relatives of domesticated crops is important, as wild plants should have useful genetic diversity to the development of more productive, resilient, and nutritious crop varieties (Castañeda-Álvarez et al. 2016). Indeed, wild relatives have long been used in genetic breeding programs contributing to a wide range of beneficial agronomic and nutritional traits (Castañeda-Álvarez et al. 2016). Among Neotropical crops, annatto (*Bixa orellana*, Bixaceae) is of great economic interest due to the production of a dye applied in cosmetics and food. Annatto was presumably domesticated in Amazonia from populations of *Bixa orellana* var. *urucurana* (Moreira et al. 2015), which is a shrub that occurs in open forests and managed landscapes and is always associated with riparian environments.

With the aim of supporting reasonable decisions about sampling the genetic diversity of the species, Dequigiovanni et al. (2018) evaluated ten populations of *B. orellana* var. *urucurana* from three Brazilian Amazonian states, using 16 nuclear SSR markers. Intermediate levels of genetic diversity were found across the studied populations, with the mean number of alleles per SSR locus ranging from 1.8 to 5.8, and the estimates of genetic diversity (measured as the expected heterozygosity) ranged from 0.32 to 0.63. Either way, the genetic diversity of these populations was higher than the germplasm maintained at the Agronomic Institute of Campinas (São Paulo, Brazil) (mean number of alleles = 4.5, genetic diversity = 0.47), which was evaluated by the same authors with the same SSR markers (Dequigiovanni et al. 2017). These results are relevant for the conservation of annatto germplasm. Wild populations of annatto in the Brazilian Amazonia may indeed be a source of new alleles, which might be useful to increase the genetic basis of annatto in breeding programs and for crop conservation.

The incorporation of wild genetic materials to germplasm banks of annatto gains more importance in the face of other interesting genetic results observed by the authors. Nine out of the ten populations showed a significant inbreeding coefficient, which ranged from 0.047 to 0.565. This means that very genetically related individuals make up these populations. With those results, populations are more prone to lose genetic diversity because of inbreeding. High levels of inbreeding were also observed for the germplasm of cultivated annatto (Dequigiovanni et al. 2017). Although most of the genetic diversity (79%) was observed within populations, a considerable proportion of genetic diversity (21%) was observed among populations. This means that the studied wild annatto populations are genetically divergent, probably because gene flow is very limited between them. Wild annatto is pollinated by bees, and fruits are dispersed through water streams. Considering pollination and seed dispersal combined with the fact that most of the studied populations were sampled from distant locations may explain these results. However, the authors also performed ecological niche modeling (ENM) (see Chap. 11) to predict the potential areas of distribution for the species, which indicated that their populations might be very isolated in the Brazilian Amazonia. In other words, ENM suggested that the disjunct distribution of the populations may have an ecological explanation. This has important implications for the conservation of wild annatto because besides the disjunct distribution, most of the studied populations were within the "arc of deforestation", a region where the agricultural frontier advances towards the forest (INPE 2002). Therefore, additionally to the natural isolated distribution, as modeled by ENM, the deforestation of the Amazon might be increasing the geographical isolation of wild annatto populations. The small and isolated populations of wild annatto across Amazonia may explain the high levels of intrapopulation inbreeding, suggesting the human action is contributing to the reduction of the genetic diversity.

When wild annatto grows near commercial annatto fields, farmers tend to remove the wild plants because they naturally cross with cultivated plants, which is associated with the perception of production losses in the cultivated annatto. This practice may also contribute to the reduction of wild annatto populations and to increase the geographical isolation of their populations, which in turn contributes to the loss of their genetic diversity. The intermediate estimates of genetic diversity, and the high levels of intrapopulation inbreeding and between-population genetic divergence highlight that urgent measures toward genetic conservation of wild annatto are needed. One way to contribute to this goal is, for example, sampling the extant populations of wild annatto across the Brazilian Amazonia and incorporating them in *ex situ* germplasm collections of cultivated annatto. Because all wild populations seem to be genetically valuable, this would enrich the current genetic variability already available for this crop.

14.3.2 Genetic Diversity Within Conservation Units

Example 3 Patterns of genetic diversity of juçara palm (*Euterpe edulis*) within a conservation unit in the Atlantic Forest in Brazil (Novello et al. 2018).

The original Atlantic Forest extended in a continuum landscape from the Northeastern Atlantic coast of Brazil to Southern Paraguay and Northern Argentina, but it was reduced to about 11% of the original coverage (Ribeiro et al. 2009) (sources vary in percentages, as other chapters have already shown, but any estimate provided in undoubtedly low). Therefore, the Atlantic Forest is considered one of the top priority areas for biodiversity conservation in the world (Myers et al. 2000, Fundação SOS Mata Atlântica and INPE 2021). This biome consists of many phytophysiognomies that harbor an impressive diversity of species, many of which are endemic and threatened with extinction (Tabarelli et al. 2003; Laurance 2009). The landscape is currently highly fragmented, and it is also the region with the highest human population density in Brazil. Unfortunately, this biome still undergoes intense anthropic pressures leading to the replacement of its natural environments. Among the most iconic species of this biome is *Euterpe edulis*, popularly known as juçara palm. *E. edulis* is a keystone species because their fruits are an important source of food for many animals (Reis et al. 2000; Galetti et al. 2013) and because of their economic interest to obtain heart-of-palm. Predatory overexploitation led *E. edulis* to be added in the list of species threatened with extinction, mainly due to the fact that extracting the heart-of-palm requires the entire plant be cut and no further sprouting occurs, leading to the death of the individual.

Novello et al. (2018) investigated the genetic diversity of *E. edulis* populations from the Serra do Mar State Park, a conservation unit currently located within the largest continuous patch of Atlantic Forest in southeastern Brazil. The authors used 14 nuclear SSR to compare the genetic diversity of *E. edulis* from protected (presumably undisturbed) areas, naturally regenerated secondary forests, and agroforests (tree plantations mixed with crops). The analyses included adult trees and juveniles that originated from seeds obtained after pulp removal of juçara fruits harvested in agroforests and secondary forests. These juveniles have also been used for reintroductions within protected areas.

High levels of genetic diversity (expressed as expected heterozygosity for adults—ranging from 0.75 to 0.81; and for juveniles—ranging from 0.76 to 0.83) and allelic richness (adults—ranging from 6.1 to 7.5; juveniles—ranging from 7 to 8.4) were observed for all studied populations. There was no statistical difference between the estimates among protected areas, secondary forests, and agroforests. These results indicate that human intervention is not reducing the levels of genetic diversity in managed areas (agroforests and secondary forest) in comparison to that of protected areas. On the other hand, despite the high levels of genetic diversity, the authors observed high estimates of inbreeding, ranging from 0.10 to 0.34. The estimates of inbreeding were not significantly different for protected areas, secondary forests, and agroforests, suggesting that the past overexploitation might have severely decreased the number of *E. edulis* in the studied areas, contributing to a genetic signature of crossings between relatives.

Overall, the highest proportion of genetic variation was observed within populations (88.3% within adult populations and 89.3% within juvenile populations). When comparing managed with non-managed populations, 97.5% of the genetic variation was found within adult populations, and 96.7% of the genetic variation was found within juvenile populations. These results suggest that low to intermediate genetic divergence occurs among most of the studied populations. Although some populations were more divergent in relation to others, high gene flow could be inferred among most populations. This scenario is positive because it suggests that the cultivation and management of *E. edulis* by local communities conserved similar levels of genetic diversity as the observed for protected forests, presumably free of human interference. In this example, Novello et al. showed that the conservation of managed *E. edulis* populations is a valuable complementation to the strict protection of natural environments. Because the seeds obtained during pulp removal had comparable genetic diversity as *E. edulis* in protected areas, the community-management of fruits could also contribute to the recovery of disturbed areas. Nonetheless, the authors highlight that this scenario can change: if juçara fruit collection intensifies in response to the market, fruit availability to frugivores may be reduced, which can compromise gene flow and the maintenance of high genetic diversity in the conservation unit.

14.3.3 Conservation and Use of Agrobiodiversity

Example 4 Genome-wide diversity of manioc (*Manihot esculenta*) varieties conserved at the Luiz de Queiroz College of Agriculture germplasm bank (Alves-Pereira et al. 2022).

Maintaining genetic diversity within crops is essential to keep their evolutionary potential which is used by farmers to adapt their traditional varieties to a changing world, select for their current needs, and is also important to develop elite material through genetic breeding (Gepts 2006; McCouch et al. 2013). In this context, regions with high socio- and biodiversity, such as Brazil, are of great importance as many crop varieties have been selected by traditional communities. Manioc (*Manihot esculenta*, Euphorbiaceae), also known as cassava, is a remarkable example of the immense agrodiversity in Brazil. Cultivated manioc (*M. esculenta* subsp. *esculenta*) was domesticated from *M. esculenta* subsp. *flabellifolia* in southwestern Amazonia, possibly as early as 10,000 years before present, and it has been spread throughout the tropics (Olsen and Schaal 1999, 2001; Olsen 2004; Leotárd et al. 2009). The largest *ex situ* manioc collections are maintained at the International Center for Tropical Agriculture (CIAT, Colombia), with about 6500 accessions *in vitro*, and at the International Institute of Tropical Agriculture (IITA, Nigeria) with about 3700 accessions. The Brazilian Agricultural Research Corporation (Embrapa) conserves approximately 4000 accessions, and there are probably more than 10,000 unique accessions conserved *ex situ*, in more than 70 cassava genebanks worldwide (Genbank Platform and CGIAR 2024). However, all the original genetic diversity of the crop can be traced back to Brazil.

Alves-Pereira et al. (2022) evaluated the genetic diversity conserved at the manioc germplasm bank of the Luiz de Queiroz College of Agriculture (State University of São Paulo) using single nucleotide polymorphisms (SNP) markers. The genetic profiles of 78 unique manioc accessions, originated from four different Brazilian biomes: Amazonia, Cerrado, Atlantic Forest and Pantanal, conserved at this germplasm bank were contrasted with eight samples of the wild relative *M. esculenta* subsp. *flabellifolia*. A total of 10,917 SNP markers revealed that the cultivated accessions had high genetic diversity (expected heterozygosity = 0.315—this is considered high because the theoretical highest estimate of genetic diversity for SNP markers is 0.5) and no inbreeding. The same patterns were observed when the manioc accessions were grouped by biomes, with the genetic diversity ranging from 0.285 to 0.312, and with inbreeding ranging from −0.22 to 0.062. These estimates suggest that the number of heterozygotes in the germplasm bank is high: because manioc is clonally propagated, it means that genetic diversity is conserved within the crop due to the selection of heterozygous plants.

When comparing cultivated manioc and its wild relative, most of the genetic variation was found within groups (67.5%), but a significant proportion was observed between these groups, suggesting that they are very genetically distinct. This is expected because of the ancient history of manioc domestication: farmers have been selecting cultivated manioc for a long time, increasing the genetic divergence from its wild relatives. Otherwise, when comparing the groups of accessions from different biomes, most of the genetic variation was observed within groups (96.6%). This result can be explained, for a sexual reproducing species, by a high level of gene flow among biomes. In the case of manioc, the interpretation is that different communities of farmers that rely on manioc cultivation, although selecting for different crop traits in response to distinct preferences in variable social and environmental contexts, have also been exchanging their local varieties with other communities from different regions. These exchange networks promote gene flow between distinct varieties at local and broader geographical scales, somewhat what is also observed for sorghum (*Sorghum bicolor*) and sweet potato (*Ipomoea batatas*), that are traditionally cultivated by smallholder farmers (Mutegi et al. 2011; Roullier et al. 2011).

The high levels of genetic diversity observed in the manioc varieties from different Brazilian biomes reinforces the importance of *ex situ* conservation strategies, such as the germplasm banks. In this example, the high genetic diversity is very desirable for genetic breeding and the manioc varieties may be also used as sources for reintroductions into smallholder communities where there are losses of cultivated material.

Now, a little disclaimer to close our discussion. Much effort has been done to study the genetic diversity of plant population species from the Neotropics, although not always these studies emphasized the important aspects for the conservation of such genetic resources. Obviously, the examples brought in this section are far from representing the different taxonomic groups, research teams, and approaches that are possible to be done. And this was not the objective of this chapter. For convenience, a few examples were chosen to contextualize the contents here described. Obviously, every single study that has been performed (and all the ones that are being performed) matter and deserve attention. Certainly, the knowledge about the genetics of neotropical species is not the only thing we need to protect and conserve. However, these approaches are powerful and may effectively contribute to this goal.

Key Points

- Biodiversity has been severely affected by human activity. Efforts to counteract the negative impacts of our own intervention in the natural landscapes include different strategies of *in situ* and *ex situ* conservation of various species;
- While *in situ* conservation is based on the maintenance of the species in their natural environment for their integral protection or rational use, *ex situ* conservation is based on bringing species to controlled environments outside their natural environments;
- In practice, population genetic studies are aimed at quantifying the genetic diversity within and among populations. These studies can be very informative to support planning and effective conservation actions.

Questions

1. What is the major advantage of *in situ* conservation of plants?
2. Why *ex situ* techniques should not be viewed as independent strategies from *in situ* conservation of plants?
3. Why within population genetic diversity and low inbreeding is important for the conservation of plant species?
4. What may be the implications of high levels of genetic divergence between plant populations in the context of conservation strategies?

References

Abdon MM, Silva JSV, Souza IM et al (2007) Desmatamento no bioma Pantanal até o ano 2002: Relações com a fitofisionomia e limites municipais. Rev Bras Cartografia 59:17–24. https://doi.org/10.14393/rbcv59n1-43959

Allendrof FW, Luikart G, Aitken SN (2013) Conservation and the genetics of populations, 2nd edn. Wiley-Blackwell, Oxford, p 624

Alves FM, Sartori ÂLB, Zucchi MI et al (2018) Genetic structure of two *Prosopis* species in Chaco areas: a lack of allelic diversity diagnosis and insights into the allelic conservation of the affected species. Ecol Evol 8:6558–6574. https://doi.org/10.1002/ece3.4137

Alves-Pereira A, Zucchi MI, Clement CR et al (2022) Selective signatures and high genome-wide diversity in traditional Brazilian manioc (*Manihot esculenta* Crantz) varieties. Sci Rep 12:1268. https://doi.org/10.1038/s41598-022-05160-8

Borém A, Lopes MTG, Clement CR et al (eds) (2012) Domestication and breeding Amazonian species. Editora Universidade Federal de Viçosa, Viçosa, p 479

Brasil (2000) Law 9.985, July 18th 2000. Regulamenta o art. 225, § 1o, incisos I, II, III e VII da Constituição Federal, institui o Sistema Nacional de Unidades de Conservação da Natureza e dá outras providências

Castañeda-Álvarez N, Khoury C, Achicanoy H et al (2016) Global conservation priorities for crop wild relatives. Nature Plants 2:16022. https://doi.org/10.1038/nplants.2016.22

Cristo-Araújo M, Rodrigues DP, Astolfi-Filho S et al (2015) Peach palm core collection in Brazilian Amazonia. Crop Breed Appl Biotechnol 15:18–25. https://doi.org/10.1590/1984-70332015v15n1a3

Dequigiovanni G, Ramos SLF, Alves-Pereira A et al (2018) Highly structured genetic diversity of *Bixa orellana* var. *urucurana*, the wild ancestor of annatto, in Brazilian Amazonia. PLoS One 13(6):e0198593. https://doi.org/10.1371/journal.pone.0198593

Dequigiovani G, Ramos SLF, Alves-Pereira A et al (2017) Genetic diversity and structure in a major Brazilian annatto (*Bixa orellana*) germplasm bank revealed by microsatellites and phytochemical compounds. Genet Resour Crop Evol 64:1775–1788. https://doi.org/10.1007/s10722-017-0535-z

Eira MTS, Ribeiro VS, Fuzuoli LC et al (2007) Coffee seed cryopreservation: current research progress. In: International conference on coffee science. Association for Science and Information on Coffee, Montpellier. 1 CD-ROM

Ellis RH, Hong TD, Roberts EH (1990) An intermediate category of seed storage behaviour? I Coffee J Exp Bot 41:1167–1174. https://doi.org/10.1093/jxb/41.9.1167

Ellis RH, Hong TD, Roberts EH (1991) An intermediate category of seed storage behaviour? II. Effects of provenance, immaturity and imbibition on desiccatio-tolerance in coffee. J Exp Bot 42:653–657. https://doi.org/10.1093/jxb/42.5.653

Ferreira ME, Grattapaglia D (1998) Introdução ao uso de marcadores moleculares, 3ª edn. Embrapa-Cenargen, Brasília, p 220

Frankham R, Ballou JD, Briscoe DA (2010) Introduction to conservation genetics, 2nd edn. Cambridge University Press, Cambridge, p 644

Fundação SOS Mata Atlântica, Instituto Nacional De Pesquisas Espaciais (2021) Atlas dos remanescentes florestais da Mata Atlântica, Período 2019–2020. Fundação SOS Mata Atlântica, São Paulo, p 73

Galetti MGR, Côrtes MC, Fadini R et al (2013) Functional extinction of birds drives rapid evolutionary changes in seed size. Science 340(6136):1086–1090. https://doi.org/10.1126/science.1233774

Genebank Platform, CGIAR (2024). https://www.genebanks.org/resources/crops/cassava/. Accessed 15 Jan 2024

Gepts P (2006) Plant genetic resources conservation and utilization: the accomplishments and future of a societal insurance policy. Crop Sci 46:2278–2292. https://doi.org/10.2135/cropsci2006.03.0169gas

Goedert CO (2006) Histórico e Avanços em Recursos Genéticos no Brasil. In: Nass LL (ed) Recursos genéticos vegetais. Embrapa Informação Tecnológica, Brasília, pp 21–58

Griffiths AJF, Wessler SR, Carroll SB et al (2015) Introduction to genetic analysis, 11th edn. W.H. Freeman, p 896

Hartl DL, Clark AG (2007) Principles of population genetics, 4th edn. Sinauer Assoc, Sunderland, p 652

Hueck K (1972) As florestas da América do Sul: Ecologia, composição e importância econômica. Polígono, São Paulo, p 466

INPE (2002) Monitoring of the Brazilian Amazonian forest by satellite, 2000–2001. http://www.inpe/br/informacoes_Eventos/amazonia.htm Accessed 15 Jan 2024

Jaramillo S, Baena M (2000) Material de apoyo a la capacitación em conservación *ex situ* de recursos fitogenéticos. Instituto Internacional para os Recursos Fitogenéticos, Cali, p 221

Laurance WF (2009) Conserving the hottest of the hotspots. Biol Conserv 142:1137. https://doi.org/10.1016/j.biocon.2008.10.011

Léotard G, Duputié A, Kjellberg F et al (2009) Phylogeography and the origin of cassava: new insights from the northern rim of the Amazonian basin. Mol Phylogenet Evol 53:329–334. https://doi.org/10.1016/j.ympev.2009.05.003

Mansur E, Pacheco G, Vieira MLC (2012) *In vitro* germplasm conservation. In: Borém A, Lopes MTG, Clement CR, Noda H (eds) Domestication and breeding Amazonian species. Editora Universidade Federal de Viçosa, Viçosa, pp 169–190

McCouch S, Baute GJ, Bradeen J et al (2013) Feeding the future. Nature 499:23–24. https://doi.org/10.1038/499023a

Moreira PA, Lins J, Dequigiovanni G et al (2015) The domestication of annatto (*Bixa orellana*) from *Bixa urucurana* in Amazonia. Econ Bot 69:127–135. https://doi.org/10.1007/s12231-015-9304-0
Mutegi E, Sagnard F, Semagn K et al (2011) Genetic structure and relationships within and between cultivated and wild sorghum (*Sorghum bicolor* (L.) Moench) in Kenya as revealed by microsatellite markers. Teor Appl Genet 122:989–1004. https://doi.org/10.1007/s00122-010-1504-5
Myers N, Mittermeier R, Mittermeier C et al (2000) Biodiversity hotspots for conservation priorities. Nature 403:853–858. https://doi.org/10.1038/35002501
Novello M, Viana JPG, Alves-Pereira A et al (2018) Genetic conservation of a threatened neotropical palm through community-management of fruits in agroforests and second-growth forests. For Ecol Manag 407:200–209. https://doi.org/10.1016/j.foreco.2017.06.059
Olsen KM, Schaal BA (1999) Evidence on the origin of cassava: Phylogeography of *Manihot esculenta*. Proc Natl Acad Sci USA 96:5586–5591. https://doi.org/10.1073/pnas.96.10.5586
Olsen KM, Schaal BA (2001) Microsatellite variation in cassava (*Manihot esculenta*, Euphorbiaceae) and its wild relatives: further evidence for a southern Amazonian origin of domestication. Am J Bot 88:131–142. https://doi.org/10.2307/2657133
Olsen KM (2004) SNPs, SSRs and inferences on cassava's origin. Plant Mol Biol 56:517–526. https://doi.org/10.1007/s11103-004-5043-9
Primack RB, Rodrigues E (2001) Biologia da conservação. Editora Planta, Londrina, p 328
Ramos LCS (2015) A conservação *in vitro* de germoplasma. In: Veiga RFA, Queiróz MA (eds) Recursos fitogenéticos: a base da agricultura sustentável no Brasil. Editora UFV, Viçosa, pp 237–241
Reis MS, Fantini AC, Nodari RO et al (2000) Management and conservation of natural populations in Atlantic rain forest: the case study of palm heart (*Euterpe edulis* Martius). Biotropica 32:894–902. https://doi.org/10.1111/j.1744-7429.2000.tb00627.x
Ribeiro MC, Metzger JP, Martensen AC et al (2009) The Brazilian Atlantic forest: how much is left, and how is the remaining forest distributed? Implications for conservation. Biol Conserv 142:1141–1153. https://doi.org/10.1016/j.biocon.2009.02.021
Roberts EH (1973) Predicting the storage life of seeds. Seed Sci Technol 1:499–514
Roullier C, Rossel G, Tay D et al (2011) Combining chloroplast and nuclear microsatellites to investigate origin and dispersal of New World sweet potato landraces. Mol Ecol 20:3963–3977. https://doi.org/10.1111/j.1365-294X.2011.05229.x
Secretariat of the Convention on Biological Diversity (2011) Convention on biological diversity: text and annexes. United Nations Environment Programme, Montreal, p 32
Silva JSV, Abdon MM (1998) Delimitação do Pantanal brasileiro e suas sub-regiões. Pesq Agrop Brasileira 33:1703–1711
Stanbury PF, Whitaker A, Hall SJ (2017) Culture preservation and inoculum development. In: Stanbury PF, Whitaker A, Hall SJ (eds) Principles of fermentation technology, 3rd edn. Butterworth-Heinemann, Oxford, pp 335–399
Tabarelli M, Pinto LP, Silva JMC et al (2003) Endangered species and conservation planning. In: Galindo-Leal C, Câmara IG (eds) The Atlantic Forest of South America: biodiversity status, trends, and outlook. CABI, Washington DC, pp 86–94

Chapter 15
What are the Strategies for Genetic Conservation in Animal Populations in the Neotropics?

Larissa S. Arantes, Fernanda J. Trindade, Marisol Domínguez, and Vanessa Paes da Cruz

Contents

Abstract The Neotropical region is renowned for its high biodiversity but also faces numerous threats to their species, with many of them at risk of extinction. Population decline of several species are often due to human-driven factors such as habitat loss and degradation, overfishing, and illegal wildlife trafficking. Conservation genetics has the potential to aid information in assessing the conservation status, monitoring populations through non-invasive methods, delineating con-

L. S. Arantes
Leibniz Institute for Zoo and Wildlife Research (IZW), Berlin, Germany

Berlin Center for Genomics in Biodiversity Research, Berlin, Germany

F. J. Trindade
Instituto Tecnológico Vale Desenvolvimento Sustentável, Belém, PA, Brazil

M. Domínguez
Unit of Evolutionary Biology/Systematic Zoology, Institute for Biochemistry and Biology, University of Potsdam, Potsdam, Germany

V. P. da Cruz (✉)
Departamento de Biologia Estrutural e Funcional, UNESP/IBB, Botucatu, SP, Brazil

M. V. Bohrer Monteiro Siqueira et al. (eds.), *Population Genetics in the Neotropics*,
https://doi.org/10.1007/978-3-031-83685-5_15

servation units within species, and unraveling the biodiversity obscured by cryptic species. Moreover, it can provide important information applicable to translocation and ex-situ conservation initiatives, and relevant for combating illegal wildlife traffic. Throughout this chapter, we present a range of conservation genetic strategies employed in diverse efforts to preserve animal populations within the neotropical region. We also examine their limitations and propose the integration of recent genomic advances and interdisciplinary approaches to pave the way for more innovative conservation approaches.

Keywords Conservation units · Ex-situ conservation · Non-invasive monitoring · Strategies · Wildlife DNA forensic

Learning Objectives
This chapter aims to provide readers with an understanding of the following:

1. Connection between genetic analyses and conservation strategies;
2. Understand the role of genetics in delineating conservation units;
3. Revealing cryptic species and resolving taxonomic uncertainties;
4. The use of non-invasive sampling methods;
5. The role of wildlife DNA forensics in combating illegal trafficking;
6. Analyze translocation strategies;
7. The importance of captive breeding and *ex situ* conservation.

15.1 Introduction

Conservation genetics strategies applied to neotropical species have been the focus of several studies, which have highlighted the importance of genetic diversity for adapting to environmental change and estimating conservation status (Quintela et al. 2022). These studies have shown that the neotropical region is highly biodiverse, but also that many species are threatened with extinction or are experiencing declines in population growth (García-Merchán et al. 2021). Furthermore, various studies have identified habitat loss and degradation as important threats to terrestrial species, while fishing activity and hunting pose major threats to marine and freshwater species (Pelicice et al. 2017; Freitas et al. 2021).

Torres-Florez et al. (2018) analyzed trends in conservation genetics research in Latin America and found that the most of the studies are focused on animal species, particularly mammals and bony fishes, with a notable bias towards vertebrates (88%). In contrast, insects, despite their biological importance and rich biodiversity, accounted for only about 3% of studies. Moreover, there is also a tendency to prioritize the study of iconic and "charismatic" species that capture greater public attention.

To date, most of the studies published in conservation genetics have used traditional genetic markers, such as mitochondrial DNA (mtDNA), microsatellites, or a few nuclear markers. In recent years, though, new approaches have been incorporated, facilitating the study of genome-wide variation patterns in a diverse set of species. As a result, questions regarding inbreeding depression, local adaptation (including climate change and other anthropogenic effects), demographic history, natural and human-induced hybridization, captive breeding, and assisted migration can be more effectively addressed (Allendorf 2010).

In the field of *conservation biology*, the integration of genetics has become indispensable, yet a disconnection persists between genetic analyses and practical conservation strategies. This chapter explores the diverse strategies employed in genetic conservation efforts for animal populations within the neotropical region. We present examples of studies using genetic methods for assessing species conservation status based on genetic metrics that provide crucial insights into the health and resilience of species. Additionally, we examine how genetic approaches aid in the identification of threats to conservation. Moving forward, we discuss the role of conservation genetics in proposing conservation units, highlighting how recent advances in genomics enhance our ability to define fine-scale population structures and identify adaptive loci, thus refining the identification of conservation priorities. We also showcase the pivotal role of genetics in revealing cryptic species and resolving taxonomic uncertainties, offering valuable insights into biodiversity conservation efforts.

The chapter also explores the use of non-invasive sampling techniques for population monitoring, emphasizing the growing significance of this alternative sampling method in assessing ecosystem health, detecting invasive species and pathogens, and revealing biodiversity and community composition. Moreover, we highlight the indispensable role of wildlife DNA forensics in combating illegal trafficking of neotropical species, and analyze translocation strategies, discussing how genetic factors such as gene flow and connectivity can inform decisions aimed to minimize inbreeding and loss of genetic diversity. We also examine the importance of captive breeding and ex situ conservation programs in maintaining genetic variation and supporting wild populations. Thus, this chapter provides a comprehensive overview of genetic conservation strategies (Fig. 15.1), underscoring their crucial role in safeguarding the rich biodiversity of the Neotropics.

15.2 Assessing Conservation Status

In the field of conservation biology, assessing the conservation status of species is crucial for determining appropriate conservation measures to prevent their extinction. Faced with challenges in obtaining long-term demographic and life-history data for monitoring and conservation of wild populations, population genetic approaches emerge as valuable tools for quickly providing insights into crucial metrics such as genetic diversity, inbreeding, effective population size, population structure and genetic load (Schmidt et al. 2024). It is also useful for diagnosing main threats for species conservation, such as hybridization and introgression.

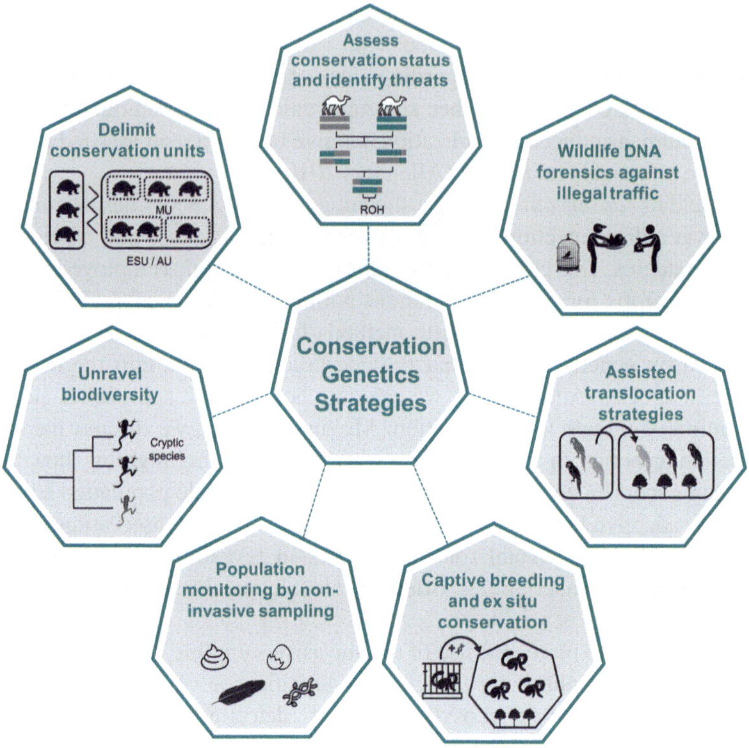

Fig. 15.1 Diverse strategies employed in conservation genetics to contribute to species protection. Each one of these topics is explored in the following sections of this chapter

Genetic diversity is the basis for biodiversity and conservation planning, particularly for threatened and endangered species, as low genetic diversity hampers their adaptive potential to environmental changes (Spielman et al. 2004). This significance is underscored by the explicit acknowledgment of the genetics of biodiversity in the Convention on Biological Diversity, which establishes the conservation mechanisms, indicators, and targets to monitor and preserve genetic diversity (Hoban et al. 2020). The motivation for coordinated efforts arised from the acknowledgment of human activities as playing a primary role in the decline of genetic diversity. Leigh et al. (2019) estimated a 5.4–6.5% decline within-population genetic diversity of wild organisms since the industrial age. Moreover, species with higher extinction risk status tended to have lower genetic diversity (Willoughby et al. 2015), although these relationships are sometimes weak and vary across taxa (Schmidt et al. 2023). For example, the harpy eagle (*Harpia harpyja*), a near-threatened species inhabiting the Amazonian and Atlantic forests, revealed lower genetic diversity in highly deforested regions compared to more preserved areas (Banhos et al. 2016). Temporal analysis also showed a concerning trend that the genetic diversity of the harpy eagle has diminished over time in these highly impacted habitats. These findings should be taken into account when designing the conservation plan for the species to ensure the implementation of effective conservation measures.

Declining populations are more strongly affected by genetic drift and higher rates of inbreeding, which can reduce the efficiency of selection, lead to fitness declines, and hinder species capacities to adapt to environmental change (Willi et al. 2006). Genetic data can be used to estimate the effective population size (*Ne*), which is the size of an idealized population that has experienced the same loss of diversity through drift as the population under consideration (Husemann et al. 2016). Thus, *Ne* is a measure for estimating population sizes with long-term viability of the species. The South American fur seal (*Arctocephalus australis*) has been experiencing a population decline due to indiscriminate commercial hunting over the past centuries. Even more severe was the bottleneck that the fur seal populations from the north-central Chilean coast and the central Peruvian coast underwent during 1993 and 1994, driven by constraints on food availability due to fluctuations in water temperature and richness associated with El Niño events (Cárcamo et al. 2021). Oliveira et al. (2012) estimated the effective population size (*Ne*) for the Peruvian population, considering the population fluctuations caused by El Niño events and also the unequal sex-ratio. They estimated a *Ne* of 2153 specimens, which should be integrated into future management plans to ensure the conservation of the species along the Peruvian coast.

In small and isolated populations, the mating between closely related individuals is more likely to occur, a phenomenon known as inbreeding. This can result in a reduction in genetic diversity and an increase in harmful recessive alleles, ultimately compromising the overall fitness and adaptability of the population (Bosse et al. 2022). Monitoring and managing inbreeding levels allows conservationists to target strategies to mitigate the negative effects, such as introducing individuals from different populations to increase gene flow and diversity.

Genomic advances have enabled us to assess inbreeding levels in a finer resolution. Runs of homozygosity (ROH) analysis assess stretches of the genome where an individual has inherited identical DNA segments from both parents due to common ancestry (Box 15.1). Longer ROH suggests more recent inbreeding, potentially leading to reduced genetic diversity and risk of expressing deleterious traits. On the other hand, shorter ROH suggests older inbreeding events, as there has been time for recombination to disrupt them. Pallotti et al. (2023) utilized ROH analysis to estimate the inbreeding levels on South American Camelids. The ROHs analysis indicated a similar inbreeding among alpacas, llamas, vicugnas, and guanacos, and suggested a loss of genetic diversity from a historical founder effect or genetic bottleneck, potentially due to the extensive hunting during the last century. Moreover, longer ROHs were found exclusively in alpacas, suggesting a more recent inbreeding compared to other species.

Saremi et al. (2019) also used ROH analysis to evaluate inbreeding effects in puma populations from North and South America. They found varying levels of inbreeding across populations, with North American isolated populations experiencing more significant inbreeding effects. Interestingly, they observed that tracts of homozygosity are rarely shared among populations, hinting at the potential assisted gene flow to restore local genetic diversity. In this study, they analyzed one individual from Florida that descended from translocated Central American individuals. Surprisingly, they identified long tracts of homozygosity, even in the presence of outbreeding. This intriguing finding implies that although translocations can initially introduce genetic diversity, the long-term maintenance of diversity in small, isolated populations need either recurrent translocations or the restoration of landscape connectivity to improve gene flow. These genomic findings have a direct implication on conservation management decisions of this highly threatened species that suffered a drastic population decline.

Box 15.1: What are runs of homozygosity (ROH) and how do they account for inbreeding levels? How does this metric contribute to informing conservation studies?

The analysis of runs of homozygosity (ROH) is a genomic tool used to assess levels of inbreeding by identifying continuous stretches of homozygous segments within an individual's genome (Lavanchy and Goudet 2023). These segments arise when identical haplotypes are inherited from both parents, indicating a common ancestry and, consequently, inbreeding. The extent and distribution of ROH within a genome can provide a quantitative measure of an individual's inbreeding level, often expressed as the ROH-based inbreeding coefficient (FROH) (Chen et al. 2023). This metric is particularly useful in populations where information is limited or unavailable, allowing for the assessment of genetic diversity and inbreeding without historical breeding records (Heweet et al. 2023; Ziegler et al. 2023). In wild populations, ROH can reveal the effects of demographic events such as bottlenecks, founder effects, and isolation, which may increase inbreeding and reduce genetic diversity (Harder et al. 2022; Heweet et al. 2023; Martin et al. 2023). This approach has been instrumental in uncovering genomic regions under selection for specific traits, aiding in the conservation (Foote et al. 2021; Lavanchy and Goudet 2023). By pinpointing these genomic regions, conservation efforts can be more targeted, promoting the preservation of genetic diversity and the health of populations.

Inbreeding is difficult to avoid after a severe bottleneck and frequently leads to loss of genetic diversity and increase in genetic load. The latter refers to the accumulation of deleterious genetic variation within a population, which can negatively impact their long-term survival and adaptive potential. A management strategy to overcome the detrimental effects of inbreeding is to promote gene flow between different genetic pools within species. Conservationists applied this strategy to Mexican wolves (*Canis lupus baileyi*) by mixing three unrelated but inbreed captive lineages beginning in 1995 (Hedrick et al. 1998). These individuals were later reintroduced in Arizona and New Mexico. Fredrickson et al. (2007) considered three fitness traits and concluded that F_1 offspring presented large fitness increases compared to the founding lineages. Moreover, they observed inbreeding depression among wolves descended from F_1 individuals, suggesting a high load of deleterious alleles, particularly within one of the three lineages. Similarly, Ochoa et al. (2022) analyzed the mutation load of rescued puma populations in Florida, discussing potential deleterious alleles introduced during the rescue process, especially in cases where the effective population size remained small. Monitoring deleterious alleles in genetically rescued populations seems to be critical for long-term conservation efforts.

In a comprehensive risk assessment for any species, it is imperative to incorporate an analysis of potential hybridization and introgression events, as these can significantly influence the fitness of offspring and, in extreme cases, contribute to species extinction (Bosse et al. 2022). Such outcomes arise from genomic incompatibility, the expression of detrimental mutations, the loss of adaptive complexes, and other contributing factors. This is the case of hybrid sea turtles from Abrolhos Archipelago, which showed lower reproductive success compared to the parental species involved in the hybridization, a phenomenon known as outbreeding depression (Arantes et al. 2020).

Genetic data can serve as a valuable tool for examining the distribution of current genetic diversity within a species across its habitat and assessing the connectivity between populations and its correlation with the environmental landscape and historical geoclimate events. Within the same biome, some species have a wide distribution with long-distance gene flow, while others are restricted to specific habitats. For example, lowland rainforest bees (*Eulaema nigrita* and *E. cingulata*) showed cross-Andean dispersal and shared mitochondrial haplotypes between populations located up to 3000 km apart (Dick et al. 2004). In contrast, the spider *Gasteracantha cancriformis* showed two mitochondrial clades occurring on opposite sides of the Eastern cordillera of the Colombian Andes, with some level of gene flow likely facilitated by the presence of low-elevation passes across the Eastern Colombian cordillera (Salgado-Roa et al. 2018).

The International Union for Conservation of Nature (IUCN) provides guidelines for evaluating the extinction risk and determining conservation priorities at both global and national levels. These guidelines have been increasingly used by countries to develop their own lists of threatened species and incorporate them into national conservation strategies. The IUCN Red List criteria, assessing extinction risk based on factors such as range, population size, and habitat quality, do not currently take genetic information explicitly into account, despite several studies highlighting the importance of genetic diversity in enabling species to adapt to environmental changes. Including genetic information in conservation assessments would enhance efforts to protect biodiversity.

A recent study proposed the incorporation of genetic data as *essential biodiversity variables* (EBVs), crucial metrics for aggregating and interpreting biodiversity observations from various sources (Hoban et al. 2022). The authors proposed four Genetic EBVs: Genetic Diversity, Genetic Differentiation, Inbreeding, and Effective Population Size—ranking them based on relevance, sensitivity to change, generalizability, scalability, feasibility, and data availability. This model intricately weaves together population size and geography with genetic diversity and presents concrete steps for researchers and makers for advancing operationalization of Genetic EBVs. Ultimately, this approach that integrates genetic insights into conservation strategies is essential for the long-term viability and protection of diverse species, as illustrated by the case studies described here.

15.3 Delimiting Conservation Units

The maintenance of the genetic diversity in natural populations is crucial for maximizing evolutionary potential, as it leverages the beneficial associations of genetics with the environment. Conservation planning aims to define appropriate conservation units in an evolutionary context in order to enhance the species evolutionary potential, to prioritize taxa or populations for conservation effort, and fine-tune management action strategies to conservation (Barbosa et al. 2018). The main conservation units are known as *evolutionarily significant units* (ESUs) and *management units* (MUs) (Box 15.2). Although there are many definitions of ESUs (summarized in Funk et al. 2012), ESUs are generally intended to mirror significant breaks in genetic diversity among populations within a species, which reflects its phylogeography history (Moritz 1994). Functionally independent populations, connected by some level of gene flow but showing diverged allele frequency, are recognized as MUs and serve as the basis for the short-term management of populations (Moritz 1994). Therefore, MUs are generally of a reduced scale compared to ESUs, and it is possible to have numerous MUs within a single ESU.

Box 15.2: How do management units (MUs) differ from evolutionarily significant units (ESUs) in the context of conservation planning, and what criteria are typically used to delineate these units?

Management Units (MUs) and Evolutionarily Significant Units (ESUs) are both concepts in conservation planning, but they differ significantly in their criteria for delineation and their implications for conservation actions (Funk et al. 2012; Shaney et al. 2020; Turbek et al. 2023). MUs are primarily defined based on genetic differentiation and demographic isolation, focusing on managing populations that are genetically distinct but may not necessarily be adaptively significant on an evolutionary timescale (Funk et al. 2012). On the other hand, ESUs are delineated based on evidence of significant evolutionary divergence, including genetic, morphologic, and sometimes ecological differentiation, which indicates a potential for long-term adaptability and survival (Turbek et al. 2023). Criteria for delineating these units often include genetic diversity, effective population size, adaptive differentiation, and functional connectivity for MUs, while ESUs require evidence of significant evolutionary divergence and potential reproductive isolation (Lima et al. 2021; Turbek et al. 2023). However, the application of these criteria can vary depending on the species and the specific conservation goals, as seen in the diverse methodologies and outcomes across the studies.

Regardless of the CUs category level, the identification of population units within species holds significant importance serving as the basis for evaluating the status of a population and devising management strategies for its maintenance or recovery. Therefore, the identification of CUs represents a crucial initial step in conservation, enabling managers and policy makers to clearly delineate the boundaries of the population units they are committed to preserving (Funk et al. 2012). For example, cryptic taxa within the monkey frogs *Pithecopus rohdei* from the Atlantic Forest were recently revealed by an integrative study including molecular and morphometric data (Ramos et al. 2019). The authors suggested that the five divergent lineages identified should be considered ESUs for conservation due to the genetic population structure found. This strategy would prioritize the conservation of these genetically distinct and disconnected populations, previously considered a single widely distributed species, thereby facilitating the implementation of targeted protection measures, such as safeguarding lineages inhabiting unprotected areas and those adapted to specific climatic conditions.

Similar strategies, starting from identifying distinct evolutionary lineages that serve as significant reservoirs of genetic diversity for the species to proposing them as independent conservation units, have been recommended for several neotropical species, such as giant otter (Pickles et al. 2011; Hernández-Romero et al. 2018), foxes (Pizarro et al. 2023), sloths (Lara-Ruiz et al. 2006), rodents (Miranda et al. 2007), dolphins (Cunha et al. 2014; Caballero et al. 2018; Oliveira et al. 2019), deers (Moscarella et al. 2003), maned wolves (Mannise et al. 2017), primates (Acevedo-Garcés et al. 2021), iguanas (Shaney et al. 2020), birds (Leite et al. 2008; Lima-Rezende et al. 2019), and fishes (Carvalho et al. 2012; Vitorino et al. 2017; García-De León et al. 2020). All of them rely on the detection of differential allele frequencies among populations, based on mitochondrial genes and/or few nuclear genes, mostly microsatellites. Moreover, these studies involve native populations that are suffering great reduction in number or diversity, exhibit significant levels of differentiation, and have the potential to serve as distinctive sources of adaptation for the species, aligning with the fundamental criteria for defining CUs as proposed by Hoelzel et al. in 2023. These populations warrant prioritization for conservation because diminishing effective population sizes can result in the loss of genetic diversity, heightened genetic drift, and increased inbreeding, ultimately leading to less fit populations. This phenomenon is commonly referred to as the "extinction vortex" (Frankham 2015). When dealing with endangered species that have small fragmented populations and face threats such as habitat loss or high inbreeding levels, delimiting CUs can also help to identify suitable individuals for translocation efforts, aiming to maximize genetic diversity without promoting outbreeding depression (Moritz 1999).

The identification of CUs is especially useful in fisheries biology to guide tailored management decisions concerning independent genetic stocks. For the fish species *Piaractus brachypomus*, populations from Orinoco and Amazon basins showed morphologically and molecularly differentiation, supported by evidence from both mitochondrial and nuclear genes, leading the authors to recognize two distinct ESUs (Escobar et al. 2015). This pattern of phylogenetic relationship linked to the geological and geographic histories of river basins was observed for several freshwater fishes in South America (Reis et al. 2016). The ultimate goal is to combine this information with life-history characteristics, reproductive seasonality and other population parameters to delineate fisheries regulations that specify local harvest quotas, fishing season limits, minimum fish size for capture, and monitor habitat and population status tailored for each ESU.

Recent advances in sequencing technologies have enabled the comprehensive exploration of genetic information across the entire genome. The availability of genomic-level data has expanded the scope of defining CUs by providing more comprehensive insights into both neutral and adaptive variations within a species. This advancement has led to the identification of an additional CU category: adaptive units (AUs) (Funk et al. 2012). AUs distinguish populations presenting genomic variants that confer unique advantages under specific local environmental conditions, a phenomenon termed local adaptation (Flanagan et al. 2018). Methods for identifying candidate adaptive loci from genomic data are mainly based on FST outlier methods and genotype–environment associations (GEAs) (Flanagan and Jones 2017). Fraser and Bernatchez (2001) have advocated for the integration of neutral genetic variation and adaptive variation, proposing the concept of Adaptive Evolutionary Conservation towards a unified framework that incorporates both aspects. Future studies on Neotropical species should strive to identify variants associated with local adaptation and integrate this information into the delineation of CUs.

The red-footed tortoise (*Chelonoidis carbonarius*) inhabiting two different ecoregions separated by the Andes within Colombia present a clear differentiation with allopatric reproductive isolation, local adaptation and ecological divergence (Gallego-García et al. 2023). Using a restriction-site-associated DNA sequencing dataset, the FST outlier detection revealed that the level of genetic differentiation was close to fixation for alternative putative adaptive alleles. The presence of the Andes Mountains, serving as a barrier to gene flow, coupled with distinct selective pressures in each environment, appears to be driving the East and West lineages along different evolutionary trajectories under divergent selection. The authors suggested that these two lineages should be managed separately to preserve their unique adaptations. Furthermore, they recognize four MUs within West lineage, contemplating the clusters distributed in extremely heterogeneous environments including dry and humid forest biomes. Understanding patterns of adaptive differentiation is crucial when prioritizing which populations to focus management efforts on, particularly in scenarios of limited resources where one population should prioritized over another, and when deciding which populations to serve as sources for translocation, re-introduction, and assisted migration efforts (Funk et al. 2012).

The primary goal of Evolutionarily Significant Units (ESUs) is to identify populations within species "possessing genetic attributes significant for present and future generations" (Ryder 1986) to prioritized conservation actions. Thus, it is crucial to consider both historical and contemporary processes shaping genomic diversity. Measuring overall genetic distinctiveness might not adequately capture the extent of local adaptation when isolated populations have diverged via neutral processes (Fernandez-Fournier et al. 2021). Moreover, a phylogeography approach might not successfully identify instances of recent local adaptation and genetic isolation (Pease et al. 2009). Combining adaptive and neutral information provides a more comprehensive understanding, although identifying adaptive loci poses challenges with often unclear interpretations in a conservation context (Shafer et al. 2015).

15.4 Using Non-invasive Sampling for Population Monitoring

The use of non-invasive sampling has been popularized since the 1980s due to the necessity of changes and the need for animal handling approaches that prioritize their well-being (Pauli et al. 2010). The application of non-invasive sampling increased between 1990 and 2000 as an alternative method for characterizing biodiversity, also thanks to improvements in molecular methods, especially for vertebrates (Taberlet and Luikart 1999; Pauli et al. 2010). Given its widespread use, terminologies for sampling strategies underwent revision, leading to the emergence of various approaches such as remote sampling and minimally invasive methods (Pauli et al. 2010; Carroll et al. 2018). The non-invasive method involves sampling and documenting biodiversity without directly handling and consequently stressing the animals, as is often the case with capture procedures. This method proves to be invaluable in situations where captured individuals may experience stress, the target species is endangered or challenging to access and locate, and when the sampling approaches could lead to mortality. Non-invasive sampling's applications extend to the identifying rarely observed species, estimating population size, structure and gene flow, as well as assisting in combating animal trafficking and disease detection, rendering it relevant in conservation (Waits and Paetkau 2005).

There are several well-defined methodological approaches for sampling animal taxa, although they are still quite biased toward vertebrates. These non-invasive sampling approaches for genetics include techniques such as hair traps, collection of "lost" biological materials, biopsy darts and, more recently, environmental DNA (eDNA, Pauli et al. 2010; Rodríguez-Castro et al. 2022). Among these methods, scat or feces stands out as the most commonly employed non-invasive approach across multiple taxa. Similarly, saliva can serve as a useful sample for several animals. While it is not exactly non-invasive in cases where the animal needs to be caught, it is considered non-invasive only when saliva is found in bite areas on a dead prey. For birds, feathers and eggshells have been utilized in a handful of studies. The latter material can also be used for reptiles and other egg-laying animals. Hairs are commonly used for sampling mammals while antlers are an interesting source for cervids, enabling researchers to trace species changes over time. Scales can be used for fishes and snakes. Mucus is the only non-invasive material found for terrestrial invertebrates. Additionally, although not strictly non-invasive, roadkill can serve as an alternative sample source, utilizing material that would otherwise be discarded. Finally, eDNA is currently a hot topic in biodiversity characterization, with a couple of examples outlined below in this section. A comprehensive overview of non-invasive sample sources, methods, and examples of studies with Neotropical species can be found in Table 15.1.

Box 15.3: What are some key challenges and potential solutions associated with non-invasive sampling techniques, particularly in the context of Conservation Genetics in the Neotropical region?
One primary challenge is the reliability of these methods, especially when dealing with faecal samples from frugivorous species in tropical areas, where the success rate of mitochondrial DNA (mtDNA) PCR amplifications can be variable, and the preservation method significantly affects the outcome. Ethanol preservation has shown better results compared to silica gel, suggesting a need for protocol optimization to enhance efficiency (Peters et al. 2023). The potential solutions to these challenges include the integration of various non-invasive sampling methods, such as environmental DNA (eDNA) analysis and cuticular swabbing, which have proven effective in detecting pathogens in endangered crayfish populations. These methods, combined with optimal storage solutions like absolute ethanol, can preserve samples effectively for genetic analysis (Baciu et al. 2022). The application of non-invasive genetic sampling in conservation genetics also benefits from advancements in DNA extraction, amplification, and data analysis techniques, making it a preferred choice for studying endangered populations (Gonçalves et al. 2022). However, the successful implementation of these techniques requires familiarity with their requirements and potential limitations, such as the need for increased incubation time and modifications in DNA elution buffer volume and temperature to improve DNA yield from samples like moulted feathers (Byrne et al. 2021). While non-invasive sampling techniques face challenges such as variable success rates and the need for protocol optimization, the integration of various methods and advancements in molecular techniques offer promising solutions for conservation genetics in the Neotropical region and beyond (Balázs et al. 2020; Banks and Piggott 2022).

In the Neotropics, studies on population genetics using non-invasive sampling follow the global trend, focusing primarily on vertebrates, particularly mammals and birds, as observed on the literature review and exemplified below. By using scats, Valqui et al. (2010) identified that the endangered marine otters (*Lontra felina*) exhibit a moderate genetic variability in its Peruvian distribution, making them likely the most diverse neotropical Lontra species. This underscores the need of focusing the conservation efforts in this region and provides valuable information for translocating individuals. Furthermore, Aristizábal Duque et al. (2018) reviewed the challenges of using scats for genetic studies on otters, pointing out issues such

Table 15.1 Non-invasive sampling methods used for monitoring populations of Neotropical species

Sampled material	Taxon	Genetic marker	References
Scat/feces/ secretions	Carnivora, Psittaciformes, Artiodactyla, Perissodactyla, Rodentia	mtDNA, CHD gene, microsatellite	Franco-Gutiérrez et al. (2017), Tirelli et al. (2019), Latorre-Cardenas et al. (2020), Delgado et al. (2021), Rodriguez-Castro et al. (2022), Salom-Pérez et al. (2022), Karen Giselle et al. (2022), Cueva et al. (2024)
Hairs	Carnivora	Microsatellite, mtDNA	Peralta et al. (2020), Lorenzana et al. (2020), Cueva et al. (2024),
Feathers	Psittaciformes, Columbiformes	Microsatellite, CHD gene, mtDNA	Olah et al. (2016), Franco-Gutiérrez et al. (2017), Peters et al. (2020)
Eggs/ eggshell	Anseriformes, Columbiformes	Microsatellite, mtDNA	Maia et al. (2017), Peters et al. (2023)
Antlers	Artiodactyla	mtDNA	Venegas et al. (2020)
Mucus	Myliobatiformes	mtDNA	Domingues et al. (2019)
Swab*	Cetacean, Anura	Microsatellite, mtDNA	Farro et al. (2008), Gallardo et al. (2012)
Road-killed	Carnivora	Microsatellite, mtDNA	Rodriguez-Castro et al. (2022), Salom-Pérez et al. (2022)
eDNA	Anura, Pristiformes, Myxozoa, fish community. Lagomorpha	Metabarcoding, Metagenomics, mtDNA	Lopes et al. (2017), Bonfil et al. (2021), Baeza et al. (2023), Hilário et al. (2023), Bonfil et al. (2024).

as genotype errors and null alleles as significant concerns. Similarly, Miotto et al. (2011) found, using mainly fecal and roadkill samples, that pumas—a species facing challenges due to habitat loss from human activities—in Brazil, although exhibiting moderate variability and connectivity, showed signatures of bottleneck events. Given their biology, characterized by long-distance dispersion and long generation times, the results demonstrate the consequences of habitat loss, indicating the necessity of conservation actions to maintain connectivity between populations.

Jaguars in the Atlantic Forest face a concerning situation highlighted by Srbek-Araujo et al. (2018). Through noninvasive fecal sampling, they identified that the remnant population in the easternmost part of the forest is significantly isolated from the species' broader distribution. Additionally, this population exhibits small effective population sizes. These findings underscore the urgency for conservation efforts to safeguard this vulnerable population. Another study, using scats and road-kill animals from a Cerrado remnant protected area surrounded by an altered environment, revealed that the maned wolves exhibited a moderate level of diversity and low inbreeding. This suggests their ability to disperse and reproduce with surrounding populations, emphasizing the importance of protecting this biome (Mannise et al. 2017). Recently, Rodríguez-Castro et al. (2020) developed a panel of DNA mini-barcodes for identification of Neotropical felid and canid species from scat samples. Interestingly, this approach extends beyond native species. Rodrigues et al. (2020) developed a methodology for sequencing DNA mini-barcodes from leporid scats, enabling the monitoring of the European rabbit invader in Brazil.

By using DNA barcodes from feathers, Olah et al. (2017) showed that the red-and-green macaws have low genetic structure, and that individuals can use different clay licks for feeding. In addition to being useful for non-invasive sampling, if dedicated sampling efforts are undertaken, these clay licks locations can also be valuable for demographic estimates. The endemic and critically endangered granada dove was studied using feathers and egg shells, revealing low genetic diversity, isolation and potential for inbreeding depression, which underscore the urgent need of conservation actions (Peters et al. 2023). In the case of fishes, the utilization of scraped mucus from captured animals was tested for Neotropical batoidea species. While not entirely non-invasive, this method involved only quick handling and caused no injuries to the animals. Additionally, it resulted in good DNA quality, offering an alternative sampling method compared to, for instance, the use of harpoons (Domingues et al. 2019).

Emerging molecular techniques such as eDNA and invertebrate DNA (iDNA) have gained global recognition, particularly for the identification and characterization of biodiversity. Some studies in the Neotropical region have utilized these methods, as discussed in another chapter of this book. In the context of non-invasive sampling, these techniques serve as remarkable tools for advancing our understanding and conservation of biodiversity.

However, the use of non-invasive samples as a source for DNA material can have a couple of pitfalls (Box 15.3). In the early stages of its use, issues such as genotype errors, mainly associated with low quality of samples and consequently DNA degradation, were often overlooked. During the 1990s, challenges related to sample preservation, DNA extraction and amplification protocols, genotype reliability, and contamination were prominent (Taberlet and Luikart 1999), though these technical limitations continue to pose ongoing challenges. Byrne et al. (2021) reviewed and summarized potential solutions, addressing issues specific to the Neotropical region. These include strategies such as sampling scats during cold days and immediately after defecation, utilizing gel to remove moisture from samples, restricting use to the scat surface to avoid contaminants and inhibitors, as well as employing multiple PCR tests to confirm genotypes.

Non-invasive sampling approach carries the potential for reacquiring samples from both the same individual and its relatives, additionally posing challenges in securing high-quality DNA due to sample degradation (Olah et al. 2017). While the occurrence of recaptures and sampling relatives were concerns in all of the mentioned studies, they were not sufficiently significant to hinder or render impractical the use of such sampling methods for population monitoring. When selecting a genetic marker, it is important to consider its amplification capabilities, especially based on the sample type. For instance, mitochondrial DNA, frequently utilized in non-invasive Neotropical studies due to its superior amplification capacity, has limitations in individual identification.

Currently, with the improvement of sequencing methods and error resistant bioinformatic analyses, non-invasive sampling has become increasingly valuable even for genomic studies (Carroll et al. 2018). Despite the need for high-quality and abundant DNA for Next Generation Sequencing (NGS), various methods have been developed and are currently used for non-invasive samples. These include target sequencing, libraries enriched with endogenous DNA, SNP arrays, and reduced SNP panels (Byrne et al. 2021). Although challenging, Baus et al. (2019) strongly advocates for the significance of utilizing NGS data to address questions relevant to Conservation Genetics in birds, including studies employing non-invasive sampling methods. They also highlight the delayed implementation of genetic techniques in Neotropical research compared to other regions, attributed to funding and technical limitations. As Byrne et al. (2021) summarized, only a few conservation genetic studies have adopted non-invasive sampling despite its well-documented advantages for animal welfare.

15.5 Genetics as a Tool to Reveal Biodiversity

Cryptic species represent lineages that are distinct but too similar phenotypically to be identified as independent from each other (Bickford et al. 2007). Their existence is strongly linked to minimal variations in morphology and, sometimes, ecological differences, leading to ambiguity and challenges in determining the true independence among lineages, which poses a significant challenge in classic taxonomic studies (Struck et al. 2018). From an evolutionary perspective, cryptic species offer valuable insights into species isolation, biogeography, reproductive barriers, physiology, and other hidden phenotypes that define species as independent evolutionary units (Fišer et al. 2018; Struck et al. 2018). It also plays a role in human health, contributing to bioprospecting for useful chemicals, the development of treatments for diseases, and the creation of antidotes for venomous species. Moreover, from a conservation viewpoint, their misidentification can yield serious consequences. The discovery of greater taxonomic diversity within a species has led to shifts in ecological classification for the species. For instance, it has revealed that some species exhibit more specialized behavior rather than being generalists (Bickford et al. 2007). Misidentifying one species as another could result in overlooking the need to protect geographical areas where the species might be endemic, with no equivalent niche occupied by other taxa. As previously highlighted in this chapter, the accurate delineation of cryptic species holds substantial importance in defining conservation units.

As morphological variation often falls short in revealing the occurrence of cryptic taxa, the use of molecular markers has emerged as a powerful tool in addressing this challenge (Bickford et al. 2007). For instance, the presence of paraphyletic lineages in a phylogenetic tree serves as a key indicator of potential cryptic taxa (e.g. Carbajal Márquez et al. 2020; Trindade et al. 2021). Since the start of DNA barcode initiatives, this method has been tested and applied to uncover cryptic species across various taxa globally (Janzen et al. 2017). Another important method that has emerged as a potential option for addressing cryptic species is environmental DNA, which offers the advantage of independence from direct animal capture, as discussed in the earlier section on non-invasive approaches (Zinger et al. 2020). However, challenges such as incomplete databases for taxa identification persist, as detailed in other chapters of this book. Moreover, genomic data provides high-resolution information, enabling more comprehensive analyses for delineating distinct lineages and further understanding their relationships.

In the Neotropics, we have observed cryptic species within several animal taxa, including invertebrates and vertebrates. By using genomic and chemical markers, Meza-Lázaro et al. (2022) showed that *Ectatomma ruidum* species form a complex of at least five ants species, being one of them endemic to Mexico. They showed that this split does not follow geographic barriers, with chemical and behavioral traits being the potential source of differentiation. This group has a complex evolutionary history, with occurrences of hybridization and heteroplasmy in specific lineages. Also using complementary traits, Lima et al. (2023a, b) found, through ultra-conserved elements (UCE) and mitochondrial genomes, associated with acoustic and morphological traits, the presence of 13 species within the passerine genus *Schiffornis* instead of the currently recognized seven. They verified varying levels of differentiation using the different markers and observed the presence of contact zones between some lineages. This system illustrates the challenges of poorly described bird taxa, particularly when relying solely on morphological data, and reveals interesting patterns of song evolution within this group.

Morales-Martínez et al. (2021) discovered, using molecular and morphological markers, that the yellow-eared bat (*Vampyressa melissa)* species complex comprises four distinct lineages, contributing to a total of six lineages within the genus. Half of these lineages are considered threatened due to their highly limited distribution, forest fragmentation, declining populations and rarity. As the newly proposed species are geographically constrained and estimated to be facing ongoing habitat loss, they have not been assigned any official conservation status, making it an important conservation concern. A similar problem with new populations of unknown conservation status was demonstrated within Cyclopes. Miranda et al. (2018) showed that the species currently acknowledged as monotypic but experiencing nomenclature instability, encompasses three additional species beyond the four previously proposed designations. This conclusion was drawn from the utilization of both mitochondrial and nuclear genetic markers, as well as consideration of phenotypic traits. The study issues a noteworthy caution regarding the conservation status of some of these species, given their distribution in areas impacted by human activities within the Amazon and the highly fragmented Atlantic Forest. Another complex that has experienced taxonomic uncertainty is the tigrinas, as demonstrated by Trindade et al. (2021) utilizing genomic markers. They showed that the tigrina complex comprises at least three cryptic species, thereby characterizing the group as paraphyletic. Additionally, considering the newly recognized lineage is older, the study suggests that the complex may exhibit the plesiomorphic phenotype of the *Leopardus* genus, posing challenges in species delimitation. The research emphasizes the importance of enhancing the sampling of tigrinas in all geographic locations where they occur, to assess in more detail all lineage relationships and geographic delimitations.

Through integrative approaches incorporating genetic data along with acoustic and morphological information, Andrade et al. (2020) confirmed the hypothesis of the presence of two cryptic species within *Pithecopus nordestinus*, also known as Northeastern Brazilian Leaf Frog. The study underscores the significance of managing the primary barrier for the species, the São Francisco River, especially concerning the transposition project for the river in areas affected by severe drought, potentially affecting species contact. Carbajal Márquez et al. (2020) revealed that the Neotropical rattlesnake (*Crotalus durissus*) species complex harbors even more species than previously believed. Employing mitochondrial and nuclear markers, considering evolutionary history, biological trait cohesion, and comparison with well-accepted congeneric species delimitation, the study suggests the existence of at least three cryptic lineages. Consequently, they propose a revision of the conservation status of the species.

Domingos et al. (2017) revealed that *Tropidurus itambere*, a Cerrado endemic lizard, may actually comprise five species, as indicated by their genomic data. This species divergence is particularly significant given their occurrence within a biodiversity hotspot in the Cerrado, a region already recognized for its endemic cryptic diversity. The authors underscore the importance of employing genomic data to unveil hidden diversity, especially within Neotropical biomes characterized by high speciose genera.

An important consideration when uncovering new cryptic species is the application of species concepts, especially given the wide array of concepts available and absence of a consensus on the most suitable definition (Bickford et al. 2007). This challenge could be particularly pronounced in highly speciose groups with rapid diversity and evolution, for which we have several examples in Neotropics (Box 15.4). Although, a systematic review is needed to determine whether the perceptual bias observed in more cryptic studies towards temperate climates is a result of a research bias or reflects a genuine hidden biodiversity (Bickford et al. 2007). Also, it is discussed that the term 'cryptic species' has been used for groups that should not necessarily be considered as such (Struck et al. 2018). To overcome this, studies often document and support their findings employing multiple complementary analyses and approaches (e.g. Andrade et al. 2020; Meza-Lázaro et al. 2022). Furthermore, species situated within the so-called gray zone of speciation (i.e. still undergoing diversification, with shared alleles, and possibly shared phenotypic and behavior characteristics) may exhibit incomplete reproductive isolation. Consequently, hybridization may occur, serving both as an obstacle to identifying cryptic diversity and as a potential consequence of it. Hybridization can also pose significant conservation problems, being the cause of outbreeding depression, extinction of local lienages, genetic and phenotypic mischaracterization, among other issues (Todesco et al. 2016). Moreover, a common challenge in studies aiming to identify cryptic diversity is avoiding collection bias by ensuring thorough geographic sampling representation. Despite the potential use of non-invasive approaches, as exemplified in this chapter, studies encounter difficulties in delimiting the occurrence of species that are rare and/or challenging to sample.

Box 15.4: What strategies could be implemented to improve our understanding of hidden biodiversity, particularly in regions like the Neotropics characterized by high species diversity and cryptic taxa?
To enhance our understanding of hidden biodiversity, particularly in regions with high species diversity and cryptic taxa, a multifaceted approach integrating various innovative strategies is essential. First, employing DNA barcoding and molecular phylogeny can significantly aid in identifying and cataloging species, including those that are cryptic or previously unrecognized (Reier et al. 2020). This method has proven effective in revealing high lineage diversity within specific ecosystems and in identifying new species within known genera (Juhel et al. 2022). Additionally, environmental DNA (eDNA) metabarcoding offers a powerful tool for biodiversity assessments, capable of detecting species not observed through traditional visual surveys, thereby providing a more comprehensive species inventory (Reier et al. 2020). Photography-based quantitative methods, combined with community science records, present another promising strategy for assessing the distribution of cryptic taxa. This approach has been successfully applied to differentiate between phenotypically similar species and clarify their geographical distributions (De León et al. 2020). The integration of vocal and genetic data has also been highlighted as beneficial for addressing taxonomic and systematic issues in avian studies, which could be adapted for broader biodiversity research (Forcina et al. 2023). Furthermore, modeling of niche the current and future distribution of cryptic species complexes under climate change scenarios can inform conservation strategies, ensuring that protected areas encompass the necessary habitats for these species' survival (Milić et al. 2019). Lastly, acknowledging the importance of local disturbances and geographic isolation in shaping biodiversity patterns can guide targeted conservation efforts, especially in regions like the Neotropics where endemicity is high even at narrow spatial scales (Ramirez et al. 2017). Collectively, these strategies underscore the need for a comprehensive, multidisciplinary approach to uncover and conserve the hidden biodiversity

Considering the challenges and advancements in recent years, several strategies could contribute to improving our understanding of hidden biodiversity. First, conducting more studies focused on building comprehensive databases for eDNA approaches across various taxa would be instrumental. Second, investing in genomics methods could provide a powerful tool for reaching a finer resolution in the phylogenetic analysis to allow the identification of cryptic species. As the first step, the collection and storage of fresh samples in optimized conservation conditions should be prioritized to meet the requirements for obtaining good quality genomic data. Third, a systematic review of widely distributed species, particularly those occurring in diverse environments and rarely observed, should be prioritized, as they may present not simply structured populations, but conceal independent lineages. This is particularly relevant in areas marked by strong anthropogenic alteration and/or identified as biodiversity hotspots, as they may harbor cryptic, endemic, and threatened species which may require different managing strategies (Bickford et al. 2007). Fourth, we underscore the importance of adopting integrative approaches that combine genetics with ecology, behavior, and morphology. Utilizing various traits on which evolution may have acted could significantly aid in untangling species complexes, rather than relying solely on morphology (Bickford et al. 2007). Lastly, we emphasize the valuable role of museums and natural collections as indispensable sources of information about biodiversity, facilitating and enriching studies in this field.

15.6 Wildlife DNA Forensics Against Illegal Traffic

Wildlife DNA forensics plays a crucial role in the fight against illegal traffic in Neotropics species. The analysis of forensic DNA allows for the identification of species, individuals, and their geographical origin, which is essential in resolving crimes involving animals (Di Rocco and Anello, 2021). DNA analysis is used to identify unknown species, determine gender, match DNA profiles, and ascertain origin and relationships (Lorenzini and Garofalo 2021). PCR-based DNA amplification, sequencing, and analysis are commonly used techniques in wildlife forensic laboratories (Jabin et al. 2019). Standardization, validation, and quality check are essential for the admissibility of forensic results as scientific evidence in court (Woodcock et al. 2023).

Illegal traffic in birds in Neotropical countries, particularly in Latin America, is an important issue that threatens the survival of many species, including parrots (Pires et al. 2016). The illegal parrot trade is driven by high demand for these birds as pets, both in developed nations and in countries where parrots are native (Sas-Rolfes et al. 2019). *Ara ararauna*, also known as the blue-and-yellow macaw, is a Neotropical parrot with a broad geographic range including eastern Panama, Colombia, Bolivia, Paraguay, and Brazil. This species is described as "common" in areas of its range and is not considered to be threatened by IUCN. They are in the illegal wildlife trade due to their attractively colored plumage, charisma, and adaptability to captivity. A significant number of *Ara ararauna* are seized by the Brazilian environmental police from illegal traders.

Fernandes and Caparroz (2013) carried out genetic analysis on 13 seized macaws in northeastern Goias, Brazil, to guide their safe release back into the wild. Several studies have relied on genetic relationships and population clusters to determine suitable release locations. This study used a 403 bp sequence of the mtDNA control region to determine the genetic relationships between the seized macaws and genetically characterized wild macaw populations in Brazil. The pairwise genetic distance between these sequences ranged from 0.0 to 3.2, indicating a high level of genetic diversity. Four of the 13 macaws selected for release contained haplotypes previously observed in wild macaws. Six of the seized macaws had close genetic relationships with the macaws of northeastern Goias, indicating they are potential candidates for release in the target area. The release of seized individuals into the wild should be conducted with caution to avoid adverse effects on local populations if genetically different individuals are introduced.

The yellow cardinal is an endangered passerine endemic of South America that also faces threats from both habitat loss and the illegal caged-bird trade. Dominguez et al. (2019) guided the reintroduction of cardinals that were confiscated in urban areas back to their original conservation units by analyzing 2 molecular markers, mtDNA and microsatellites, in birds seized from the illegal pet trade. The project was successful in genetically determining the origin of confiscated individuals, releasing them into their respective wild population and evaluating the success of this conservation action with a post-release monitoring through radio-telemetry.

Molecular markers can be used to identify trafficked animals and determine their origin in the Neotropical region (Túnez et al. 2021), providing valuable information for conservation and forensic analysis. The illegal trade in tortoises, driven by a demand for exotic pets, traditional medicines, and the illegal wildlife, has had devastating consequences for tortoise populations worldwide. Giant tortoises are a group of slow-moving, long-lived, and iconic reptiles known for their impressive size, longevity, and distinctive appearance. Giant tortoises, which are a group of threatened species targeted by Illegal Wildlife Trade (IWT), are primarily found in the Galápagos Islands and the Aldabra Atoll in the Indian Ocean. These regions are isolated and have allowed different species and subspecies of giant tortoises to evolve independently. In the Galapagos Islands, there are 15 recognized species, with each species endemic to a specific Galapagos island. Furthermore, some species are housed in breeding centers located on the islands of Santa Cruz, San Cristobal, and Southern Isabela.

During a routine border inspection of a bus traveling between Ecuador and Peru, 29 turtles were found wrapped in plastic and packed in cardboard boxes (Quinzin et al. 2023), in which three individuals did not survive. The molecular markers can be used to identify trafficked tortoises and their origin, because there is a genetic reference database with mitochondrial (control region) and nuclear microsatellite markers of Galapagos species. Quinzin et al. (2023) utilized the same markers employed in building the genetic reference database to identify the origins of the confiscated giant tortoises. Thus, they assigned eight individuals to *C. chatamensis*, endemic to the island of San Cristobal, and 17 individuals to *C. guntheri* and *C. vicina,* endemic to the island Isabela. According to Quinzin et al. (2023), the genetic analyses together with captive-born registry data provide evidence that the smuggled juveniles were from two breeding centers dedicated toward conservation located on the Galapagos islands of San Cristobal and Isabela. This is the first documentation of smuggled tortoises being taken from breeding centers rather than the wild. The findings from genetic analysis provided key evidence that enabled legal investigation and enhanced the fight against IWT by identifying the trafficked species using genetic markers.

Illegal trafficking of shark and ray meat has become a significant concern in recent years, posing threats to marine biodiversity and ecosystems. This is because they play crucial roles in maintaining balance and stability within aquatic environments. However, due to factors such as overfishing, habitat degradation, and the demand for their meat and fins, many species of sharks and rays are now facing serious population declines and extinction risks.

15.7 Genetics Assists Translocation Strategies

Animal conservation using translocations, encompass the relocation of species from one habitat to another for conservation objectives. This process can be categorized into two scenarios: (1) in situ reinforcements, which involve the collection and transportation of wild organisms to another natural area to bolster the recipient conspecific population; and (2) ex situ reinforcements, which entail a captive phase lasting from less than one generation to several generations before eventual release. Genetics plays a crucial role in assisting translocation strategies for wildlife conservation. By using genomic data, scientists can gain insights into the diversity, population dynamics, and interactions of species, which can inform the implementation of translocations (Flesch et al. 2020). Genomic data can guide the selection of donor individuals/populations, understand interspecific interactions, and design population reinforcements for genetic and evolutionary rescue (Neaves et al. 2022).

Box 15.5: How do translocations differ from reintroductions in the context of animal conservation?
Translocations and reintroductions are both strategies in animal conservation, each with distinct objectives and considerations (Marino et al. 2024). Translocations encompass the broader practice of moving species from one location to another for conservation benefits, which includes reintroductions but also encompasses reinforcements and range expansions to enhance ecosystem function or assist species in adapting to environmental change (Hoffmann et al. 2015; Gross et al. 2024). Reintroductions specifically aim to restore species to areas where they have been extirpated or become extinct (IUCN/SSC 2013; Evans et al. 2023). The primary goal of reintroductions is to re-establish a viable, free-ranging population in the wild, often as part of broader conservation and ecosystem restoration efforts (Evans et al. 2023; Smith et al. 2023). While translocations can aim to increase genetic diversity and adaptability of populations by introducing individuals into existing populations or new areas, reintroductions focus on restoring ecological functions and species to their natural habitats where they have disappeared (Gaywood et al. 2022; Evans et al. 2023). The success of these efforts is influenced by various factors, including the genetic and demographic status of the species, habitat quality, and the species' behavior in the new environment (Wallen et al. 2023). Moreover, the social dynamics and ecological interactions of the species involved are importants, as these can significantly impact the success of both translocations and reintroductions (Stanley-Price 2022; Greggor et al. 2024).

Additionally, genomic approaches can help ensuring that translocated individuals represent the best candidates in terms of genetic diversity and divergence, relatedness and adaptation. Efforts to select individuals with greater genetic diversity, lower relatedness and better adapted to the habitat of the recipient population should be prioritized. These approaches can also be used to monitor the success of translocations, confirm the incorporation of genetic diversity in newly established populations, and inform decision-making in translocation management (Norman and Christidis 2021). This is particularly significant given that currently less than 10% of published translocation papers were experiments or pilots. It is important to ensure, in advance, the viability of the strategy, as for some taxa and scenarios, this process can indeed be ineffective (e.g. Lopes et al. 2021). However, most translocation actions do not yet incorporate genetic analyses (Langridge et al. 2021).

The Spix's Macaw (*Cyanopsitta spixii*) is an example of a parrot species that became extinct in the nature, before effective conservation measures could be implemented. The species is endemic to a restricted semiarid area of Brazil called Caatinga and little is known about its ecology and behavior in the wild. The existence of few captive individuals provided an opportunity for reestablishing a wild population. The Blue-Winged Macaw (*Primolius maracana*) is the most similar species to Spix's Macaw in terms of ecology and biology, making it an appropriate model species to assist in the reintroduction efforts. Captive Spix's Macaws were released alongside with Blue-Winged Macaws as heterospecific flocks (Purchase et al. 2024). This approach allowed the released Spix's Macaws to interact with and integrate into wild Blue-Winged Macaw groups at the release site. The goal of this strategy was to increase the total number of individuals in the release cohort and enhance the post-release survival and adaptation of the Spix's Macaws, which had been reared for multiple generations in captivity and released in the absence of wild conspecifics. The captive Spix's Macaws released in the program quickly acclimatized to their habitat and demonstrated the ability to reproduce, with the first breeding event occurring within 9 months of the second release. The breeding success of captive Spix's Macaws in the wild indicates that they are still capable of hatching and rearing chicks in their natural habitat.

From a pragmatic standpoint, reintroductions and translocations represent different approaches with similar aims (Box 15.5). Reintroductions involve releasing captive individuals into their original habitat distribution, while translocations entail moving individuals from one location to another. Both approaches aim to restore populations by providing the necessary resources and environment for them to thrive and develop into healthy populations (Moraes et al. 2023). Moreover, rewilding initiatives have also emerged as a tool aimed at restoring environments at both biodiversity and community levels, using both translocations and reintroductions. In the Neotropics, notable rewilding efforts are underway.

In Argentina, extensive areas have been restored and prepared to accommodate translocated individuals of cervids, macaws, rodents, and large carnivores, with the aim of restoring populations that were once abundant in the region (Zamboni et al. 2017). In the Iberá region, not only have jaguars been successfully translocated, but they have also begun reproducing, establishing a founding population. This project has already utilized genetics to aid in decision-making regarding translocation candidates for certain species, and will continue to do so as part of ongoing monitoring efforts. However, conducting extensive biodiversity translocation work comes with its challenges, including issues of public acceptance, human-wildlife conflicts arising from potential livestock predation, and the risk of introducing diseases (Donadio et al. 2022a, b). A similar initiative took place in the Atlantic Forest, involving the reintroduction of the red-humped agouti and the howler monkey, aiming to favor interaction across a broad ecological spectrum (Fernandez et al. 2017). Despite the absence of genetic monitoring before or during the process, assessing the demographic and genetic structure of the current population in comparison to pre-translocation data would be highly valuable to understand the impact on the species. Nonetheless, this initiative is already demonstrating significant impacts on seed dispersal and, potentially, forest regeneration (Genes et al. 2019).

Currently, there are several biases in the application of translocation as a management action worldwide, particularly concerning the use of genetics as both a decision-making tool and a monitoring tool. While the majority of published translocation examples are found in North America and Oceania, the Neotropics, which face significant habitat alterations and harbor a large endangered biodiversity, also warrant attention for the application of translocation strategies. Another notable bias exists in the choice of target taxa for translocation efforts. While charismatic mammals and birds dominate the literature, the conservation needs of invertebrates and reptiles, which comprise endangered biodiversity in the Neotropics, are often overlooked (Evans et al. 2023; Langridge et al. 2021; Moraes et al. 2023). However, it is important to recognize that translocation alone cannot resolve conservation challenges. Proper habitat restoration may also be necessary to facilitate gene flow and population growth (Moraes et al. 2023).

15.8 Captive Breeding and Ex Situ Conservation

Ex situ conservation is the practice of preserving species outside their natural habitats, through captive breeding, botanical gardens, and seed banks, aiming to rescue species from extinction. Such strategies are particularly important for species that are extinct in the wild or have critically low population sizes. Ideally these programs should also focus on genetic rescue to enhance the genetic diversity of endangered species, combating inbreeding depression, and increasing their evolutionary potential.

Despite the extensive implementation of captive-bred and cultivated release programs across the Neotropic (Soorae 2021), many initiatives have not explicitly integrated genetic considerations into their strategies. These projects cover a broad spectrum of species. For example, reptiles like the captive breeding programs of Orinoco crocodiles (*Crocodylus intermedius*) in Venezuela and Colombia (Saldarriaga-Gómez et al. 2023) and of Orinoco turtle (*Podocnemis expansa*) in Venezuela (Peñaloza et al. 2013) aim for reinforcement of wild populations.

Additionally, there are captive breeding programs of birds that were rescued or confiscated after being illegally collected, such as the Scarlet macaw (*Ara macao*) and the Great green macaw (*Ara ambiguus*) in Costa Rica (Toft and Wright 2015). These efforts extend to mammals and plants, including the reintroduction of the collared peccary (*Pecari tajacu*) in Argentina (Zamboni et al. 2017; Camino et al. 2019) and the restoration of threatened tree species in Brazil's Araucaria Forest by collecting seeds, producing seedlings in a nursery and reintroducing them in protected areas (Shaw 2019). However, financial constraints frequently impede long-term monitoring and comprehensive genetic analysis, which are crucial for evaluating the success of these conservation efforts.

Notably, some conservation efforts in the Neotropics have started to indirectly incorporate genetic awareness. For instance, the management plan for Darwin's rhea (*Rhea pennata pennata*) in Chile, used three geographically distinct populations as genetic sources for breeding, demonstrating an awareness of the need for genetic diversity (Saucedo et al. 2018, 2019). Similarly, the coral reef restoration efforts in the Caribe adopted an innovative approach with ex-situ gamete fertilization instead of traditional cloning methods. This project aims to enhance genetic diversity, although further genetic analyses are necessary to confirm its effectiveness in increasing the genetic diversity of wild populations (Henry et al. 2021).

Neotropical case studies where genetics actually played a central role offer insights into its potential for enhancing conservation outcomes. For example, the breeding program for the Spix's macaw (*Cyanopsitta spixii*), a species extinct in the wild, carefully considered genetic makeup to minimize inbreeding, underscoring the critical role that genetics plays in species recovery (Caparroz et al. 2001). In a recent reintroduction, released individuals were chosen to maximize the genetic diversity of the re-introduced population (Purchase et al. 2024). Similarly, for the golden lion tamarin (*Leontopithecus rosalia*) genetic analyses of re-introduced and translocated populations revealed an increase in effective population size and a decrease in inbreeding coefficients, indicating the success of these strategies in enhancing genetic diversity and population structure (Moraes et al. 2017). The Andean condor (*Vultur gryphus*), South America's most endangered vulture, represents a prime example of how genetic and movement data are essential for addressing the conservation challenges of long-lived and highly mobile species. Recent research on the Andean condors in the equatorial Andes revealed that the captive population retains considerable genetic diversity, offering potential for strengthening wild populations (Padro et al. 2023). Genetic tools can also help evaluate the success of reintroductions. Metabarcoding analysis, a molecular technique to identify multiple taxa within environmental samples, was employed to assess the diet of re-introduced Red and Green Macaws (*Ara chloropterus*) in Argentina, where they were locally extinct. The study revealed that these macaws successfully adapted to local food resources (Volpe et al. 2022). This finding highlights the utility of metabarcoding in providing critical insights into the dietary habits of re-introduced species, thereby offering a measure of the success of conservation efforts.

However, cautionary tales, such as the giant Galapagos tortoise (*Geochelone hoodensis*) repatriation program, remind us of the complexities of captive breeding.

This initiative aimed at restoring populations through captive breeding inadvertently led to a loss of genetic variation in the wild because only a few released breeders reproduced (Milinkovitch et al. 2004). This example highlights the need for careful genetic management within conservation planning. Ultimately, effective strategies require a multidisciplinary approach that incorporates genetic management, behavioral training, disease screening, habitat restoration, community engagement, and post-release monitoring, among others.

The potential of museums and seed banks in contributing to these conservation efforts is significant, yet underutilized. Museums, for example, play a pivotal role in storing tissue and DNA samples, although their utilization in conservation has yet to be fully realized. Similarly, seed banks offer a parallel strategy for plant conservation. In Colombia, the National Plant Germplasm Bank is part of an ex situ public effort for conserving many species (Valencia, Lobo and Ligarreto 2010), which currently maintains more than 30,000 different accessions from 275 species (Cerón-Souza et al. 2023).

Incorporating the emerging field of cryopreservation into genetic conservation efforts offers a promising approach for preserving the genetic diversity of species in the Neotropics. By allowing the preservation of cells, tissues, organs, or any other biological materials at very low temperatures, cryobanks represent repositories where biological materials are stored using cryopreservation techniques preserving the genetic diversity of species for future restoration efforts. While initiatives to organize cryobanks in this region are underway, the infrastructure necessary for their widespread implementation remains in development. A notable example of cryopreservation's potential in conservation is seen in the efforts to protect coral reefs in Hawaii. This project underscores the viability of cryopreservation for endangered species, aiming to safeguard the genetic diversity of coral species threatened by disease and environmental changes (Lager et al. 2023). By successful cryopreservation and revival of coral fragments, scientists hope to facilitate future restoration efforts and enhance the resilience of coral populations. Further investment in this area could unlock new conservation possibilities, offering a repository of genetic material for future research and restoration projects. As these technologies and infrastructures evolve, they promise to become integral components of conservation strategies, ensuring the genetic heritage of the Neotropics unique flora and fauna is maintained for future generations.

The Neotropics is a region rich in biodiversity facing significant conservation challenges, where insufficient economic resources have produced a slow development of genetic methods. Hence, the integration of genetics into conservation planning is crucial for the long-term success of ex situ and reintroduction programs. Despite financial and logistical constraints, these examples underscore the potential benefits of incorporating genetic considerations in enhancing species recovery and ecosystem resilience.

15.9 Final Remarks

In the face of current escalating biodiversity loss, the integration of diverse conservation strategies is essential for protecting our planet's natural heritage. Combining genetics with other conservation methods (habitat restoration, anti-poaching efforts, community engagement) in a multifaceted approach we are better prepared to tackle the challenges of species survival. There is a pressing need for more integrative research that not only focuses on genetic studies but also incorporates multiple conservation techniques.

The benefits of incorporating genetic analysis in conservation strategies have been demonstrated in the examples throughout this chapter. However, it also became evident that genetics is not being used as much as it could be within conservation frameworks. While genetic research, at least in the Neotropic, often informs the initial stages of conservation efforts by identifying species or conservation units and selecting individuals for breeding programs, its role in latter stages, which require government support and involve direct conservation actions, is limited. Hence, the integration of genetic information into the strategic planning for the latter phases of conservation work is still significantly missing. Moreover, the limited post-release monitoring of re-introduced individuals in the Neotropic underscores the necessity of funding to assess the success of reintroductions by monitoring the fitness of populations and evaluating the risk of outbreeding depression. Adding individuals to a small, inbred population can restore genetic variability by increasing heterozygosity in following generations, with the potential to mask deleterious alleles. However, this practice carries the risk of outbreeding depression, as fitness may be reduced by diluting locally adapted alleles. In addition, while captive populations can serve as a genetic reservoir, it is essential to monitor them to detect if adaptations that could have occurred during captivity can reduce fitness in the wild habitat. So, how can we better use genetics not just to study biodiversity but to actively protect it? Given the complex, transboundary threats to biodiversity, it is vital to strengthen the link between genetic research, policy-making, and conservation actions on the ground. In this sense, international collaborations are key to amplifying the effectiveness of conservation initiatives, particularly for species whose distributions span multiple countries.

Citizen science, by involving the public in scientific research, has been key for the success of many conservation projects in the Neotropic. Conservation initiatives integrating citizen participation benefit from data collected across extensive geographical areas. This involvement of society in conservation causes can yield greater and long-standing impacts on conservation efforts. Beyond its educational value, by engaging a broader community in data collection and monitoring efforts, citizen science can offer a valuable contribution to conservation. For instance, the Yellow Cardinal conservation project in Argentina demonstrated the power of citizen science by conducting extensive surveys to locate the species across locations with historical records simultaneously (Dominguez et al. 2021), a task that would have been unfeasible for a single team of scientists. This collaborative planned sampling was carried out during three consecutive years and helped discover the remaining populations of this endangered species, raising public awareness about its conservation challenges. In the Caribbean, citizen science contributes to conservation efforts by enabling local communities to report on coral tissue loss disease (Estrada-Saldívar et al. 2021, https://www.perryinstitute.org/sctld/). Similarly in Panama, the indigenous communities Emberá and Wounaan work alongside with a non-governmental organization (NGO) to assist in field work and mounting transmitters for the conservation of the Harpy Eagle (*Harpia harpyja*), a critically endangered bird in central America (https://peregrinefund.org/projects/harpy-eagle).

Bridging the gap between researchers, who generate the data, and stakeholders, who implement conservation strategies, remains a critical challenge (Taft et al. 2020). This connection is vital for ensuring that the knowledge from scientific research is effectively translated into actionable conservation policies. An harmonious relationship between science and policy is necessary so authorities can incorporate the information generated by scientists in conservation actions, and also to avoid unintended consequences of rewilding efforts without adequate evaluation (Balza et al. 2023).

As we consider the practicalities of intervening in nature to save species potentially unadapted to current conditions, we must confront a paradox at the heart of conservation. Are we counteracting evolution by obstructing natural selection with our conservation efforts? This question encourages us to ponder our impact on natural environments and consider whether the struggle of these species to persist is a direct consequence of abrupt changes caused by human actions. With habitat loss as a primary threat to biodiversity, the conservation focus could shift towards protecting environments and communities, allowing well-adapted species to flourish naturally.

This reflection leads us to an open-ended inquiry about the future of conservation. In prioritizing habitat protection and fostering ecosystems where species can naturally evolve, providing ecosystem services that enhance resilience, might we find a more sustainable approach to conserving biodiversity? Exploring these complexities in a holistic manner reveals that solutions lie not just in scientific advancement but in a collective reevaluation of our relationship with the natural world.

Key Points

- Genetic approaches provide insights into the conservation status of species by assessing genetic diversity, inbreeding, effective population size, population structure, and genetic load, aiding in identifying conservation threats such as hybridization and introgression.
- Conservation planning relies on defining appropriate Conservation Units (CUs) to enhance species evolutionary potential and prioritize conservation efforts. Evolutionarily Significant Units (ESUs) and management units (MUs) are identified based on significant breaks in genetic diversity among populations, guiding targeted protection measures and management strategies.
- Non-invasive sampling techniques provide invaluable tools for monitoring populations, estimating population size, structure, and gene flow, and combating wildlife trafficking and disease, enabling effective conservation efforts without stressing animals.

- Genetic markers play a crucial role in uncovering cryptic species, which offer insights into evolutionary processes, species isolation, and ecological niches. Molecular techniques aid in accurately delineating species boundaries, guiding conservation units delimitation, and avoiding misidentification that could lead to overlooking conservation needs.
- DNA forensics helps identifying species, individuals, and their origins, which is important for combating illegal trafficking of Neotropical species. Genetic analysis guides the safe release of confiscated animals back into the wild, ensuring genetic diversity and the success of conservation efforts.
- Genetic data informs translocation strategies by guiding the selection of donor individuals/populations, ensuring adaptively representative and diverse translocated populations, and monitoring their success. Genetic considerations are essential for maximizing the effectiveness of translocation efforts and minimizing the risk of outbreeding depression.
- Ex situ conservation programs, such as captive breeding, are vital for rescuing species from extinction, but genetic considerations are often overlooked. Integrating genetic analyses into conservation strategies enhances success by minimizing inbreeding, maximizing genetic diversity, and ensuring the adaptive potential of captive populations.

Questions

- What are runs of homozygosity (ROH) and how do they evaluate inbreeding levels? How does this metric contribute to informing conservation studies?
- How do Management Units (MUs) differ from Evolutionarily Significant Units (ESUs) in the context of conservation planning, and what criteria are typically used to delineate these units?
- What are some key challenges and potential solutions associated with non-invasive sampling techniques, particularly in the context of Conservation Genetics in the Neotropical region?
- What strategies could be implemented to improve our understanding of hidden biodiversity, particularly in regions like the Neotropics characterized by high species diversity and cryptic taxa?
- How do translocations differ from reintroductions in the context of animal conservation?

References

Acevedo-Garcés YA, Valencia LM, Di Fiore A et al (2021) Current and historical genetic structure of the White-Footed Tamarin (*Saguinus leucopus*). In: Nardelli M, Túnez JI (eds) Molecular ecology and conservation genetics of neotropical mammals. Springer International Publishing, Cham, pp 171–197

Allendorf TD (2010) A framework for the park–people relationship: insights from protected areas in Nepal and Myanmar. Int J Sustain Dev World Ecol 17(5):417–422. https://doi.org/10.1080/13504501003788180

de Andrade FS, Haga IA, Ferreira JS et al (2020) A new cryptic species of Pithecopus (Anura, Phyllomedusidae) in north-eastern Brazil. Eur J Taxon 723:108–134. https://doi.org/10.5852/ejt.2020.723.1147

Arantes LS, Ferreira LCL, Driller M et al (2020) Genomic evidence of recent hybridization between sea turtles at Abrolhos Archipelago and its association to low reproductive output. Sci Rep 10:12847. https://doi.org/10.1038/s41598-020-69613-8

Aristizábal Duque SL, Orozco-Jiménez LY, Zapata-Escobar C et al (2018) Conservation genetics of otters: review about the use of non-invasive samples. Therya 9:85–93. https://doi.org/10.12933/therya-18-515

Baciu I, Fedorca A, Ionescu G (2022) Noninvasive genetics knowledge from the brown bear populations to assist biodiversity conservation. Diversity 14:121. https://doi.org/10.3390/d14020121

Baeza JA, Ortega J, Montes-Carreto LM et al (2023) Whole and nearly complete mitochondrial genomes of an endemic and endangered neotropical rabbit (*Romerolagus diazi*) assembled using non-invasive eDNA metagenomics (field droppings). J Nat Hist 57:1220–1234. https://doi.org/10.1080/00222933.2023.2186810

Balázs G, Vörös J, Lewarne B, Herczeg G (2020) A new non-invasive in situ underwater DNA sampling method for estimating genetic diversity. Evol Ecol 34:633–644. https://doi.org/10.1007/s10682-020-10053-1

Balza U, Baldi R, Rodríguez-Planes L et al (2023) Scientific evidence does not support the translocation of guanacos in Argentina. Cons Sci and Practice 5:e13031. https://doi.org/10.1111/csp2.13031

Banhos A, Hrbek T, Sanaiotti TM et al (2016) Reduction of genetic diversity of the Harpy Eagle in Brazilian tropical forests. PLoS One 11:e0148902. https://doi.org/10.1371/journal.pone.0148902

Banks SC, Piggott MP (2022) Non-invasive genetic sampling is one of our most powerful and ethical tools for threatened species population monitoring: a reply to Lavery et al. Biodivers Conserv 31:723–728. https://doi.org/10.1007/s10531-022-02377-x

Barbosa S, Mestre F, White TA et al (2018) Integrative approaches to guide conservation decisions: using genomics to define conservation units and functional corridors. Mol Ecol 27:3452–3465. https://doi.org/10.1111/mec.14806

Baus I, Miño CI, Monge O (2019) Current trends in the application of non-invasive genetic sampling to study neotropical birds: uses, goals, and conservation potential. Avian Biol Res 12:125–132. https://doi.org/10.1177/1758155919848229

Bickford D, Lohman DJ, Sodhi NS et al (2007) Cryptic species as a window on diversity and conservation. Trends Ecol Evol 22:148–155. https://doi.org/10.1016/j.tree.2006.11.004

Bonfil R, Palacios-Barreto P, Vargas OUM et al (2021) Detection of critically endangered marine species with dwindling populations in the wild using eDNA gives hope for sawfishes. Mar Biol 168:60. https://doi.org/10.1007/s00227-021-03862-7

Bonfil R, Díaz-Jaimes P, Palacios-Barreto P et al (2024) Improved eDNA assay evidences further refugia for critically endangered smalltooth sawfish (Pristis pectinata) in Mexico. Front Mar Sci 11:1290661. https://doi.org/10.3389/fmars.2024.1290661

Bosse M, Van Loon S (2022) Challenges in quantifying genome erosion for conservation. Front Genet 13:960958. https://doi.org/10.3389/fgene.2022.960958

Byrne MS, Peralta DM, Ibañez EA et al (2021) Non-invasive sampling techniques applied to conservation genetic studies in mammals. In: Nardelli M, Túnez JI (eds) Molecular ecology and conservation genetics of neotropical mammals. Springer International Publishing, Cham, pp 63–83

Caballero S, Hollatz C, Rodríguez S et al (2018) Population structure of Riverine and Coastal Dolphins *Sotalia fluviatilis* and *Sotalia guianensis*: patterns of nuclear and mitochondrial diversity and implications for conservation. J Hered 109:757–770. https://doi.org/10.1093/jhered/esy049

Camino M, Cirignoli S, Varela D et al (2019) Pecari tajacu. In: SAyDS–SAREM (eds.) Categorización 2019 de los mamíferos de Argentina según su riesgo de extinción. Lista Roja de los mamíferos de Argentina. http://cma.sarem.org.ar

Caparroz R, Miyaki CY, Bampi MI et al (2001) Analysis of the genetic variability in a sample of the remaining group of Spix's Macaw (*Cyanopsitta spixii*, Psittaciformes: Aves) by DNA fingerprinting. Biol Conserv 99:307–311. https://doi.org/10.1016/S0006-3207(00)00196-8

Carbajal Márquez R, Cedeño-Vázquez J, Martinez A et al (2020) Accessing cryptic diversity in Neotropical rattlesnakes (Serpentes: Viperidae: Crotalus) with the description of two new species. Zootaxa 4729:451–481. https://doi.org/10.11646/zootaxa.4729.4.1

Cárcamo D, Pizarro M, Orellana M et al (2021) Long-term monitoring for conservation: closing the distribution gap of *Arctocephalus australis* in central Chile. BMC Res Notes 14:170. https://doi.org/10.1186/s13104-021-05583-y

Carroll EL, Bruford MW, DeWoody JA et al (2018) Genetic and genomic monitoring with minimally invasive sampling methods. Evol Appl 11:1094–1119. https://doi.org/10.1111/eva.12600

Carvalho DC, Oliveira DAA, Beheregaray L et al (2012) Hidden genetic diversity and distinct evolutionarily significant units in an commercially important Neotropical apex predator, the catfish *Pseudoplatystoma corruscans*. Conserv Genet 13:1671–1675. https://doi.org/10.1007/s10592-012-0402-6

Cerón-Souza I, Delgadillo-Duran D, Polo-Murcia SM et al (2023) Prioritizing Colombian plant genetic resources for investment in research using indicators about the geographic origin, vulnerability status, economic benefits, and food security importance. Biodivers Conserv 32:2221–2261. https://doi.org/10.1007/s10531-023-02599-7

Chen C, Zhu B, Tang X et al (2023) Genome-wide assessment of runs of homozygosity by whole-genome sequencing in diverse horse breeds worldwide. Genes 14:1211. https://doi.org/10.3390/genes14061211

Cueva DF, Zug R, Pozo MJ et al (2024) Evidence of population genetic structure in Ecuadorian Andean bears. Sci Rep 14:2834. https://doi.org/10.1038/s41598-024-53003-5

Cunha HA, Medeiros BV, Barbosa LA et al (2014) Population structure of the endangered Franciscana Dolphin (*Pontoporia blainvillei*): reassessing management units. PLoS One 9:e85633. https://doi.org/10.1371/journal.pone.0085633

Delgado PM, Argüelles CF, DeMatteo KE (2021) Using noninvasive techniques to monitor game species targeted by poaching in Misiones Argentina. Neotropical Biodiversity 7:78-85. https://doi.org/10.1080/23766808.2021.1897354

De León LF, Cornejo A, Gavilán RG et al (2020) Hidden biodiversity in Neotropical streams: DNA barcoding uncovers high endemicity of freshwater macroinvertebrates at small spatial scales. PLoS One 15:e0231683. https://doi.org/10.1371/journal.pone.0231683

Di Rocco F, Anello M (2021) The use of forensic DNA on the conservation of neotropical mammals. Molecular ecology and conservation genetics of neotropical mammals, 85–98

Dick CW, Roubik DW, Gruber KF et al (2004) Long-distance gene flow and cross-Andean dispersal of lowland rainforest bees (Apidae: Euglossini) revealed by comparative mitochondrial DNA phylogeography. Mol Ecol 13:3775–3785. https://doi.org/10.1111/j.1365-294X.2004.02374.x

Domingos FMCB, Colli GR, Lemmon A et al (2017) In the shadows: phylogenomics and coalescent species delimitation unveil cryptic diversity in a Cerrado endemic lizard (Squamata: Tropidurus). Mol Phylogenet Evol 107:455–465. https://doi.org/10.1016/j.ympev.2016.12.009

Domingues RR, Garrone-Neto D, Hilsdorf AWS et al (2019) Use of mucus as a non-invasive sampling method for DNA barcoding of stingrays and skates (batoid elasmobranchs). J Fish Biol 94:512–516. https://doi.org/10.1111/jfb.13919

Dominguez M, Lapido R, Gorrindo A et al (2021) A citizen science survey discloses the current distribution of the endangered Yellow Cardinal *Gubernatrix cristata* in Argentina. Bird Conserv Int 31:139–150

Donadio E, Di Martino S, Heinonen S (2022a) Rewilding Argentina: lessons for the 2030 biodiversity targets. Nature 603:225–227. https://doi.org/10.1038/d41586-022-00631-4

Donadio E, Zamboni T, Martino SD (2022b) Bringing jaguars and their prey base back to the Iberá wetlands, Argentina. In: Moehrenschlager A, Ewen JG, Gaywood MJ et al (eds) Conservation translocations. Cambridge University Press, Cambridge, pp 443–448

Escobar LMD, Andrade-López J, Farias IP et al (2015) Delimiting evolutionarily significant units of the fish, *Piaractus brachypomus* (Characiformes: Serrasalmidae), from the Orinoco and Amazon River Basins with insight on routes of historical connectivity. J Hered 106:428–438. https://doi.org/10.1093/jhered/esv047

Estrada-Saldívar N, Quiroga-García BA, Pérez-Cervantes E et al (2021) Effects of the stony coral tissue loss disease outbreak on coral communities and the benthic composition of cozumel reefs. Front Mar Sci 8:632777. https://doi.org/10.3389/fmars.2021.632777

Evans MJ, Pierson JC, Neaves LE et al (2023) Tendências na pesquisa de translocação de animais. Ecografia 2023:e06528

Farro APC, Rollo MM Jr, Silva JM Jr et al (2008) A simple protocol for a low invasive DNA accessing in *Stenella longirostris* (Cetacea: Delphinidae). Pan-Am J Aqua Sci:130–134

Fernandes GA, Caparroz R (2013) DNA sequence analysis to guide the release of blue-and-yellow macaws (*Ara ararauna*, Psittaciformes, Aves) from the illegal trade back into the wild. Mol Biol Rep 40:2757–2762. https://doi.org/10.1007/s11033-012-2294-4

Fernandez FAS, Rheingantz ML, Genes L et al (2017) Rewilding the Atlantic Forest: restoring the fauna and ecological interactions of a protected area. Perspect Ecol Conserv 15:308–314. https://doi.org/10.1016/j.pecon.2017.09.004

Fernandez-Fournier P, Lewthwaite JMM, Mooers AØ (2021) Do we need to identify adaptive genetic variation when prioritizing populations for conservation? Conserv Genet 22:205–216. https://doi.org/10.1007/s10592-020-01327-w

Fišer C, Robinson CT, Malard F (2018) Cryptic species as a window into the paradigm shift of the species concept. Mol Ecol 27:613–635. https://doi.org/10.1111/mec.14486

Flanagan SP, Jones AG (2017) Constraints on the FST–heterozygosity outlier approach. J Hered 108:561–573. https://doi.org/10.1093/jhered/esx048

Flanagan SP, Forester BR, Latch EK et al (2018) Guidelines for planning genomic assessment and monitoring of locally adaptive variation to inform species conservation. Evol Appl 11:1035–1052. https://doi.org/10.1111/eva.12569

Flesch EP, Graves TA, Thomson, J M et al (2020) Evaluating wildlife translocations using genomics: a bighorn sheep case study. Ecology and Evolution, 10(24):13687–13704

Foote AD, Hooper R, Alexander A et al (2021) Runs of homozygosity in killer whale genomes provide a global record of demographic histories. Mol Ecol 30:6162–6177. https://doi.org/10.1111/mec.16137

Forcina G, Boesman P, Jowers MJ (2023) Cryptic diversity in a neotropical avian species complex untangled by neglected genetic evidence. Stud Neotropical Fauna Environ 58(1):130–137. https://doi.org/10.1080/01650521.2021.1915674

Franco-Gutiérrez LJ, Álvarez-Cardona J, Soto-Calderón ID (2017) Sex identification of neotropical macaws (*Ara* spp.) from invasive and non-invasive samples. Ornitología Colombiana 16:1–07

Frankham R (2015) Genetic rescue of small inbred populations: meta-analysis reveals large and consistent benefits of gene flow. Mol Ecol 24:2610–2618. https://doi.org/10.1111/mec.13139

Fraser DJ, Bernatchez L (2001) Adaptive evolutionary conservation: towards a unified concept for defining conservation units. Mol Ecol 10:2741–2752

Fredrickson RJ, Siminski P, Woolf M et al (2007) Genetic rescue and inbreeding depression in Mexican wolves. Proc Biol Sci 274:2365–2371. https://doi.org/10.1098/rspb.2007.0785

Freitas C, Villegas-Ríos D, Moland E et al (2021) Sea temperature effects on depth use and habitat selection in a marine fish community. J Anim Ecol 90:1787–1800. https://doi.org/10.1111/1365-2656.13497

Funk WC, McKay JK, Hohenlohe PA et al (2012) Harnessing genomics for delineating conservation units. Trends Ecol Evol 27:489–496. https://doi.org/10.1016/j.tree.2012.05.012

Gallardo CE, Correa C, Morales P et al (2012) Validation of a cheap and simple nondestructive method for obtaining AFLPs and DNA sequences (mitochondrial and nuclear) in amphibians. Mol Ecol Resour 12(6):1090–1096. https://doi.org/10.1111/1755-0998.12013

Gallego-García N, Vargas-Ramírez M, Shaffer HB (2023) The importance of cryptic diversity in the conservation of wide-ranging species: the red-footed tortoise *Chelonoidis carbonarius* in Colombia. Mol Ecol 32:4531–4545. https://doi.org/10.1111/mec.17052

García-Merchán VH, Calle CA, Sosa-Arango CC et al (2021) Conservation genetics in the neotropical carnivorous mammals: a systematic review. Mol Ecol Conserv Genet Neotrop Mamm:277–295. https://doi.org/10.1007/978-3-030-65606-5_12

Gaywood MJ, Ewen JG, Hollingsworth PM et al (eds) (2022) Conservation translocations. Cambridge University Press

Genes L, Fernandez FAS, Vaz-de-Mello FZ et al (2019) Effects of howler monkey reintroduction on ecological interactions and processes. Conserv Biol 33:88–98. https://doi.org/10.1111/cobi.13188

Giselle RK, Gemesio LF, Cavalcanti AF et al (2022) Human highly modified landscapes restrict gene flow of the largest neotropical canid, the maned wolf. Biodivers Conserv 31:1229–1247. https://doi.org/10.1007/s10531-022-02385-x

Gonçalves CF, Rodriguez-Castro KG, Dib LV et al (2022) Non-invasive genetic sampling reveals a habitat use extension of Chrysocyon brachyurus and Leopardus guttulus inside a protected area of Southeastern Brazil. Biota Neotrop 22:e20221346. https://doi.org/10.1590/1676-0611-BN-2022-1346

Greggor AL, Merrick MJ, Wisinski CL et al (2024). Conservation breeding and translocation. Behavioural Responses to a Changing World, 206.

Gross IP, Wilson AE, Wolak ME (2024) The fitness consequences of wildlife conservation translocations: a meta-analysis. Biol Rev 2:348–371. https://doi.org/10.1111/brv.13025

Harder AM, Kirksey KB, Mathur S et al (2022) Detectability of runs of homozygosity is influenced by analysis parameters as well as population-specific demographic history. BioRxiv 2022-09. https://doi.org/10.1101/2022.09.29.510155

Hedrick PW, Miller PS, Geffen E et al (1998) Genetic evaluation of the three captive mexican wolf lineages. Zool Biol 16:47–69

Henry JA, O'Neil KL, Pilnick AR et al (2021) Strategies for integrating sexually propagated corals into Caribbean reef restoration: experimental results and considerations. Coral Reefs 40:1667–1677. https://doi.org/10.1007/s00338-021-02154-2

Hernández-Romero PC, Gutiérrez-Rodríguez C, Valdespino C et al (2018) The role of geographical and ecological factors on population divergence of the neotropical otter *Lontra longicaudis* (Carnivora, Mustelidae). Evol Biol 45:37–55. https://doi.org/10.1007/s11692-017-9428-5

Hewett AM, Stoffel MA, Peters L et al (2023) Selection, recombination and population history effects on runs of homozygosity (ROH) in wild red deer (*Cervus elaphus*). Heredity 130:242–250. https://doi.org/10.1038/s41437-023-00602-z

Hilário HO, Mendes IS, Guimarães Sales N et al (2023) DNA metabarcoding of mock communities highlights potential biases when assessing Neotropical fish diversity. Environ DNA 5:1351–1361. https://doi.org/10.1002/edn3.456

Hoban S, Bruford M, D'Urban Jackson J et al (2020) Genetic diversity targets and indicators in the CBD post-2020 Global Biodiversity Framework must be improved. Biol Conserv 248:108654. https://doi.org/10.1016/j.biocon.2020.108654

Hoban S, Archer FI, Bertola LD et al (2022) Global genetic diversity status and trends: towards a suite of Essential Biodiversity Variables (EBVs) for genetic composition. Biological Reviews, 97(4):1511–1538

Hoelzel AR (2023) Where to now with the evolutionarily significant unit? Trends Ecol Evol. https://doi.org/10.1016/j.tree.2023.07.005

Hoffmann M, Duckworth JW, Holmes K et al (2015) The difference conservation makes to extinction risk of the world's ungulates. Conserv Biol 29:1303–1313. https://doi.org/10.1111/cobi.12519

Husemann M, Zachos FE, Paxton RJ et al (2016) Effective population size in ecology and evolution. Heredity 191–192. https://doi.org/10.1038/hdy.2016.75

IUCN/SSC (2013) Guidelines for reintroductions and other conservation translocations. IUCN Species Survival Commission, Gland

Jabin G, Dewan Y, Khatri H et al (2019) Identifying the tick Amblyomma javanense (Acari: Ixodidae) from Chinese pangolin: generating species barcode, phylogenetic status and its implication in wildlife forensics. Experimental and Applied Acarology, 78:461–467

Janzen DH, Burns JM, Cong Q et al (2017) Nuclear genomes distinguish cryptic species suggested by their DNA barcodes and ecology. Proc Natl Acad Sci 114:8313–8318. https://doi.org/10.1073/pnas.1621504114

Juhel JB, Marques V, Utama RS et al (2022) Estimating the extended and hidden species diversity from environmental DNA in hyper-diverse regions. Ecography 2022:e06299. https://doi.org/10.1111/ecog.06299

Lager C, Perry R, Daly J et al (2023) Metrics of coral microfragment viability. bioRxiv, 2023-01. https://doi.org/10.1101/2023.01.03.522625

Langridge J, Sordello R, Reyjol Y (2021) Existing evidence on the outcomes of wildlife translocations in protected areas: a systematic map. Environ Evid 10:29. https://doi.org/10.1186/s13750-021-00236-w

Lara-Ruiz P, Lopez GG, Santos FR, Soares L.S (2006) Extensive hybridization in hawksbill turtles (Eretmochelys imbricata) nesting in Brazil revealed by mtDNA analyses. Conservation Genetics 7:773–781

Latorre-Cardenas MC, Gutiérrez-Rodríguez C, Rico Y (2020) Estimating genetic and demographic parameters relevant for the conservation of the Neotropical otter, *Lontra longicaudis*, in Mexico. Conserv Genet 21:719–734. https://doi.org/10.1007/s10592-020-01283-5

Lavanchy E, Goudet J (2023) Effect of reduced genomic representation on using runs of homozygosity for inbreeding characterization. Mol Ecol Resour 4:787–802. https://doi.org/10.1111/1755-0998.13755

Leigh DM, Hendry AP, Vázquez-Domínguez E et al (2019) Estimated six per cent loss of genetic variation in wild populations since the industrial revolution. Evol Appl 12:1505–1512. https://doi.org/10.1111/eva.12810

Leite KCE, Seixas GHF, Berkunsky I et al (2008) Population genetic structure of the blue-fronted Amazon (*Amazona aestiva*, Psittacidae: Aves) based on nuclear microsatellite loci: implications for conservation. Genets Mol Res 7:819–829. https://doi.org/10.4238/vol7-3gmr474

Lima CS, Magalhães RF, Santos FR (2021) Conservation issues using discordant taxonomic and evolutionary units: a case study of the American manatee (*Trichechus manatus*, Sirenia). Wildl Res 48:385–392. https://doi.org/10.1071/WR20197

Lima RD, Bocalini F, Silveira LF (2023a) Integrative revision of species limits in the genus *Schiffornis* (Aves: Tityridae) reveals cryptic diversity in the Neotropics. Zool J Linn Soc. https://doi.org/10.1093/zoolinnean/zlad108

Lima J, Rocha Alexandre B et al (2023b) Genetic management applied to conservation of reduced and fragmented wild populations. In: Galetti PM Jr (ed) Conservation genetics in the neotropics. Springer International Publishing, Cham, pp 227–249

Lima-Rezende CA, de Souza RO, Caparroz R (2019) The spatial genetic structure of the White-banded Tanager (Aves, Passeriformes) in fragmented Neotropical savannas suggests two evolutionarily significant units. Biotropica 51:234–244. https://doi.org/10.1111/btp.12623

Lopes CM, Sasso T, Valentini A et al (2017) eDNA metabarcoding: a promising method for anuran surveys in highly diverse tropical forests. Mol Ecol Resour 17:904–914. https://doi.org/10.1111/1755-0998.12643

de Lopes JM, Alves CBM, Peressin A et al (2021) Dazed and confused: behavioural constraints impose major challenges to fish passage in the neotropics. Aquat Conserv Mar Freshw Ecosyst 31:3403–3415. https://doi.org/10.1002/aqc.3716

Lorenzana G, Heidtmann L, Haag T et al (2020) Large-scale assessment of genetic diversity and population connectivity of Amazonian jaguars (*Panthera onca*) provides a baseline for their conservation and monitoring in fragmented landscapes. Biol Conserv 242:108417. https://doi.org/10.1016/j.biocon.2020.108417

Lorenzini R, Garofalo L (2021) Wildlife forensics: DNA analysis in wildlife forensic investigations. In Forensic DNA analysis (pp. 357–384). Apple Academic Press

Moraes AM, Lima JS, Alexandre BR et al (2023) Genetic management applied to conservation of reduced and fragmented wild populations. In Conservation genetics in the neotropics. Cham: Springer International Publishing, pp 227–249. https://doi.org/10.1007/978-3-031-34854-9_10

Maia TA, Vilaça ST, Silva RS et al (2017) DNA sampling from eggshells and microsatellite genotyping in rare tropical birds: Case study on Brazilian Merganser. Gen and Mol Biol 40:808–812. https://doi.org/10.1590/1678-4685-GMB-2016-0297

Mannise N, Cosse M, González S et al (2017) Maned wolves retain moderate levels of genetic diversity and gene flow despite drastic habitat fragmentation. Endanger Species Res 34:449–462. https://doi.org/10.3354/esr00859

Marino F, McDonald RA, Crowley SL et al (2024) Rethinking the evaluation of animal translocations. Biol Conserv 292:110523. https://doi.org/10.1016/j.biocon.2024.110523

Martin CA, Sheppard EC, Illera JC et al (2023) Runs of homozygosity reveal past bottlenecks and contemporary inbreeding across diverging populations of an Island-colonizing bird. Mol Ecol 32:1972–1989. https://doi.org/10.1111/mec.16865

Meza-Lázaro RN, Peña-Carrillo KI, Poteaux C et al (2022) Genome and cuticular hydrocarbon-based species delimitation shed light on potential drivers of speciation in a Neotropical ant species complex. Ecol Evol 12:e8704. https://doi.org/10.1002/ece3.8704

Milić D, Radenković S, Ačanski J et al (2019) The importance of hidden diversity for insect conservation: a case study in hoverflies (the Merodon atratus complex, Syrphidae, Diptera). J Insect Conserv 23:29–44. https://doi.org/10.1007/s10841-018-0111-7

Milinkovitch MC, Monteyne D, Gibbs JP et al (2004) Genetic analysis of a successful repatriation programme: giant Galapagos tortoises. Proc Roy Soc London Series B: Biol Sci 271:341–345. https://doi.org/10.1098/rspb.2003.2607

Miotto RA, Cervini M, Figueiredo MG et al (2011) Genetic diversity and population structure of pumas (*Puma concolor*) in southeastern Brazil: implications for conservation in a human-dominated landscape. Conserv Genet 12:1447–1455. https://doi.org/10.1007/s10592-011-0243-8

Miranda GB, Andrades-Miranda J, Oliveira LFB et al (2007) Geographic patterns of genetic variation and conservation consequences in Three South American Rodents. Biochem Genet 45:839–856. https://doi.org/10.1007/s10528-007-9122-x

Miranda FR, Casali DM, Perini FA et al (2018) Taxonomic review of the genus *Cyclopes* Gray, 1821 (Xenarthra: Pilosa), with the revalidation and description of new species. Zool J Linnean Soc 183:687–721. https://doi.org/10.1093/zoolinnean/zlx079

Moraes AM, Ruiz-Miranda CR, Ribeiro MC et al (2017) Temporal genetic dynamics of reintroduced and translocated populations of the endangered golden lion tamarin (*Leontopithecus rosalia*). Conserv Genet 18:995–1009. https://doi.org/10.1007/s10592-017-0948-4

Morales-Martínez DM, Rodríguez-Posada ME, Ramírez-Chaves HE (2021) A new cryptic species of yellow-eared bat *Vampyressa melissa* species complex (Chiroptera: Phyllostomidae) from Colombia. J Mammal 102:90–100. https://doi.org/10.1093/jmammal/gyaa137

Moritz C (1994) Defining 'Evolutionarily Significant Units' for conservation. Trends Ecol Evol 9:373–375. https://doi.org/10.1016/0169-5347(94)90057-4

Moritz C (1999) Conservation units and translocations: strategies for conserving evolutionary processes. Hereditas 130:217–228. https://doi.org/10.1111/j.1601-5223.1999.00217.x

Moscarella RA, Aguilera M, Escalante AA (2003) Phylogeography, population structure, and implications for conservation of White-Tailed Deer (*Odocoileus virginianus*) in Venezuela. J Mammal 84:1300–1315. https://doi.org/10.1644/BRB-028

Neaves LE, Ogden R, Hollingsworth PM (2022) Genomics 9t and Conservation Translocations. Conservation translocations, 9:271

Norman JA, Christidis L (2021). A spatial genetic framework for koala translocations: where to?. Wildlife Research, 48(3):193–201.

Ochoa A, Onorato DP, Roelke-Parker ME et al (2022) Give and take: effects of genetic admixture on mutation load in endangered Florida panthers. J Hered 113:491–499. https://doi.org/10.1093/jhered/esac037

Olah G, Heinsohn RG, Brightsmith DJ et al (2016) Validation of non-invasive genetic tagging in two large macaw species (*Ara Macao* and *A. chloropterus*) of the Peruvian Amazon. Conserv Genet Resour 8:499–509. https://doi.org/10.1007/s12686-016-0573-4

Olah G, Heinsohn RG, Brightsmith DJ et al (2017) The application of non-invasive genetic tagging reveals new insights into the clay lick use by macaws in the Peruvian Amazon. Conserv Genet 18:1037–1046. https://doi.org/10.1007/s10592-017-0954-6

de Oliveira LR, Fraga LD, Majluf P (2012) Effective population size for South American sea lions along the Peruvian coast: the survivors of the strongest El Niño event in history. J Mar Biol Assoc U K 92:1835–1841. https://doi.org/10.1017/S0025315411001871

de Oliveira LR, Fraga LD, Ott PH et al (2019) Population structure, phylogeography, and genetic diversity of the common bottlenose dolphin in the tropical and subtropical southwestern Atlantic Ocean. J Mammal 100:564–577. https://doi.org/10.1093/jmammal/gyz065

Pallotti S, Picciolini M, Antonini M et al (2023) Genome-wide scan for runs of homozygosity in South American Camelids. BMC Genomics 24:470. https://doi.org/10.1186/s12864-023-09547-3

Pauli JN, Whiteman JP, Riley MD et al (2010) Defining noninvasive approaches for sampling of vertebrates. Conserv Biol 24:349–352. https://doi.org/10.1111/j.1523-1739.2009.01298.x

Pease KM, Freedman AH, Pollinger JP et al (2009) Landscape genetics of California mule deer (Odocoileus hemionus): the roles of ecological and historical factors in generating differentiation. Molecular Ecology, 18(9):1848–1862

Pelicice FM, Azevedo-Santos VM, Vitule JRS et al (2017) Neotropical freshwater fishes imperilled by unsustainable policies. Fish Fish 18:1119–1133. https://doi.org/10.1111/faf.12228

Peñaloza CL, Hernández O, Espín R et al (2013) Harvest of endangered Sideneck River Turtles (Podocnemis spp.) in the Middle Orinoco, Venezuela. Copeia 2013(1):111–120. https://doi.org/10.2307/41827125

Peralta DM, Ibañez EA, Lucero S et al (2020) A new minimally invasive and inexpensive sampling method for genetic studies in pinnipeds. Mamm Res 65:11–18. https://doi.org/10.1007/s13364-019-00453-2

Peters C, Nelson H, Rusk B et al (2020) A novel method to optimise the utility of underused moulted plumulaceous feather samples for genetic analysis in bird conservation. Conserv Genet Resour 12:457–467. https://doi.org/10.1007/s12686-019-01117-8

Peters C, Geary M, Hosie C et al (2023) Non-invasive sampling reveals low mitochondrial genetic diversity for a critically endangered island endemic species. https://www.researchsquare.com. Accessed 19 Oct 2023

Pickles RSA, Groombridge JJ, Zambrana Rojas VD et al (2011) Evolutionary history and identification of conservation units in the giant otter, *Pteronura brasiliensis*. Mol Phylogenet Evol 61:616–627. https://doi.org/10.1016/j.ympev.2011.08.017

Pires DE, Blundell TL, Ascher DB (2016) mCSM-lig: quantifying the effects of mutations on protein-small molecule affinity in genetic disease and emergence of drug resistance. Scientific reports, 6(1):29575

Pizarro EJ, Julio-Kalajžić B, Sallaberry-Pincheira N et al (2023) Species delimitation and intraspecific diversification in recently diverged South American foxes. Mamm Res. https://doi.org/10.1007/s13364-023-00717-y

Purchase C, Lugarini C, Purchase C et al (2024) Reintroduction of the extinct-in-the-wild Spix's Macaw (*Cyanopsitta spixii*) in the Caatinga forest domain of Brazil. Diversity 16:80. https://doi.org/10.3390/d16020080

Quintela FM, Lopes CM, Gonçalves GL (2022) Genetics applied to mammal conservation. Front Genet 13:974631. https://doi.org/10.3389/fgene.2022.974631

Quinzin MC, Bishop AP, Miller JM et al (2023) Galapagos giant tortoise trafficking case demonstrates the utility and applications of long-term comprehensive genetic monitoring. Anim Conserv. Animal Conservation, 26(6):826–838. https://doi.org/10.1111/acv.12870

Ramirez JL, Birindelli JL, Carvalho DC et al (2017) Revealing hidden diversity of the underestimated neotropical ichthyofauna: DNA barcoding in the recently described genus *Megaleporinus* (Characiformes: Anostomidae). Front Genet 8:149. https://doi.org/10.3389/fgene.2017.00149

Ramos EKS, de Magalhães RF, Marques NCS et al (2019) Cryptic diversity in Brazilian endemic monkey frogs (Hylidae, Phyllomedusinae, Pithecopus) revealed by multispecies coalescent and integrative approaches. Mol Phylogenet Evol 132:105–116. https://doi.org/10.1016/j.ympev.2018.11.022

Reier S, Sattmann H, Schwaha T et al (2020) Unravelling the hidden biodiversity–the establishment of DNA barcodes of fish-parasitizing Acanthocephala Koehlreuther, 1771 in view of taxonomic misidentifications, intraspecific variability and possible cryptic species. Parasitology 147:1499–1508. https://doi.org/10.1017/S0031182020001316

Reis RE, Albert JS, Di Dario F et al (2016) Fish biodiversity and conservation in South America. J Fish Biol 89:12–47. https://doi.org/10.1111/jfb.13016

Rodrigues NT, Saranholi BH, Angeloni TA et al (2020) DNA mini-barcoding of leporids using noninvasive fecal DNA samples and its significance for monitoring an invasive species. Ecol Evol 10:5219–5225. https://doi.org/10.1002/ece3.5863

Rodríguez-Castro KG, Saranholi BH, Bataglia L et al (2020) Molecular species identification of scat samples of South American felids and canids. Conserv Genet Resour 12:61–66. https://doi.org/10.1007/s12686-018-1048-6

Rodriguez-Castro KG, Lemos FG, Azevedo FC et al (2022) Human highly modified landscapes restrict gene flow of the largest neotropical canid, the maned wolf. Biodivers Conserv 31. https://doi.org/10.1007/s10531-022-02490-x

Ryder OA (1986) Species conservation and systematics: the dilemma of subspecies. Trends in Ecology and Evolution, 1:9–10

Saldarriaga-Gómez AM, Ardila-Robayo MC, Medem F et al (2023) Hope is the last thing lost: Colombian captive-bred population of the critically endangered Orinoco crocodile (*Crocodylus intermedius*) is a genetic reservoir that could help to save the species from extinction. Nat Conserv 53:85–103. https://doi.org/10.3897/natureconservation.53.104000

Salgado-Roa, Fabian C et al (2018) Gene flow and Andean uplift shape the diversification of Gasteracantha cancriformis (Araneae: Araneidae) in Northern South America. Ecology and Evolution 8.14:7131–7142

Salom-Pérez R, Wultsch C, Adams JR et al (2022) Genetic diversity and population structure for ocelots (*Leopardus pardalis*) in Costa Rica. J Mammal 103:68–81. https://doi.org/10.1093/jmammal/gyab146

Saremi NF, Supple MA, Byrne A et al (2019) Puma genomes from North and South America provide insights into the genomic consequences of inbreeding. Nat Commun 10:4769. https://doi.org/10.1038/s41467-019-12741-1

Sas-Rolfes M, Challender DW, Hinsley A et al (2019) Illegal wildlife trade: Scale, processes, and governance. Annual Review of Environment and Resources, 44(1):201–228

Saucedo C, Saavedra A, Herrera P (2018) Darwin's Rhea conservation and recovery at Patagonia National Park, Chile. In: Ratite review of the AZA Struthioniformes Taxon Advisory Group Newsletter, pp 9–10

Saucedo C, Saavedra A, Herrera P (2019) News from the Darwin's Rhea (*Rhea pennata*) recovery and conservation program in Patagonia National Park, Chile. In: Ratite review of the AZA Struthioniformes Taxon Advisory Group Newsletter, pp 36–39

Schmidt C, Hoban S, Hunter M et al (2023) Genetic diversity and IUCN red list status. Conserv Biol 37:e14064. https://doi.org/10.1111/cobi.14064

Schmidt TL, Thia JA, Hoffmann AA (2024) How can genomics help or hinder wildlife conservation? Annu Rev Anim Biosci 12:45–68. https://doi.org/10.1146/annurev-animal-021022-051810

Shafer ABA, Wolf JBW, Alves PC et al (2015) Genomics and the challenging translation into conservation practice. Trends Ecol Evol 30:78–87. https://doi.org/10.1016/j.tree.2014.11.009

Shaney KJ, Diaz-Ramirez LG, Espindola S et al (2020) Defining intraspecific conservation units in the endemic Cuban Rock Iguanas (*Cyclura nubila nubila*). Sci Rep 10:21607. https://doi.org/10.1038/s41598-020-78664-w

Shaw TE (2019) Species diversity in restoration plantings: important factors for increasing the diversity of threatened tree species in the restoration of the Araucaria forest ecosystem. Plant Divers 41:84–93. https://doi.org/10.1016/j.pld.2018.08.002

Smith MM, Knife CG, Eklund D et al (2023) Predicting and prioritizing genetic diversity outcomes of animal translocations. Conserv Sci Pract 5:e12937. https://doi.org/10.1111/csp2.12937

Soorae PS (ed) (2021) Global conservation translocation perspectives, 2021: case studies from around the globe. IUCN SSC Conservation Translocation Specialist Group, Environment Agency... & Calgary Zoo

Spielman D, Brook BW, Frankham R (2004) Most species are not driven to extinction before genetic factors impact them. Proc Natl Acad Sci 101:15261–15264. https://doi.org/10.1073/pnas.0403809101

Srbek-Araujo AC, Haag T, Chiarello AG et al (2018) Worrisome isolation: noninvasive genetic analyses shed light on the critical status of a remnant jaguar population. J Mammal 99:397–407. https://doi.org/10.1093/jmammal/gyy007

Stanley-Price M (2022) Species translocations, taxon replacements, and rewilding. In Routledge Handbook of Rewilding (pp. 68–80). Routledge.

Struck TH, Feder JL, Bendiksby M et al (2018) Finding evolutionary processes hidden in cryptic species. Trends Ecol Evol 33:153–163. https://doi.org/10.1016/j.tree.2017.11.007

Taberlet P, Luikart G (1999) Non-invasive genetic sampling and individual identification. Biol J Linn Soc 68:41–55. https://doi.org/10.1111/j.1095-8312.1999.tb01157.x

Taft HR, McCoskey DN, Miller JM et al (2020) Research–management partnerships: an opportunity to integrate genetics in conservation actions. Conserv Sci Pract 2:e218. https://doi.org/10.1111/csp2.218

Tirelli FP, de Freitas TR, Michalski F et al (2019) Using reliable predator identification to investigate feeding habits of Neotropical carnivores (Mammalia, Carnivora) in a deforestation frontier of the Brazilian Amazon. Mammalia 83:415–427. https://doi.org/10.1515/mammalia-2018-0106

Todesco M, Pascual MA, Owens GL et al (2016) Hybridization and extinction. Evol Appl 9:892–908. https://doi.org/10.1111/eva.12367

Toft CA, Wright TF (2015) Parrots of the wild: a natural history of the world's most captivating birds. University of California Press, Oakland

Torres-Florez JP, Johnson WE, Nery MF et al (2018) The coming of age of conservation genetics in Latin America: what has been achieved and what needs to be done. Conserv Genet 19:1–15. https://doi.org/10.1007/s10592-017-1006-y

Trindade FJ, Rodrigues MR, Figueiró HV et al (2021) Genome-wide SNPs clarify a complex radiation and support recognition of an additional cat species. Mol Biol Evol 38:4987–4991. https://doi.org/10.1093/molbev/msab222

Túnez JI, Ibañez EA, Nardelli M et al (2021) The use of molecular markers in neotropical mammal conservation. Molecular Ecology and Conservation Genetics of Neotropical Mammals, 35–62

Turbek SP, Funk WC, Ruegg KC (2023) Where to draw the line? Expanding the delineation of conservation units to highly mobile taxa. J Hered 114:300–311. https://doi.org/10.1093/jhered/esad011

Valencia RA, Lobo M, Ligarreto GA (2010) Estado del arte de los recursos genéticos vegetales en Colombia: Sistema de Bancos de Germoplasma. Cien Tecnol Agropecuaria 11:85–94

Valqui J, Hartl GB, Zachos FE (2010) Non-invasive genetic analysis reveals high levels of mtDNA variability in the endangered South-American marine otter (Lontra felina). Conserv Genet 11:2067–2072. https://doi.org/10.1007/s10592-010-0069-9

Venegas C, Varas V, Vásquez JP et al (2020) Non-invasive genetic sampling of deer: a method for DNA extraction and genetic analysis from antlers. Gayana 84:67–74. https://doi.org/10.4067/S0717-65382020000100075

Vitorino CA, Nogueira F, Souza IL et al (2017) Low genetic diversity and structuring of the Arapaima (Osteoglossiformes, Arapaimidae) population of the Araguaia-Tocantins basin. Front Genet 8:277092. https://doi.org/10.3389/fgene.2017.00159

Volpe NL, Thalinger B, Vilacoba E et al (2022) Diet composition of reintroduced Red-and-Green Macaws reflects gradual adaptation to life in the wild. Condor 124:duab059. https://doi.org/10.1093/ornithapp/duab059

Waits LP, Paetkau D (2005) Noninvasive genetic sampling tools for wildlife biologists: a review of applications and recommendations for accurate data collection. J Wildl Manag 69:1419–1433. https://doi.org/10.2193/0022-541X(2005)69[1419:NGSTFW]2.0.CO;2

Wallen J, Norén K, Angerbjörn A et al (2023) Context-dependent demographic and genetic effects of translocation from a captive breeding project. Anim Conserv 26:412–423. https://doi.org/10.1111/acv.12831

Willi Y, Van Buskirk J, Hoffmann AA (2006) Limits to the adaptive potential of small populations. Annu Rev Ecol Evol Syst 37:433–458. https://doi.org/10.1146/annurev.ecolsys.37.091305.110145

Willoughby JR, Sundaram M, Wijayawardena BK et al (2015) The reduction of genetic diversity in threatened vertebrates and new recommendations regarding IUCN conservation rankings. Biol Conserv 191:495–503. https://doi.org/10.1016/j.biocon.2015.07.025

Woodcock L, Gooch J, Wolff K et al (2023) Fingermarks in wildlife forensics: A review. Forensic Science International, 111781

Zamboni T, Di Martino S, Jiménez-Pérez I (2017) A review of a multispecies reintroduction to restore a large ecosystem: the Iberá Rewilding Program (Argentina). Perspect Ecol Conserv 15:248–256. https://doi.org/10.1016/j.pecon.2017.10.001

Ziegler TE, Molina A, Ramón M et al (2023) Analysis of the genomic landscape of inbreeding in two divergent groups of Spanish Florida goats. J Anim Breed Genet 140:316–329. https://doi.org/10.1111/jbg.12759

Zinger L, Donald J, Brosse S et al (2020) Chapter Nine - Advances and prospects of environmental DNA in neotropical rainforests. In: Dumbrell AJ, Turner EC, Fayle TM (eds) Advances in ecological research. Academic, pp 331–373

Chapter 16
What Can We Say About Plant Domestication in the Neotropics?

Elizabeth Ann Veasey, Flaviane Malaquias Costa, Wellington Ferreira do Nascimento, Santiago Linorio Ferreyra Ramos, and Charles R. Clement

Contents

E. A. Veasey (✉) · F. M. Costa
Departamento de Genética, Escola Superior de Agricultura "Luiz de Queiroz", Universidade de São Paulo, Piracicaba, São Paulo, SP, Brazil
e-mail: eaveasey@usp.br

W. F. do Nascimento
Centro de Ciências de Chapadinha, Universidade Federal do Maranhão, Chapadinha, Maranhão, Brazil

S. L. F. Ramos
Instituto de Ciências Exatas e Tecnologia, Universidade Federal do Amazonas, Itacoatiara, AM, Brazil

C. R. Clement
Instituto Nacional de Pesquisas da Amazônia, Manaus, Amazonas, AM, Brazil

M. V. Bohrer Monteiro Siqueira et al. (eds.), *Population Genetics in the Neotropics*, https://doi.org/10.1007/978-3-031-83685-5_16

Abstract Plant domestication is a coevolutionary process in which plants are selected to make them more useful to humans and adapted to the environment created by humans, while at the same time becoming less adapted to their original habitat in nature. This process is often considered to be related to the origins of agriculture, which occurred later but simultaneously in several regions of the earth, including the Neotropics. However, plant domestication in the Neotropics occurred within agroecological systems that focused on individual plants rather than populations, as is typical of agriculture. In this chapter we define the concepts of population domestication and landscape domestication, as well as different levels of plant population domestication, such as incidentally co-evolved, incipiently domesticated, semi-domesticated, and domesticated plants, and centers of domestication, including single and multiple independent centers of plant domestication in this part of the American continent. We examined the roles of archaeology, archaeobotany, paleobiolinguistics, genetics and genomics, and ethnobotany for a greater understanding of plant domestication in the Neotropics. We also define the concepts of domestication syndromes, genetic drift, selection signatures, among others, related to domestication in the Neotropics. Numerous plant species were domesticated in the Neotropics, such as *Capsicum* spp., *Manihot esculenta*, *Theobroma* spp., *Phaseolus* spp., *Bertholletia excelsa*, *Annona cherimola*, *Bixa orellana*, *Cucurbita* spp., among others, and case studies were presented for *Ananas comosus*, *Zea mays*, *Solanum lycopersicum* and *Bactris gasipaes*. Finally, we present future directions in the study of plant domestication in the Neotropics.

Keywords Agricultural origins · Archaeology · Center of domestication · Genetics · Plant domestication

Learning Objectives

This chapter will help readers to understand the following:

1. Plant domestication, degrees of domestication and landscape domestication;
2. Domestication syndromes;
3. Agricultural origins and its relation to plant domestication;
4. Centers of domestication;
5. Single and multiple independent centers of domestication;
6. The roles of archaeology, especially archaeobotany, paleobiolinguistics, genetics and genomics, and ethnobotany in the study of plant domestication in the Neotropics.

16.1 Introduction

The Neotropical region comprises the areas of South America, Central America, southern and central Mexico, and the Antilles between the Tropics of Cancer and of Capricorn (Morrone 2018). In the Neotropics, more than 6500, 1206 and 6261 native plant species have been used in Mesoamerica, the Central Andes, and lowland South America, respectively, and at least 1148, 428 and 600 are cultivated in these three regions (Clement et al. 2021a). According to these authors, fully domesticated crops are more numerous in Mesoamerica (251), followed by the Andes (124) and the South American lowlands (45), meaning that domestication is more common in the Neotropics than previously recognized and started much earlier than reliance on food production systems.

This chapter will present the concepts of plant domestication, domestication syndrome and degrees of domestication; the origins of agriculture and its relation to plant domestication; the centers of domestication, and the comparison among single and multiple independent plant domestications in the Neotropics; the roles of archaeology, archaeobotany, paleobiolinguistics, ethnobotany, genetics and genomics, and other disciplines that have contributed to understanding plant domestication in the Neotropics. We will also describe some case studies to illustrate the fascinating evolutionary history of plants cultivated in this part of the American continent.

16.2 Plant Domestication, Including Degrees of Domestication, Landscape Domestication and Domestication Syndromes

Darwin (1859) used the idea of domestication to introduce his own ideas about natural selection because everyone has some idea about what domestication is. Curiously, however, Darwin never provided a clear definition. Rather, he concentrated on process [(selection, both unconscious (similar to natural selection) and methodical (similar to modern plant or animal breeding)], preconditions (the availability of variation) and results (more variation). Although Darwin considered the domestication of both animals and plants, we will concentrate on plants. Perhaps because Darwin never defined it, domestication has received numerous definitions, such as: '… a co-evolutionary process by which human selection on the phenotypes of promoted, managed or cultivated plant populations results in changes in the population's genotypes that make them more useful to humans and better adapted to human intervention' (Clement 1999); 'A process that is based on genetic change in domesticated populations making them more adapted to the environment created by cultivation, but at the same time poorly adapted to their original habitat in nature' (Ladizinsky 1998); 'The process that causes the loss of characteristics (mainly reproductive ones) in populations of cultivated plants necessary for their survival in nature in primary environments, i.e., the process which makes plant populations dependent on human intervention for its reproduction' (Hillman and Davies 1990); or 'The genetic architecture that makes populations dependent on both the interference from man and the environment created by man' (Harlan 1992).

Observe that most of the definitions include the term 'process', which is a series of actions taken in order to achieve a particular end. Thus, a process has a beginning in time and can arrive at numerous intermediate stages, including what exists today, but only has an end if the plant population goes extinct. Observe also that most authors are interested in a specific result: a loss of adaptation to natural environments. This is important because if a plant population loses this adaptation, it becomes dependent on humans; if humans abandon it, it goes extinct.

Since domestication is a process, a plant population involved in the process may be classified into different degrees of domestication (Clement 1999), such as *wild* (a naturally evolved population whose genotypes and phenotypes have not been modified by human intervention); *incidentally co-evolved* (a population that volunteers and adapts in a human disturbed environment, possibly undergoing genetic change, but without human selection, typical of weeds); *incipiently domesticated* (a population that has been modified by human selection and intervention, but whose average phenotype is still within the range of variation found in the wild population for the trait(s) subject to selection; *semi-domesticated* (a population that is significantly modified by human selection and intervention so that the average phenotype may diverge from the range of variation found in the wild population for the trait(s) subject to selection; and *domesticated* (a plant population similar to the semi-domesticated one but whose ecological adaptability has been reduced to the point that it can only survive in human-created environments, specifically in cultivated landscapes) (Harlan 1992). Like evolution, domestication is a process that occurs in populations. It is common, however, to speak of domesticated species, even though these do not exist. The term domesticated species means a species with domesticated populations. There are 727 incipiently domesticated species in Mesoamerica, 517 in lowland South America, while the Central Andes has no available data yet. As for semi-domesticated and domesticated species, Mesoamerica has 170 and 251, the Central Andes 304 and 124, while lowland South America has 38 and 45, respectively (Clement et al. 2021a).

The environment in which the domestication process occurs is as important as the plant population and the humans. Hence, another important concept is that of landscape domestication, "a conscious process by which human manipulation of the landscape results in changes in landscape ecology and in the demographics of its plant and animal populations, resulting in a landscape more productive and congenial for humans" (Clement 1999). Human manipulation can be classified into a small set of categories of relatively simple practices: (1) removal of non-useful plants, (2) protection of useful plants, (3) attraction of non-human animal dispersers, (4) transportation (dispersal) of useful plants, (5) selection of phenotypes, (6) fire management, (7) planting of useful plants, and (8) soil improvement (Levis et al. 2018), where only category 3 is not conscious—or intentional. The outcome of these practices in a patch of forest or other ecosystem transforms that patch into a domesticated landscape, which might be a forest (Levis et al. 2018), a savanna (Ferreira et al. 2022) or another ecosystem (Clement 2024). The domesticated landscapes that everyone is familiar with include gardens, farming plots, pastures, and agro-industrial monocultures, but many people have difficulty recognizing a forest as a domesticated landscape (Clement et al. 2021b).

Returning to plant domestication, a frequent question is: how can we know/decide if a plant population is domesticated, semi-domesticated, or incipiently domesticated? A common practice among researchers interested in plant domestication is to compare non-managed or wild populations with managed populations and look for signs of the domestication syndrome, and of lower genetic diversity, mostly due to founder effects (bottleneck) or directional selection, which are typical of plants manipulated by humans. For example, a study conducted with piquiá (*Caryocar villosum*), a widely distributed Amazonian forest fruit tree, comparing homegarden populations, more subjected to anthropic action, with forest populations, that have much less interaction with humans, in the Tapajós National Forest, observed lower genetic diversity parameters for the homegarden populations, with lower number of effective alleles, allelic richness and observed heterozygosity, thus demonstrating piquiá's incipient domestication (Francisconi et al. 2021). Another study conducted with purple passion fruit (*Passiflora edulis* f. *edulis*) compared landraces (local varieties traditionally managed and reproduced by family farmers), commercial cultivars and genebank accessions from Colombia with Single Nucleotide Polymorphism (SNP) markers, and found higher average expected heterozygosity for the landraces (0.78) than the commercial cultivars and germplasm accessions (0.41 and 0.43, respectively) (Castillo et al. 2021). The authors discussed these results in terms of a possible genetic bottleneck for purple passion fruits during domestication or adaptive pressure for highland production in the Andes compared to Amazonia and the lowlands where yellow passion fruit (*P. edulis* f. *flavicarpa*) is grown.

Box 16.1 What is a domestication syndrome?

Darwin (1859) observed that crops are different from their wild ancestors in those traits that are of interest to, hence selected by humans. The domestication syndrome is defined as the suite of traits that marks a crop's divergence from its wild ancestor(s) (Harlan 1971, 1992; Meyer et al. 2012). The main elements of the domestication syndrome, when comparing a domesticated with a non-domesticated plant are: a reduction/loss of dispersal mechanisms, such as the non-shattering trait, reduction of seed dispersal aids, fragile rachis, dehiscent pods, reduction/loss of dormancy, more compact growth habit, precocity, gigantism (the increase in the size of seeds, fruits, tubers), with tastier and more colorful fruits, insensitivity to photoperiod, synchronous tillering and ripening, loss of secondary metabolites, reduction/loss of toxic compounds (alkaloids, etc.), change in reproductive system, life cycle shift, change in ploidy, among others (Harlan 1992; Meyer et al. 2012; Olsen and Wendel 2013).

An example of the use of a domestication syndrome is a study comparing wild and domesticated populations of Amazon tree grape (*Pourouma cecropiifolia*) populations cultivated by the Ticuna Indigenous People in Western Amazonia, using traditional approaches (differences in morphological traits) and ecological approaches (Pedrosa et al. 2018). Fruits of domesticated plants had 20 times greater mass and twice as much edible pulp than wild plants, even those from the nutrient rich floodplains. Also, although domesticated plants had heavier seeds, the domesticated fruits had proportionally (46%) smaller seed mass compared to wild fruits. The authors concluded that the high light availability and poor soils of cultivated landscapes may have contributed to seed mass reduction, while human selection promoted seed mass decrease in correlation with fruit mass increase. In chili pepper (*Capsicum annuum*), variations in shape and increase in fruit size are the main changes observed in domesticated fruits, accompanied by change in fruit orientation (from erect fruit in the wild ancestor to pendant fruit in domesticated forms) (Lopez-Moreno et al. 2023). The same kinds of changes in the domestication syndrome are common in four other domesticated chili peppers (*C. baccatum, C. chinense, C. frutescens, C. pubescens*) (Pickersgill 2007). Ferrer et al. (2020) examined the domestication syndrome of native Mesoamerican trees in Mayan homegardens. They found that individuals growing in homegardens, compared to those growing in the forest, have larger foliar area and higher volume of inflorescence in *Brosimum alicastrum* and of the flower in *Cordia dodecandra*, and greater foliar area and fruit weight in *Spondias purpurea*.

A typical example of domestication syndromes is the domestication of peanuts (*Arachis hypogeae*) revealed through the morphological characteristics observed in the most popularly cultivated peanuts (Fávero and Valls 2009): (i) the increased tenacity of the 'peg' (structure that pushes the ovary into the ground, enabling the underground formation of fruits), which facilitates the harvesting of a greater quantity of seeds by simple plucking, unlike wild species whose 'peg' collapses upon maturation; (ii) reduction of the isthmus between the fruit segments, resulting in the occurrence of two or more seeds in a single segment; (iii) plant habit, which is no longer prostrate, as in wild species, to more erect, facilitating uprooting; (iv) concentration of fruits at the base of the plant, unlike wild species, which tend to release their seeds as far away as possible from the mother plant, thus avoiding competition; (v) proliferation of ovules in some varieties, where more than two and even six seeds are observed within the same pod—in wild species, there are generally two, or very rarely, three seeds; and (vi) annual cycle, unlike most wild species, which have a perennial habit, with some exceptions.

Another example of domestication syndrome involves the peach palm (*Bactris gasipaes*), domesticated in southwestern Amazonia. Wild peach palms are found across southern Amazonia in open forest and their fruits are rich in oils, considered an important source of energy. The fruits of wild peach palms are very small, varying in size from 1 to 2 grams. Domesticated peach palm populations have increased fruit sizes (10 to 70 g, rarely 200 g), greater amounts of pulp (mesocarp), greater quantity of starches and smaller quantity of oils in the mesocarp, as well as variation in colors (yellow, orange, red and their gradations) (Fig. 16.1).

Fig. 16.1 Illustration showing fruits of peach palm (*Bactris gasipaes*); (**a**) From left to right, wild, incipiently domesticated, and domesticated fruits. Bar: 1 cm. (**b**) From the left to right, fruit sizes showing the increase in the size of the mesocarp, characteristic of microcarpa, mesocarpa and macrocarpa peach palms. Bar: 1 cm. Illustration by Clarissa Alves Stefanello. (Adapted from Vieira et al. (2021))

16.3 Origins of Agriculture and Its Relation to Plant Domestication

In the standard narrative of Euroamerican dominance of the planet, plant domestication leads to the origins of agriculture, which in turn allows sedentary populations and the emergence of civilizations (Graeber and Wengrow 2021; Bogaard et al. 2021). In this regard, many questions such as where, when, how, and why humans began to practice agriculture have been raised, and numerous scientists have tried to answer these questions. Following these, another question arises: how were the plants modified during this process? And what about the dispersal of domesticated plants to other parts of the world? To understand these questions, numerous sources of evidence have been explored (Harlan 1971), such as living plants (biosystematics analyses, including genetics, cytogenetics, morphology, ecology, etc. of the crops and their near relatives); plants of the past (archaeology, archaeobotany, palynology, paleobotany, etc); living humans (linguistics, oral traditions, attitudes towards crops in culture, tradition, magic, etc); and humans of the past (archaeology, history, art, etc.). In this chapter, we illustrate several studies including such methodology to understand the origins of plants in the Neotropics.

Before we proceed, a proper definition of agriculture is, according to the Oxford English Dictionary, the theory or practice of cultivating the soil to produce crops; or the practice of growing crops, rearing livestock, and producing animal products (such as milk and eggs), regarded as a single sphere of activity. It is also defined as farming of plants and husbandry of animals. Originally, however, agriculture meant the culture or tending of fields (*agri* in Latin) and is therefore associated with waving fields of grains, such as wheat (*Triticum* spp.) or barley (*Hordeum vulgare*), in southwest Asia. Other cultures include horticulture (gardens), arboriculture (trees) and silviculture (woods or forests).

Most students of domestication consider that plant domestication for food production began globally 12,000–11,000 years ago at the end of the most recent ice age and during the transition to the present interglacial period, the Holocene, answering the question of when plant domestication began. Two major chronological periods are of greatest interest: the transition to the Holocene from about 12,000–9000 B.P., and the middle Holocene between 7000 and 4000 B.P. (Larson et al. 2014), although humans domesticated plants throughout the Holocene (Meyer et al. 2012). Simultaneously, in various locations of the world, a change from hunting-gathering to agricultural economies took place, which had profound consequences for humankind, our way of life, and the environment (Gepts and Papa 2002; Graeber and Wengrow 2021). The main question that has intrigued many scientists for so long is why human beings abandoned the hunter-gatherer lifestyle that successfully sustained us for nearly 300,000 years to adopt agriculture, which is more labor intensive and yields lower quality diets? This transition from hunter-gatherer to agriculture was not abrupt and happened slowly. To answer 'why', many theories and models have been raised by many authors, but a common element is sedentary lifestyles (Ladizinsky 1998). However, not all humans abandoned the hunter-gatherer lifestyle and adopted agriculture; there is a large middle ground (Denham and Donohue 2022). It is only during the last couple of millennia that some urban societies have come to depend on agriculture to the near exclusion of hunting and gathering (Graeber and Wengrow 2021).

To answer the question where it happened, at least 11 regions of the Old and New World have been proposed as independent centers of origin of agriculture (Larson et al. 2014; Fuller et al. 2014; Kraft et al. 2014). According to these authors, some of these regions were the source of major domesticates that spread to adjacent regions, while others involved more regionally important species often regarded as "minor" crops today. However, the theory of centers of origin of agriculture is questioned today by many scholars (Nelson and Ladiges 2001; Clement et al. 2021a; Mellanby et al. 2024), as agriculture has no single and no simple origin. According to Mellanby et al. (2024) "a wide variety of plants and animals have been independently domesticated at different times and in numerous places." The practice of intensive agriculture may have happened in some areas but in many others more complex methods to manage plants and animals occurred. The narrative of linking plant domestication to agriculture origins is no longer accepted (Clement et al. 2021a).

In the Neotropics, for a long time Mesoamerica and the Andean region were considered the main areas of origin of agriculture, but more recently Amazonia was shown to be similarly important (Clement et al. 2015; Levis et al. 2017; Watling et al. 2018; Casas et al. 2019; Iriarte et al. 2020; Lombardo et al. 2020). Unlike Mesoamerica and the Andes, which are relatively small areas geographically and clearly delimited culturally, Amazonia is a continental sized area within South America with enormous cultural diversity. In Amazonia, the origins of domestication and of farming occurred in various areas at various times throughout the Holocene (Clement et al. 2010).

However, it is important to understand that there was no agriculture in Amazonia before European conquest (Clement et al. 2012; Shepard Jr. et al. 2020), nor in the rest of the Neotropics (Clement et al. 2021a). According to Clement (2024), "in Amazonia, forests were domesticated before gardens expanded into swiddens, and plant populations were domesticated both in forests and in gardens and swiddens". Plants are treated in a different manner in gardens, with important consequences for domestication (Ames 1939; Clement 2024). Human selection in gardens can identify and propagate any difference that might be useful—or just different. The result is dozens, or hundreds, or even thousands of varieties typical of gardens, such as observed in manioc (*Manihot esculenta*) and maize (*Zea mays*) (Alves-Pereira et al. 2018; Costa et al. 2023; Clement 2024). Anyone can easily identify what is planted in a garden or a swidden, but it takes experience to identify domesticated plants in a forest. Hence, in Amazonia we need to consider at least two domestications: domestication of landscapes and domestication of populations of species (Clement 1999, 2024; Clement et al. 2021a), as described above (sub-item 16.2). In Mesoamerica and the Andes, centralized states arose, and the term agriculture is generally attributed to their food production systems, whereas in Amazonia there are more diverse societies, without the presence of states. However, all these societies always exchanged plants, such as maize, tomatoes (*Solanum lycopersicum*), manioc, and many others. This interaction between human societies and ecosystems resulted in the domestication of many plants in the Americas without necessarily developing a form of agriculture as we know it today (Vieira et al. 2021).

16.4 Centers of Domestication

A center of domestication represents the region where a given species was domesticated, that is, where it was adapted for human use. This process occurs mainly through the selection, cultivation and maintenance of plants that present phenotypes with characteristics 'desirable' to humans, such as larger fruits, a greater quantity of fruits, better organoleptic properties (aroma, flavor, texture, and appearance) of the fruits and other edible parts of the plant (seeds, roots, and tubers), among other characteristics of the domestication syndrome (Naithani 2021). When the genetic variability of a domesticated species is compared with their wild relatives, it is observed that cultivated populations have high genetic diversity in their center of domestication, because the domestication process in this location occurred with numerous morphological types intermediate between derivatives and wild types, and because wild populations continue to hybridize with cultivated populations

(Vavilov 1992). It is also observed that most domesticated populations are more vulnerable to parasites, pests, and pathogens, and are less resilient under unfavorable environmental conditions. However, these lost characteristics may be present in populations from regions where the crop was first domesticated, that is, in incipient domesticates, wild progenitors and even related species. Thus, useful traits can be reintroduced into crops through genetic improvement programs (Naithani 2021).

A major center of crop domestication in the Neotropics is Amazonia, with at least 83 native species containing populations domesticated to some degree (Clement 1999; Clement et al. 2010, 2015, 2016; Hilbert et al. 2017; Watling et al. 2018; Iriarte et al. 2020). The upper Madeira River basin is an important area of plant domestication for many crops, such as manioc, peach palm, lady's finger peppers (*Capsicum baccatum*) (Clement et al. 2010, 2016). Other important crops domesticated in Amazonia are pineapple (*Ananas comusus*), coca (*Erythroxylum coca*), tobacco (*Nicotiana tabacum*), guarana (*Paullinia cupana*), and less important ones include cubiu (*Solanum sessiliflorum*), assai-do-pará (*Euterpe oleracea*), other hot peppers (*Capsicum chinense* and *C. frutescens*), inga (*Inga edulis*), Brazil nut (*Bertholletia excelsa*), cupuaçu (*Theobroma grandiflorum*), genipap (*Genipa americana*), leren (*Calathea allouia*), sweet potato (*Ipomoea batatas*), cocoyam (*Xanthosoma sagittifolium*), yam (*Dioscorea trifida*), mapati (*Pourouma cecropiifolia*), abiu (*Pouteria caimito*), bacuri (*Platonia insignis*), cashew (*Anacardium occidentale*), sapota (*Quararibea cordata*), biriba (*Rollinia mucosa*), guava (*Psidium guajava*), and annatto (*Bixa orellana*) (Clement et al. 2010, 2015, 2016).

Biological, archaeological, and ethnographic information about annatto, integrated with ethnobotanical observations of cultivated (*B. orellana* var. *orellana*) and non–cultivated populations, revealed that domestication occurred in northern South America and identified the wild ancestor of cultivated annatto (*B. orellana* var. *urucurana*) (Moreira et al. 2015). The proximity between wild and cultivated annatto samples in the upper Madeira River basin, using microsatellite markers and chloroplast DNA, provided evidence that annatto domestication began in this region from var. *urucurana* (Dequigiovanni 2017).

16.4.1 Single and Multiple Independent Plant Domestications in the Neotropics

Apart from a single event of domestication, where a crop is domesticated in a certain area and then dispersed to other areas, multiple and independent domestications are possible. Multiple domestications may arise from the same wild species occurring in different geographic areas (Ladizinsky 1998). An example of independent domestication from the same wild species in the Neotropics is coca. Known as the "Divine Leaf" by the Inka, coca has been cultivated for over 8000 years, representing a significant pharmaceutical plant in South America (White et al. 2021). It is the source of the alkaloid 'cocaine', an insecticide and local anesthetic of large impact in Western Medicine, as well as the illegal drug well-known around the world. This crop comprises four closely related varieties classified into two species (Amazonian and Huánuco varieties within *Erythroxylum coca*, and Colombian and Trujillo varieties within *E. novogranatense*). By using a phylogenomic approach, White et al. (2021) demonstrated a novel and robust hypothesis of multiple independent origins of different coca varieties from *E. gracilipes*, a widespread, wild species comprised of at least two main clades. The Colombian and Trujillo coca varieties are descended from a single, ancient domestication event in northwestern South America. Huánuco coca was domesticated more recently, possibly in southeastern Peru. And Amazonian coca either shares a common domesticated ancestor with Huánuco coca, or it was the product of a third and more recent independent domestication event in the western Amazon basin.

Another interesting example of independent domestication in the Neotropics is the genus *Phaseolus*. Both common bean (*P. vulgaris*) and lima bean (*P. lunatus*) have undergone two independent domestications. Together with single independent domestication events for *P. coccineus*, *P. dumosus* and *P. acutifolius*, that occurred in Mesoamerica, at least seven independent and isolated processes of domestication have occurred for *Phaseolus* from different wild species (Bitocchi et al. 2017). For the common bean, two major domesticated gene pools originated, one in Mesoamerica and another in the Andes (Bitocchi et al. 2013). A reduction in genetic diversity in both gene pools was found, which was threefold greater in the Andes compared with the Mesoamerica, due to a bottleneck that occurred before domestication in the Andes. A similar scenario has been observed for *P. lunatus*. One of its gene pools was in the Andes, which gave rise to the large-seeded landraces collectively known as the 'big Lima' cultivars. The other was in Mesoamerica, which gave rise to a great variety of small-seeded Mesoamerican landraces (Bitocchi et al. 2017).

The Cucurbitaceae family has interesting examples of independent domestications (Chomicki et al. 2020). The *Cucurbita* genus offers four main crops, each one with a different history of domestication. *Cucurbita pepo* is native to North America (Northeast Mexico, and the Southeast and Central United States (Paris et al. 2012). Independent domestication occurred in Eastern North America (Sanjur et al. 2002), originating the forms now classified as *C. pepo* subsp. *pepo* and *C. pepo* subsp. *ovifera* (Paris et al. 2012). A second crop is *C. maxima*, with the subspecies *maxima* and *andreana*. The *maxima* subspecies (Hubbard squash, buttercup squash, giant pumpkin) was cultivated as early as 4000 BP along the coast of Peru, while subspecies *andreana* appears to have been domesticated in what is today Argentina (Chomicki et al. 2020). The third domesticated species of *Cucurbita*, *C. moschata* (the butternut squash) may have originated in Northern South America (Sanjur et al. 2002). However, nuclear sequence data cast doubt on a South American origin of this species (Kates et al. 2017). The fourth crop, *C. argyrosperma* is native to Mexico (Sanjur et al. 2002; Barrera-Redondo et al. 2021). By sequencing libraries for 192 individuals of the wild and domesticated subspecies of *C. argyrosperma* across Mexico, Barrera-Redondo et al. (2021) found a monophyletic origin of this domesticated crop in the lowlands of Jalisco, in Mexico.

16.5 Plant Domestication in the Neotropics

Understanding the complex interactions between biological and cultural factors involved in plant domestication requires the integration of multiple disciplines, such as archaeology, archaeobotany, linguistics, genetics and genomics, and ethnobotany (De Candolle 1883; Vavilov 1926). In the following we will present some approaches used in plant domestication studies in the Neotropics.

16.5.1 The Role of Archaeology and Archaeobotany

Archaeology plays an important role in reconstructing the history of plant domestication and dispersal, providing evidence that helps researchers understand how the relationship between humans and domesticated plants evolved over time. It is the discipline dedicated to the study of the human past through the analysis of material remains left by past societies. These remains may include artifacts, architectural structures, ceramics, tools related to agriculture, human, animal, and plant remains in archaeological sites. Archaeologists use specialized techniques to unearth, preserve, analyze, and interpret these remains to reconstruct the history and culture of past human societies (Archila et al. 2008; Piperno 2011; Pearsall 2015; Watling et al. 2018; Iriarte et al. 2020; Lombardo et al. 2020; Peripato et al. 2023).

A new study has just predicted, through image analysis that detected ancient earthwork structures (geoglyphs, ditches, lagoons, and wells), that there are still between 10,000 and 24,000 archaeological sites to be discovered in Amazonia. This study also evaluated the relationship between the occurrence and abundance of 53 domesticated species close to archaeological sites, where the abundance of many of these species increased, such as the Brazil nut, *annatto*, pitanga (*Eugenia uniflora)* and macaúba (*Acrocomia aculeata)* (Peripato et al. 2023). The association between Brazil nut populations and Amazonian *Terra Preta* (also known as *Terra Preta de Índio*), as well as other archaeological sites, is now well known (Shepard Jr. and Ramirez 2011; Levis et al. 2012; Thomas et al. 2015), demonstrating the interface between the species and ancient human populations. It follows that the abundance and distribution of useful species in Amazonia indicates areas involved with anthropogenic activities in the past (Levis et al. 2012, 2017, 2018; Maezumi et al. 2018).

Archaeology provides material and contextual evidence for plant domestication studies that help researchers understand *where, when, how* and *why* the domestication of a given species occurred, mentioned above in Sect. 16.3, involving specific geographic contexts (Archila et al. 2008; Piperno 2011; Pearsall 2015; Watling et al. 2018; Iriarte et al. 2020; Lombardo et al. 2020). Dating refers to the process of determining the chronological age of archaeological remains, which allows establishment of a chronology for archaeological events, to define *where* and *when* the species was present, and to suggest *when* domestication occurred and when the species was dispersed. There are several dating methods used in archaeology, of which radiocarbon dating (carbon-14) is one of the most widely used by the scientific community.

Archaeobotany, also called paleoethnobotany, is the study of botanical remains in an archaeological context, aiming at cultural interpretations about past human populations, their multiple relationships with plants and the evolutionary processes involving plants (Piperno 2006; Archila et al. 2008; Piperno 2011; Pagan-Jimenez Jr. 2015; Pearsall 2015). Species identification can be performed through the analysis of archaeobotanical remains, which can be classified into two groups: (i) Macro-remains, such as fruits, seeds, tubers, roots, wood, charcoal, leaves, and fibers, which can be seen with the naked eye; and (ii) Micro-remains, such as phytoliths, starch grains and pollen, whose visualization depends on chemical preparation and the use of microscopes. Phytoliths (from the Greek, "plant stone") are rigid structures, made of silica, found in some plant tissues and which persist after the decomposition of the plant (Piperno 2006; Pearsall 2015). These structures present specific patterns in terms of shape and size, which sometimes allow the identification of species (Piperno 2006; Wesolowski et al. 2010; Piperno 2011; Pagan-Jimenez Jr. 2015; Pearsall 2015; Hilbert et al. 2017; Watling et al. 2018, 2020a, b).

The dating and distribution of numerous species throughout the Neotropics has been reported in different domestication studies (Piperno 2011; Kistler et al. 2018; Watling et al. 2018; Iriarte et al. 2020; Lombardo et al. 2020) and has also been useful in investigating dispersal routes between regions, such as maize (Kistler et al. 2018, 2020; Costa et al. 2022). Hypotheses of maize dispersal in the lowlands of South America were proposed through the compilation of archaeological records of the species in this part of the continent, genomic and morphological analyzes of modern local varieties and the distribution of indigenous linguistic families in the region. The species' dispersal routes were associated with different types of maize grain and the migrations of native ancestral Indigenous Peoples (Costa et al. 2022) (Fig. 16.2).

Starch molecules from maize, yam (*Dioscorea* spp.), pine nut (*Araucaria angustifolia,* Paraná pine) and sweet potato were identified in human dental arches found in four sambaquis along the coast of Santa Catarina, Southern Brazil. Sambaquis are deposits of shells built by humans, made up of organic and calcareous materials that, deposited over time, are exposed to the weather. Some indigenous groups use them as a sanctuary, burying their dead, while others choose it as a special location to build their houses. In this study, 53 archaeological human teeth were analyzed, dating back to 1390 BP, indicating that these species were present in the diet of the inhabitants of these middens (Wesolowski et al. 2010). Another archaeobotanical study analyzed phytoliths and starch in ceramics (3000 to 1500 years BP) found in the upper Madeira River basin, in Southwestern Amazonia (Watling et al. 2020a). Twenty different starch morphotypes were observed, but only eight could be taxonomically identified to species, and some of the morphotypes were from maize. The species most represented in the ceramics was maize, found in 12 of the 17 ceramic fragments analyzed. Starch grains produced by maize with soft (floury) and hard (popcorn) endosperms were found in these residues, implying that varieties with different types of endosperm were being used during this period. The damage

Fig. 16.2 Hypotheses of maize dispersal (arrows) suggested by Costa et al. (2022) on map of neotropical biomes, and the archaeological records of maize (in years BP)

present in the starch grains found in the pots indicates that the plants underwent different mechanical processes (e.g., milling) and heating (toasting and boiling), which indicates the production and processing of maize flour (Watling et al. 2020a). There is evidence of maize being used to make drinks, which may have been consumed collectively at ceremonies (Kater et al. 2020). Ethnobotanical research identified different typical uses of maize to date, such as the use of maize as a fermented drink (*chicha)* in indigenous and riverside communities in Southwestern Amazonia (Costa et al. 2021).

Morphological analyzes of plant structures allow archaeologists to investigate how domesticated species have been modified over time because of human selection during domestication. Changes in certain morphological patterns over time and/or involving a geographic context can assist in the investigation of *how* and *why* domestication occurred, as was observed in populations of wild rice (*Oryza* spp.) in southwestern Amazonia (Hilbert et al. 2017). An increase in the size of rice husk phytoliths was observed, providing evidence of the domestication to increase grain size (Hilbert et al. 2017). Until this study was performed, rice domestication processes had only been documented in Asia (*Oryza sativa*) and Africa (*O. glaberrima*), and this study now includes the Americas.

In the North of Pará, in northeastern Amazonia, archaeological records of Brazil nut, such as charred fruits and wood fragments, were found in the Pedra Pintada Cave, occupied by ancient hunter-gatherers since 11,000 years BP (Roosevelt et al. 1996). In Acre, around 2000 years BP, archaeological evidence demonstrated that the Brazil nut produced smaller fruits than those found today (Pärssinen et al. 2021). Of the 151 fruits collected at the archaeological site, it was possible to measure the diameter of 69 fruits, whose average diameter was 5.7 cm (minimum and maximum diameters of 3.9 and 6.6 cm) (Pärssinen et al. 2021). Currently, Brazil nut fruits measure 11–15 cm, with more and larger seeds (Shepard Jr. and Ramirez 2011). This evidence also suggests that selection processes aimed at increasing the productivity of Brazil nut were performed by Indigenous Peoples in the past (Pärssinen et al. 2021).

16.5.2 The Role of Paleobiolinguistics

Linguistic studies can be useful in domestication studies in several ways, as language plays a fundamental role in communication and can reveal interactions between humans and domesticated plants. Paleobiolinguistics (PBL) is an interdisciplinary field that combines linguistics, anthropology, evolutionary biology, and paleontology. PBL employs the comparative method of historical linguistics to reconstruct the biodiversity known by human groups of the past. Studies in this area seek to understand how human language developed over time and how it relates to biological evolution. Linguistic histories have also been used to expand knowledge about the distribution and dispersal of species (Brown 2006; Shepard Jr. and Ramirez 2011; Brown et al. 2013a, b, c, 2014). PBL, for example, was used to establish *when* and *where* manioc, hot pepper, squash and maize developed importance for different prehistoric human groups native to Americas (Brown et al. 2013a, b, c, 2014).

In the case of maize, the determinations identified by PBL agree, in general, with previous information regarding the time and space of the domestication and dispersal of the species (Brown et al. 2014). The oldest linguistic reconstruction referring to maize (6591 years BP) (Brown et al. 2014) is located near the region of early domestication in the Balsas River valley, Mexico (Matsuoka et al. 2002; Piperno et al. 2009). In lowland South America, the oldest linguistic record was 4400 years BP, in Amazonia, for the Arawak language family. In Rondônia, Southwestern Amazonia, the oldest linguistic records were dated to 2000 years BP, where the presence of five different indigenous linguistic groups was concentrated, including Tupi (Brown et al. 2014). Observe, however, that maize arrived in Amazonia much earlier than these reconstructions. The reason is that indigenous languages in Amazonia diversified in the last 5000 years and the languages present before this time cannot be reconstructed.

Linguistic analyzes for the word maize showed strong links between the Southwestern Amazonia and Eastern Brazil, suggesting a cultural expansion of maize traditions (Kistler et al. 2018). A loanword for maize with origins from the Arawak languages—originating in Southwestern Amazonia (Walker and Ribeiro 2011)—to the languages of the Macro-Jê group in the Brazilian savanna and on the Atlantic coast (Kistler et al. 2018) was observed, corroborating with the maize dispersal hypotheses presented by Costa et al. (2022).

Results for the Brazil nut suggested that the three main language families involved with the distribution of the species were Arawak, Carib and Tupi (Shepard Jr. and Ramirez 2011). However, terms for Brazil nut are not reconstructed in proto-Tupi (ancestral language), which suggests that approximately five millennia ago, when the Tupi language emerged in the Southwestern Amazonia (Walker et al. 2012; Ramallo et al. 2013; Iriarte et al. 2017; and Silva et al. 2020), the species may not have been important to this indigenous group at the time (Shepard Jr. and Ramirez 2011). The Arawak linguistic family is important for loaning words referring to Brazil nuts to other linguistic families in Amazonia. Northern Amazonia presents the greatest linguistic diversity of terms for Brazil nuts (Shepard Jr. and Ramirez 2011), which indicates the importance of the species in this region for indigenous populations in the past.

16.5.3 The Role of Genetics and Genomics

Genetics, a discipline that studies the genes, as well as the variation and heredity of characteristics, has provided important insights into the domestication of plants through its concepts, tools, and techniques. Among the various techniques available, the analysis of morphological, biochemical, and molecular markers has played a fundamental role in elucidating the evolutionary history of numerous species. Furthermore, in recent decades, the genomic era has made it possible to sequence genomes and obtain more consistent information about cultivated plants and their respective wild relatives (Shi and Lai 2015).

Based on the genetic markers and techniques mentioned above, several crops had their domestication process elucidated in the Neotropics. For example, there were controversies over the evolutionary history of the cultivated peanut, *Arachis hypogaea* and *A. villosulicarpa*. Initially, it was believed that these two species were native to Paraguay and Mato Grosso do Sul, in Brazil (Krapovickas and Gregory 1994). However, morphological and molecular evidence indicated that cultivated peanuts originated in northwestern Argentina and southern Bolivia, where their wild ancestors are found: *A. duranensis*, *A. ipaensis* and *A. monticola* (Krapovickas and Gregory 1994). Probably, the Indigenous People of this region began the domestication of peanuts at least 6000–7000 years BP (Krapovickas 1996). Genetic analyzes carried out using microsatellite markers (SSRs), single nucleotide polymorphisms (SNPs), and chloroplast genome sequencing confirmed that *A. duranensis, A. ipaensis* and *A. monticola* are close ancestors of *A. hypogaea*. Using the polymorphism of non-coding regions of cpDNA and the 5S rDNA of the wild tetraploid *A. monticola* and the nine diploid species proposed as peanut's closest relatives, Grabiele et al. (2012) concluded that the maternal genome belongs to *A. duranensis*, from Argentina, and the paternal to *A. ipaensis*, from Bolivia. Complementing this study, a phylogenetic tree obtained by SSR indicated that *A. duranensis* and *A. ipaensis* are close ancestors of *A. hypogaea* and *A. monticola* (Moretzsohn et al. 2013). Molecular aspects associated with agronomic traits related to peanut domestication were analyzed using 158 accessions and a total of 17,338 SNPs, through sequencing of specific fragments/loci (Zhang et al. 2017). Using 11 agronomic traits from a Genome Wide Association Study (GWAS), highly differentiated genomic regions were found between accessions of *A. hypogaea* and *A. fastigiata* using F_{ST} values. Among the 1429 genes, 662 were located on chromosome A3, suggesting the presence of key selective signatures caused by human selection during the domestication process. Such discoveries provide promising insights into the genetic architecture of traits related to peanut domestication (Zhang et al. 2017).

Box 16.2 What does "selective signature" mean?

Selective signatures, also known as selective sweeps or genetic hitchhiking, are terms that explain how alleles for favorable adaptations, and alleles near them on the chromosome, become more frequent in a population due to natural and human selection. Natural selection favors alleles for adaptation to the environment. The more favorable the allele for the environment, the more likely the individuals that possess that allele will live long enough to reproduce and pass that desirable trait on to their offspring. Eventually, undesirable traits will be bred out of the population and only the favorable alleles will be left. After a particularly strong selection for a trait that is the most desirable, a selective sweep will happen. Not only will the alleles that code for the favorable adaptation increase in frequency and be seen more often in the population, other traits that are controlled by genes that are close to those favorable genes on the chromosome will also be selected for, whether they are good or bad adaptations. That explains the term "genetic hitchhiking", as these genes and their alleles come along for the selection ride (Scoville 2020). Selective signature, which is involved in a reduction of nucleotide variation resulting from strong selective pressures acting on particular loci, have also been observed in genes associated with domestication or diversification phenotypes. The best example of a domestication-related selective sweep occurs in maize at the gene *tb1*, suggesting that *tb1* was targeted by human selection. Much less nucleotide diversity is found in maize relative to teosinte (the ancestor of maize) compared to genes that have not been targeted by human selection (Purugganan and Fuller 2009; Tsiantis 2011).

The origin of manioc had been debated for a long time, and finally it was better understood with the use of molecular markers. Manioc consists of three subspecies: *Manihot esculenta* ssp. *esculenta* (cultivated form), *M. esculenta* ssp. *flabellifolia* (the closest wild relative) and *M. esculenta* ssp. *peruviana* (another close wild relative probably not involved in manioc domestication). Together with *M. pruinosa*, these form the primary gene pool (Allem 1994). The gene pool concept was proposed by Harlan and de Wet (1971) to provide a practical guide for placing existing classifications into genetic perspective. The primary gene pool includes species and subspecies that can be crossed easily, usually belonging to the same species, and produce fertile hybrids. The secondary and the tertiary gene pools comprise species that are difficult to cross with cultivated species (secondary gene pool), or extremely difficult to cross with cultivated species (tertiary gene pool).

Three different molecular markers [sequencing of the nuclear gene glyceraldehyde 3-phosphate dehydrogenese (G3pdh), SNPs and SSRs] were used to evaluate the relationships among cultivated varieties of manioc and wild populations of *M. esculenta* ssp. *flabellifolia* and *M. pruinose* (Olsen and Schaal 1999, 2001; Olsen 2004). The authors concluded that the latter species did not contribute to the gene pool of cultivated manioc and that the genetic variability found in cultivated manioc is a subset of the genetic variability found in the populations of ssp. *flabellifolia* that occur in southwestern Amazonia. Domestication of manioc may have started as early as 10,000 years BP, in the upper Madeira River basin (Olsen and Schaal 1999). Another study (Léotard et al. 2009) using the G3pdh gene to examine cultivated manioc with a wider geographic sample of ssp. *flabellifolia* and other potentially hybridizing *Manihot* species, confirmed the previous results, suggesting that manioc was domesticated only once from *M. esculenta* ssp. *flabellifolia* populations occurring in northern Mato Grosso, Rondônia and Acre states, in Brazil, and adjacent areas of northern Bolivia. A recent study identified signatures of selection putatively associated with resistance genes, plant development and response to abiotic stresses that might have been important for the crop's domestication and diversification resulting from cultivation in different environments (Alves-Pereira et al. 2022).

One last example is the domestication history of cherimoya (*Annona cherimola*). Researchers believed that its center of domestication was in the Andean valleys of southern Ecuador and northern Peru. However, more recent research has shown that its center of domestication is, in fact, Mesoamerica (Larranaga et al. 2017). These authors assessed a large data set of 1765 accessions of cherimoya collected from six different countries across the Neotropics with nine highly informative microsatellite markers. Six scenarios were tested to examine three different areas of origin and subsequent migration histories, and the best supported scenario was that of a Mesoamerican origin and migration first to Ecuador and Northern Peru and from there to Southern Peru and Bolivia.

16.5.4 The Role of Ethnobotany

Prehistoric inhabitants developed ways of life that were completely integrated with natural dynamics, where human societies constructed the landscapes in which they lived. This cultural construction can be identified historically by the domestication of plants and landscapes. For example, prior to the arrival of European colonizers in Brazil, there were already millennia of forest management and plant domestication (Bacci et al. 2022). At present, rural communities, with their traditional knowledge about the use of plants and the management of their natural resources, preserve plants *in situ* and use their products for their subsistence, valuing local biodiversity (Lamarca et al. 2020). This interaction, which explores the relationship between popular and scientific knowledge of plants, is known as Ethnobotany (Martin 1995).

Ethnobotany means research aimed at understanding what people know about plants, how plants form parts of their systems of beliefs and conceptions of the world, and how humans use and manage plants for reproducing their social and cultural lives (Casas et al. 2016). For example, the discoveries of anthropic soils, also known as *terra preta de Índio* (TPI), demonstrate that people domesticated soils as well. In the Amanã Sustainable Development Reserve, in Amazonas, Brazil, more than 200 pre-Columbian burial urns were discovered in TPI, indicating ancient occupation (Shepard Jr. and Ramirez 2011), where high densities of useful, large-seeded plants were found, including Brazil nut. Brazil nut trees have also been found in association with geoglyphs—square or circular man-made trenches dated between 1000 and 2500 years ago (Pärssinen et al. 2009) in the state of Acre in Brazil, and in Riberalta, Bolivia, near the confluence of the Beni and Madre de Dios rivers (Shepard Jr. and Ramirez 2011). Observations, reports of the culture and behavior of the Indigenous Peoples of Amazonia, facilitated the recruitment of Brazil nuts in different parts of the region. The Wari's (Indigenous People) funeral customs involved the complete destruction of all belongings of the deceased, and when families return to their usual Brazil nut groves, they burn the undergrowth and discard fruit capsules from previous years' collection seasons, altering the appearance of the forest and eliminating physical reminders of past moments shared with the deceased (Shepard Jr. and Ramirez 2011).

Another ethnobotany study including TPI was carried out in three riverside communities located in the municipality of Manicoré, along the middle Madeira River, Amazonas state, Brazil, based on interviews with 62 residents (Junqueira et al. 2011). Residents mentioned more ethnospecies on TPI (mean ± standard deviation: 19.5 ± 8.9) than on non-anthropogenic soils (NAS) (17.4 ± 8.5), and the use value of the environment for the residents was higher on TPI (19 ± 5.7) than on NAS (16.2 ± 6.0). This study shows that the current residents value secondary forests on anthropogenic soils, having incorporated these domesticated landscapes into their knowledge and practices.

Ethnobotany has also played an important role documenting the main cultural and biological factors influencing human selection and other evolutionary processes guided by humans to domesticate species and landscapes in their territories (Lins

Neto et al. 2014). The management of plant populations may cause phenotypic changes in the characteristics of a plant that is targeted by human selection over time, which is the domestication process. Sousa Júnior et al. (2018) were interested in assessing this process in *Caryocar coriaceum*, that produces a fruit called pequi, in the Chapada do Araripe, Northeast Brazil. The authors used both morphometric and ethnobotanical approaches. Three different management regimes were observed: cultivation, *in situ* management (collection) and incipient management (the tolerance and protection of individuals). The fruits from individuals grown under cultivation were significantly different from those of individuals that were managed *in situ* and from those under incipient management. The study showed that local people perceive great morphological diversity among the study populations, including taste, color, and fruit size, consistent with the findings of the morphometric analyses. Fruits from cultivated areas were significantly larger than from other systems, corroborating the local perception of morphological variation related to the distinct management regimes in use. Fruit size is a preferred feature for humans (Meyer et al. 2012), who select this feature and keep it in cultivated populations.

Ethnobotany is an essential tool to detect the interaction between humans and germplasm diversity generated by the continued use and maintenance of crops, especially when practiced in traditional production systems, considering that these systems usually involve human selection of different varieties, interaction between wild and cultivated plants, leading to new varieties. Casas et al. (1996) studied different forms of management of wild, weedy, and domesticated plants carried out by the Nahua and the Mixtee Indigenous People in the Balsas River basin, Mexico, conducting inventories of plant resources, gathering ethnobotanical information about plant uses, forms of management, ethno classification, social and cultural roles of plants, and ecological data, such as distribution, phenology, etc. They also conducted specific interviews to obtain detailed information about the forms of management of the most important edible plants, and some evaluations of their roles in subsistence. The authors discuss these forms of plant management as examples of current routes of plant domestication. They observed that the types of plants managed and the ease of management, the requirements satisfied by those plants, the morphological features of plants used by humans and the ease of human selection, as well the preferences for particular colors, flavors, odors, shapes, or textures, are all aspects modulated by culture that influence the intensity of the human-plant interaction, which influence domestication trends through time.

Using ethnobotany data obtained through field surveys conducted in 70 rural communities involving 2049 farms, Costa et al. (2017) detected a microcenter of maize diversity in western Santa Catarina. Microcenters of diversity are defined as restricted geographical areas within which significant diversity of genetic resources is accumulated (Clement 1999). Costa et al. (2017) found extensive landrace richness and the presence of wild relative species, associated with local human activities and sociocultural aspects. Also using ethnobotany data from 129 interviewed farmers, as well as morphological and molecular markers, Costa et al. (2023) identified four microcenters and four regions of maize diversity in different biomes (Amazonia, Cerrado, Caatinga, Atlantic Forest, and Pampa) of lowland South America, in Brazil and Uruguay. The uses and preferences attributed to the landraces were associated with the characteristics that farmers consider important, especially for food, which influence the selection of their landraces. Results of this study showed high genetic richness and diversity of maize landraces (mostly exclusive to the respective areas) and races, as well as the presence of active landrace diversification due to current human action as diagnosed by ethnobotanical indicators.

Another study with maize was conducted in Mexico, in the Sierra Sur of Oaxaca, a cultural region represented by five municipalities inhabited by the Zapotec Indigenous People, who have lived there for millennia. The study was conducted in seven communities of two municipalities, where 20 semi-structured interviews from each community (N = 140) inquired about cultural practices, with questions about maize seed, ear, and cob management and selection and maize cultivation (Aragón-Martinez et al. 2023). Thirty maize plants from 28 fields (*milpa*) were assessed with ISSR markers. Milpa is a traditional farming system made up of a polyculture, which constitutes a dynamic space of genetic resource management. Its main species is maize, accompanied by various species of beans, squash, chili peppers, tomatoes, and many others depending on the region; the combination of corn-bean-squash is known as "the Mesoamerican triad" (Conabio 2016). Milpa is the Indigenous Peoples' laboratory where they select better ears, cultivate different varieties in time and place, alone or mixed, retain the main characteristics that characterize each landrace by endogamy, and promote their isolation or introgression, regardless of their geographical proximity (Aragón-Martinez et al. 2023). The results of this study indicated that traditional management of maize in this region, particularly as it relates to the selection of ears and seeds, and control of the farming cycle, maintains a balance between morphological differentiation, moderate levels of genetic diversity and gene flow, and low levels of population differentiation. These types of studies show that traditional management shapes genetic diversity to produce dozens of landraces and agronomic races, transforming them into genetic and morphological reservoirs that can be used in the present and future (Aragón-Martinez et al. 2023).

16.6 Case Studies of Plant Domestication in the Neotropics

16.6.1 Pineapple

Pineapple (*Ananas comosus* var. *comosus*; Bromeliaceae) is native to and was domesticated in South America. According to Bertoni (2019), cited by Chen et al. (2019), the genus name Ananas means 'excellent fruit' in the Guaraní language of Paraguay. Pineapple domestication started >6000 years BP with archaeobotanical remains dated 3500 years BP in South America, and later distributed to Mesoamerica >2500 years BP (Coppens d'Eeckenbrugge and Duval 2009). Pineapple was dispersed to several American countries as an exchange between tribes. However, after the discovery of America, it became known worldwide, when it was taken to Europe, Asia and Africa (Crestani et al. 2010).

Most of the pineapple cultivars belong to *A. comosus*, a diploid species (2n = 2x = 50), although there are also polyploids in the genus *Ananas* (Cotias de Oliveira et al. 2000). *A. comosus* is self-incompatible and mostly vegetatively propagated from axillary suckers and the fruit crown. *A. comusus* has five botanical varieties: the edible pineapple, var. *comosus*; the smooth-leaved var. *erectifolius* and the spiny var. *bracteatus*, two fiber cultigens exploited as ornamentals; and two wild varieties (var. *microstachys* and var. *parquazensis*) (Coppens d'Eeckenbrugge et al. 2018).

Coppens d'Eeckenbrugge and Duval (1997) and Coppens d'Eeckenbrugge and Leal (2003), cited by Clement et al. (2010), proposed that the Guiana shield is where the pineapple was domesticated. The Guiana shield, especially the eastern part of it, contains considerable phenotypic and genetic diversity, including wild phenotypes with relatively large and fleshy fruits and a wide range of large-fruited cultivars. A common practice in eastern Guiana is the picking of clones in the wild and transferring them to home gardens. While the Guianas constitute the likely center of pineapple domestication from wild and intermediate forms, the ample western Amazonian cultivar diversity suggests a secondary center where sexual recombination mostly involved domesticated forms (Coppens d'Eeckenbrugge et al. 2018). The wild species, *A. comosus* var. *microstachys* from northern Amazonia, is a likely ancestor of cultivated pineapple (Coppens d'Eeckenbrugge et al. 2018). It has long, narrow leaves, small fruits, with white or yellow fibrous pulp, and a high sugar and acidity content, in addition to having numerous seeds (Vieira et al. 2021). One of the domesticated varieties mentioned above, the red pineapple (var. *bracteatus*), was once cultivated for fiber, fruit juice and as a living hedge, and now is a pantropical ornamental. The bracteatus plant presents bright pink-to-red colored fruit and long bracts, with long leaves, coarse spines, and abundant suckers (Baker and Collings 1939, cited by Chen et al. 2019).

Among the characteristics that differentiate domesticated pineapple from wild pineapple is the presence of a greater number of wider and shorter leaves. This modification had a relevant impact in obtaining energy through photosynthesis by these plants, which began to use greater quantities of the sugar produced during photosynthesis in the development of larger and sweeter fruits. It is also noteworthy that with the decrease in the production of viable seeds in domesticated pineapple varieties, vegetative (asexual) propagation gained more importance as it is a mode of reproduction that does not depend on seeds, and this considerably reduced its reproductive success under natural conditions. This decline has made domesticated pineapple varieties much more dependent on human management. These modifications represent the pineapple domestication syndrome (Vieira et al. 2021). Also, greatly reduced diversity in cultivars relative to their wild progenitor suggests a severe domestication bottleneck experienced by this clonally propagated crop (Crestani et al. 2010), as mentioned above in sub-item 16.2.

16.6.2 Maize

Hypotheses about maize (*Zea mays* var. *mays*) domestication have been the subject of discussion since the end of the nineteenth century. The unicentric theory has been widely accepted by the scientific community in recent years due to genetic (Matsuoka et al. 2002; Vigouroux et al. 2008; van Heerwaarden et al. 2011) and archaeological (Piperno et al. 2009) evidence, and the fact that the endemic distribution of the *Z. mays* var. *parviglumis*, the wild progenitor, is in the Balsas River region (Fukunaga et al. 2005). However, the discussion about the domestication of maize was reignited with new evidence that reveals a complex history of domestication. Kistler et al. (2018) suggested that maize dispersed from Mexico and arrived in lowland South America semi-domesticated and continued to diversify in the region. According to the maize stratified theory (Kistler et al. 2018), the domesticate was selected from populations of *Z. mays* var. *parviglumis* in Mexico, ~9000 years ago, crossing Central America 7500 years ago (Piperno et al. 1985) and dispersed throughout the lowlands of South America ~6500 years BP (Bush et al. 1989; Brugger et al. 2016). Thus, structured maize populations with stable domestication syndromes evolved in South America from partially domesticated ancestral populations originating from Mexico (Kistler et al. 2018; Costa et al. 2023). Based on genomic, linguistic, archaeological, and paleontological evidence, Southwestern Amazonia was proposed as a probable secondary center of maize diversification in South America, within which the process of domestication of the populations continued (Kistler et al. 2018). This theory does not contradict the unicentric theory, which considers that the domestication process began in Mexico. However, it highlights that domestication was not complete when maize dispersed from Mexico to other regions. A new hypothesis posits a second dispersal event of maize that hybridized with *Z. mays* var. *mexicana*, arriving in South America after 6000 years BP. The hypothesis suggests that this hybridization may have conferred agronomic advantages to maize, such as increased ear size and number of grains per row (Yang et al. 2023).

The domestication process resulted in a series of morphological changes between wild teosinte and cultivated maize. Human intervention was an indispensable condition for this transformation and the origin of the cultivated populations. These changes ranged from changing the plant's architecture (Doebley et al. 1997, 2006; Studer and Doebley 2012) and characteristics associated with the species' reproductive organs (Iltis 1983; Beadle 1972, 1980; Doebley et al. 1990; Dorweiler and Doebley 1997). The morphological differences between the female inflorescences of maize and teosinte occurred due to a series of mutations which involved five genes (Beadle 1972, 1980; Doebley et al. 1990; Dorweiler and Doebley 1997). This event altered the distribution of nutrients in the plant and caused drastic morphological change, which was captured by human selection (Iltis 1983). Among the various consequences of maize domestication, the transformation of a small ear of teosinte, with easily dispersed grains, into an ear of maize with a large quantity of grains tightly attached to the rachis is the most dramatic. Teosinte has a large number of small and fragile ears, each ear with six to 12 seeds, in one or two rows. Maize has one or more large ears, with many rows, not fragile (Hake and Ross-Ibarra 2015). The revolutionary transition of the two-row teosinte to the maize we know today, with the high number of rows, dramatically improved crop yields (Doebley 2004; Bommert et al. 2013; Janzen et al. 2019), becoming an important source of food for different human societies in the Americas and around the world.

16.6.3 Tomato

Tomato domestication history involves two major transitions: one from wild *Solanum pimpinellifolium* to a semi-domesticated intermediate, *S. lycopersicum* var. *cerasiforme* (SLC) in South America, and a second transition from SLC to fully domesticated *S. lycopersicum* var. *lycopersicum* (SLL) in Mesoamerica (Razifard et al. 2020). The first stage involves selection, mainly for fruit size (going from the size of a blueberry to that of a cherry) and other secondary characteristics, occurring in Peru and Ecuador; and the second with a migration of semi-domesticated *S. lycopersicum* var. *cerasiforme*, through two population genetic bottlenecks, from Ecuador to Central America and later to Mexico, where a more intensive domestication of adaptive characteristics occurred in local agronomic environments, originating the varieties of *S. lycopersicum* var. *lycopersicum* (Blanca et al. 2012, 2015; Peralta et al. 2008; Zhu et al. 2018; Razifard et al. 2020). The main population genetic bottleneck in tomato domestication occurred during the migration from Peru to Ecuador, due to the high number of exclusive alleles that are found only in accessions of *S. pimpinellifolium* present in Peru, granting these accessions the title of basal group of the red-fruited tomato species (Blanca et al. 2015). Despite recent discoveries, most authors classify Peru and Ecuador as centers of tomato origin and not as centers of domestication, as the intensity of wild tomato domestication in these countries is still not very clear (Bergougnoux 2014; Saavedra et al. 2017). Mexico, specifically in the Vera Cruz—Puebla region, is accepted as the center of tomato domestication and diversification, due to the high morphological variability and diversity of varieties found in this region for the species *S. lycopersicum* (Peralta et al. 2008), and Peru as the center of diversity for wild relatives (Blanca et al. 2012, 2015).

During tomato domestication, changes occurred in a wide range of morphological and physiological characteristics of the crop compared to its wild ancestor (Bergougnoux 2014). The number of fruits per plant did not increase considerably, however, the size, circumference and weight of individual fruits did (Zhu et al. 2018). There was also selection for greater biomass and yield (Bergougnoux 2014). Razifard et al. (2020) describe fruit size increases from the southern to the northern region of the Americas, this being a consequence of the selection processes for this characteristic during tomato domestication. Other radical changes compared with wild tomato phenotypes are fruits that on average have thicker pericarps, more locules, lower dry weight, lower levels of beta-carotene, soluble solids, and citric acid, and higher levels of malic acid (Razifard et al. 2020). The wild species *S. pimpinellifolium* presents fruits with low deviations from the round phenotype, while cultivated varieties present a wide range of phenotypes in terms of fruit shape, including shapes similar to pears, torpedoes, bells, and rounded and flattened forms (Tanksley 2004). The seeds of wild plants are smaller and lighter, and this change in increased weight and size would be associated with a response to selection pressure for uniform germination and seedling vigor during domestication (Doganlar et al. 2000). Another modification was the increase in the number of locules in cultivated tomatoes, whose number of locular cavities can vary from two to ten or more in cultivated tomatoes, compared to wild tomatoes, which have only two cavities, that is, they are bilocular (Muñoz et al. 2011).

16.6.4 Peach Palm

Peach palm (*Bactris gasipaes*) was once an important crop and is the only Neotropical palm with domesticated populations, whose fruit is used primarily for food (Kerr and Clement 1980; Clement et al. 2010) and is currently an agribusiness for hearts-of-palm production (Clement et al. 2010). Patiño (2002) hypothesized that domestication began with selection for its stem wood, preferred for the manufacture of tools. Subsequently, there was selection of fruits with varying colors (yellow, orange, red and its variations, larger pulp (mesocarp) and larger sizes (Viera et al. 2021). The oily small fruits of wild var. *chichagui* were selected for size, which reduced oil content and increased starch content, as the latter is energetically easier for the plant to produce. This phenotypic variability in fruits and uses is observed in different regions, including the selection of fruits to be consumed cooked, fruits with larger amounts of starch to make a fermented drink, or consumption of palm hearts. These changes in the fruits are the domestication syndrome, as mentioned in the sub-item 16.2.

The domestication of peach palm started in southwestern Amazonia (Clement et al. 2017), from where it was dispersed along the Ucayali River northwards, leading to the formation of populations of plants that produced fruits with high starch and low oil contents, often used to produce fermented beverages. This dispersal continued throughout western Amazonia and western northern South America, and over the Andes to the Ecuadorian and Colombian Pacific coast, and finally arrived in Honduras in Central America. The second dispersal occurred along the Madeira River, creating populations of plants that produce fruits with a balance in the oil and starch contents, appreciated as snacks (Clement et al. 2017).

16.7 Final Remarks

Plant domestication in the Neotropics, and elsewhere, is an ongoing and dynamic process. Much information is available about plant domestication around the world. Despite significant advances in recent decades, in the Neotropics a better understanding of the origin and processes of plant domestication are still needed for numerous species, including the reconstruction of plant exchanges and human migration routes based on the geographic distribution of plants. To help understand the evolutionary history of plant species in this region, efforts from different disciplines are essential to provide new techniques and multiple perspectives on the relationship between plants and humans in the Americas, including the disciplines of archaeology, archaeobotany, paleobiolinguistics, genetics, genomics, ethnobotany, among others. New methods in each of these areas are being used, such as new generation sequencing leading to advances in genomics, technologies, and better image resolutions for analyzing archaeological microremains, such as pollen and phytoliths, software and analytical approaches for ethnobotanical and paleobiolinguistic data, ecological niche modeling and advanced analysis of spacial images and geographic information systems, among other tools. By using this knowledge, we believe it is possible to advance research to obtain more accurate data on the factors involved in the domestication of plants in the context and peculiarities of this region.

It is worth mentioning that the neotropical region, with emphasis on Amazonia, has an immense diversity of plants in different degrees of domestication, many of which are of great importance to traditional communities and urban populations. Many are already completely domesticated, such as maize, beans, tomato, potatoes, sweet potatoes, manioc, pineapple, squash, papaya (*Carica papaya*), avocado (*Persea americana*), guava, quinoa (*Chenopodium quinoa*), peanuts, rubber trees (*Hevea brasiliensis*), cashew, passion fruit, jaboticaba (*Plinia cauliflora*), peach palm, guarana, and hot peppers, among others. Some are still in the process of domestication (semi-domesticated), such as cacao (*Theobroma cacao*), yellow mombin (*Spondias mombin*); (and incipiently domesticated), such as Brazil nuts, piquiá, cupuaçu, among others. In this context, more detailed studies on the different aspects of these crops are necessary, including the development of conservation and management strategies. Considering multidisciplinary knowledge and modern techniques for obtaining and interpreting data, the need for research aimed at understanding the factors involved in the evolution of domesticated plants is increasingly relevant. In our understanding, there is still a lot to be said about the domestication of plants in the Neotropical region.

Key Points

- The center of domestication of a crop is found within the distribution of the wild populations of the same species. The center of domestication often has greater genetic diversity because of the presence of both wild and domesticated populations, but the regions where the crop shows maximum differences from its wild ancestor may be in regions to which it was dispersed and where it diversified.

- Amazonia is a major center of crop domestication, with at least 83 native species containing populations showing different degrees of domestication. Lowland South America, including Amazonia, contains at least 600 species with populations showing some degree of domestication.
- Landscapes started to be domesticated before their plant populations, by management of selected plants whose abundance was enhanced before morphological changes (the domestication syndrome) were evident.
- Also, in Amazonia, forests were domesticated before gardens expanded into swiddens, and plant populations were domesticated both in forests and in gardens and swiddens.
- Different research areas have provided important evidence on the origins of plant domestication in the Neotropics, such as archaeology, archaeobotany, paleobiolinguistics, genetics and genomics, and ethnobotany, among others.

Questions

- What is a domestication syndrome? Provide examples of domestication syndromes in two crops domesticated in the Neotropics.
- What does landscape domestication mean?
- What does 'the center of plant domestication' mean?
- Where were maize, tomato and peach palm domesticated?
- Why are the research areas of archaeology and genetics important to understand the origins of plant domestication?

References

Allem AC (1994) The origin of *Manihot esculenta* Crantz (Euphorbiaceae). Genet Resour Crop Evol 41:133–150. https://doi.org/10.1007/BF00051630

Alves-Pereira A, Clement CR, Picanço-Rodrigues D et al (2018) Patterns of nuclear and chloroplast genetic diversity and structure of manioc along major Brazilian Amazonian rivers. Ann Bot 121:625–639. https://doi.org/10.1093/aob/mcx190

Alves-Pereira A, Zucchi MI, Clement CR et al (2022) Selective signatures and high genome-wide diversity in traditional Brazilian manioc (*Manihot esculenta* Crantz) varieties. Sci Rep 12:1268. https://doi.org/10.1038/s41598-022-05160-8

Ames O (1939) Economic annuals and human cultures. Botanical Museum of Harvard University, Cambridge

Aragón-Martínez M, Serrato-Díaz A, Rocha-Munive MG et al (2023) Traditional management of maize in the Sierra Sur, Oaxaca, maintains moderate levels of genetic diversity and low population differentiation among landraces. Econ Bot 77:282–304. https://doi.org/10.1007/s12231-023-09580-z

Archila S, Giovannetti M, Lema V (2008) Arqueobotánica y teoría arqueológica: discusiones desde Suramérica. Universidad de Los Andes, Facultad de Ciencias Sociales, Departamento de Antropología, CESO, Ediciones Uniandes, p 282

Bacci MB, Valdati J, Pimenta LHF (2022) Pessoas, plantas e territórios: Relações entre etnobotânica e fitogeografia da Mata Atlântica no entorno do parque estadual da Serra do Tabuleiro, SC. Geosul 37:141–165. https://doi.org/10.5007/2177-5230.2022.e83985

Baker KF, Collins JL (1939) Notes on the distribution and ecology of Ananas and Pseudananas. Am J Bot 26:697–702. https://doi.org/10.2307/2437017

Barrera-Redondo J, de la Vega GS, Aguirre-Liguori JA et al (2021) The domestication of *Cucurbita argyrosperma* as revealed by the genome of its wild relative. Hortic Res 8:109. https://doi.org/10.1038/s41438-021-00544-9

Beadle GW (1972) Mystery of maize. Field Mus Nat Hist Bull 43:9–11. https://core.tdar.org/document/111371/mystery-of-maize

Beadle GW (1980) The ancestry of corn. Sci Am 242:112–119. https://doi.org/10.1038/scientificamerican0180-112

Bergougnoux V (2014) The history of tomato: from domestication to biopharming. Biotechnol Adv 32:170–189. https://doi.org/10.1016/j.biotechadv.2013.11.003

Bertoni MS (2019) Contributions à l'étude Botanique des Plantes Cultivées. Ex Sylvis, Puerto Bertoni, Alto Parana.

Bitocchi E, Bellucci E, Giardini A et al (2013) Molecular analysis of the parallel domestication of the common bean (*Phaseolus vulgaris*) in Mesoamerica and the Andes. New Phytol 197:300–313. https://doi.org/10.1111/j.1469-8137.2012.04377.x

Bitocchi E, Rau D, Bellucci E et al (2017) Beans (*Phaseolus* spp.) as a model for understanding crop evolution. Front Plant Sci 8:722. https://doi.org/10.3389/fpls.2017.00722

Blanca J, Cañizares J, Cordero L et al (2012) Variation revealed by SNP genotyping and morphology provides insight into the origin of the tomato. PLoS One 7:e48198. https://doi.org/10.1371/journal.pone.0048198

Blanca J, Montero-Pau J, Sauvage C et al (2015) Genomic variation in tomato, from wild ancestors to contemporary breeding accessions. BMC Genomics 16:257. https://doi.org/10.1186/s12864-015-1444-1

Bogaard A, Allaby R, Arbuckle BS et al (2021) Reconsidering domestication from a process archaeology perspective. World Archaeol 53:56–77. https://doi.org/10.1080/00438243.2021.1954990

Bommert P, Nagasawa NS, Jackson D (2013) Quantitative variation in maize kernel row number is controlled by the FASCIATED EAR2 locus. Nat Genet 45:334–337. https://doi.org/10.1038/ng.2534

Brown CH (2006) Glottochronology and the chronology of maize in the Americas. In: Staller JE, Tykot RH, Benz BF (eds) Histories of maize: multidisciplinary approaches to the prehistory, biogeography, domestication, and evolutions of maize. Elsevier, San Diego, pp 648–663. https://doi.org/10.1016/B978-012369364-8/50299-0

Brown C, Clement C, Epps P et al (2013a) The paleobiolinguistics of domesticated manioc (*Manihot esculenta*). Ethnobiol Lett 4:61–70. https://doi.org/10.14237/ebl.4.2013.5

Brown C, Clement C, Epps P et al (2013b) The paleobiolinguistics of domesticated chili pepper (*Capsicum* spp.). Ethnobiol Lett 4. https://doi.org/10.14237/ebl.4.2013.2

Brown CH, Luedeling E, Wichmann S et al (2013c) The paleobiolinguistics of domesticated squash (*Cucurbita* spp.). In: Quinlan M, Lepofsky D (eds) Explorations in ethnobiology: the legacy of Amadeo Rea. Society of Ethnobiology, Boston, pp 132–161

Brown CH, Clement CR, Epps P et al (2014) The paleobiolinguistics of maize (*Zea mays* L.). Ethnobiol 5:52–64. https://doi.org/10.14237/ebl.5.2014.130

Brugger SO, Gobet E, van Leeuwen JFN et al (2016) Longterm man-environment interactions in the Bolivian Amazon: 8000 years of vegetation dynamics. Quat Sci Rev 132:114–128. https://doi.org/10.1016/j.quascirev.2015.11.001

Bush MB, Piperno DR, Colinvaux PA (1989) A 6,000 year history of Amazonian maize cultivation. Nature 340:303–305. https://doi.org/10.1038/340303a0

Casas A, del Vfizquez MC, Viveros JL et al (1996) Plant management among the Nahua and the Mixtec in the Balsas River Basin, Mexico: an ethnobotanical approach to the study of plant domestication. Hum Ecol 24:455–478. https://doi.org/10.1007/BF02168862

Casas A, Blancas J, Lira R (2016) Mexican ethnobotany: interactions of people and plants in Mesoamerica. In: Lira R, Casas A, Blancas J (eds) Ethnobotany of Mexico. Ethnobiology. Springer, New York. https://doi.org/10.1007/978-1-4614-6669-7_1

Casas A, Ladio AH, Clement CR (2019) Editorial: ecology and evolution of plants under domestication in the Neotropics. Front Ecol Evol 7:231. https://doi.org/10.3389/fevo.2019.00231

Castillo NCR, Wu X, Chacón MI et al (2021) Genetic diversity of purple passion fruit, *Passiflora edulis* f. *edulis*, based on single-nucleotide polymorphism markers discovered through genotyping by sequencing. Diversity 13:144. https://doi.org/10.3390/d13040144

Chen LY, VanBuren R, Paris M et al (2019) The bracteatus pineapple genome and domestication of clonally propagated crops. Nat Genet 51:1549–1558. https://doi.org/10.1038/s41588-019-0506-8

Chomicki G, Schaefer H, Renner SS (2020) Origin and domestication of Cucurbitaceae crops: insights from phylogenies, genomics and archaeology. New Phytol 226:1240–1255. https://doi.org/10.1111/nph.16015

Clement CR (1999) 1492 and the loss of Amazonian crop genetic resources. I. The relation between domestication and human population decline. Econ Bot 53:188–202. https://doi.org/10.1007/BF02866498

Clement CR (2024) Agrobiodiversity in Amazonia. In: Scheiner SM, Encyclopedia of biodiversity, vol 4, 3rd edn. Elsevier, Inc, Oxford, pp 228–238. https://doi.org/10.1016/B978-0-12-822562-2.00170-5

Clement CR, Cristo-Araújo M, d'Eeckenbrugge GC et al (2010) Origin and domestication of native Amazonian crops. Diversity 2:72–106. https://doi.org/10.3390/d2010072

Clement CR, Borém A, Lopes MTG (2012) From plant domestication to breeding. In: Borém A, Lopes MTG, Clement CR et al (eds) Domestication and breeding: Amazonian species. Editora da Univ. Fed, Viçosa, pp 11–37

Clement CR, Denevan WM, Heckenberger MJ et al (2015) The domestication of Amazonia before European conquest. Proc R Soc B282:20150813. https://doi.org/10.1098/rspb.2015.0813

Clement CR, Picanço-Rodrigues D, Alves-Pereira A et al (2016) Crop domestication in the upper Madeira River basin. Bol Mus Para Emílio Goeldi, sér. Ciências Humanas 11:193–205. https://doi.org/10.1590/1981.81222016000100010

Clement CR, Cristo-Araújo M, d'Eeckenbrugge GC et al (2017) Origin and dispersal of domesticated peach palm. Front Ecol Evol 5:148. https://doi.org/10.3389/fevo.2017.00148

Clement CR, Casas A, Parra-Rondinel FA et al (2021a) Disentangling domestication from food production systems in the Neotropics. Quat 4:4. https://doi.org/10.3390/quat4010004

Clement CR, Levis C, Cabral de Oliveira J et al (2021b) Naturalness is in the eye of the beholder. Front For Glob Change 4:800294. https://doi.org/10.3389/ffgc.2021.800294

CONABIO (2016) La milpa. Comisión Nacional para el Conocimiento y Uso de la Biodiversidad, Cd. de México México https://www.biodiversidad.gob.mx/diversidad/sistemas-productivos/milpa. Accessed at 05/02/2024

Coppens d'Eechenbrugge G, Duval M-F, Leal F (2018) The pineapple success story: from domestication to pantropical diffusion. In: Ming R (ed) Genetics and genomics of pineapple. Plant genetics and genomics: crops and models, vol 22. Springer, Cham. https://doi.org/10.1007/978-3-030-00614-3_1

Coppens d'Eeckenbrugge G, Duval MF (2009) The domestication of pineapple: context and hypotheses. Pnappl News 16:15–27

Costa FM, Almeida Silva NC, Ogliari J (2017) Maize diversity in southern Brazil: indication of a microcenter of *Zea mays* L. Genet Resour Crop Evol 64:681–700. https://doi.org/10.1007/s10722-016-0391-2

Costa FM, Silva NCA, Vidal R, Clement CR, Freitas FO, Alves-Pereira A, Petroli CD, Zucchi MI, Veasey EA (2022) Maize dispersal patterns associated with different types of endosperm and migration of indigenous groups in lowland South America. Ann Bot 129:737-751. https://doi.org/10.1093/aob/mcac049

Costa FM, Silva NCA, Vidal R et al (2021) Entrelaçado, a rare maize race conserved in Southwestern Amazonia. Genet Resour Crop Evol 68:51–58. https://doi.org/10.1007/s10722-020-01008-0

Costa FM, Almeida Silva NC, Vidal R et al (2023) A new methodological approach to detect microcenters and regions of maize genetic diversity in different areas of Lowland South America. Econ Bot 77:345–371. https://doi.org/10.1007/s12231-023-09588-5

Cotias-de-Oliveira ALP, Assis JGA, Bellintani MC et al (2000) Chromosome numbers in Bromeliaceae. Gen Mol Biol 23:173–177. https://doi.org/10.1590/S1415-47572000000100032
Coppens d'Eeckenbrugge G, Leal F (2003) Morphology, anatomy and taxonomy. In: Bartholomew DP, Paul RE, Rohrbach KG (eds) The pineapple: botany, production and uses. CABI, Wallingford, Oxford, pp 13–32.
Crestani M, Barbieri RL, Hawerroth FJ et al (2010) Das Américas para o Mundo - origem, domesticação e dispersão do abacaxizeiro. Cienc Rural 40:1473–1483. https://doi.org/10.1590/S0103-84782010000600040
Darwin C (1859) On the origin of species by means of natural selection, or the preservation of the favoured races in the struggle for life. John Murray, London
De Candolle A (1883) Origine des plantes cultivées. Librairie Germer Baillière, Paris
Denham T, Donohue M (2022) Mapping the middle ground between foragers and farmers. J Anthropol Archaeol 65:101390. https://doi.org/10.1016/j.jaa.2021.101390
Dequigiovanni G (2017) Genetic structure, mating system and domestication of annatto (*Bixa orellana* L.) using molecular markers. PhD Thesis, Luiz de Queiroz College of Agriculture, University of São Paulo, Piracicaba. https://doi.org/10.11606/T.11.2018.tde-26102017-142451
Doebley F (2004) The genetics of maize evolution. Annu Rev Genet 38:37–59. https://doi.org/10.1146/annurev.genet.38.072902.092425
Doebley JF, Stec A, Wendel J et al (1990) Genetic and morphological analysis of a maize-teosinte F2 population: implications for the origin of maize. Proc Natl Acad Sci USA 87:9888–9892. https://doi.org/10.1073/pnas.87.24.9888
Doebley JF, Stec A, Hubbard L (1997) The evolution of apical dominance in maize. Nature 386:485–488. https://doi.org/10.1038/386485a0
Doebley JF, Gaut BS, Smith BD (2006) The molecular genetics of crop domestication. Cell 127:1309–1321. https://doi.org/10.1016/j.cell.2006.12.006
Doganlar S, Frary A, Tanksley SD (2000) The genetic basis of seed-weight variation: tomato as a model system. Theor Appl Genet 100:1267–1273. https://doi.org/10.1007/s001220051433
Dorweiler JE, Doebley JF (1997) Developmental analysis of teosinte glume architecture 1: a key locus in the evolution of maize (Poaceae). Am J Bot 84:1313–1322. https://doi.org/10.2307/2446130
Fávero AP, Valls JFM (2009) Domesticação e melhoramento de amendoim. In: Borem A, Lopes MTG, Clement CR (eds) Domesticação e melhoramento: espécies amazônicas. Universidade Federal de Viçosa, Viçosa
Ferreira MJ, Levis C, Chaves L et al (2022) Indigenous and traditional management creates and maintains the diversity of ecosystems of South American tropical savannas. Front Environ Sci 10:809404. https://doi.org/10.3389/fenvs.2022.809404
Ferrer MM, Montañez-Escalante PI, Ruenes-Morales MR et al (2020) Growing out of the tropical forests: domestication syndrome of native Mesoamerican trees in Mayan homegardens. Genet Resour Crop Evol 67:587–604. https://doi.org/10.1007/s10722-019-00833-2
Francisconi AF, Alves RP, Clement CR et al (2021) Genetic structure and diversity identify incipient domestication of Piquiá [*Caryocar villosum* (Aubl.) pers.] along the lower Tapajós River, Brazilian Amazonia. Genet Resour Crop Evol 68:1487–1501. https://doi.org/10.1007/s10722-020-01078-0
Fukunaga K, Hill J, Vigouroux Y et al (2005) Genetic diversity and population structure of teosinte. Genetics 169:2241–2254. https://doi.org/10.1534/genetics.104.031393
Fuller DQ, Denham T, Arroyo-Kalin M et al (2014) Convergent evolution and parallelism in plant domestication revealed by an expanding archaeological record. Proc Natl Acad Sci USA 111:6147–6152. https://doi.org/10.1073/pnas.1308937110
Gepts P, Papa R (2002) Evolution during domestication. In: Encyclopedia of Life Sciences. Nature Publishing Group, London. https://doi.org/10.1038/npg.els.0003071
Grabiele M, Chalup L, Robledo G et al (2012) Genetic and geographic origin of domesticated peanut as evidenced by 5S rDNA and chloroplast DNA sequences. Plant Syst Evol 298:1151–1165. https://doi.org/10.1007/s00606-012-0627-3

Graeber D, Wengrow D (2021) The dawn of everything: a new history of humanity. Penguin, London
Hake S, Ross-Ibarra J (2015) Genetic, evolutionary and plant breeding insights from the domestication of maize. elife 4:e05861. https://doi.org/10.7554/eLife.05861
Harlan JR (1971) Agricultural origins: centers and noncenters. Science 174:468–474. https://doi.org/10.1126/science.174.4008.468
Harlan JR (1992) Crops and man. Crop Science Society of America, Madison
Harlan JR, de Wet JMJ (1971) Toward a rational classification of cultivated plants. Taxon 20:509–517. https://doi.org/10.2307/1218252
van Heerwaarden J, Doebley J, Briggs WH (2011) Genetic signals of origin, spread, and introgression in a large sample of maize landraces. Proc Natl Acad Sci USA 108:1088–1092. https://doi.org/10.1073/pnas.1013011108
Hilbert L, Neves EG, Pugliese F et al (2017) Evidence for mid-Holocene rice domestication in the Americas. Nat Ecol Evol 1:1693–1698. https://doi.org/10.1038/s41559-017-0322-4
Hillman GC, Davies MS (1990) Domestication rates in a wild-type wheats and bartley under primitive cultivation. Biol J Linn Soc Lond 39:39–78. https://doi.org/10.1111/j.1095-8312.1990.tb01611.x
Iltis HH (1983) From teosinte to maize: the catastrophic sexual transmutation. Science 222:886–894. https://doi.org/10.1126/science.222.4626.886
Iriarte J, Smith R, Gregorio de Souza J et al (2017) Out of Amazonia: Late-Holocene climate change and the Tupi-Guarani trans-continental expansion. The Holocene 27:967–975. https://doi.org/10.1177/0959683616678461
Iriarte J, Elliott S, Maezumi SY et al (2020) The origins of Amazonian landscapes: plant cultivation, domestication and the spread of food production in tropical South America. Quat Sci Rev 248:106582. https://doi.org/10.1016/j.quascirev.2020.106582
Janzen GM, Wang L, Hufford MB (2019) The extent of adaptive wild introgression in crops. New Phytol 221:1279–1288. https://doi.org/10.1111/nph.15457
Junqueira AB, Shepard GH Jr, Clement CR (2011) Secondary forests on anthropogenic soils of the Middle Madeira River: valuation, local knowledge, and landscape domestication in Brazilian Amazonia. Econ Bot 65:85–99. https://doi.org/10.1007/s12231-010-9138-8
Kater T, Almeida FO, Mongeló GZ et al (2020) Variabilidade estratigráfica e espacial dos contextos cerâmicos no Sítio Teotônio. Rev Arqueol 33:198–220. https://doi.org/10.24885/sab.v33i1.685
Kates HR, Soltis PS, Soltis DE (2017) Evolutionary and domestication history of Cucurbita (pumpkin and squash) species inferred from 44 nuclear loci. Mol Phylogenet Evol 111:98–109. https://doi.org/10.1016/j.ympev.2017.03.002
Kerr WE, Clement CR (1980) Práticas agrícolas de conseqüências genéticas que possibilitaram aos índios da Amazônia uma melhor adaptação às condições ecológicas da região. Acta Amazon 10:251–261. https://doi.org/10.1590/1809-43921980102251
Kistler LM, Maezumi SY, de Souza JG et al (2018) Multi-proxy evidence highlights a complex evolutionary legacy of maize in South America. Science 362:1309–1313. https://doi.org/10.1126/science.aav0207
Kistler L, Thakar HB, VanDerwarker AM et al (2020) Archaeological Central American maize genomes suggest ancient gene flow from South America. PNAS 117:33124–33129. https://doi.org/10.1073/pnas.2015560117
Kraft KH, Brown CH, Nabhan GP et al (2014) Multiple lines of evidence for the origin of domesticated chili pepper, *Capsicum annuum*, in Mexico. Proc Natl Acad Sci USA 111:6165–6170. https://doi.org/10.1073/pnas.1308933111
Krapovickas A (1996) Agricultura indígena en las llanuras de la Cuenca del Plata. Acad Nac de Agron y Vet 49:18–26
Krapovickas A, Gregory WC (1994) Taxonomía del género *Arachis (Leguminosae)*. Bonplandia, pp 81–186. https://doi.org/10.30972/bon.81-43559
Ladizinsky G (1998) Plant evolution under domestication. Springer, Dordrecht, p 254. https://doi.org/10.1007/978-94-011-4429-2_2

Lamarca EV, Oliveira-Júnior CJFD, Barbedo CJ (2020) Etnobotânica na conservação de espécies com sementes sensíveis à dessecação: o exemplo da *Eugenia brasiliensis* Lam. Hoehnea 47:e372019. https://doi.org/10.1590/2236-8906-37/2019

Larranaga N, Albrtazzi FJ, Fontecha G et al (2017) A Mesoamerican origin of cherimoya (*Annona cherimola* Mill.): implications for the conservation of plant genetic resources. Mol Ecol 26:4116–4130. https://doi.org/10.1111/mec.14157

Larson G, Piperno DR, Allaby RG et al (2014) Current perspectives and the future of domestication studies. PNAS 111:6139–6146. https://doi.org/10.1073/pnas.1323964111

Léotard G, Duputié A, Kjellberg F et al (2009) Phylogeography and the origin of cassava: new insights from the northern rim of the Amazonian basin. Mol Phylogenet Evol 53:329–334. https://doi.org/10.1016/j.ympev.2009.05.003

Levis C, de Souza PF, Schietti J et al (2012) Historical human footprint on modern tree species composition in the Purus-Madeira interfluve, Central Amazonia. PLoS One 7:e48559. https://doi.org/10.1371/journal.pone.0048559

Levis C, Costa FRC, Bongers F et al (2017) Persistent effects of pre-Columbian plant domestication on Amazonian forest composition. Science 355:925–931. https://doi.org/10.1126/science.aal0157

Levis C, Flores BM, Moreira PA et al (2018) How people domesticated Amazonian forests. Front Ecol Evol 5:171. https://doi.org/10.3389/fevo.2017.00171

Lins Neto EMF, Peroni N, Casas A et al (2014) Brazilian and Mexican experiences in the study of incipient domestication. J Ethnobiol Ethnomed 10:33. https://doi.org/10.1186/1746-4269-10-33

Lombardo U, Iriarte J, Hilbert L et al (2020) Early Holocene crop cultivation and landscape modification in Amazonia. Nature 581:190–193. https://doi.org/10.1038/s41586-020-2162-7

Lopez-Moreno H, Basurto-Garduño AC, Torres-Meraz MA et al (2023) Genetic analysis and QTL mapping of domestication-related traits in chili pepper (*Capsicum annuum* L.). Front Ecol Evol 7:231. https://doi.org/10.3389/fgene.2023.1101401

Maezumi SY, Alves D, Robinson M et al (2018) The legacy of 4,500 years of polyculture agroforestry in the eastern Amazon. Nat Plants 4:540–547. https://doi.org/10.1038/s41477-018-0205-y

Martin GJ (1995) Ethnobotany, conservation and community development. In: Ethnobotany. Springer, Boston. https://doi.org/10.1007/978-1-4615-2496-0_8

Matsuoka Y, Vigouroux Y, Goodman MM et al (2002) A single domestication for maize shown by multilocus microsatellite genotyping. Proc Natl Acad Sci USA 99:6080–6084. https://doi.org/10.1073/pnas.052125199

Mellanby K, Crawford GW, Gray AW et al (2024) How agriculture and domestication began. In: Origins of agriculture. Encyclopaedia Britannica. Available at: https://www.britannica.com/topic/agriculture/How-agriculture-and-domestication-began

Meyer RS, DuVal AE, Jensen HR (2012) Patterns and processes in crop domestication: an historical review and quantitative analysis of 203 global food crops. New Phytol 196:29–48. https://doi.org/10.1111/j.1469-8137.2012.04253.x

Moreira PA, Lins J, Dequigiovanni G et al (2015) The domestication of annatto (*Bixa orellana*) from *Bixa urucurana* in Amazonia. Econ Bot 69:127–135. https://doi.org/10.1007/s12231-015-9304-0

Moretzsohn MC, Gouvea EG, Inglis PW et al (2013) A study of the relationships of cultivated peanut (*Arachis hypogaea*) and its most closely related wild species using intron sequences and microsatellite markers. Ann Bot 111:113–126. https://doi.org/10.1093/aob/mcs237

Morrone JJ (2018) Evolutionary biogeography of the Andean region. CRC Press, Taylor and Francis Group, Boca Raton

Muños S, Ranc N, Botton E et al (2011) Increase in tomato locule number is controlled by two single-nucleotide polymorphisms located near WUSCHEL. Plant Physiol 156:2244–2254. https://doi.org/10.1104/pp.111.173997

Naithani S (2021) History and science of cultivated plants. Pressbooks, Corvallis

Nelson G, Ladiges PY (2001) Gondwana, vicariance biogeography and the New York School revisited. Aust J Bot 49:389–409. https://doi.org/10.1071/BT00025

Olsen KM (2004) SNPs, SSRs and inferences on cassava's origin. Plant Mol Biol 56:517–526. https://doi.org/10.1007/s11103-004-5043-9

Olsen KM, Schaal BA (1999) Evidence on the origin of cassava: phylogeography of *Manihot esculenta*. Proc Natl Acad Sci USA 96:5586–5591. https://doi.org/10.1073/pnas.96.10.5586

Olsen KM, Schaal BA (2001) Microsatellite variation in cassava (*Manihot esculenta*, Euphorbiaceae) and its wild relatives: further evidence for a Southern Amazonian origin of domestication. Am J Bot 88:131–142. https://doi.org/10.2307/2657133

Olsen KO, Wendel JF (2013) Crop plants as models for understanding plant adaptation and diversification. Front Plant Sci 4:290. https://doi.org/10.3389/fpls.2013.00290

Pagan-Jimenez JR (2015) Almidones: guia de material comparativo moderno del Ecuador para los estudios paleoetnobotanicos en el Neotropico, vol 1. Instituto Nacional de Patrimonio Cultural, Quito

Paris HS, Lebeda A, Kristkova E et al (2012) Parallel evolution under domestication and phenotypic differentiation of the cultivated subspecies of *Cucurbita pepo* (Cucurbitaceae). Econ Bot 66:71–90. http://www.jstor.org/stable/41493901

Pärssinen M, Schaan D, Ranzi A (2009) Pre-Columbian geometric earthworks in the upper Purús: a complex society in western Amazonia. Antiquity 83:1084–1095. https://doi.org/10.1017/S0003598X00099373

Pärssinen M, Ferreira E, Virtanen PK et al (2021) Domestication in motion: macrofossils of pre-colonial Brazilian nuts, palms and other Amazonian planted tree species found in the upper Purus. Environ Archaeol 26:309–322. https://doi.org/10.1080/14614103.2020.1765295

Patiño VM (2002) Historia y dispersión de los frutales nativos del neotrópico. Centro Internacional de Agricultura Tropical, Cali

Pearsall DM (2015) Paleoethnobotany: a handbook of procedures, 3rd edn. Routledge, London/New York

Pedrosa HC, Clement CR, Schietti J (2018) The domestication of the Amazon tree grape (*Pourouma cecropiifolia*) under an ecological lens. Front Plant Sci 9:203. https://doi.org/10.3389/fpls.2018.00203

Peralta IE, Spooner DM, Knapp S (2008) Taxonomy of wild tomatoes and their relatives (*Solanum* sect. *Lycopersicoides*, sect. *Juglandifolia*, sect. *Lycopersicon*; Solanaceae). Syst Bot Monogr 84:186. https://doi.org/10.2307/25027972

Peripato V, Levis C, Moreira GA et al (2023) More than 10,000 pre-Columbian earthworks are still hidden throughout Amazonia. Science 382:103–109. https://doi.org/10.1126/science.ade2541

Pickersgill B (2007) Domestication of plants in the Americas: insights from Mendelian and molecular genetics. Ann Bot 100:925–940

Piperno DR (2006) Phytoliths: a comprehensive guide for archaeologists and paleoecologists. Altimira Press, Oxford

Piperno DR (2011) The origins of plant cultivation and domestication in the New World tropics: patterns, process, and new developments. Curr Anthropol 52:453–470. https://doi.org/10.1086/659998

Piperno DH, Clary KH, Cooke RG et al (1985) Preceramic maize in Central Panama: phytolith and pollen evidence. Am Anthropol 87:871–878. https://repository.si.edu/handle/10088/18804

Piperno DR, Ranere AJ, Holst I et al (2009) Starch grain and phytolith evidence for early ninth millennium b.p. maize from the Central Balsas River Valley, Mexico. Proc Natl Acad Sci USA 106:5019–5024. https://doi.org/10.1073/pnas.0812525106

Purugganan MD, Fuller DQ (2009) The nature of selection during plant domestication. Nature 457:843–848. https://doi.org/10.1038/nature07895

Ramallo V, Bisso-Machado R, Bravi C et al (2013) Demographic expansions in South America: enlightening a complex scenario with genetic and linguistic data. Am J Phys Anthropol 150:453–463. https://doi.org/10.1002/ajpa.22219

Razifard H, Ramos A, Della Valle AL et al (2020) Genomic evidence for complex domestication history of the cultivated tomato in Latin America. Mol Biol Evol 37:1118–1132. https://doi.org/10.1093/molbev/msz297

Roosevelt AC, Lima Da Costa M, Lopes Machado C et al (1996) Paleoindian cave dwellers in the Amazon: the peopling of the Americas. Science 272:373–384. https://doi.org/10.1126/science.272.5260.373

Saavedra TM, Figueroa GA, Cauih JGD (2017) Origin and evolution of tomato production *Lycopersicon esculentum* in México. Ciênc Rural 47:3. https://doi.org/10.1590/0103-8478cr20160526

Sanjur OI, Piperno DR, Andres CT et al (2002) Phylogenetic relationships among domesticated and wild species of Cucurbita (Cucurbitaceae) inferred from a mitochondrial gene: implications for crop plant evolution and areas of origin. Proc Natl Acad Sci USA 99:10923–10928. https://doi.org/10.1073/pnas.012577299

Scoville H (2020) What is selective sweep? ThoughtCo. https://www.thoughtco.com/what-is-selective-sweep-1224718

Shepard GH Jr, Ramirez H (2011) "Made in Brazil": human dispersal of the Brazil nut (*Bertholletia excelsa*, Lecythidaceae) in Ancient Amazonia. Econ Bot 65:44–65. https://doi.org/10.1007/s12231-011-9151-6

Shepard JRGH, Clement CR, Lima HP et al (2020) Ancient and traditional agriculture in South America: Tropical lowlands. In: Hazlett R (ed) Encyclopedia of agriculture and the environment. Oxford University Press, New York, pp 1–48. https://doi.org/10.1093/acrefore/9780199389414.013.597

Shi J, Lai J (2015) Patterns of genomic changes with crop domestication and breeding. Curr Opin Plant Biol 24:47–53. https://doi.org/10.1016/j.pbi.2015.01.008

Silva MAC, Nunes K, Lemes R et al (2020) Genomic insight into the origins and dispersal of the Brazilian coastal natives. Proc Natl Acad Sci USA 117:2372–2377. https://doi.org/10.1073/pnas.1909075117

Sousa Júnior JR, Collevatti RG, Lins Neto EMF et al (2018) Traditional management affects the phenotypic diversity of fruits with economic and cultural importance in the Brazilian Savanna. Agrofor Syst 92:11–21. https://doi.org/10.1007/s10457-016-0005-1

Studer A, Doebley JF (2012) Evidence for a natural allelic series at the maize domestication locus *teosinte branched1*. Genetics 191:951–958. https://doi.org/10.1534/genetics.112.138479

Tanksley SD (2004) The genetic, developmental, and molecular bases of fruit size and shape variation in tomato. Plant Cell 16:S181–S189. https://doi.org/10.1105/tpc.018119

Thomas E, Alcázar Caicedo C, Mcmichael CH et al (2015) Uncovering spatial patterns in the natural and human history of Brazil nut (*Bertholletia excelsa*) across the Amazon Basin. J Biogeogr 42:1367–1382. https://doi.org/10.1111/jbi.12540

Tsiantis M (2011) A transposon in *tb1* drove maize domestication. Nat Genet 43:1048–1050. https://doi.org/10.1038/ng.986

Vavilov NI (1926) Studies on the origin of cultivated plants. Bull Appl Bot Plant Breed 16:139–248

Vavilov NI (1992) Origin and geography of cultivated plants. Cambridge University Press, New York

Vieira LN, Fraga HPF, Serbin GM et al (2021) Domesticação de plantas: um processo co-evolutivo. Genét esc 16:356–367. https://doi.org/10.55838/1980-3540.ge.2021.394

Vigouroux Y, Glaubitz J, Matsuoka Y et al (2008) Population structure and genetic diversity of new world maize races assessed by DNA microsatellites. Am J Bot 95:1240–1253. https://doi.org/10.3732/ajb.0800097

Walker RS, Ribeiro LA (2011) Bayesian phylogeography of the Arawak expansion in lowland South America. Proc Biol Sci 278:2562–2567. https://doi.org/10.1098/rspb.2010.2579

Walker RS, Wichmann S, Mailund T (2012) Cultural phylogenetics of the Tupi language family in lowland South America. PLoS One 7:e35025. https://doi.org/10.1371/journal.pone.0035025

Watling J, Shock MP, Mongelo GZ et al (2018) Direct archaeological evidence for Southwestern Amazonia as an early plant domestication and food production centre. PLoS One 13:e0199868. https://doi.org/10.1371/journal.pone.0199868

Watling J, Almeida F, Kater T et al (2020a) Archaeobotanical data from ceramic occupations of the Cachoeira do Teotônio. Bol Mus Para Emílio Goeldi 15:e20190075. https://doi.org/10.1590/2178-2547-bgoeldi-2019-0075

Watling J, Castro MT, Simon MF et al (2020b) Phytoliths from native plants and surface soils from the Upper Madeira river, SW Amazonia, and their potential for paleoecological reconstruction. Quat Int 550:85–110. https://doi.org/10.1016/j.quaint.2020.03.045

Wesolowski V, de Souza SMFM, Reinhard KJ et al (2010) Evaluating microfossil content of dental calculus from Brazilian sambaquis. J Archaeol Sci 37:1326–1338. https://doi.org/10.1016/j.jas.2009.12.037

White DM, Huang JP, Jara-Muñoz OA et al (2021) The origins of coca: museum genomics reveals multiple independent domestications from progenitor *Erythroxylum gracilipes*. Syst Biol 70:1–13. https://doi.org/10.1093/sysbio/syaa074

Yang N, Wang Y, Liu X et al (2023) Two teosintes made modern maize. Science 382(6674). https://doi.org/10.1126/science.adg8940

Zhang X, Zhang J, He X et al (2017) Genome-wide association study of major agronomic traits related to domestication in peanut. Front Plant Sci 8:1611. https://doi.org/10.3389/fpls.2017.01611

Zhu G, Wang S, Huang Z et al (2018) Rewiring of the fruit metabolome in tomato breeding. Cell 172:249–261.e12. https://doi.org/10.1016/j.cell.2017.12.019

Chapter 17
Strategies for Breeding Animal Populations in the Neotropics

Luciana C. A. Regitano, Concepta McManus, Felipe Pimentel, and Adriana M. G. Ibelli

Contents

L. C. A. Regitano (✉)
Embrapa Pecuária Sudeste, São Carlos, São Paulo, SP, Brazil

Universidade Federal de São Carlos, São Carlos, SP, Brazil
e-mail: luciana.regitano@embrapa.br

C. McManus
Universidade de Brasília (UNB), Instituto de Biologia, Brasília, DF, Brazil

F. Pimentel
Universidade Anhembi Morumbi, São Paulo, SP, Brazil

A. M. G. Ibelli
Embrapa Pecuária Sudeste, São Carlos, São Paulo, SP, Brazil

Universidade Estadual do Centro-Oeste, Guarapuava, Paraná, PR, Brazil

M. V. Bohrer Monteiro Siqueira et al. (eds.), *Population Genetics in the Neotropics*,
https://doi.org/10.1007/978-3-031-83685-5_17

Abstract In this chapter, we highlight that the biodiversity of animal genetic resources (AnGR) is essential for breeding, aiming food and agriculture in the Neotropics. Furthermore, we discuss how AnGR are paramount to facing climate change and the consequent disruptions to the food chain supply, such as increased disease outbreaks, that are introducing new challenges for animal production, a scenario where resilience is desired. In this context, we describe basic concepts, as well as methods and strategies currently used in animal breeding to improve productivity, product quality and novel and desirable phenotypes. We also provide an overview on how omics have been contributing with genetic selection in different livestock species.

Keywords Livestock · Genomic selection · Breeding programs · Genomics · Tropical adaptation

Learning Objectives

1. Understand the use of animal genetic resources (AnGR) in production systems in the Neotropics;
2. Types of information available for the conservation and breeding of AnGR;
3. How can different technologies be used to improve animal production in the Neotropics?

17.1 Introduction

Animal genetic resources (AnGR) for food and agriculture are an essential stock of biodiversity. They are defined as those animal species that are used or may be used for food and agriculture and the populations within each of them. During the colonisation of the Neotropics, settlers potentially brought farm animals from the Iberian Peninsula and West Africa. These funding populations have naturally adapted to low-input agricultural environments in these regions, creating locally adapted breeds. Furthermore, multiple waves of importation to the western hemisphere have occurred, including product-specialised breeds, mainly from the Northern Hemisphere, which lack adaptation to the tropical environment and thus, are highly dependent on environmental control practices, such as chemical protection from parasites and acclimation of production facilities.

The neotropical region is recognized as one of the most important biodiversity hotspots in the world, housing a significant portion of the planet's species diversity. Countries within this region often share similar environmental challenges, including deforestation, habitat fragmentation, climate change, and biodiversity loss. Efforts to conserve and sustainably manage the natural resources of the Neotropics are essential for the well-being of both local communities and the global ecosystem.

AnGR are fundamental to facing climate change and the consequent disruptions to food supplies. For instance, increased disease outbreaks introduce new challenges for animal production, a scenario where resilience is necessary. Resilience is defined by the Intergovernmental Panel on Climate Change (IPCC) as the ability of a system and its parts to anticipate, absorb, accommodate, or recover from the effects of a hazardous event in a timely and efficient manner.

Animal breeding is the controlled reproduction of selected domestic animals to improve desirable traits to suit human needs. Animal breeding involves knowledge of quantitative and molecular genetics, statistics, reproductive physiology, and computer science. As the main fuel for genetic selection is diversity, the balance between the exploitation of resources for food and agriculture and the conservation of ecosystems is crucial for achieving the Sustainable Development Goals (SDGs) of the United Nations—2030 Agenda, particularly SDG 2 (zero hunger), SDG113 (climate action) and 15 (life on land), which aim to enhance the delivery of ecosystem services for all types of environments.

Over the past thirty years, substantial efforts and progress have been made to conserve these resources (Paiva et al. 2016). Across continents and regions, national programs, nongovernmental organisations (NGOs) (e.g., breeding associations), and producer-based activities have been initiated for *in situ* and *ex situ* (gene banking) conservation. Exemplary efforts include developing *in situ* conservation of rare breeds across all regions and recognising locally adapted breeds by governments.

These resources include more than 7000 distinct local (reported in only one country) and 1000 transboundary (reported by several countries) livestock breeds of approximately 40 species found worldwide (FAO 2015). Some of those breeds can be considered locally adapted because they have been in a country for sufficient time to be genetically adapted to one or more of the traditional production systems in the country. In the case of the neotropical countries, most farm animal species arrived with colonisers around the year 1500, with new breeds introduced over time. Tropical environments pose unique challenges for animal production due to high temperatures, humidity, and disease pressure. In recent years, our understanding of global climate change has enhanced stress on genetic resources for food and agriculture. Raising temperatures and rainfall enable disease vectors and invasive species to extend their range, which can adversely affect the health of livestock and the livelihoods of those who depend on them. Over time, selection pressures have been imposed to increase productivity through animal breeding and reproductive technologies. However, this has often also reduced animal fitness, disease resistance, and tolerance to many of these changes. However, with careful selection and breeding, it has been possible to develop animal breeds that are better adapted to these conditions and thrive in tropical environments.

Livestock production also contributes to climate change. For instance, it is estimated that about 12% of global greenhouse gas emissions are attributed to the livestock sector via land use, land-use change, manure management, and enteric fermentation (FAO 2023). Meanwhile, farmers are under pressure to increase productivity to feed growing populations and reduce the environmental impact of this production. In this way, several studies have attempted to predict models, such as those related to impacts on poverty, methane emissions, and animal diseases. However, agricultural production systems are complex, as many livestock species were domesticated in the same region, and many of these breeds are now globally distributed and overlaid by various production systems. This is why some researchers believe that climate change will have a greater effect on developing countries, where smallholders and pastoralists mostly carry out production (Morton 2007). Among the climatic variables, heat stress is the main factor that negatively affects livestock production, reducing growth, reproductive performance, milk production and milk components (milk proteins, minerals, and fats). The incidence of endo- and ectoparasites is also a concern (Marchioretto et al. 2023; Romero et al. 2023).

Globally, both well-established and newly formed national gene banks operate and provide genetic security for commercial and rare breeds. Additionally, a more formal structure among nations leveraging their mutual interests in AnGR conservation through the direction given by FAO became prominent worldwide. Global drivers of population growth, income, and climate change already impact how AnGR are used. Increased animal productivity (unit of output/animal) should be a primary format for sustainably addressing food security needs. As the climate is expected to become highly dynamic and extreme, impacting food security, either new or existing genotypes of interest will have to be developed, requiring access and utilisation of a broad array of genetic variation. Beyond that, the ever-increasing amount of information on livestock genomes in the last decades and tools that can provide genomic knowledge to address these previously stated problems is juxtaposed with these challenges. While genotyping costs have greatly decreased over time, perhaps the more important driver is our ability to manage, manipulate, store and understand the big data being generated.

Indiscriminate crossbreeding and introduction of exotic breeds are two threats to maintaining adapted local breeds and genetic conservation. Important information for conservation programs in the Neotropics includes population size and structure, genetic diversity between and within breeds, spatial dissemination, and production environments. Population parameters can be calculated from pedigree and molecular data and can aid in decision-making at the farm, breed, and national levels.

Monitoring changing inbreeding rates and genetic diversity within a population are key parameters to guide for breeding programs as they have an impact on sustainable genetic gain and performance (i.e., inbreeding depression). Some parameters used in conservation programs, with data from both pedigrees and molecular markers, the latter being used to monitor pedigree errors, include:

17.1.1 Pedigree Completeness

The degree to which a family pedigree contains all available genetic linkages between family members. Completeness can be affected by factors such as missing or incomplete records, or limited access to historical data. It is essential for estimating inbreeding, avoiding inbreeding, determining the risk of extinction for a population and determining the proportion of ancestors known per parental generation (number of equivalents to discrete generations (t)) for each individual in a pedigree, which is the sum of $(1/2)^n$, where n is the number of generations separating the individual from each known ancestor (Boichard et al. 1997).

17.1.2 Generation Interval

Refers to the average age of the parents at birth of their offspring, used to measure turnover and genetic gain.

17.1.3 Inbreeding Coefficient

Represents the probability that two alleles at any locus are identical by descent, i.e., an individual will inherit two identical alleles, which are copies of one allele from the same ancestor. It depends on the average relatedness between mating pairs, which is low when an animal shares a low percentage of genes with the rest of the population. Various Wright's inbreeding coefficients (Wright 1965) (Fig. 17.1) can be calculated (F_{IT}, F_{ST}, and F_{IS}) to describe the partition of genetic variation within and among populations.

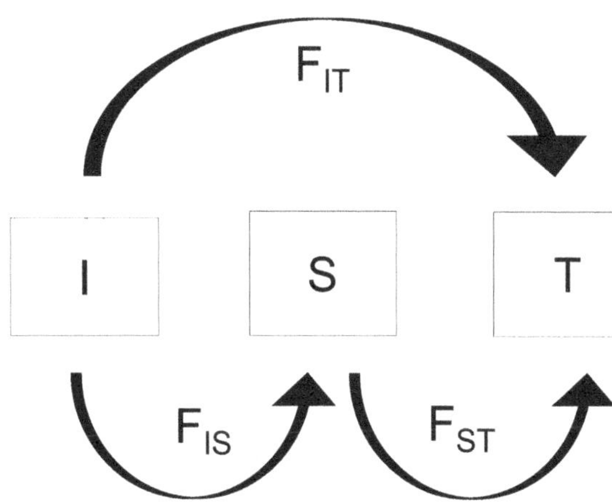

Fig. 17.1 Wright's Inbreeding coefficients and their use *I* Individual, *S* Subpopulation, *T* whole population

17.1.4 Effective Number of Founders (f$_e$)

The number of idealised individuals who, in a population or breeding programme, would contribute the same genetic diversity as the actual founders contributed or the number of founders required to produce the observed genetic diversity under the condition that all founders contributed equally to the genetic diversity of the population.

$$f_e = \frac{1}{\sum_1^N q_k^2}$$

where f_e is the effective number of founders, q_k is the genetic contribution of the kth founder to the reference population and N is the number of founders. A lower effective number of founders translates to reduced genetic diversity and potentially an increased risk of inbreeding depression and loss of genetic variation over time.

17.1.5 Effective Number of Ancestors (f$_a$)

The effective number of ancestors is the minimum number of ancestors, not necessarily founders, required to account for the genetic diversity observed in a reference population (Boichard et al. 1997), according to the expression:

$$f_a = \frac{1}{\sum_1^N \left(q_k\left(1-\sum_1^{(n-1)} a_i\right)\right)^2}$$

where f_a is the effective number of ancestors, q_k is the total genetic contribution of the kth ancestor to the reference population, a_i was the component of the genetic contribution of the kth ancestor that has already been accounted for by the contribution of the ith ancestor. The number of ancestors was denoted by N, and the number of ancestors related to another ancestor was denoted by n. A low number indicates overuse of a low number of sires (male reproducer).

17.1.6 Founder Genome Equivalent (F$_g$)

The number of founder genomes required to account for the total genetic diversity of the reference population without loss of founder alleles due to genetic drift (MacCluer et al. 1986; Lacy 1989).

$$F_g = \frac{1}{2\sum_1^{2f} f_k^2}$$

where f is the number of founder genes and f_k is the gene frequency of the kth gene.

A Genetic conservation index (GCI) was defined by Alderson et al. (1992):

$$GCI = \frac{1}{\sum p_i^2}$$

where P_i is the proportion of genes of the founder i in the pedigree of an animal and assumes that the conservation program wishes to maintain the maximum number of alleles possessed by the base population. This parameter is useful for the selection of an individual sire within an overall breed selection policy and in the formulation of group breeding programs. Problems arise with this parameter as there is no accounting for concentration of breeding to non-founder animals in subsequent generations and cannot be used without a pedigree.

17.1.7 Heterozygosity, a Measure of Genetic Variability

A locus is considered polymorphic when the most common allele has a frequency below 0.9. The heterozygosity of a marker is the probability of an individual being heterozygous at the marker locus and depends on the number of alleles and their frequency in the population (Nei 1978).

17.1.8 Effective Population Size (N_e)

Frequently calculated as $N_e = \frac{1}{2\Delta F_L}$, where N_e *is* is the effective population size, ΔF_y the yearly change in mean inbreeding within the population, and L is the generation interval in years. For N_e estimation, variation in allele frequency in a population is determined by its actual size.

It is important to highlight that the real number of individuals in a population is rarely its effective size; actual population size is the number of individuals who contribute uniformly to the gene pool; some individuals have a higher reproduction rate than others (they have greater reproductive performance or fitness); the distribution of males and females may result in difficulty for some individuals to mate or, due to inbreeding, reduce their individual capacity; the effective population size is a theoretical measure that compares the genetic behaviour of a population with the behaviour of an "ideal" population (Wright 1978). As the effective population size decreases, the probability of the allele frequency changing due to chance (drift) increases. In order to clarify parameters to help the understanding of species at risk, the FAO has classified "at-risk categories" (FAO 2022) of animal breeds depending on their reproductive capacity (Table 17.1).

Table 17.1 Population size thresholds for the assignment of breeds to current risk categories

	Species reproductive capacity			
	High[a]		Low[b]	
Risk category	Lower threshold	Upper threshold	Lower threshold	Upper threshold
Critical	0	≤100	0	≤300
Endangered	>100	≤1000	>300	≤3000
Vulnerable	>1000	≤2000	>3000	≤6000
Not at risk	>2000	∞	>6000	∞

[a]Cassowary, Chicken, Chilean Tinamou, Dog, Duck, Emu, Goose, Guinea Fowl, Guinea Pig, Nandu, Ostrich, Partridge, Peacock, Pheasant, Pig, Pigeon, Quail, Rabbit, Swallow, Turkey
[b]Alpaca, Ass, Bactrian camel, Buffalo, Cattle, Deer, Dromedary, Goat, Guanaco, Horse, Llama, Sheep, Vicuña, Yak

17.2 Breeding Strategies and Programs

In the context of farm animals in the Neotropics, various breeding strategies are employed to improve livestock productivity and adaptability to the local environment. Additionally, artificial insemination and embryo transfer technologies are also often used to facilitate the spread of superior genetics within livestock populations. Furthermore, conservation breeding programs for indigenous or creole livestock breeds are sometimes implemented to safeguard genetic diversity and preserve important cultural and agricultural resources. Sustainable breeding practices also promote animal welfare and minimise environmental impacts within farming systems. Where an individual farmer does not have a sufficient number of animals, community-based participatory breeding programs provide alternative solutions to maximise the conservation and use of this locally adapted breed while integrating contemporary animal breeding methods with popular knowledge to support the socioeconomic development of regional communities.

Breeding is by definition changing the genetic makeup of a population of animals, such as a breed or ecotype. The selection of breeding goals and the design of an effective breeding program are important. Breeding goals are dictated by market demand, nonetheless as goals can take many generations to achieve, it may not be possible to predict what consumers will prefer several years in advance. For example, consumers today want leaner and more tender meat than they did, for example, in the 1950s. This also means changes in the use of technologies to evaluate the traits, such as ultrasound to determine the fat and lean content of live animals, hastening changes in carcass quality to meet consumer demands.

Breeding methods depend on heritability and genetic correlations for desirable traits. The heritability (see Chap. 8) is the proportion of genetic variation (V_G) relative to phenotypic variation (V_P) and can be calculated as follows:

$$H^2 = \frac{V_G}{V_P} \text{ or } h^2 = \frac{V_A}{V_P}.$$

where H^2 is broad-sense heritability, V_G is genetic variability, V_P is phenotypic variation, h^2 is narrow-sense heritability, and V_A is additive genetic variation.

$$V_P = V_G + V_E$$

$$V_G = V_A + V_D\left(+V_I\right),$$

where V_D is the dominance variance, and V_I is the interaction between V_D and V_I. In general, $h^2 < H^2$.

For selection purposes, only narrow-sense heritability is considered, as only the additive effects of alleles are passed from one generation to the other, while dominant and epistatic effects are unpredictable as they depend upon the diploid combination of gametes. Heritability applies to a specific trait measured within a particular population of animals at a specific point in time. This approach is essential because there can be no genetic change in the population if the phenotypic variation is not partially explained by genetic variation. Genetic change can be rapid if heritability is high, and simple means of selection, like selection based solely on the individual's phenotype, are effective. Using a scale from 0 to 1, a heritability of 0.45 means that 45% of the total variance in a trait is controlled by additive gene action or, in other words, the variability in the trait in a population is largely due to additive genetic differences among animals. Single traits with high heritabilities (>0.40) can be used in effective breeding programs. It is essential to keep in mind that heritability is a population parameter, and thus cannot be generalised among different populations, but in general traits related to fertility or disease resistance consistently have low heritabilities (<0.15); therefore, most of the variation is due to the environment. Examples include days to first breeding, number of inseminations to conceive and mastitis incidence. Traits related to production, such as milk yield, have intermediate heritabilities. Other examples include daily weight gain and the number of eggs chickens lay. Traits such as stature and milk quality traits, such as fat, protein, and lactose percentages, tend to have greater heritabilities. Therefore, heritability is one of the most important considerations when designing breeding programs.

The genetic correlation (rg) is the proportion of variance that two traits share due to genetic causes. This correlation arrives from the fact that many genes affect two or more traits as well as because genes affecting one trait may be in linkage disequilibrium with genes affecting another trait (linkage disequilibrium is the non-random association of alleles at different loci in a given population). Genetic correlations can be positive or negative and can range from +1 to −1.

The closer to 1 or −1, the stronger the correlation (positive or negative) between traits. For breeding purposes, this means that one trait will also be selected along with the other, even though no deliberate selection exists for the second trait. For example, selection for increased milk production also increases protein production, and selection for improved weight gain in broiler chickens also increases carcass fat content. With negative genetic correlations, it is challenging to select both characteristics simultaneously. For example, as milk production increases in dairy cows through genetic selection, it becomes more difficult for high-producing cows to conceive. This negative correlation is partly due to the partitioning of the cows' nutrients between production and reproduction, prioritising production in early lactation, as this has a significant metabolic demand. In other words, nutrients are used for this purpose instead of for reproduction (resource allocation theory—trade-offs of maintenance, growth, reproduction, and immune function). Thus, selecting for improved fertility may result in a reduction in milk production and vice versa.

17.2.1 Selective Breeding

Several genetic and genomic methods have been used in the breeding programs in the neotropical species, including selective breeding. It involves selecting animals with desirable traits and breeding them to produce offspring with increased value for those traits. This is traditionally carried out using quantitative genetics and uses the natural variations in traits among members of any population. Genetic improvement requires understanding the two primary sources of variation: genetics and the environment. Differences in animal performance (growth, reproduction, and productivity) may be caused by the environment, such as the quantity and quality of feed, management, and even the climate. These are not transmitted to the next generation. For example, only approximately 30% of the variation in milk production in dairy cattle is due to additive genetic effects and is therefore transmitted to the next generation, with the remainder being due to non-additive genetic variation or environmental effects. Thus, environmental factors must be controlled when selecting breeding stock to reveal the variation due to genetic effects.

The genetic gain per year (ΔG) depends on several balancing factors:

$$\Delta G = \left(\mathrm{A}\sigma \mathrm{gi}\right) / L,$$

where A is the accuracy of selection (depending on the depth of the pedigree), σg is the standard deviation of the additive genetic variation for the trait, i is the selection intensity (proportion of the population selected for further breeding), and L is the generation interval. Except for σg, the other values in the equation can be changed by management practices. More complete pedigree records on candidates for selection can increase the accuracy of selection by allowing a better estimate of the individual's genetic value. If selection is made only after candidates reach full maturity (to have better genetic data), this will increase the generation interval. Selection intensity can also be increased by reducing the proportion of the population used in breeding, but this should be balanced with the generation interval and inbreeding. Because the generation interval is the denominator in the equation, all else being equal, anything that increases it reduces the annual genetic progress (ΔG).

Furthermore, the ranking of animals for breeding relies on statistical and genetic knowledge, equating their performance records to that of a statistical model. The basic animal model for phenotypic measurement yi of an individual i ($1 \leq i \leq n$) is

$$Y_i = \mu + a_i + e_i$$

with the population mean μ, environmental component $e_i \sim N(0,\sigma_E^2)$ with environmental variance σE2, and breeding values distributed as $a^T = (a_1,\ldots,a_n)T \sim N(0,\sigma_A^2)$ with additive genetic variance σ_A^2 and additive genetic relatedness matrix A, which represents the relatedness among individuals. This model often includes fixed effects (such as year or age) and additional random effects that account for permanent environmental conditions. The underlying assumption is that all animals are from the same genetic population and that the breeding values (ai) are measured as the deviation from the mean of this population, and therefore have a mean of zero. Best linear unbiased prediction, or BLUP, is a technique for predicting random effects, usually an animal's genetic merit.

The first task is to eliminate environmental effects statistically, which involves setting up a system of equations to solve for all the genetic effects of dams and sires simultaneously. All available relative information is included in *g* and increases the accuracy of the evaluation. This so-called animal model is used for evaluating cattle, sheep, goats, chickens, and pigs. A pedigree is also valuable when a trait's heritability is relatively low and is expressed later in life or only in one sex.

Genetic gains are permanent and cumulative from generation to generation. Additional complications arise from simultaneously trying to improve multiple traits, as some traits are genetically correlated, either positively or negatively; they may still be complementary or antagonistic. These can be combined in a selection index:

$$I_i = a_1 EPD_{i1} + a_2 EPD_{i2} + \ldots + a_n EPD_{in}$$

I Index value for animal *I*; a_1, a_2, etc.—economic weights and *EPD*—expected progeny differences.

$$EPD = h^2 \left(P - \underline{P}\right)$$

where h^2 = trait heritability, *P* = phenotypic value of a characteristic and $\underline{P}$ is the contemporary group mean.

Economic selection indices are based on maternal, terminal, and all-purpose production systems. Maternal indices are focused on improving profitability in herds that retain females and sell calves at weaning. They place heavy weights on Expected Progeny Differences (EPDs) for cow longevity, fertility, and maintenance while improving calf weaning weights. Terminal indices are designed for commercial producers that retain ownership of their calves and market their carcass quality.

17.2.2 Crossbreeding

It involves crossing different breeds to produce offspring that have desirable traits from both parental breeds (complementation). For example, crossing *Bos indicus* breeds such as Brahman with *Bos taurus* breeds such as Angus or Hereford can result in offspring with improved heat tolerance, disease resistance, and meat quality. Another consideration in crossbreeding is heterosis, or hybrid vigour, which occurs when offspring performance exceeds the average performance of the parent breed. This is a common phenomenon in which crossbred offspring display increased size, growth rate, and fertility, especially when the breeds are more genetically distant. Crossbreeding is the main strategy in swine and poultry breeding, mostly due to negative correlation between performance and reproductive traits. In this case, male and female inbred lines are developed with specific traits being improved within lines, followed by crossbreeding between them to generate the hybrid animals that will be found in the commercial farms and will supply the markets worldwide (Neeteson et al. 2023). This inbred line selection and crossbreeding is controlled by few large commercial companies, and the success of the developed genetic material in the tropics is highly dependent on technologies to reproduce the ideal environment for them. In these species, production based on local breeds is restricted to small family-based systems.

Crossbreeding can additionally be used in introgression of monogenic or oligogenic traits. This is a quite useful process for introducing climate and disease resistance in tropical production systems and can significantly benefit from the use of genomics in the process of identifying the desired combination of both genomes (see Sect. 17.2.3 Genomic selection).

17.2.3 Genomic Selection (GS)

Genomic selection involves the use of a sufficient number of DNA markers to cover all genomic segments to predict an animal's genetic value for a particular trait. By enabling the control of the Mendelian segregation of alleles into gametes, this method can be used to identify animals with desirable characteristics more quickly and accurately and can help accelerate the breeding process.

The use of DNA/genomic information in breeding programs has become possible due to the advances in sequencing methods that occurred between the last two decades. Until 2010, Sanger sequencing was the only approach available to perform sequencing, but with next-generation sequencing (NGS) becoming available commercially (post-2005), the analysis throughput was highly improved and genomic centres have emerged worldwide. Livestock (poultry, cattle, horses, sheep, pigs, salmon and goats), were among the first species for which reference genomes were available, and their quality, representativeness and annotation are continuously being improved (Clark et al. 2020; Gong et al. 2023). Technological developments have reduced costs, increased throughput and provided much greater detail, completely revolutionising how genomic information is obtained and used. Recently, the quality of these materials has improved with the generation of new scaffolds using long-read sequences (such as PacBio), optical mapping (such as BioNano Genomics), and chromatin conformation capture. Pangenomes, representing a variety of individuals, are being built. Genomics differs from 'classical genetics' in that it considers an organism's full complement of hereditary material rather than one gene or gene product at a time. Moreover, genomics focuses on interactions between loci and alleles within the genome, as well as other interactions, such as epistasis (for example, when one gene locus masks or modifies the phenotype of another), pleiotropy (when the same gene influences two or more seemingly unrelated phenotypic traits) and heterosis.

Genome strategies are typically constructed using tens to hundreds of individuals of different breeds, representing genetic diversity and demographic history (review Chap. 10). In this way, millions of genetic variants were uncovered, and many single-nucleotide polymorphisms (SNPs) became available in livestock species for the first time. Millions of SNP markers have been developed and mined, generating many genome-wide SNP arrays. These SNP arrays can be classified as high-density arrays (600 k to 800 k), medium-density arrays (70 k to 80 k), or low-density arrays (2 k to 5 k) and allowed the development of genome-wide association studies (GWAS) and Genomic selection (GS). For example, at least three million cattle and millions of poultry and pigs have already been genotyped with this kind of SNP array. Using genomics to identify adaptive and resilient traits related to climate change will enable us to address these challenges more effectively by characterising the genetic resources available for food, agriculture and adaptation. Genomic tools have come to the fore in allowing us to evaluate functional genomic regions and their associations with adaptation. Genome coverage with SNPs and cost-effective SNP genotyping arrays have provided GS programs that are revolutionising animal breeding, with predictions of doubling genetic gains compared to traditional selection.

Genomic selection enables us to predict the genetic value of animals from genome-wide SNP markers, which are usually estimated as genomic best linear unbiased predictors (gBLUP). While much progress has been made in easily measuring production traits, such as egg or milk production, few advances have been made in understanding traits that are becoming more important, such as those related to climate change and its associated challenges, including higher temperatures, changes in rainfall, more extreme and frequent weather events, disruptions to feed supplies, and disease outbreaks, although promising results are emerging for tick resistance in beef cattle. One suggested approach is the introgression of adaptation genes from locally adapted breeds into high-productivity commercial breeds or vice versa. Introgression strategies assume that one or more alleles in genes of interest are present in one population and absent in the other. The main goal is to select individuals who carry these favourable alleles in the donor population and introduce them to the recipient population by cross-breeding. Next, back-crossing to the recipient population and selection for the introduced trait would result in fixing the favourable alleles for increased adaptation in our example while maintaining the minimal possible remaining donor genome. This process has made great contributions from genomics, which allows for the identification and monitoring of population-specific haplotypes. Accordingly, the use of gBLUP outperformed the use of conventional BLUP in simulated divergent populations.

Genomic selection presumably works through indirect association because the variants of interest, i.e. the SNPs associated to the selected trait, are in linkage disequilibrium with one or several variants genotyped. There are problems with this method in that the accuracy of genomic values is based on being able to predict the effects of large chromosome segments that segregate within closely related animals. Due to recombination, these segments break up over generations, so the prediction equation's accuracy rapidly decays over time. Whole genome sequence data may be used to overcome these difficulties, with the imputation of data (haplotype reconstruction) for descendants that have been genotyped with fewer SNPs. This imputed data would be used to predict genomic values, especially in populations with fewer sires. As sequencing costs decrease, livestock geneticists are resequencing the genomes of many animals.

Several other challenges exist in breeding for adaptation and sustainability, such as a clear definition of phenotypes. There is also a need for cross-border protocols (for example, the simultaneous measurement of appropriate correlated and productive traits that could serve as proxies); protocols for the collection and storage of data; and criteria for data sharing, combining and analysis so that relevant comparisons can be made between different countries and studies. In other words, we must develop inexpensive and efficient methods to accurately phenotype large numbers of animals for traits like methane emission, if genomic studies are to become useful in mitigating the impacts of global climate change.

Several genes are important for animal improvement in the tropics, and their specific relevance depends on the traits and production systems being targeted. Methods for the identification of useful genes include the comparison of genomes between adapted/non/adapted breeds/species, the modelling of genetic components under stress conditions using reaction norms, differential gene expression analyses and association studies for polymorphisms in specific genes and GWAS. Genomic studies have identified genes that act at the physiological (transitional regulation, heat shock proteins, endocrine function, immune response, oxidative stress, energy production/metabolism and electrolyte/water regulation) and morphological (body size, growth, development, skin and hair traits) levels.

17.3 Breeding and Conservation Genetics and Its Impact on the Traits of Interest

Several traits are important for livestock breeding in the Tropics. These traits can be broadly categorised into productivity traits, adaptive traits, and health traits. Productivity traits are those that contribute to the production of high-quality meat, milk, or other animal products. Examples include growth rate, feed efficiency, carcass weight, and meat quality. As stated previously these are the major traits used in breeding programs helping to improve the quantitative traits in several livestock species. In the last century, those strategies have allowed the development of most of the commercial breeds and lines of poultry, pig, fish, cattle, and others available nowadays in the global markets. However, the adaptive traits, those that enable animals to thrive in challenging tropical environments, are becoming essential to be studied and included as new phenotypes in breeding programs, especially with the climate changes expected in the upcoming years (FAO 2015). Examples of adaptive traits are heat tolerance, the ability to digest low-quality forage, health traits, which contributes to the overall health and welfare of animals, *e.g.* resistance to parasites and infectious diseases, and predisposition to metabolic diseases. Notably, the relative importance of these traits may vary depending on the specific production system and market demands. For example, in a grass-fed production system, traits such as sensitivity to forage quality and forage utilisation may be more important than feed efficiency expressed under high-energy diets. Similarly, temperament and handling ease may be more important than growth rate or carcass weight in markets where consumers prioritise animal welfare. The adaptability and resilience of livestock to climate change depend on many traits, including those that influence fitness, such as longevity and disease resistance. These traits typically have low heritability and have been negatively selected for as a correlated effect of selection for production traits. In this context, it is also interesting to highlight the efforts to characterise several indigenous and local breeds worldwide, especially in Asia and Africa, to understand and to identify molecular markers related to adaptation and disease resistance (FAO 2015).

17.3.1 Beef and Dairy Cattle

Incidence of ticks and of tick-borne diseases is an important concern for cattle production in tropical areas, with massive economic impacts. Breeding for these traits is challenging, as obtaining accurate measures of the incidence of ticks (manually counting adult female ticks in one side of the animal body) is laborious while determination of tick-born hemoparasites levels is based on costly laboratory assays. Using SNP data from nearly 4 thousand Braford and Hereford cattle assayed for tick (*Boophilus microplus*) incidence, from which over 1850 were assayed for infection with the tick-born hemoparasites *Babesia bovis*, *Babesia bigemina* and *Anaplasma marginale*, heritabilities for the hemoparasites were found to be lower than 0.1 with no phenotypic correlation between tick burden and hemoparasite levels (Romero et al. 2023). Thus, according to these authors, even though these hemoparasites are transmitted by ticks, selection for a lower number of ticks will not reduce the level of hemoparasites. In Brazil, breeding programmes of Hereford, Braford and Angus breeds successfully implemented tick resistance in their selection criteria. Moreover, the combination of tick infestation data from different breeds and countries in a multi-trait model, as phenotypes collected in different countries may behave as different traits, is a promising strategy to overcome small sizes of phenotyped populations and increase selection accuracy (Cardoso et al. 2021). This is a promising approach for genomic selection for tick resistance, as many small data sets exist from tropical areas, including the neotropics.

In dairy cattle, several efforts have been made to produce crossbreds of taurine cattle with *Bos taurus indicus* cattle to improve heat stress tolerance, disease resistance while maintaining milk production. The Australian Friesian Sahiwall and Australian milking zebu composite breeds, in Australia, produced more than 6000 liters/lactation, which is approximately two times more than the *Bos indicus* ancestors, and showed higher tick resistance when compared with the taurine breeds (Marchioretto et al. 2023). In Brazil, the Girolando breed, a crossbred of Holstein × Gyr can be highlighted as a successful strategy for milk production. An organised Girolando genetic breeding program was established between producers and Brazilian Agricultural Research Corporation (Embrapa), improving, in about 20 years, 63% of milk production at 305 days (Silva et al. 2022), leveraging Brazil to be the fifth largest milk producer in the world (FAO 2021).

17.3.2 Poultry

Poultry is considered one of the main affordable types of meat, acceptable almost worldwide. In the last 50 decades, genetic selection's main breeding goals were related to feed efficiency, live weight, and yields. Commercial crossbreds were developed by both governmental institutions and private companies. Nowadays, more than 90% of the lines used in poultry production around the world come from about a few breeding companies for broilers (Laughlin 2007) and, also for layers. The main poultry breeding companies have been using genomic selection in the last decades, as well as, expanding their breeding goals, considering health and welfare traits to produce more balanced animals (Neeteson et al. 2023).

Among the governmental efforts, the one started by Embrapa in the 1990s developing broilers and layers lines could be highlighted. Embrapa Swine and Poultry has maintained their poultry lines for over 30 years, and in the last decade has initiated partnerships with small companies to disseminate their lines. The alternative laying hens (named Embrapa 051), a crossbred of Rhode Island Red and White Plymouth Rock are brown egg laying hens and stand out by being rustic and well adapted to less intensive systems. This product has been commercialised by

Gramado Avicultura and is about 4% of the Brazilian brown eggs market share. In a similar effort, Embrapa Swine and Poultry also signed a licence agreement with West Aves for the multiplication and commercialization of its poultry lines: industrial (Embrapa 021) and colonial broiler (Embrapa 041) lines and brown eggs industrial (Embrapa 031) and white eggs industrial (Embrapa 011) laying hens (Embrapa 2024). Those products are well adapted to tropical environments with good productivity. Some of these lines have now been exported to Angola, which needs to expand its production with competitive prices. Furthermore, the characterization of indigenous breeds has been also used to improve selection in dual-purpose chickens and to improve resistance to Newcastle disease in Africa (Habimana et al. 2021).

17.3.3 Swine

The swine breeding programs evolved similarly to poultry production, with a small number of breeding companies having stock products available worldwide which are, in general, developed in temperate zones and with highly environmental control (Hörtenhuber et al. 2020). There are two rising concerns on pig production systems: one related to environmental impact, mainly due to waste manure management and land occupation for feed production, and the other related to negative effects of heat stress and health status that could affect efficiency and growth rate of the animals. Although most of pig systems are based in highly confined and controlled systems, a special concern applies to less technological pig farms and with limited resources, which will be heavily affected by climate changes (Monteiro et al. 2021). Thinking about the environmental impact, the use of feed conversion rate as a selection criterion has been suggested to be associated with lowering the environmental impact of pig fattening. When adaptation traits are considered, the selection for heat tolerance and resilience, associated with high growth rates have been discussed to improve the welfare, sustainability, while maintaining the production rates in commercial pigs. As in poultry, there are several studies also attempting to characterise indigenous breeds, specially, those from China, to improve genetic resistance to african swine fever, porcine reproductive and respiratory syndrome (PRRS) and other infectious diseases.

17.3.4 Fish

Breeding programs are key strategies in advancing aquaculture within tropical areas, since this production activity is facing several challenges, especially those related to climate changes, such as the fluctuating water temperatures and disease outbreaks. In general, these programs aim to develop fish and shellfish species that grow faster, in about 4–20% per generation, and are more resistant to diseases, thus leading to more efficient aquaculture (Olesen et al. 2003). This is especially important in regions where food security can be a challenge. Another concern in aquaculture is overfishing, a major threat to many tropical fish populations, as well as competition with introduced exotic species. Aquaculture breeding programs can act as a powerful tool for rehabilitation of threatened species. By raising these species in controlled environments, they can be reintroduced into the wild to replenish dwindling stocks (Sae-Lim et al. 2017). Among the species that could be highlighted are the shrimp and tilapia. As in other species, the breeding programs started with mass selection for growth and evolved to genomics selection aiming to find candidate genes to disease resistance, processing yields, meat quality, thermotolerance within others.

17.4 The Use of New Technologies on Livestock Breeding Programs and Conservation

Advances in genomics and breeding technologies are helping to identify and target genes more efficiently and are leading to the fast development of new livestock breeds that are better adapted to the challenges of tropical production systems. By identifying and targeting genes important for desirable traits, it is also possible to accelerate genetic progress through genome manipulation (transgeny, gene editing, etc.). Furthermore, in the last decade, commensal microorganisms, besides the pathogenic ones, have become notably important. Studies on their role in the host physiology have emerged, since microbiome composition could be beneficial to several host's traits, especially but not limited to those related to feed efficiency and methane emission (Kaur et al. 2023).

17.4.1 Metagenomics

Metagenomics, the study of all the genes present in the ecosystem, has become incredibly fast and inexpensive over the past decade, employing high-throughput DNA sequencing. All animals co-exist with microbial communities colonising their bodies and their environments. The effects of this associated microbiota are as diverse as remediating toxins, extracting energy from food, modulating the immune system, and affecting the development of the nervous system in mammals.

A good example of the relevance of microbiota in animal breeding is its role in methane production. Methane is the result of microbial fermentation in the gastrointestinal tract. From the point of view of climate change, methane is a greenhouse gas with a global warming potential 28 times greater than that of carbon dioxide. To date, attempts to use genetic selection to decrease methane production by cattle to mitigate emissions have been unsuccessful, mainly due to the lack of massive phenotype records. Implementing selection for methane emission is expected to result in effective genetic progress in dairy cows, as recent studies point to moderate heritabilities (Beauchemin et al. 2020).

Metagenomics allows the abundance of all genes present in the animal microbiota to be determined and metabolic pathways to be predicted. Understanding how these microbial communities change over time and affect the health and production of animals can help us better understand disease resistance and methane production to mitigate global warming.

As the abundance of many microorganisms is affected by the host genotype and thus can be considered an inherited trait, there is also an opportunity to select better microbiome profiles. Including microbiome data was shown to increase the accuracy of gBLUP predictions for many traits (Pérez-Enciso et al. 2021). Metagenomics can be a useful technology for tracking the microbial community and its activity by identifying genes and metabolic pathways, as well as genes related to antibiotic synthesis and antitumor substances, among other uses. In addition, metatranscriptomics allows the identification of messenger RNA (mRNA), and non-coding RNAs (ncRNAs) which represent genes or RNA molecules that are switched on and actively expressed in an environment. These methodologies depend on correctly annotating diverse genes and ncRNA sequences that may not have homologues in current databases.

Nevertheless, the costs of shotgun sequencing of entire metagenomes are still limited for its general use in breeding programs. A less expensive method, the metabarcoding, is the targeted sequencing of hypervariable regions of the genes encoding the 16S subunit of bacterial/archeal ribosomal RNA, the 18S subunit of eukaryotic ribosomal RNA or the internal transcribed spacer (ITS) of fungi, which produces phylogenetic signals for taxonomic profiling. Although this technique enables the taxonomic description, it does not reveal genes and metabolic potential of the microbiome. Moreover, the depth of the taxonomic classification is usually low, not allowing for species and strains differentiation.

17.4.2 Gene Editing

Genome editing (also called gene editing) refers to a group of technologies used to change an organism's DNA. With the use of clustered, regularly interspaced, short palindromic repeats (CRISPR) and CRISPR-associated nuclease 9 (Cas9) gene-editing technology, the number of target genes and multiplex genomic modifications have increased, and the production of 'genetically manipulated animals' has become a reality.

Gene editing plus genetic and genomic breeding may accelerate the process of coping with the challenges of climate change which will put more pressure on animal production, requiring ever more efficient and sustainable performance in this area. Livestock genome editing is advancing rapidly as a technical field, offering an opportunity to make specific changes to an animal's genome to increase productivity and disease resistance. For example, the myostatin gene leads to increased muscle growth, and to date, this gene has been edited in cattle, sheep, goats, and channel catfish. In addition to production traits, resistance to diseases such as African swine fever and porcine reproductive and respiratory syndrome has also become a target for genome editing. The sleek hair mutation in cattle and the naked neck in chicken are examples of major genes (genes that have a great impact on polygenic traits) that could be targets for adaptation to global warming. For instance, as ticks use the fur to climb and attach to the host, the sleek hair mutation benefits both the adaptation to higher temperatures and to cattle ticks.

Genome editing can also be applied in research to understand underlying biological processes, which may provide improved models of human diseases and speed the progress in producing animals for xenotransplantation. For their organ sizes, pigs are a natural focus of xenotransplantation research.

The use of genome-edited animals for human consumption will depend on legislative modifications and appropriate classification of different genetically modified animals. Precise and accurate genome-editing techniques differ greatly from transgenics. Nonetheless, research needs to target the proper traits, species, and regions to achieve optimal food security benefits. This leads to the question of whether farmers will adopt genome-edited animals. It is possible, for instance, that edited disease-resistant or disease-tolerant animals could be kept in the same herds as disease-susceptible animals and we must understand the epidemiological implications of this coexistence. Genome editing for improved adaptation to environmental challenges may be desirable under some circumstances. Nevertheless, it may also be criticised as an increase in technological interventions by humans in natural systems. In fact, a single strategy will not resolve climate change adaptation. Besides genetic changes in animals (within and across species), modifications must also occur in animal housing, reproduction, nutrition, and health care.

17.4.3 Landscape Genomics

Landscape genomics studies the relationship between adaptive genetic imprints in genomes and environmental heterogeneity or system production among natural populations or livestock/crop species. It is thus a combination of landscape ecology and population genetics, aimed at providing information on the interaction between geographical coordinates and their associated biotic and abiotic environment traits, as well as evolutionary processes, including gene flow, genetic drift, adaptation and selection.

This discipline has emerged as a useful technology in conservation studies (see more details Chap. 12). It may reveal attributes such as genetic adaptation to specific environmental stressors, including diseases, parasites, extreme heat, vegetation type and drought, or any combination of these stressors. The results have been used as a further criterion to optimise on-site and off-site conservation schemes and predict areas to preserve or expand animal production.

The adaptation of livestock to local climates has shaped the spatial distribution of various genes, which may help us understand adaptive variation. Understanding why certain AnGRs are adapted to a given environment is important. Such an understanding would also be a reliable indicator of what other environments these animals can survive and flourish in. Therefore, characterising AnGRs and the production environments in which they thrive is a priority. At present, breeds reared in harsh environments are usually less studied but are potentially the most useful sources for future climate change adaptations. For this reason, detailed information on the local management of a breed and the environment to which it is adapted, in conjunction with extensive data on the local climate, vegetation, water resources and soil characteristics, is fundamental. An economic study on setting priorities for the Sustainable Development Goals showed that investing monetary resources in research is much more important for increasing livestock production and crop yield than achieving the 2 °C target threshold for global warming (Ivanovich et al. 2023). Although populations can evolve in response to climate change, this response is limited by the adaptive potential and genomic vulnerability found among populations. Metagenomics can provide insight into the coadaptation of AnGRs with other organisms in the same production environment.

Genomic selection can be used in both pure-breeding and crossbreeding programs for adaptation, assuming that phenotypes are available. Thus, programs are needed to record animal production performance in developing countries. Given the importance of landscapes and the environment in this approach, developing new tools and methods to improve the integration of geographical information, as well as its organisation, storage, and dissemination to stakeholders, will be vital. Several applications of these techniques are currently being studied, such as the AdaptMap goats project, an international coordinated effort to genotype and resequence goat breeds, focusing on the genetics of adaptation. This project indicates the relationship between geographical origin and genomic diversity (Stella et al. 2018). A more comprehensive understanding of the dynamics of adaptive genomic attributes across and among collections of organisms that coexist in the same ecosystems and combining those data with geographical information technologies has led to the emerging field of landscape genomics. NEXTGEN is a European Union-funded programme that concentrates on a comparative analysis of intraspecific genomic data of cattle, sheep, and pigs. Landscape genomics was used to identify genomic regions associated with disease resistance and adaptation (https://archiveweb.epfl.ch/nextgen.epfl.ch/index.html%3Fp=21.html). The Yak Genome Database (yakgenomics.com), an internet-based resource, is a further example of the increasing use of genomic technologies. Landscape genomics is also being applied to studies of genetic resources, including trees and grasses, that will help researchers to understand how best to use these tools.

17.5 Final Remarks

Livestock production is already being affected by climate change. Although many existing technologies in animal breeding systems will play a crucial role in mitigation and adaptation, genomic approaches are changing how researchers deal with extensive amounts of data and phenotypes. Genomic selection, gene editing and metagenomics, among other strategies, could be successfully used to identify and select resilient and adapted animals to mitigate the impact of climate change. Moreover, genomics tools help increase livestock production, which faces challenges such as the need to improve resistance to diseases and parasites, the need to decrease farming's impact on the environment, the need to reduce methane emissions from ruminants and increased competition for land and water resources. In their attempts to better use these high-precision genomic technologies, researchers are now seeking high-precision phenotypes that better characterise adaptation to the challenges of climate change associated with livestock production.

Key Points

- Animal production in the Neotropics can be an invaluable source of genetic material in the face of climate change;
- Animal breeding in the Neotropics should consider traits related to fitness as well as production;
- Along with quantitative genetics, genomics and biotechnology can be used to increase productivity.

Questions

1. What are the major traits of interest for selecting AnGR in the Neotropics?
2. What different types of data are available for animal breeding in the neotropics, and what are their limitations?
3. How do different types of information affect our ability to design a breeding program?
4. Some examples of genetic programs that consider climatic factors are provided.

Acknowledgments We thank CAPES (Brazil, Finance Code 001) for the free access to the journals used in the literature review. A.M.G.I., C.M. and L.C.A.R. are recipients of a productivity fellowship from the National Council of Scientific and Technological Development (CNPq).

References

Alderson G, Alderson L, Bodó I (1992) A system to maximize the maintenance of genetic variability in small populations. In: Conservation of domestic livestock. CAB International, Wallingford, pp 18–29

Beauchemin KA, Ungerfeld EM, Eckard RJ et al (2020) Review: fifty years of research on rumen methanogenesis: lessons learned and future challenges for mitigation. Animal 14(Suppl 1):s2–s16. https://doi.org/10.1017/S1751731119003100

Boichard D, Maignel L, Vender É (1997) The value of using probabilities of gene origin to measure genetic variability in a population. Genet Sel Evol 29:5–23. https://doi.org/10.1186/1297-9686-29-1-5

Cardoso FF, Matika O, Djikeng A et al (2021) Multiple country and breed genomic prediction of tick resistance in beef cattle. Front Immunol 12:620847. https://doi.org/10.3389/fimmu.2021.620847

Clark EL, Archibald AL, Daetwyler HD et al (2020) From FAANG to fork: application of highly annotated genomes to improve farmed animal production. Genome Biol 21:285. https://doi.org/10.1186/s13059-020-02197-8

Embrapa (2024) Brasil exporta primeira remessa de genética Embrapa para angola. Available at: https://www.embrapa.br/suinos-e-aves/busca-de-noticias/-/noticia/86514480/brasil-exporta-primeira-remessa-de-genetica-embrapa-para-angola?p_auth=fLA6UAWc. Accessed 25 Apr 2024

FAO (2023) Pathways towards lower emissions—a global assessment of the greenhouse gas emissions and mitigation options from livestock agrifood systems. FAO, Rome. https://doi.org/10.4060/cc9029en. Accessed on 4 June, 2024

FAO (2022) Monitoring the diversity of animal genetic resources. Intergovernmental technical working group on animal genetic resources for food and agriculture. Available at: https://www.fao.org/3/nl061en/nl061en.pdf. Accessed on 25 Apr 2024

FAO (2021) Dairy Market Review: emerging trends and outlook.10. Available from: https://www.fao.org/3/cb7982en/cb7982en.pdf. Accessed on 25 Apr 2024

FAO (2015) The second report on the state of the world's animal genetic resources for food and agriculture. FAO Commission on Genetic Resources for Food and Agriculture Assessments, Rome, p 562

Gong Y, Li Y, Liu X et al (2023) A review of the pangenome: how it affects our understanding of genomic variation, selection and breeding in domestic animals? J Animal Sci Biotechnol 14:73. https://doi.org/10.1186/s40104-023-00860-1

Habimana R, Ngeno K, Okeno TO et al (2021) Genome-wide association study of growth performance and immune response to Newcastle disease virus of indigenous chicken in Rwanda. Front Genet 12:723980. https://doi.org/10.3389/fgene.2021.7239803

Hörtenhuber SJ, Schauberger G, Mikovits C et al (2020) The effect of climate change-induced temperature increase on performance and environmental impact of intensive pig production systems. Sustain For 12:9442. https://doi.org/10.3390/su12229442

Ivanovich CC, Sun T, Gordon DR et al (2023) Future warming from global food consumption. Nat Clim Chang 13:297–302. https://doi.org/10.1038/s41558-023-01605-8

Kaur H, Kaur G, Gupta T et al (2023) Integrating omics technologies for a comprehensive understanding of the microbiome and its impact on cattle production. Biology (Basel) 2023(9):1200. https://doi.org/10.3390/biology12091200

Lacy RC (1989) Analysis of Founder Representation in Pedigrees: Founder Equivalents and Founder Genome Equivalents. *Zoo Biology, 8*, 111–123. https://doi.org/10.1002/zoo.1430080203

Laughlin KF (2007) The evolution of genetics, breeding and production. Temperton fellowship; Harper Adams University: Newport, 15:1–55. Available at: https://citeseerx.ist.psu.edu/document?repid=rep1&type=pdf&doi=2f9c5d50b3fdd28b7923137846ebe149c9152d56. Accessed on 22 Apr 2024

MacCluer JW, Vande Berg JL, Read B, Ryder OA (1986) Pedigree analysis by computer simulation. Zoo Biol 5(2):147–160. https://doi.org/10.1002/ZOO.1430050209

Marchioretto PV, Chanaka Rabel RA, Allen CA et al (2023) Development of genetically improved tropical-adapted dairy cattle. Anim Front 13(5):7–15. https://doi.org/10.1093/af/vfad050

Monteiro ANTR, Brossard L, Gilbert H et al (2021) Environmental impacts and their association with performance and excretion traits in growing pigs. Front Vet Sci 8:677857. https://doi.org/10.3389/fvets.2021.677857

Morton JF (2007) The impact of climate change on smallholder and subsistence agriculture. PNAS 104(50):19680–19685. https://doi.org/10.1073/pnas.070185510
Neeteson AM, Avendaño S, Koerhuis A et al (2023) Evolutions in commercial meat poultry breeding. Animals 13:3150. https://doi.org/10.3390/ani13193150
Nei M (1978) Estimation of average heterozygosity and genetic distance from a small number of individuals. Genetics 89:583–590. https://doi.org/10.1093/genetics/89.3.583
Olesen I, Gjedrem T, Bentsen HB et al (2003) Breeding programs for sustainable aquaculture. J Appl Aquac 13(3–4):179–204. https://doi.org/10.1300/J028v13n03_01
Paiva SR, McManus CM, Blackburn H (2016) Conservation of animal genetic resources—a new tact. Livest Sci 193:32–38. https://doi.org/10.1016/j.livsci.2016.09.010
Pérez-Enciso M, Zingaretti LM, Ramayo-Caldas Y et al (2021) Opportunities and limits of combining microbiome and genome data for complex trait prediction. Genet Sel Evol 53:65. https://doi.org/10.1186/s12711-021-00658-7
Romero ARS, do Nascimento AV, Oliveira MCS et al (2023) Genetic parameters and multi-trait genomic prediction for hemoparasites infection levels in cattle. Livest Sci 273:105259. https://doi.org/10.1016/j.livsci.2023.105259
Sae-Lim P, Kause A, Mulder HA et al (2017) Breeding and genetics symposium: climate change and selective breeding in aquaculture. J Anim Sci 95(4):1801–1812. https://doi.org/10.2527/jas.2016.1066
Silva MVGB, Ferreira Júnior E, Panetto JCC, et al (2022) Programa de Melhoramento Genético da Raça Girolando—Sumário de Touros—Resultado do Teste de Progênie (Avaliação Genética/Genômica)—Junho/2022 [Genetic Improvement Program for the Girolando Breed—Summary of Bulls—Result of the Progeny Test (Genetic/Genomic Evaluation)—June/2022]. Embrapa Gado de Leite [Embrapa Dairy Cattle], 248. Available at: https://ainfo.cnptia.embrapa.br/digital/bitstream/doc/1144061/1/DOC-266-Programa-de-Melhoramento-Genetico-da-Raca-Girolando.pdf. Accessed on 4 June 2024
Stella A, Nicolazzi EL, Van Tassell CP et al (2018) AdaptMap: exploring goat diversity and adaptation. Genet Sel Evol 50:61. https://doi.org/10.1186/s12711-018-0427-5
Wright S (1965) The interpretation of population structure by F-statistics with special regard to systems of mating. Evolution 19:395–420. https://doi.org/10.1111/j.1558-5646.1965.tb01731.x
Wright S (1978) The relation of livestock breeding to theories of evolution. J Anim Sci 46(5):1192–1200. https://doi.org/10.2527/jas1978.4651192x

Chapter 18
How did Evolutionary Processes Influence the Genetic Structure of Humans in the Neotropics?

Tábita Hünemeier, Kelly Nunes, and Marcos Araújo Castro e Silva

Contents

T. Hünemeier (✉)
Departamento de Genética e Biologia Evolutiva, Universidade de São Paulo, São Paulo, SP, Brazil

Institute of Evolutionary Biology, IBE/CSIC-UPF, Barcelona, Spain
e-mail: hunemeier@usp.br

K. Nunes
Departamento de Genética e Biologia Evolutiva, Universidade de São Paulo, São Paulo, SP, Brazil

M. A. Castro e Silva
Departament de Medicina i Ciències de la Vida (MELIS-UPF), Universitat Pompeu Fabra, Barcelona, Spain

M. V. Bohrer Monteiro Siqueira et al. (eds.), *Population Genetics in the Neotropics*,
https://doi.org/10.1007/978-3-031-83685-5_18

Abstract The Neotropics, encompassing Central and South America, stand out for their extraordinary biodiversity and cultural richness. Indigenous Americans exhibit remarkable cultural and linguistic diversity, but their genetic diversity is among the lowest among human populations. Environmental factors, including altitude, temperature, and biodiversity, have influenced the genetic makeup of Indigenous Americans since the earliest peopling of the America. Additionally, human impact, such as hunting, harvesting, and wildlife management, has significantly affected neotropical biodiversity. Niche construction, driven by plant and animal domestication and landscape modification, has played a crucial role in agriculture and has heavily influenced the genetic diversity of the region. The interplay between humans and their environment has shaped human genetic adaptations concerning diet and susceptibility to infectious diseases, affecting contemporary indigenous and admixed populations. Furthermore, migrations have played a significant role in shaping the genetic structure of neotropical human populations. Over time, multiple migration waves have contributed to the genetic diversity observed in present-day populations. The peopling of the America involved successive migrations dating back at least 15,000 years, followed by later post-contact migrations from various world regions. The study of genetic variation in neotropical populations has provided insights into their evolutionary history, demographic changes, and the impact of historical events like colonization and the transatlantic slave trade. Understanding the complex interplay of genetic, environmental, and historical factors is vital for conserving the biodiversity of humans and ecosystems. This chapter discusses the importance of migrations, demographic changes and natural selection in shaping the genetic diversity of neotropical human populations.

Keywords Genetic diversity · Indigenous Americans · Migration · Demographic Changes · Natural Selection

Learning Objectives

1. Understand the origins of Indigenous populations from America and how the peopling of the American continent have shaped the genetic diversity of ancient and present-day human populations.
2. Examine the genetic evidence, in the light of the archaeological and linguistic data regarding human occupation, population movements and demographic dynamics in this region, as well as their lasting effects on genetic diversity.
3. Explore the unique genetic profiles of Neotropical human populations, examining adaptations to diverse environments, diets, and pathogens.
4. Analyze the practical significance of studying human genetic diversity, considering its impact on human health.

18.1 Introduction

The Neotropical region is characterized by its rich and varied geography, including dense rainforests, expansive savannas, mountain ranges, and extensive river systems in the tropical areas of the American continent, spanning from Mexico to Argentina (Palma-Silva et al. 2022). This region comprised an astonishing array of plant and animal species, accounting for almost one third of global biodiversity (Rull 2008, 2011; Palma-Silva et al. 2022). Also, the Neotropics area includes the Amazon Rainforest, the biggest tropical rainforest in the world, and home to an enormous variety of plants, insects, mammals, and birds (Hoorn et al. 2010; Antonelli et al. 2018).

Regarding human history, Neotropics have a rich cultural scenario, with ancient civilizations that flourished in these regions during at least 15,000 years before European contact. Pre-contact civilizations diversified over this time, giving rise to various peoples, such as the great Mesoamerican and Andean empires (e.g., the Maya and Inca), as well as numerous different cultures of the lowlands of Central and South America (e.g., Tupi and Taino). These Indigenous peoples left behind a legacy of advanced agricultural practices, monumental architecture, intricate material culture, landscape modifications, and a vast number of domesticated, and semi-domesticated, species of plants, along with a few animals (Larson et al. 2014; Clement et al. 2015; Neves and Heckenberger 2019; Iriarte et al. 2020). These societies also held considerable knowledge about diverse topics including astronomy, engineering, medicine, writing, mathematics, among others (Graeber and Wengrow 2021).

Humans have been modifying and shaping the Neotropics region since their first arrival. One striking example suggests the human influence on the Late Pleistocene extinctions, especially of megafaunal species, not only in America, but also globally (Bergman et al. 2023). The extent to which this process was influenced by various other factors, including climate change and the repercussions of the end of the last glacial period, remains a subject of ongoing debate. During the Holocene humans also started to perform different types of niche construction, generating considerable environmental changes, such as the construction of domesticated landscapes, as represented by the Amazonian dark earths mostly concentrated along riverside areas, and also earthworks (particularly geoglyphs) which are currently estimated to number in the thousands in the Amazonian area (Iriarte et al. 2020; Peripato et al. 2023). Notably, this also includes the domestication of at least 83 plant species within the Amazon, making it one of the world's major crop domestication centers (Clement et al. 2015).

From a scientific perspective, the Neotropics provide a valuable setting for research in various disciplines, including ecology, biology, anthropology and genetics. Thus, regarding the study of human population genetics, first and foremost, the Neotropical region is characterized by unique and diverse ecosystems, leading to distinct genetic profiles among its human populations (Rull 2008, 2011; Adhikari et al. 2017; Castro E Silva et al. 2022; Palma-Silva et al. 2022). Understanding this genetic diversity offers insights into the historical processes of human migration, adaptation, health and the complex interactions between populations and their environments over time.

In this perspective, the purpose of this chapter is to provide an overview of some of the key questions surrounding the genetic diversity, history and evolution of humans in the Neotropics. In the first section we start by looking at how humans first came to be in this region and then explore some of the complex processes that have shaped the genetic makeup of people throughout the Neotropics over time. We also examine how human genetic diversity is distributed across the Neotropics, analyzing the connections between various populations and evaluating evidence of long-standing patterns of genetic continuity. Furthermore, we explore the genetic adaptations caused by natural selection in Neotropical environments in greater detail in the second section. We also highlight a few instances where native populations have adapted to the Neotropical environments, pathogens and food resources. In its third section, the chapter also considers the consequences arising from the genetic diversity of present-day Neotropical peoples, exploring the implications on both native and non-native populations. This chapter concludes with a consideration of the wider ramifications and importance of the knowledge that has been generated about the genetic diversity of humans in the Neotropics.

18.2 Neotropical Human Genetic Diversity

18.2.1 Peopling of the America

The peopling of the American continent trace back its origins to human migrations from East and Northeast Asia. Human history in Northeast Asia predates the Last Glacial Maximum (LGM; 26.5 to 19 thousand years ago) (Lambeck et al. 2014), by the presence of the Ancient North Siberians (Sikora et al. 2019) as evidenced by archaeological findings at sites like Yana River (31,600 years ago) and Mal'ta (24,000 years ago) (Raghavan et al. 2014; Graf and Buvit 2017). Although the Ancient North Siberians did not persist, their genetic legacy endured through admixture with East Asians around 20–18,000 years ago, likely giving rise to the ancestors of the Indigenous Americans peoples, or Native Americans (Figs. 18.1 and 18.2).

According to the Beringian standstill hypothesis, the ancestors of Native Americans experienced a period of isolation, either preceding or during their initial migration to America (Bonatto and Salzano 1997; Tamm et al. 2007; Hoffecker et al. 2023). This isolation is estimated to have endured for a minimum of 4600 years (Pinotti et al. 2019), though likely lasted more than 10,000 years (Tamm et al. 2007; Hoffecker et al. 2023). The precise reasons for this isolation, whether ecological barriers or Beringia acting as a bioclimatic refugium during the Last Glacial Maximum, are still not understood (Figs. 18.1 and 18.2) (Bonatto and Salzano

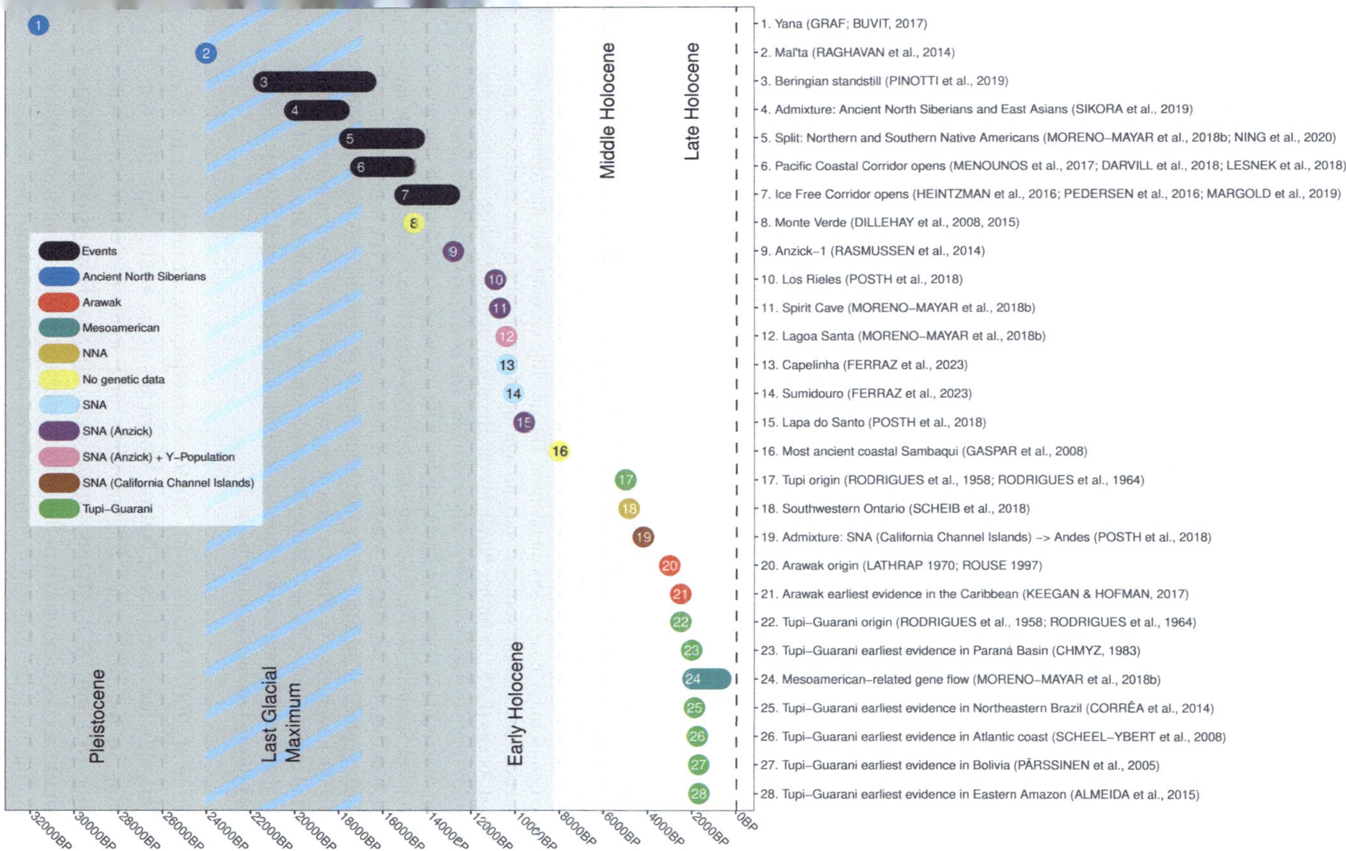

Fig. 18.1 Temporal overview of key events shaping neotropical human diversity. The timeline depicts major milestones in the history of Neotropical indigenous populations, as discussed in the text. These events are organized chronologically according to start date (confidence intervals are not shown), assigned numerical identifiers, and accompanied by concise descriptions and references. The back panel of the plot incorporates shades of gray to denote geological periods, with bright blue stripes representing the Last Glacial Maximum. Specific genetic or ethnolinguistic affiliations are color-coded, as indicated in the legend on the right of the plot. The events and their numerical identifiers presented in this figure are further detailed and geographically contextualized in Figs. 18.2 to 18.4

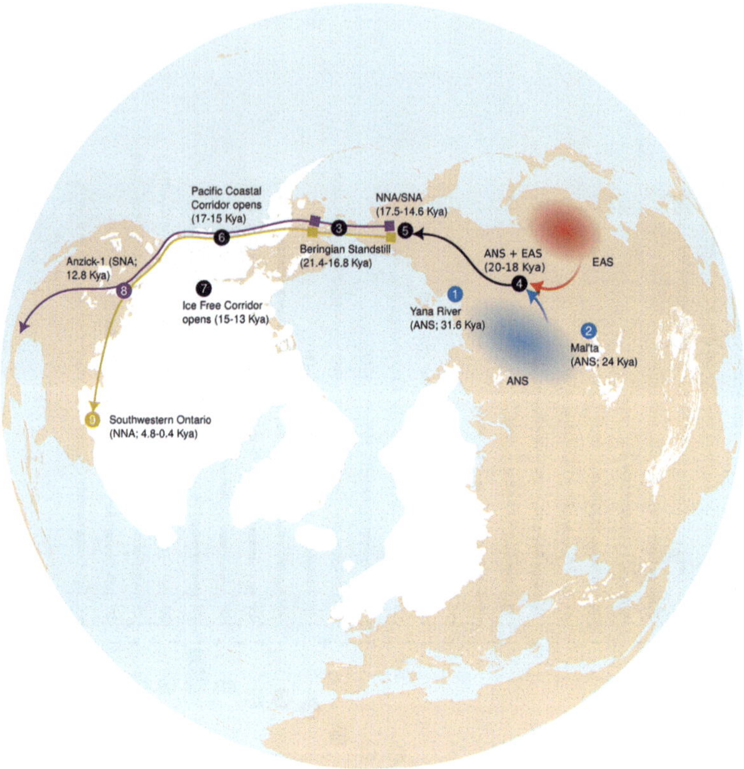

Fig. 18.2 Initial peopling of the America. This schematic representation illustrates the key milestones in the origin of Indigenous American peoples. The timeline encompasses the formation of Indigenous American ancestors in Northeast Asia through the admixture of Ancient North Siberians (ANS) and East Asians (EAS). It further highlights the Beringian standstill, which prompted the development of distinctive genetic diversity, followed by the fast post-Last Glacial Maximum dispersion into the American continent. The figure finishes with the major population split that gives rise to Northern and Southern Native American lineages (NNA and SNA, respectively). The ice sheet distribution during the Last Glacial Maximum is indicated in white. The map was plotted using the 'sf' and 'ggplot2' R packages, using the Orthographic projection centering on the North Pole (Latitude = 90; Longitude = 0). Points represent the approximate location of important archeological sites and events cited in the text, while the arrows simply indicate the direction of hypothetical dispersal routes. The numerical sequence of points has been arranged chronologically, referencing the primary timeline depicted in Fig. 18.1

1997; Tamm et al. 2007; Sikora et al. 2019). Subsequently, around 17.5 to 14.6 thousand years ago, the ancestors of Native Americans underwent genetic differentiation, leading to the divergence into northern and southern Native American branches (Figs. 18.1 and 18.2) (Moreno-Mayar et al. 2018b). This period of genetic divergence marks a crucial moment in the early history of Native American populations, shaping the foundational genetic structure of northern and southern branches that persisted through subsequent millennia.

Throughout the Last Glacial Maximum, extensive continental glaciers obstructed an inland route to the American continent from Beringia (Meltzer 2010). However, following the Last Glacial Maximum, an ice-free corridor opened along the Rocky Mountains, offering a feasible passage for human migration approximately 15–13 thousand years ago (Figs. 18.1, 18.2, and 18.3) (Perego et al. 2009; Potter et al. 2018). An alternative migratory route is the Pacific coastal corridor, supported by evidence of human arrival dating back to 17–15 thousand years ago (Figs. 18.1, 18.2, and 18.3). The latter pathway is currently considered the most plausible based on a convergence of multiple lines of evidence, spanning archaeological findings to genetic analyses (Perego et al. 2009; Lesnek et al. 2018). Although, both routes might have been used at different times, as the glaciers in Northern North America began to melt, creating pathways for our species to enter the new continent (Figs. 18.1, 18.2, and 18.3) (Lambeck et al. 2014; Potter et al. 2018). Then, likely around 16,000 years ago, humans started to migrate from Beringia into the American continent, marking the first time any hominin set foot on this continent (Skoglund and Reich 2016; Willerslev and Meltzer 2021; Silva et al. 2022).

Indigenous Americans, as revealed by extensive studies, predominantly trace their ancestry to either the Northern Native American (NNA) or Southern Native American (SNA) branches, with limited exceptions (Posth et al. 2018; Moreno-Mayar et al. 2018b). Notably, South American populations exhibit exclusive SNA ancestry, indicating that their initial dispersion occurred subsequent to the divergence of the SNA and NNA branches. While some populations in Central and South America display contributions from both lineages, it is inferred that these admixture events took place after the initial settlement (Scheib et al. 2018). However, distinctive cases are observed among speakers of Na-Dené and Eskimo-Aleut languages in northern North America, requiring additional gene flow from East Asian groups to explain their genetic diversity (Rasmussen et al. 2010; Reich et al. 2012; Moreno-Mayar et al. 2018a; Flegontov et al. 2019).

Other exceptions are several contemporary Indigenous groups in the Amazon (Karitiana and Suruí), as well as the Xavante from the Brazilian central plateau, along with an ancient individual from the Lagoa Santa site in Minas Gerais dating back to 10,000 years ago, which display an increased genetic affinity with Australasian populations (Figs. 18.1 and 18.3) (Raghavan et al. 2015; Skoglund et al. 2015; Moreno-Mayar et al. 2018b; Campelo dos Santos et al. 2022). Recent findings have expanded this connection to include two additional modern-day groups from the Central-West Brazilian region (Guaraní Kaiowá) and the northern Peruvian Pacific coast (Chotuna), indicating a wider distribution of this ancestry contribution (e Silva et al. 2021). This apparent excess affinity with Australasians in certain South American populations has been attributed to the contribution of an unsampled population, known as "*Ypikuéra*" (meaning "ancestral" in Tupi language) or "Y." This discovery implies a more intricate population history, involving either an additional influx from Beringia or a significant genetic structure in the ancestors of Native Americans. However, the percentage of this additional ancestry in the identified groups remains relatively low, ranging from 1% to 3% of the total genetic makeup (Skoglund et al. 2015; e Silva et al. 2021).

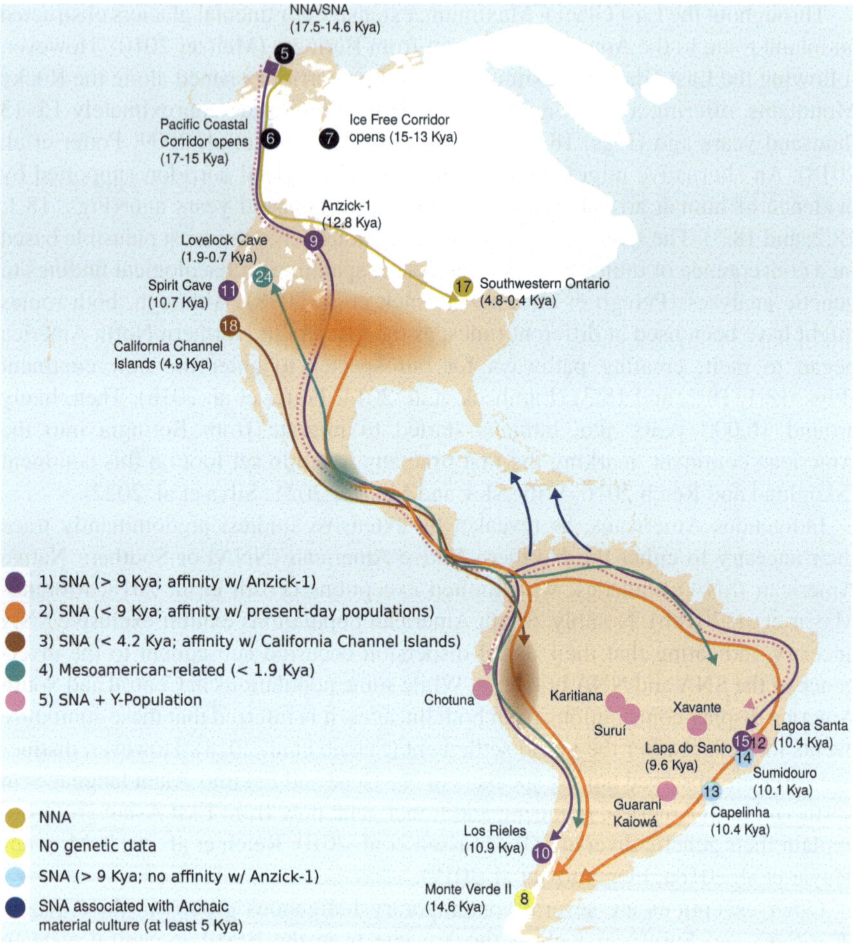

Fig. 18.3 Origins of Caribbean, Central, and South American natives. This figure shows the four main ancestry components that formed South American Indigenous populations. After splitting from the Northern Native American (NNA; green) lineage, the initial wave of settlers in the region corresponds to a Southern Native American (SNA) group genetically linked to Anzick-1 (purple). Subsequently, a second SNA influx (orange), distinct in its lack of this genetic affinity to Anzick-1 and higher affinity to present-day Indigenous populations, at least partially replaced the first group starting approximately 9000 years ago, however some exceptions contradict this model (red). Beyond these two primary dispersals, a distinctive genetic contribution emerged after 4.2 thousand years ago and geographically confined to Central Andes. This component is genetically linked to SNA ancient individuals from the California Channel Islands (brown). A fourth ancestry is detected as an excess affinity between some contemporary and ancient individuals and Australasian populations (pink), modeled as the contribution of an unsampled group called Y-population. At least 5 thousand years ago an SNA group associated with so-called Archaic material culture started to appear in Caribbean Islands (dark green). The map was plotted using the 'sf' and 'ggplot2' R packages, using the Orthographic projection centering on the Caribbean (Latitude = 20; Longitude = −80). The ice sheet distribution during the Last Glacial Maximum is indicated in white. The numerical sequence of points has been arranged chronologically, referencing the primary timeline depicted in Fig. 18.1

An intriguing hypothesis posits that the genetic connection observed between certain Indigenous American and Australasian populations may stem from shared archaic introgression, potentially involving Denisovans, Neanderthals, or another unidentified archaic hominin. The widespread signal of low Denisovan ancestry across Eastern Eurasian and Native American populations raises questions (Qin and Stoneking 2015; Peyrégne et al. 2023). Notably, Oceania displays a heightened Denisovan ancestry compared to Eastern Eurasians and Indigenous Americans. In Oceania, this ancestry correlates with a New Guinean ancestry rather than an Australian ancestry, suggesting recent gene flow from New Guinea as a contributing factor. Conversely, Denisovan ancestry in Eastern Eurasians and Indigenous American populations correlates equally with both New Guinea and Australian ancestry, pointing toward a common source for Denisovan ancestry in these groups (Qin and Stoneking 2015). If this shared Denisovan ancestry plays a role in the observed genetic affinity between certain Indigenous Americans and Australasian populations is still an open question. To date, there is no supporting evidence indicating a correlation between Denisovan ancestry in Asians or Native Americans and Australasian ancestry (Peyrégne et al. 2023).

In conclusion, in South America, existing data strongly suggest the occurrence of three distinct dispersals of SNA populations into the continent (Posth et al. 2018). The initial group, closely related to Anzick-1, gave way around 9000 years ago to a second SNA influx lacking this genetic affinity but demonstrating increased affinity with contemporary Indigenous populations, hinting at a scenario of at least partial demographic replacement (Figs. 18.1 and 18.3). Furthermore, a third population influx, with genetic ties to individuals from the California Channel Islands, contributed to populations in the central Andes before 4200 years ago, potentially associated with the diffusion of agriculture from Mesoamerica (Figs. 18.1 and 18.3) (Sutter 2021). Notably, another study (Moreno-Mayar et al. 2018b (https://www.science.org/doi/10.1126/science.aav2621)) identified a significant and recent influx of Mesoamerican-related ancestry into Indigenous populations spanning both North and South America (Figs. 18.1 and 18.3). While this genetic signal underscores a shared demographic history across geographically distant regions, including the Lovelock Cave site in North America and contemporary South American Indigenous populations, its potential relationship to the aforementioned evidence of North American gene flow into Andean populations remains unclear. This admixture likely reflects the movement of relatively small groups that, while contributing genetically to local populations, did not lead to their complete demographic or cultural replacement. The predominant genetic and archaeological evidence supports a settlement of the continent after the Last Glacial Maximum. However, a handful of potential archaeological sites dating back to the Last Glacial Maximum era, or even earlier, introduces the possibility of a more prolonged human presence on the continent, albeit the evidence supporting this possibility remains very limited (Silva et al. 2022).

18.3 Holocene Population Dynamics in the Neotropics

18.3.1 Amazonia and Andes

South America experienced a rapid human dispersal with the arrival of initial populations approximately 15–16 thousand years ago (Prates et al. 2020). Early populations from this region likely separated into two groups, independently migrating along the west and east coasts (Bodner et al. 2012; Goldberg et al. 2016; Brandini et al. 2018; Gómez-Carballa et al. 2018; Prates et al. 2020). An alternative hypothesis proposes that the Andes, Amazon, and coastal areas in South America were colonized by three distinct lineages that diverged before entering the continent (Rothhammer and Dillehay 2009), with the possibility that the Andean region was settled through a later split from the Pacific coastal branch (Skoglund and Reich 2016). Population history models for indigenous Peruvians support the latter idea, with an inferred split date of around 12,000 years ago, aligning with the trifurcation hypothesis (Harris et al. 2018). These findings indicate that significant lineage splits occurred early in the settlement of South America, and their dispersion was rapid. This is supported by the earliest evidence of human presence in the Southern Cone, which comes from Monte Verde II archeological site, dated to 14.6 thousand years ago (Figs. 18.1 and 18.3) (Dillehay 2009; Dillehay et al. 2015) and the discovery of ancient human remains dating back more than 9000 years ago on both sides of the continent. Noteworthy locations include Cuncaicha in Peru and Los Rieles in Chile on the Pacific side, as well as Lapa do Santo and Lagoa Santa in Brazil on the Atlantic side (Figs. 18.1 and 18.3) (Posth et al. 2018; Moreno-Mayar et al. 2018b). Notably, there is evidence of long-term genetic continuity within significant continental areas. For instance, permanent settlements in the Andes began around 9000 years ago, while genetic analyses suggest enduring continuity in the Lake Titicaca region, possibly dating back to 3.8 or even 7 thousand years ago, persisting in present-day Aymara and Quechua-speaking peoples (Lindo et al. 2018).

Eastern South America boasts some of the earliest human remains in America, found at sites in the Lagoa Santa region in southeastern Brazil, dating back to 10.4–9.6 thousand years ago (Box 18.1). Individuals from these sites are inferred to be descendants of the first southern Native American population influx (Posth et al. 2018; Moreno-Mayar et al. 2018b). As discussed before, around 9000 years ago, a new southern Native American group began to arrive, partially displacing the early migrants, as evidenced by their reduced affinity with Anzick1 and increased affinity with modern South American indigenous peoples (Posth et al. 2018; Moreno-Mayar et al. 2018b). However, a recent study (Ferraz et al. 2023) analyzing 34 ancient individuals from coastal and riverine Sambaqui sites (Box 18.1) has revealed individuals older than 9000 years ago lacking substantial affinity to Anzick-1 (Capelinha and Sumidouro; Figs. 18.1 and 18.3), contradicting the narrative of two sequential waves of migration into South America, one with Anzick-1-related ancestry and another without (Ferraz et al. 2023).

Box 18.1: Lagoa Santa peoples and Sambaqui mound builders

Lagoa Santa sites are famous for their well-preserved early Holocene human skeletons. The sites comprise caves with a sheltered area and have a rich history of human occupation spanning three distinct phases: early, middle, and late Holocene (Strauss et al. 2020). Excavations since 2001 have uncovered 39 human burials, predominantly from the late phase of the early Holocene (Posth et al. 2018; Moreno-Mayar et al. 2018b), placing Lagoa Santa as one of the most densely populated sites in this period. Ritualistic treatment of the dead, including mutilation and defleshing was practiced around 9600–9400 calBP, depicting the ancient case of ritualistic funeral in the neotropics (Strauss et al. 2015). Sediments indicate repeated combustion activities, suggesting an early Archaic economy focused on staple carbohydrates and hunting. The lithic assemblage consists mainly of small flakes and cores, with a shift towards locally available crystal quartz as the dominant raw material around 9900 calBP. Also, low mobility is supported by isotopic and anthropological studies (Strauss et al. 2020). The individuals excavated from the archaeological sites of Lagoa Santa exhibit Paleoamerican craniofacial morphology (Neves and Hubbe 2005), meaning they are morphologically different from present-day Native Americans. However, genomic studies have shown that the skeletons recovered from this region, although morphologically distinct, originate from the early Beringian populations that populated South America, genetically closely related to present-day Indigenous peoples (Posth et al. 2018; Moreno-Mayar et al. 2018b). In this sense, both the emblematic Luzia and her contemporaries from Lagoa Santa are believed to originate from a dispersal through Beringia, as all the earliest inhabitants of the American continent. The morphological differences between Paleoamericans and other ancient and modern individuals from America may be explained by microevolutionary factors following the arrival of the first Americans on the continent (González-José et al. 2008).

The Sambaqui, a term used by Tupi speakers, are shell mounds produced by diverse fisher-gatherer communities, along the Atlantic coast, as well as riverside, and lake areas, starting at least 8000 years ago (Gaspar et al. 2008). These groups likely occupied the Brazilian coast until the arrival of Tupi-Guarani and Macro-Jê speakers, suggested by the presence of ceramics from Tupiguarani and Taquara/Itararé traditions in the uppermost layers of some Sambaqui (Gaspar et al. 2008). Due to the limited number of individuals studied, the relationship between the builders of Sambaqui mounds and contemporary indigenous populations remains largely unexplored. Nevertheless, recent data suggests a stronger affinity between the Sambaqui builders and present-day Jê speakers when compared to other indigenous groups in eastern South America. It is crucial to interpret this finding cautiously, given that the analyzed Jê-speaking communities exhibit a high degree of endogamy (Castro E Silva et al. 2020, 2022; Ferraz et al. 2023). The observed high level of endogamy can result in an accelerated genetic drift effect, potentially introducing biases in inferences related to genetic similarity (Moorjani and Hellenthal 2023).

After the initial population dispersals, South America's diverse climates and environments led to a demographic and evolutionary history characterized by significant variation across time and space. In this sense, a comprehensive analysis incorporating calibrated radiocarbon datings from 13 to 2 thousand years ago and spanning 5464 datings from 1147 archaeological sites, reveals two main phases in this demographic history (Goldberg et al. 2016). The first phase, from 13 to 5.5 thousand years ago, witnessed rapid geographic expansion followed by density-dependent population growth. Population sizes increased swiftly until reaching carrying capacity, after which they would have remained relatively constant between 9 and 5.5 thousand years ago. Additionally, a recent analysis of high-quality curated radiocarbon dates supports that demographic stability was achieved by 11 thousand years ago (Prates et al. 2020). Around 5.5 thousand years ago, the adoption of a sedentary lifestyle and intensified food production marked the onset of a new phase characterized by exponential population growth, particularly in cultural centers, notably those situated in the central and northern Andes. According to this model, more than half of the population growth occurred during this second stage (Goldberg et al. 2016). However, studies indicate that population growth rates varied significantly between regions and over time. In the Andes, expansion began around 9000 years ago, earlier than in other eastern regions such as Patagonia, where a more gradual and late expansion initiated between 7.5 and 5 thousand years ago (Perez et al. 2016, 2017; Prates et al. 2020).

The shift in human population growth rates aligns with a Middle Holocene climatic change, transitioning from previously dry and variable precipitation to consistently increased precipitation in the Southern Hemisphere's tropical forests (Iriarte et al. 2017). In this sense, an increase in precipitation expanded tropical rainforests between 5 and 1 thousand years ago, fostering human population growth and movement, especially in the southern Amazon forest (Iriarte et al. 2017). This shift to wetter climates also led to increased sedentism and the prominence of agriculture, evidenced by an elevated rate of landscape modifications during the Late Holocene (Iriarte et al. 2020). Plant domestication in South America broadly coincided with the extinction of the last megafauna species around 9–8 thousand years ago, with some plant species fully domesticated by 6 thousand years ago. However, the initial use of various plants dates back to the Late Pleistocene or Early Holocene (Iriarte et al. 2020). Although the boost in food production did not directly translate into an increased population growth rate, it mainly occurred in specific cultural centers, especially in the Andes, where intensive agricultural systems took precedence as the primary subsistence strategy (Goldberg et al. 2016; Perez et al. 2017).

Some studies propose that this second phase of exponential growth continued until the arrival of Europeans, especially in Amazonia (Arroyo-Kalin 2017). In contrast, other research suggests a slowdown or even a decline in population size in certain areas, potentially due to reaching carrying capacity, autochthonous diseases, or environmental and social changes (Arroyo-Kalin and Riris 2021; Bush et al. 2021). Nevertheless, as expected the majority of genetic and archaeological evidence indicates the most significant population decrease rate occurred after the arrival of Europeans in America (Browning et al. 2018; Castro E Silva et al. 2022).

The Andes-Amazon divide model, historically utilized to understand various archaeological, ethnolinguistic, genetic, and demographic patterns in South America, more recently has been criticized for limiting and biasing the study of indigenous peoples, particularly regarding genetic diversity (Pearce et al. 2020). This model prompts an evolutionary framework where opposing dynamics have historically shaped the Andes and Amazonia (Tarazona-Santos et al. 2001). The Andes, characterized by long-standing large populations and intensive food production, led to highly hierarchical and interconnected societies with preserved genetic diversity within populations and reduced genetic differentiation across populations. In contrast, Amazonia, inhabited by small and isolated hunter-gatherer groups in diverse environments, exhibited low within-population genetic diversity and high among-population genetic differentiation due to limited gene flow.

The reassessment of cultural and demographic complexity in the Andes and Amazonia is ongoing as evidence of denser occupation and complex cultures in the Amazon grows (Heckenberger et al. 2003; Heckenberger and Neves 2009; Roosevelt 2013; Piperno et al. 2015; Clement et al. 2015; Pearce et al. 2020). Recent findings reveal highly populated permanent settlements along major rivers, establishing Amazonia as a major center for crop domestication with at least 83 domesticated species, thousands of geoglyphs and even cities within the Amazon (Iriarte et al. 2020; Peripato et al. 2023; Rostain et al. 2024). These developments involved extensive environmental modifications, including the creation of Amazonian dark earths (ADEs), crucial for plant domestication and food production, contributing to increased population growth rates (Clement et al. 2015; Neves and Heckenberger 2019; Iriarte et al. 2020).

18.3.2 Central America and the Caribbean

Certain regions, shaped by unique geographical factors, exhibit distinctive genetic variation patterns, such as Central America, the Caribbean, and the Southern Cone. Central America, was the exclusive entry point for the dispersal of humans into South America and the crucial link between distinct genetic lineages from North and South America, witnessing through time extensive population movements and the exchange of resources, ideas, and populations (Capodiferro et al. 2021). The archaeological and paleoecological records strongly suggest a continuous human occupation of the Isthmo-Colombian region dating back 16 thousand years (Ranere and Cooke 2021). Whereas, patterns of regional genetic continuity in some indigenous communities in continental Central America have been inferred through the analysis of mitochondrial DNA data and nuclear genomes (Reich et al. 2012). While the examination of ancient individuals predating the arrival of Europeans (around 1.4–0.6 thousand years ago) reveals a remarkable genetic continuity spanning nearly a thousand years, with these ancient lineages persisting in the Indigenous

American component present in contemporary admixed populations from the region (Capodiferro et al. 2021).

The analysis of ancient individuals from the Isthmo-Colombian region also unveiled the presence of a unique ancestry component distinct from other areas of the continent. This distinctive ancestry component can be modeled as the result of an admixture event between two distinct SNA lineages originating in North America. Notably, one of these lineages involves an additional contribution from another unsampled population, referred to as unsampled population I (Capodiferro et al. 2021). In turn, the oldest archaeological sites discovered in the Caribbean Islands date back approximately 8 thousand years before the present and are located on the island of Trinidad near the northern coast of South America (Napolitano et al. 2019). Dating between 5.8 and 2.5 thousand years ago, archaeological sites showcasing stone tool technologies from the Lithic and Archaic eras are dispersed across Barbados, Cuba, Hispaniola, Puerto Rico, St. Martin, and Tobago (Napolitano et al. 2019). Comparatively, the oldest archaeological sites identified in Jamaica, San Salvador, and St. Lucia date back to at most 1800 years ago.

The genetic composition of Caribbean human populations is thought to have been shaped by at least two major dispersal events occurring at different times. The first, dating back at least 6 thousand years ago, brought individuals with so-called Archaic stone tool technologies (Fig. 18.3) (Napolitano et al. 2019), although earlier studies proposed two separate population dispersals from South America during this period (Schroeder et al. 2018; Nägele et al. 2020). Analyses suggest that individuals associated with the Archaic period in Cuba derive from a single source (Fernandes et al. 2021), contradicting previous evidence of additional migration with an affinity to North American individuals (Nägele et al. 2020). Approximately 2500 years ago, a new wave of people from the Ceramic Age, originating from Arawak-speaking populations in northern South America, is believed to have arrived (Figs. 18.1 and 18.4). The dispersal of the Ceramic Age ancestry component into the Caribbeans will be further explored in the following section.

18.3.3 Southern Cone

Whereas, in the Southern Cone, the Monte Verde II site in Patagonia stands as the earliest indisputable evidence of human activity, dating back to 14.6 thousand years ago (Figs. 18.1 and 18.3) (Dillehay 2009). Additionally, the region has yielded the most ancient DNA ever genotyped in the continent (Los Rieles site in Chile), dating back 12 thousand years (Posth et al. 2018). These findings highlight an exceptionally rapid colonization of the Southern Cone following the initial arrival on the continent around 16–15 thousand years ago (Prates et al. 2020). Even Tierra del Fuego, the continent's southernmost point, was inhabited before 8000 years ago, a time when it remained connected to South America due to substantially lower sea levels (Morello et al. 2012).

Fig. 18.4 Late Holocene population expansions. This figure summarizes the hypothesis for the Arawak and Tupi-Guarani demic diffusion based on linguistic, archaeological and genetic data, as discussed in the text. Green points on the map pinpoint the approximate location of the earliest Tupiguarani material culture occurrences in the limits of their geographic expansion. While the Caribbean Islands, highlighted in red, indicate where the most ancient evidence of Saladoid-Barrancoid culture was found. Meanwhile, arrows are used to illustrate the hypothetical dispersal routes. Shaded areas represent the center of origin of the Arawak, Tupi and Tupi-Guarani linguistic groups. Important rivers mentioned in the text are labeled and highlighted in a darker tone of blue. The map was generated using the 'sf' and 'ggplot2' R packages, employing the Robinson projection. The numerical sequence of points has been arranged chronologically, referencing the primary timeline depicted in Fig. 18.1

During the Middle Holocene, the ancestors of Southern Cone populations underwent divergence into three primary lineages, as evidenced by ancient individuals from the Argentinean Pampas, Patagonia, and the Center-Southern region of Chile. The lineage originating from the Center-Southern region of Chile represents a group of ancestors to contemporaneous Mapuche population, while the Patagonian lineage corresponds to ancestors of present-day peoples in the Patagonia region (Nakatsuka et al. 2020). Notably, present-day populations from the Southern Cone exhibit elevated genetic affinities among themselves and with contemporary indigenous communities in Central-Southern Chile (Raghavan et al. 2015; la Fuente et al. 2018).

18.4 Population Expansions of the Late Holocene

A transformational time marked by unprecedented population growth along with major worldwide population movements arose during the Late Holocene (Loog et al. 2017; Stoneking et al. 2023). Evidence indicates this demographic growth was sustained by the intensification of food production, impulsed by long processes of landscape modification, the transition to sedentary lifestyles, advancements in technology, but most importantly by plant and animal domestication. The confluence of these factors played a pivotal role in creating demographic expansions and large-scale migratory patterns that have left enduring imprints on the patterns of genetic and cultural diversity distribution across the world (Sokal et al. 1991; Cordaux et al. 2004; Wen et al. 2004; de Filippo et al. 2012; Ammerman and Cavalli-sforza 2014).

These processes also significantly increased the impact of infectious diseases, due to the close proximity between humans and their domesticated animals. Also, the concentration of human populations in densely inhabited areas enhanced pathogen transmission between populations due to migrations (Stoneking et al. 2023). Many infectious diseases, including tuberculosis, plague, herpes, and chickenpox, have their origins in this period (Sikora et al. 2023). The increased selective pressure originated by this wide range of diseases has been recently implicated in the evolution of immune genes, likely contributing to an increased genetic risk of autoimmune diseases such as multiple sclerosis (Barrie et al. 2024).

Furthermore, according to the agricultural expansion hypothesis, between approximately 10,500 and 4500 years ago, the development of food production, centered on the domestication of a limited number of wild plant and animal species, occurred independently in at least nine agriculture and herding homelands, distributed across the world, leading to the transition from hunter-gatherer to farming (and/or herding) subsistence strategy in these places (Diamond and Bellwood 2003).

In line with this hypothesis, the agricultural expansions can be attributed to three key advantages acquired by farmers over hunter-gatherers. Firstly, farmers achieved higher food yields per unit of productive land. Secondly, while usually hunter-gatherer societies are mobile, food-producing societies adopted a sedentary lifestyle, allowing them to accumulate stored food surpluses. Lastly, epidemic infectious diseases that originated from domestic animals provided an additional advantage when facing hunter-gatherer groups. These combined advantages facilitated the displacement, replacement or assimilation of hunter-gatherers in the areas of expansion of these early farmers/herders (Diamond and Bellwood 2003). Numerous examples of this phenomenon have been documented across the world, including the expansion of the Bantu in Africa, the Austronesian in South Asia, Oceania, Polynesia, and Madagascar, and most famously the Indo-European in Eurasia (Stoneking et al. 2023).

By the Late Holocene populations in Central and South America are believed to have broadly settled into more or less cohesive regions, showing prolonged continuity for several thousand years prior to the European arrival (Willerslev and Meltzer 2021). Despite the emergence of Andean civilizations (e.g. Norte Chico, Nazca, Wari, Tiwanaku, and Inca) the population movements during this time are supposed to have been relatively smaller in scale compared to earlier periods of the continent. In this sense, the territorial expansion of these populations over large areas may not have entailed extensive, widespread movements of people.

In spite of that, South America's linguistic landscape reveals compelling correlations with its material culture traditions, particularly the material culture traditions of the Late Holocene. Major language families from South American lowlands, namely Arawak, Karib, Jê (a linguistic family of the Macro-Jê stock), and Tupi-Guarani (a linguistic family of the Tupi stock), are posited to have origins and dispersal centers in the Amazonian region (Dixon and Aikhenvald 2006; Neves 2011). While lacking complete correspondence, existing evidence supports the hypothesis that the geographic distribution of these language families aligns at least partially with some Late Holocene material culture traditions (Noelli 2008; Castro E Silva et al. 2020; Gregorio de Souza et al. 2020; Nägele et al. 2020). Demic diffusion is proposed as the mechanism through which these languages and material culture were propagated together (Box 18.2) (Gregorio de Souza et al. 2020).

Box 18.2: Demic or cultural diffusion?

What are the mechanisms governing the diffusion of farming practices and languages? Did these phenomena occur through migrations of farming communities, thereby introducing their lifestyle and languages (referred to as demic diffusion), or did local hunter-gatherer groups adopt farming and language from nearby farmers (characterized as cultural diffusion)? Moreover, in the context of demic diffusion, the fate of local hunter-gatherer groups becomes crucial: were they entirely displaced/replaced, or did they undergo partial assimilation into the expanding farmer communities?

Although archaeology, biological anthropology, and linguistics were the only sources of evidence in the past, currently the ultimate resolution usually lies within the realm of genetics. Assuming the identification of the homeland of the original farming group and recognizing genetic distinctions between these individuals and the groups into whose territories they expanded, genetic investigations can elucidate the degree to which contemporary populations possess ancestry derived from farmers versus indigenous hunter-gatherer populations (Stoneking et al. 2023).

Specifically, evidence suggests that the Arawak, Karib, Jê, and Tupi-Guarani linguistic families respectively overlap to the Saladoid-Barrancoid, Incised-Punctate, Una, and Tupiguarani material culture traditions (Noelli 2008; Corrêa 2015). Notably, only the Saladoid-Barrancoid and Tupiguarani cultural complexes exhibit a broader geographical presence outside de boundaries of the Amazonian basin. The Saladoid-Barrancoid culture extended its influence to the Caribbean islands of Puerto Rico and Hispaniola, while the Tupiguarani complex spanned over 5000 kilometers across eastern South America, encompassing diverse ecosystems such as the Brazilian highlands, Caatinga, Atlantic forests, and the Argentine pampas (Keegan 1995; Noelli 1998, 2008). Subsequently, we undertake a more intricate exploration of the Tupi Expansion, which is the most well-known from a genetic standpoint. Additionally, we provide a brief discussion of the Late Holocene dispersals into the Caribbean islands, representing the only portion of the Arawak expansion studied using genetic data. The aim is to present a comprehensive illustration of how genetics is being used to infer population dynamics in the Neotropics throughout the Late Holocene.

18.4.1 Tupi Expansion

Tupi speakers represent a group of Amazonian peoples which originated and adapted specifically to the challenges of forest life, particularly thriving in riverine forest environments. The widespread geographical distribution of the Tupi people has been a subject of considerable interest for archaeologists and linguists for a long time. Over the years, there has been a prevailing hypothesis suggesting that the Tupi people originated and expanded from a unique common origin in the Amazon (Noelli 1998, 2008; Castro E Silva and Hünemeier 2023). Several lines of evidence converge to support that the Tupi's center of origin lies in the region between the Madeira and Guaporé rivers in Southwestern Amazonia (Figs. 18.1 and 18.4). One compelling line of evidence is the higher linguistic diversity among Tupi speakers in this area, notably characterized by the presence of the most deeply diverged languages within the Tupi language family (Campbell and Grondona 2012; Walker et al. 2012). Additionally, genetic diversity in populations from this region exceeds that found in populations from other areas (Ramallo et al. 2013; Santos et al. 2015). Particularly in the case of the Tupi-Guarani, the most widespread of the 10 Tupi linguistic families, evidences suggest a center of dispersal in a region of eastern Amazonia between the Xingu and Tocantins rivers (Figs. 18.1 and 18.4) (Almeida and Neves 2015; Souza Mello and Kneip 2017; O'Hagan et al. 2019).

Aligned with the agricultural expansion hypothesis, the Tupi people developed an agricultural food production system, however diverging significantly from the prevalent strategies adopted by agriculturalists worldwide. In contrast to the common practice of monocultures involving one or more cereal species in highly anthropized environments, the Tupi developed a system known as polyculture agroforestry (Neves 2013; Iriarte et al. 2020). This distinctive approach seamlessly integrated the cultivation of domesticated plants with the management of semi and non-domesticated plants within forest environments. The Tupi's agricultural practices showcase their cultural adaptation to the intricate ecosystems of the Amazon rainforest, reflecting a profound understanding of sustainable and resilient food production systems tailored to their unique environmental context.

The available evidence, including genetic analyses, support a demic diffusion model aligning with a longstanding hypothesis positing that the dispersal of the Tupi-Guarani population from Amazonia was propelled by continuous demographic expansion sustained by the development of their agriculturalist food production systems (Brochado 1984; Corrêa 2015; Castro E Silva et al. 2020). Moreover, insights from paleoecological and paleoclimatic investigations suggest that a more humid climate during the Late Holocene led to forest expansions in the southern Hemisphere around 3 to 2 thousand years ago, coinciding with the onset of the Tupi Expansion (Iriarte et al. 2017). The expansion of riverine forests, in particular, would have presented an ecological opportunity for the Tupi-Guarani population to expand, as it provided the essential environmental conditions conducive to their adapted polyculture agroforestry practices for food production. This highlights how the confluence of ongoing cultural evolution of Indigenous communities in the region, encompassing aspects such as plant domestication, landscape modification, and the development of food production systems with opportune environmental changes, resulted in heightened rates of population growth and, subsequently, territorial expansion in some instances.

The Tupiguarani material culture tradition can be internally classified into three primary sub traditions: Guaraní, Amazon Tupinambá, and Atlantic Forest Tupinambá. These subtraditions are primarily distributed in the Paraná Basin, southeastern Amazonia, and the Atlantic coast, respectively (Almeida and Neves 2015). Current archeological evidence indicates the initial appearance of Tupiguarani ceramics in these regions occurred approximately between 1600 and 2010 years ago (Figs. 18.1 and 18.4) (Chmyz 1983; Pärssinen et al. 2005; Scheel-Ybert et al. 2008; Corrêa 2015; Almeida and Neves 2015). The model that emerges for the expansion of the Tupi-Guarani, with support from the current genetic evidence, is that a branch of the Tupi-Guarani migrated southward, settling in the Paraná basin and eventually giving rise to the Guarani (Fig. 18.4). Simultaneously, another branch radiated towards the mouth of the Amazon river, subsequently dispersing along the Atlantic coast, extending down to present-day Southeastern Brazil, forming the Tupi coastal populations, collective denoted as Tupinambá (Fig. 18.4) (Brochado 1984; Corrêa 2015; Castro E Silva et al. 2020).

18.4.2 Arawak Expansion into the Caribbean Islands

In turn, a connection between the Saladoid-Barrancoid ceramics and Arawakan languages was first proposed by American archeologist Donald Lathrap (Lathrap 1970). According to existing evidence, it is posited that the Arawak population originated from a dispersion center located in the northwestern Amazonian region, spanning from the Upper Amazon river in Brazil to the Middle Orinoco river in Venezuela (Figs. 18.1 and 18.4) (Lathrap 1970; Rouse 1997). Linguistic analysis suggests that approximately 3000 years ago, Arawak-speaking communities experienced a rapid dispersion across the Negro and Orinoco floodplains. This dispersion extended through the Amazon, reaching as far as the Caribbean and Guyana coasts, and finally the Antilles (Figs. 18.1 and 18.4) (Hill and Santos-Granero 2002; Heckenberger 2013).

In recent years, genetic studies have played a pivotal role in shedding light on the intricate details of human settlement in the Caribbean islands by Arawak speaking populations. These investigations have provided a more nuanced understanding of the population dynamics and migratory patterns that shaped the genetic landscape of the region. By delving into the genomic data of present-day populations and analyzing ancient DNA from archaeological sites, researchers have been able to trace the footsteps of these ancestral communities.

The image that emerges from this research is that, circa 2500 years ago, Saladoid material culture started to appear in Puerto Rico and the northern Lesser Antilles marking the start of the Ceramic Age in the Caribbean (Figs. 18.1 and 18.4) (Keegan and Hofman 2017). The Ceramic Age newcomers would have encountered Archaic Age peoples and interacted, yet little to no evidence of admixture has been identified (Napolitano et al. 2019; Nägele et al. 2020). In the majority of the Greater Antilles, the Archaic Age ancestry component underwent a substantial replacement (>98%) by Ceramic Age ancestry. Contrarily, in Cuba, the Archaic Age ancestry persisted with minimal admixture for over 2500 years (Fernandes et al. 2021). Most Ceramic Age individuals are inferred to be largely genetically homogenous and linked to a northwestern South America origin (Fernandes et al. 2021), the hypothesized Arawak homeland (Lathrap 1970; Rouse 1997). Finally, present-day individuals from Puerto Rico and Cuba exhibit approximately 14% of their ancestry attributed to the Ceramic Age ancestors.

18.5 Human Genetic Adaptations to the Neotropical Environments

Neotropical America, with its vast expanses of tropical rainforests, high mountains, savannas, and deserts, is home to a rich diversity of ethnic groups and cultures. Over millennia, evolutionary processes, especially natural selection, have played a crucial role in shaping the genetic diversity and differentiation of these populations. This selective force has led to genetic (Fig. 18.5), physiological and cultural adaptations in response to the region's diverse environmental, climatic and pathogenic conditions. In this diverse landscape, human evolution unfolded, with local conditions and resources playing a fundamental role in driving natural selection.

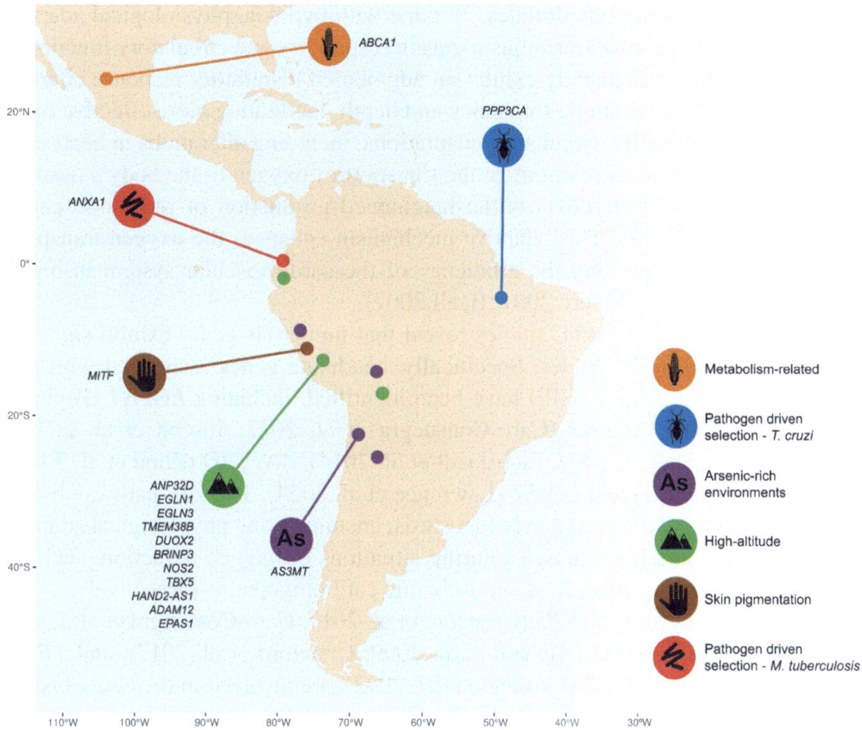

Fig. 18.5 Genetic Adaptations to the Neotropical Environments. This figure highlights some genetic signatures of natural selection found in the Neotropical environment. The dots in orange represent signs of selection in genes related to metabolism, in blue, selection directed by the pathogen *T. cruzi*, in red by *M. tuberculosis*, in purple the arsenic-rich environment, and in brown and green in response to high altitudes. (Modified from Hünemeier (2024))

18.5.1 Adaptations to the Andean highlands

Continuously inhabited for at least 12,000 years by modern humans (Rademaker et al. 2014), the Andes highlands stand as one of the most extreme environments in the Neotropical region, imposing a myriad of challenges on organisms that necessitate physiological and ecological adaptations for survival (Espinoza-Navarro et al. 2011). With average altitudes reaching 4000 meters, the Andes present low oxygen concentrations, extreme temperatures, harsh weather conditions, and harmful doses of ultraviolet radiation, all compounded by limited resources. These local characteristics have spurred numerous studies to understand how organisms respond to this seemingly inhospitable environment.

Undoubtedly, the primary challenge posed by the Andean highlands is the reduction in atmospheric pressure, leading to a diminished availability of oxygen. This condition, referred to as hypoxia, significantly impacts the health and well-being of individuals residing at high altitudes. To cope with hypoxia, physiological adaptations come into play, involving adjustments in respiratory and circulatory functions. Andean populations frequently exhibit an augmented respiratory response characterized by increased breathing frequency and depth, facilitating more effective oxygen uptake. Additionally, circulatory adaptations, including alterations in heart rate and cardiac output, work to optimize the transport of oxygen to the body's tissues. A notable response to hypoxia is the heightened production of red blood cells, known as polycythemia. This adaptive mechanism enhances the oxygen transport capacity, thereby improving the efficiency of the cardiovascular system in high-altitude environments (Moore 2001; Beall 2007).

From a genetic standpoint, studies reveal that numerous genes exhibit signs of positive selection in the Andes. Specifically, candidate genes associated with the *Hypoxia Inducible Factor* (HIF) have been identified, including *EGLN1* (Bigham et al. 2009, 2010), *EGLN3* (Caro-Consuegra et al. 2022; Joseph et al. 2023), *PRKAA1* (Bigham et al. 2009; Eichstaedt et al. 2014), *ANP32D* (Zhou et al. 2013; Eichstaedt et al. 2014) and *EPAS1* (Lawrence et al. 2024). The HIF pathway holds significance in regulating oxygen homeostasis, encompassing physiological adaptations in skeletal and heart muscles during situations of oxygen reduction, such as those induced by muscular effort and ischemic cardiomyopathy, respectively.

Other genes, such as *BRINP3* (Crawford et al. 2017; Caro-Consuegra et al. 2022; Joseph et al. 2023), *NOS2* (Bigham et al. 2009; Crawford et al. 2017), and *TBX5* (Crawford et al. 2017; Caro-Consuegra et al. 2022), are involved in the Nitric Oxide pathway (NOS), which at high altitudes acts as a vasodilator, relaxing blood vessels to enhance oxygen transport to tissues. The NOS pathway plays a pivotal role in regulating blood flow, immune responses, and cardiovascular health. Positive selection signals have been also identified in the *DUOX2* gene (Jacovas et al. 2018; Borda et al. 2020; Caro-Consuegra et al. 2022), associated with the generation of Reactive Oxygen Species (ROS). Exposure to reduced atmospheric pressure and oxygen levels at high altitudes induces oxidative stress, surpassing the body's detoxification capacity and leading to an increased production of ROS. Organisms acclimated to high altitudes exhibit adaptive responses, enhancing antioxidant defense mechanisms to counteract excessive ROS and mitigate oxidative stress. While excessive ROS levels can be detrimental, moderate levels play a pivotal role in cellular signaling pathways associated with adaptive responses, including angiogenesis (Schieber and Chandel 2014).

Additionally, numerous candidate genes exhibit a TP53 pathway selection signal, centered around the p53 protein, which safeguards genomic stability at high altitudes by orchestrating DNA repair mechanisms in response to elevated ultraviolet radiation, oxidative stress, and reduced oxygen levels (Eichstaedt et al. 2014; Jacovas et al. 2015, 2018). The TP53 pathway is also involved in regulating apoptosis versus cell survival, influencing tissue homeostasis in the face of stressors like hypoxia. The pathway's involvement in mitigating oxidative stress and collaborating with other signaling pathways highlights its central role in cellular adaptation to the unique challenges of high-altitude environments.

Beyond hypoxia, other signs of selection have been identified. In Andean Peruvians, a notable signal was found in *MITF*, a gene involved in the development of melanocytes, which is associated with both pigmentation and the vitamin D receptor (Caro-Consuegra et al. 2022). Additionally, a recent study uncovered selective signals in genes associated with the immune response. Specifically, the study identified signs of positive selection in genomic regions linked to the immune response against tuberculosis (*Mycobacterium tuberculosis*) (Joseph et al. 2023). The candidate genes *ANXA1*, *CD44*, and *ADGRE*1 were highlighted for their roles in the direct recognition of bacterial antigens by innate immune macrophages, while the genes *KCNA3* and *FCRL4* were implicated in the pro-inflammatory adaptive immune response. This suggests that, alongside adaptations to hypoxia, there are evolutionary changes in genes related to immune responses in high-altitude environments, particularly in the context of combating infectious diseases.

The dynamic interactions between indigenous people and their environment in the Andes offer a wealth of opportunities to investigate not only physiological adaptations, but also the underlying genetic, molecular, and ecological processes that contribute to their resilience. Although there are other regions of the world where people live at high altitudes, such as Tibet and Ethiopia, most of the selective signals observed in the Andean highlands are unique. This uniqueness underscores the importance of studying Andean populations as a distinct and valuable resource for understanding human adaptation to extreme environments.

18.5.2 Adaptations to the Andean Arsenic-rich Environments

One of the most intriguing examples of natural selection is observed in arsenic-rich Andean environments. Arsenic contamination of groundwater is a prevalent issue in many parts of the Andean region, often stemming from natural sources of arsenic in geological formations or mining activities (Cooke and Abbott 2008). This contamination can result in elevated levels of arsenic in rivers and drinking water, posing a significant threat to the health of local populations (Smith et al. 2006; Van Den Bergh et al. 2010). Prolonged exposure to high levels of arsenic through drinking water is associated with various health problems, including skin lesions and cancer (Martinez et al. 2011; Sage et al. 2017). Over thousands of years, human communities in the Andean highlands have coped with drinking water tainted by arsenic, prompting the question of whether these populations have developed adaptations over time to thrive in their potentially toxic environment.

Initial insights emerged from studies conducted on indigenous women in the northern Argentine Andes who regularly consumed drinking water with elevated arsenic concentrations. These women displayed genetic variations in the *AS3MT* gene, a pivotal player in arsenic metabolism. Individuals carrying the arsenic tolerance haplotype can metabolize arsenic more rapidly. Consequently, by minimizing exposure to toxic metabolites, they may have gained a selective advantage in environments with high arsenic content (Schlebusch et al. 2015).

Since the initial report of arsenic adaptation in humans, signatures of natural selection in *AS3MT* gene have been identified in studies involving indigenous populations from various regions of the Andes, including Argentina, Chile, Peru, and Bolivia (Eichstaedt et al. 2015; Apata et al. 2017; Jacovas et al. 2018; De Loma et al. 2022). These findings strongly support the hypothesis that arsenic exposure has been a driving force for human adaptation, leading to enhanced tolerance through more efficient arsenic detoxification in diverse Andean populations.

In the realm of environmental adaptations, numerous organisms have demonstrated the ability to tolerate toxic chemicals in their surroundings. Surprisingly, our understanding of human adaptations to such toxic chemicals remains limited. The discoveries regarding protective variants of the *AS3MT* gene open a new avenue for exploring how human populations might develop genetic resilience to environmental toxins, shedding light on the intricate mechanisms of adaptation within the human genome not only in the Neotropics but in other environments around the world.

18.5.3 Adaptations to the Amazon Rainforest

The Amazon rainforest is frequently acknowledged as one of the most biodiverse areas on the planet, given its extensive variety of plant, animal, and microbial species. The wealth of biodiversity in the Amazon rainforest is contrasted by the formidable challenges it presents, rendering it a harsh environment for long-term human habitation. Despite its nutrient-rich characteristics, the Amazon introduces distinctive obstacles to sustained human survival, characterized by the unpredictable availability of food resources (Bailey et al. 1989; Headland and Bailey 1991), restricted light penetration resulting from the dense canopy (Ratnam et al. 2011), and a diverse array of pathogens (Guernier et al. 2004). Notably, tropical forests, including the Amazon, are estimated to harbor approximately 70% more pathogens compared to temperate forests (Guernier et al. 2004). This elevated pathogenic diversity makes pathogens and tropical diseases potential contributors to local selective pressures.

Regardless of the unique characteristics of the Amazon environment, there remains a scarcity of studies that have assessed natural selection within this region. In 2015, Amorim et al. conducted a study evaluating convergent evolution between populations in the Amazonian and African rainforest. Identifying shared selective signals, the study primarily pinpointed genes associated with the immune system (*CCL28, CWH43, SCP2*), lipid metabolism (*CWH43, SCP2*), and thermoregulation (*HSF2*), among others (Amorim et al. 2015). This research provides insights into the genetic adaptations of populations in rainforests, shedding light on specific pathways influenced by natural selection in response to the challenges posed by these diverse and intricate environments.

Another study, focused on indigenous populations from the Peruvian Amazon and the Amazon lowlands, revealed indications of selection in the *PTPRC* gene. This gene plays a crucial role in the production of the CD45 protein, which is involved in the recognition of pathogens, particularly viral ones (Borda et al. 2020). Furthermore, in another investigation, indications of selection were noted in the *CBLB* and *CARD8* genes among indigenous individuals from the Peruvian Amazon. The *CBLB* gene encodes a negative regulator of adaptive immune responses, while *CARD8* mediates inflammasome activation in response to pathogens and other signals (Caro-Consuegra et al. 2022).

Finally, in a recent genomic and functional exploration within indigenous populations of the Brazilian Amazon, a prominent signal of natural selection was identified in a cluster of genes associated with the immune response to *Trypanosoma cruzi* infection (*PPP3CA* and *DYNC1I1*) and to the mosquito bite reaction (*NOS1AP*). *T. cruzi* is the parasite responsible for Chagas disease, endemic to America. The study uncovered that a variant of the *PPP3CA* gene provided a protective effect against Chagas disease infection. Through comprehensive functional studies, it was demonstrated that this gene plays a role in reducing the infection rate among carriers, unveiling a potential mechanism contributing to resistance against Chagas disease in Amazonian populations (Couto-Silva et al. 2023). This study underscores the potential significant role of Chagas disease in driving natural selection within native Amazonian populations. The findings highlight the imperative for more comprehensive exploration of human adaptations within the intricate environmental dynamics of the Amazon rainforest.

18.6 Adaptations Gene-culture Coevolution

Human evolution is not solely driven by the natural environment, climate, and pathogens; rather, it is profoundly shaped by cultural practices, knowledge, and innovations that influence how societies adapt to their surroundings. One of the most compelling examples of natural selection involving cultural factors in human populations is linked to the emergence of agriculture. The development of agriculture was a pivotal milestone in human history, often signifying a transition from nomadic, foraging societies to settled communities engaged in farming and animal husbandry.

Genetic adaptations might have occurred in response to the new dietary and environmental challenges introduced by agriculture (Balaresque et al. 2007; O'Brien and Laland 2012). For example, the increased reliance on cereal grains and domesticated animals could have led to genetic changes related to the metabolism of new food sources (Perry et al. 2007), lactose tolerance (Bersaglieri et al. 2004; Voight et al. 2006; Tishkoff et al. 2007), and resistance to certain diseases associated with agricultural settlements (Balaresque et al. 2007; O'Brien and Laland 2012).

In the Neotropical region, the most striking example of selection driven by agricultural practices comes from the indigenous populations of Mexico. Agricultural practices originated in Mexico between 10,000 and 7000 yBP, and the region is considered, along with the Andes, one of the primary centers for plant domestication and the development of agricultural practices in America (Pope et al. 2001; Zizumbo-Villarreal and Colunga-GarcíaMarín 2010). The cultivation of crops such as beans, squash, and especially maize formed the foundational basis for early agricultural societies in Mesoamerica (Zizumbo-Villarreal and Colunga-GarcíaMarín 2010).

A previous study revealed that Native American populations exhibit a natural selective signature on an autochthonous allele of the *ABCA1* gene (rs9282541) from America. These days, this allele is associated with low HDL cholesterol levels and obesity-related comorbidities. However, in the past, this variant may have favored intracellular cholesterol retention and energy storage (Acuña-Alonzo et al. 2010).

Upon further study of this variant, Hünemeier et al. discovered a significant correlation between the frequency of the *ABCA1*230Cys*-derived allele and the distribution of *Zea* pollen relics in Mesoamerican populations (Hünemeier et al. 2012). Using a Bayesian approach, the authors estimated the age of the derived allele at 7500 yBP, aligning with the age of domestication of maize in southwestern Mexico, approximately 10,000–6200 yBP. These discoveries reveal how a constructed agricultural niche and an autochthonous Native American allele became the focal points of an ongoing directional selective sweep as a consequence of the origin and spread of maize culture in ancient Mesoamerica.

This is an example of niche construction, showcasing how human beings, rather than remaining passive to their environment, actively engage in shaping it. Niche construction underscores the active role organisms play in molding their ecological and cultural surroundings. The subsequent modifications to this environment can, in turn, influence natural selection and evolutionary processes by favoring traits that prove advantageous within the altered niche. However, according to the thrifty genotype hypothesis, some authors argue that evolutionary history did not end there. Some of these genetic variants, which played a crucial role during the agricultural food transition process, are now associated with the development of a range of metabolic diseases in our contemporary society, characterized by an abundance of food resources and ultra-processed foods (see Box 18.3) (Neel 1962; Acuña-Alonzo et al. 2010; Yang and Ye 2021; Johnson et al. 2022).

Box 18.3: Thrifty Genotype Hypothesis and Domestication of Maize in Mesoamerica

The thrifty genotype hypothesis, proposed by geneticist James V. Neel in 1962, suggests that genetic variants advantageous in times of food scarcity have undergone positive selection throughout human evolutionary history. These variants promoted efficient energy utilization in ancient times, aiding adaptation to fluctuating food availability. However, in modern societies with abundant food resources, these same variants may lead to metabolic disorders.

The introduction of maize cultivation in Mesoamerica brought significant shifts in settlement patterns, with communities heavily reliant on maize as a primary food source. Despite expectations, agricultural and sedentary lifestyles did not guarantee food stability, as evidenced by periodic crop failures (Wells and Stock 2020). In this context, carriers of the ABCA1 variant (rs9282541) may have experienced selective advantages. This variant, associated with reduced cholesterol efflux, likely facilitated intracellular cholesterol and energy storage, benefiting adipose tissue in various functions such as energy regulation, reproduction, and immune support. In current Neotropical native populations, this variant is observed with an average frequency of 12% (ranging from 0% to 31%) (Acuña-Alonzo et al. 2010). However, in contemporary societies, these variants may contribute to metabolic diseases like low HDL levels, obesity (Villarreal-Molina et al. 2007; Acuña-Alonzo et al. 2010), cardiovascular disease and type 2 diabetes (Ochoa-Guzmán et al.

2020; Peña-Espinoza et al. 2024). This juxtaposition highlights the dual nature of the thrifty genes, wherein beneficial adaptations in ancestral environments can predispose individuals to health risks in modern contexts of excess resources (Fig. 18.6).

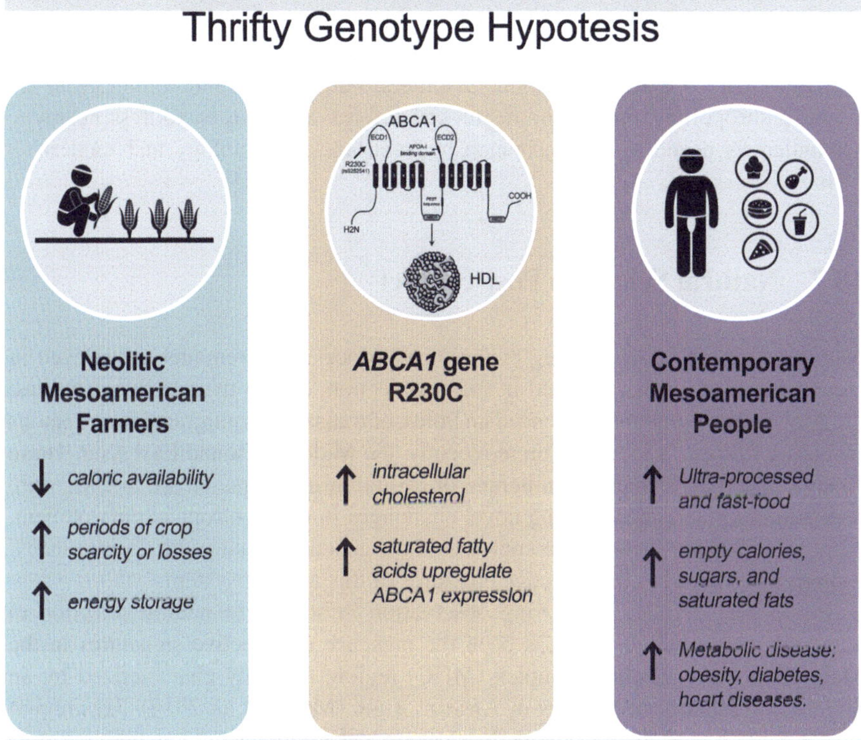

Fig. 18.6 Thrifty Genotype Hypothesis applied to autochthonous *ABCA1 R230C* variant (rs92822541) in Neotropical Mesoamerica. *ABCA1* is a protein playing a fundamental role in cholesterol homeostasis, mediating the formation of high-density lipoproteins and facilitating cholesterol efflux. Functional studies indicate that the *R230C* variant (rs9282541) demonstrates reduced cholesterol efflux. The presence of this variant favors the accumulation of 30% more intracellular cholesterol and may impact energy storage. This energy reserve could have conferred an adaptive advantage to Neolithic Mesoamerican populations during the transition from hunter-gatherer to maize farming lifestyles, particularly due to low caloric availability following the advent of agriculture and periods of crop scarcity or losses. Conversely, numerous studies conducted in the current Mesoamerican population have linked this variant to metabolic diseases such as obesity, cardiovascular disease, and diabetes. It's recognized that recent decades have witnessed a significant dietary transition marked by the prevalence of industrialized and ultra-processed foods, which are rich in empty calories, sugars, and saturated fats. Recent research indicates that diets high in saturated fatty acids upregulate *ABCA1* expression. Furthermore, the R230C variant interacts with dietary macronutrients, impacting plasma lipid levels, BMI, HDL-C, and serum adiponectin levels, which are often associated with metabolic diseases. Thus, this example may support the hypothesis of thrifty genotypes in Neotropical Mesoamerica.

(*continued*)

Box 18.3: (Continued)

The cultivation of maize in Mesoamerica not only shaped genetic adaptations but also influenced patterns of human disease. The transition from traditional maize-based diets to modern diets rich in processed foods may have increased metabolic disorders among indigenous populations in the region (Mendoza-Caamal et al. 2020). Understanding the relationship between maize cultivation, genetic adaptations, and human disease offers insights into the evolutionary origins of metabolic disorders. Furthermore, integrating genetics, anthropology, and public health can inform strategies to address health challenges posed by the mismatch between ancestral biology and modern lifestyles.

18.7 Natural Selection Post-contact

Natural selection is an ongoing evolutionary process that remains active even in recent times. As further detailed in the next section, during the last five centuries, the American continent experienced an influx of transatlantic migrants from regions including Europe, Africa, and more recently, the Middle East and East Asia. These foreign genomes, shaped by millennia of evolution in diverse environments, were now admixing and encountering novel challenges in the Neotropical environment. Now, scientists are striving to comprehend how genetic variants with diverse ancestral origins respond to the selective pressures of this new ecosystem.

In the past decade, a recurring observation in studies of natural selection in admixed Latino populations has been the presence of selective signatures in the Major Histocompatibility Complex (MHC) region, typically characterized by an increased frequency of variants of African origin (Meyer et al. 2018). This region stands out as one of the most gene-dense areas in the human genome, with a majority of its genes closely associated with the immune response. Within this complex, the HLA gene family holds particular prominence, playing a critical role in the immune system. These genes encode proteins that are integral to recognizing foreign substances (antigens), including pathogens and toxins, showcasing the vital role of the HLA in shaping immune responses and reflecting its significance in the genetic landscape of admixed Latino populations.

Two additional studies, focusing on identifying signs of selection across the genomes of admixed Latino populations, observed signals in genes related to the immune system, metabolism, and diseases. In one study encompassing populations from Colombia, Mexico, Peru, and Puerto Rico, genes associated with the immune system, particularly in pathways related to cytokine receptor interaction, T cell receptor signaling, and antigen presentation, were identified. The analysis also highlighted genes related to metabolism, including drug and xenobiotic metabolism, and steroid hormone biosynthesis. Furthermore, the study discovered genes associated with specific diseases such as type I diabetes, Alzheimer's disease, and

Leishmaniasis (Norris et al. 2018). The other study, involving populations from Brazil, Chile, Colombia, Mexico, and Peru, identified 15 genes associated with the development or regulation of the immune system and 9 genes related to dietary practices through their association with metabolism-related phenotypes or anthropometric measurements (Mendoza-Revilla et al. 2022).

In the exploration of post-contact natural selection within Latin populations, researchers have unearthed genetic variants stemming from different ancestral origins, encompassing African, European, and Native American roots. This discovery underscores the intriguing possibility that alleles originating from various genetic backgrounds may confer adaptive advantages in the context of Neotropical America. It is noteworthy that the observed selection signals may exhibit commonality across multiple populations, indicating shared adaptive responses. However, the specificity of these signals to particular populations adds an additional layer of complexity, highlighting the intricate interplay of genetic diversity and adaptation within the diverse mosaic of Latin American populations.

18.8 Genetic Distinctiveness of Contemporary Neotropical Populations

The last five centuries, the American continent has witnessed unprecedented population dynamics, movements, and demographic shifts, beginning with the European contact in 1492. The colonization process initiated a profound demographic transformation, leading to the extermination of over 90% of the indigenous populations, numbering in the millions at the time (Adhikari et al. 2017). Various factors contributed significantly to this decimation process, encompassing epidemics, enslavement, extermination through wars of conquest, manipulation of conflicts between rival indigenous groups, forced displacement from ancestral territories, habitat destruction, disruption of traditional subsistence strategies, and the erosion of indigenous knowledge, among other impactful elements (Thornton 1987; Stannard 1992). Some evidence suggests that the depopulation was more intense later during the colonization period in more isolated areas (Jones et al. 2021).

The colonization era also witnessed extensive admixture among Indigenous populations, European colonizers, and enslaved Africans forcibly brought to the continent, shaping the genetic landscape in unprecedented ways. The unique combination of these factors highlights the distinctiveness of the Western Hemisphere, particularly in terms of human genetic history of Latin America. This region experienced a much more widespread admixture process compared to North America, where the segregation between Indigenous Americans, European, and African populations was more pronounced (Adhikari et al. 2017). Hence, the genetic study of these populations presents a valuable opportunity to explore the effects of large-scale admixture and population reduction events on human genetic diversity patterns and evolution.

In this sense, the historical trajectory of America has resulted in the contemporary genetic diversity of the region encompassing a substantial portion of the world's human genetic variability, as previously highlighted by others (Adhikari et al. 2017). Although the predominant ancestry of present-day populations in America can be traced to the three primary continental ancestral populations (Native Americans, Europeans, and Africans) and the contribution from ancestors originating in Asia, Oceania, and other global regions has been minimal and largely confined to specific geographic areas. Furthermore, the genetic makeup of contemporary American populations is a consequence of the varying contributions from these three major continental ancestry components. These components were intermixed in diverse proportions and timelines, influenced by various socioeconomic and demographic factors.

Another phenomenon contributing to the distinctive contemporary human genetic diversity in America and the Neotropical region is the pronounced non-random mating dynamics in the admixture processes that shaped these populations, including more evidently the sex-biased admixture, most prevalent during the initial period of the colonization (Adhikari et al. 2017). This non-random mating primarily occurred between European men and Indigenous American and African women (Pena 2002; Kehdy et al. 2015). Several factors led to this pattern, including the predominance of European male settlers, their dominant socioeconomic roles, the elevated mortality rate among Indigenous and African men, who were also prevented from leaving descendants, along with the sexual violence against women from these latter groups.

Analyses of mitochondrial DNA and Y chromosome have unequivocally demonstrated that the majority of paternal genetic lineages trace back to European ancestors, while the predominant maternal lineages can be linked to Indigenous American or African origins (Carvajal-Carmona et al. 2000; Bedoya et al. 2006; Adhikari et al. 2016). The investigation of the X chromosome, as well as autosomal genetic variation, has consistently corroborated these findings (Adhikari et al. 2016; Webster and Wilson Sayres 2016). An additional important factor influencing the patterns of genetic diversity of American populations is the ancestry-assortative mating, which happens when individuals select mates with comparable amounts of ancestry (Kehdy et al. 2015). Socioeconomic factors are likely important underlying causes of this phenomenon, given that they are often linked with ancestry proportions, influencing mating probabilities and genetic diversity patterns (Mas-Sandoval et al. 2023).

In general, recognizing the genetic diversity of present-day Neotropical populations necessitates an understanding that it is shaped by a complex interplay of past demographic events and the proportional contributions from continental ancestral populations. This insight is paramount for the formulation of effective public health policies and the development of tailored health treatments that account for the specific genetic risks inherent in these diverse populations. By unraveling this intricate genetic history, geneticists pave the way for targeted interventions that address the unique health challenges faced by Neotropical communities, ultimately fostering more equitable and effective healthcare strategies.

18.9 Final Remarks

The intricate history of the peopling of America, shaped by human migrations originating from Asia, reveals a multifaceted evolutionary and demographic history. This process encompasses the enduring genetic legacy arising from the contact and intermixing dynamics of Northeast and East Asian ancestral populations, the emergence of a distinctive genetic diversity within the Beringian ancestors, the rapid dispersal of the ancestors of Native Americans following the Last Glacial Maximum, and the subsequent successive population turnovers. This extensive history unfolds against the backdrop of an intricate process of biological and cultural adaptation, prompted by the immensely diverse environments encountered across the continent.

This process of adaptation has produced in the Neotropics not only incredibly rich cultural diversity, including multiple strategies of food production integrated to their ecosystems, but also the evolution of genetic adaptations to the local environmental challenges, encompassing adaptations to consumption of the food resources and to respond to pathogens. The study of this complex history provides profound insights into how human demographic dynamics and evolution has shaped their genetic diversity and impact the health of present-day populations.

The study of human occupation in the Neotropical region also provides a compelling illustration of the profound influence of past climate changes on human demography. This encompasses critical aspects such as the distribution, growth rates, and movement patterns of populations. These past climatic shifts altered the selective pressures within environments, sometimes even creating ecological opportunities that facilitated the thriving and expansion of certain groups. These findings underscore how historical climatic shifts have shaped human history, offering valuable insights into the complex interplay between environmental factors and human demographic patterns.

The long-term consequences of this genetic history can be found in both modern native and non-native populations throughout the region, where they had a considerable influence on phenotypic diversity and, more importantly, health outcomes. This impact includes genetic risks and adaptations linked with traditional and modern industrialized diets, susceptibility to infectious and auto-immune diseases, the genetic burden of deleterious mutations introduced by diverse global populations and propelled to higher frequencies by genetic drift, among other health-related aspects. The complex relationship between historical genetic legacies and present-day health underscores the significant impact of the human recent and distant past on the well-being of populations in the region.

Key Points

- Indigenous American populations trace their ancestry back to ancestors who lived in Northeast and East Asia during the Late Pleistocene.
- Distinct ancestry components were formed within the American continent and several population turnovers and demographic shifts shaped and reshaped the genetic makeup of the continent.
- Serial population bottlenecks and founder events led to an increased effect of genetic drift, both reducing the genetic diversity and increasing the proportion of deleterious variants in present-day indigenous and admixed American populations.
- The European colonization of America dramatically changed the genetic diversity patterns of human populations, causing intense population reductions and widespread admixture of at least three continental ancestry components (African, European and Indigenous American).
- Human populations, both in the past and present, adapted to Neotropical environments, diets and pathogens. Several examples have been documented, including resistance to Chagas disease, adaptations to high-altitude and arsenic-rich environments, among others.

Questions

1. How do Indigenous American populations relate to other human populations, and what distinguishes the genetic diversity of Indigenous Americans from that of other groups?
2. Analyze the profound demographic transformation initiated by European contact in 1492, including the population bottlenecks and admixture events. How have these factors influenced the genetic diversity of human populations in the American continent?
3. What significance does the study of human genetic diversity hold in enhancing our understanding of human diseases? Furthermore, why is it particularly important to focus on the study of underrepresented populations?
4. In what ways have evolutionary processes, particularly natural selection, influenced the genetic diversity of populations in the Neotropical America? Can you provide examples of human genetic adaptations evolved in response to Neotropical environments?

References

Acuña-Alonzo V, Flores-Dorantes T, Kruit JK et al (2010) A functional ABCA1 gene variant is associated with low HDL-cholesterol levels and shows evidence of positive selection in Native Americans. Hum Mol Genet 19(14):2877–2885. https://doi.org/10.1093/hmg/ddq173

Adhikari K, Mendoza-Revilla J, Chacón-Duque JC et al (2016) Admixture in Latin America. Curr Opin Genet Dev 41:106–114. https://doi.org/10.1016/j.gde.2016.09.003

Adhikari K, Chacón-Duque JC, Mendoza-Revilla J et al (2017) The genetic diversity of the Americas. Annu Rev Genomics Hum Genet 18:277–296. https://doi.org/10.1146/annurev-genom-083115-022331

de Almeida FO, Neves EG (2015) Evidências arqueológicas para a origem dos Tupi-Guarani no leste da Amazônia. Mana 21:499–525. https://doi.org/10.1590/0104-93132015v21n3p499

Ammerman AJ, Cavalli-sforza LL (2014) The neolithic transition and the genetics of populations in Europe. Princeton University Press

Amorim CEG, Daub JT, Salzano FM et al (2015) Detection of convergent genome-wide signals of adaptation to tropical forests in humans. PLoS One 10:e0121557. https://doi.org/10.1371/journal.pone.0121557. eCollection 2015

Antonelli A, Zizka A, Carvalho FA et al (2018) Amazonia is the primary source of Neotropical biodiversity. Proc Natl Acad Sci USA 115:6034–6039. https://doi.org/10.1073/pnas.1713819115

Apata M, Arriaza B, Llop E, Moraga M (2017) Human adaptation to arsenic in Andean populations of the Atacama Desert. Am J Phys Anthropol 163:192–199. https://doi.org/10.1002/ajpa.23193

Arroyo-Kalin M (2017) Human niche construction and population growth in pre-Columbian Amazonia. Archaeol Int 20:122–136

Arroyo-Kalin M, Riris P (2021) Did pre-Columbian populations of the Amazonian biome reach carrying capacity during the Late Holocene? Philos Trans R Soc Lond Ser B Biol Sci 376:20190715. https://doi.org/10.1098/rstb.2019.0715

Bailey RC, Head G, Jenike M et al (1989) Hunting and gathering in tropical Rain Forest: is it possible? Am Anthropol 91:59–82. https://doi.org/10.1525/aa.1989.91.1.02a00040

Balaresque PL, Ballereau SJ, Jobling MA (2007) Challenges in human genetic diversity: demographic history and adaptation. Hum Mol Genet 16 Spec No. 2:R134–R139. https://doi.org/10.1093/hmg/ddm242

Barrie W, Yang Y, Irving-Pease EK et al (2024) Elevated genetic risk for multiple sclerosis emerged in steppe pastoralist populations. Nature 625:321–328. https://doi.org/10.1038/s41586-023-06618-z

Beall CM (2007) Two routes to functional adaptation: Tibetan and Andean high-altitude natives. Proc Natl Acad Sci USA 104(Suppl 1):8655–8660. https://doi.org/10.1073/pnas.0701985104

Bedoya G, Montoya P, García J et al (2006) Admixture dynamics in Hispanics: a shift in the nuclear genetic ancestry of a South American population isolate. Proc Natl Acad Sci USA 103:7234–7239. https://doi.org/10.1073/pnas.050871610L

Bergman J, Pedersen RØ, Lundgren EJ et al (2023) Worldwide Late Pleistocene and Early Holocene population declines in extant megafauna are associated with Homo sapiens expansion rather than climate change. Nat Commun 14:7679. https://doi.org/10.1038/s41467-023-43426-5

Bersaglieri T, Sabeti PC, Patterson N et al (2004) Genetic signatures of strong recent positive selection at the lactase gene. Am J Hum Genet 74:1111–1120. https://doi.org/10.1086/421051

Bigham AW, Mao X, Mei R et al (2009) Identifying positive selection candidate loci for high-altitude adaptation in Andean populations. Hum Genomics 4:79–90. https://doi.org/10.1186/1479-7364-4-2-79

Bigham A, Bauchet M, Pinto D et al (2010) Identifying signatures of natural selection in Tibetan and Andean populations using dense genome scan data. PLoS Genet 6:e1001116. https://doi.org/10.1371/journal.pgen.1001116

Bodner M, Perego UA, Huber G et al (2012) Rapid coastal spread of First Americans: novel insights from South America's Southern Cone mitochondrial genomes. Genome Res 22:811–820. https://doi.org/10.1101/gr.131722.111

Bonatto SL, Salzano FM (1997) A single and early migration for the peopling of the Americas supported by mitochondrial DNA sequence data. Proc Natl Acad Sci USA 94:1866–1871. https://doi.org/10.1073/pnas.94.5.1866

Borda V, Alvim I, Mendes M et al (2020) The genetic structure and adaptation of Andean highlanders and Amazonians are influenced by the interplay between geography and culture. Proc Natl Acad Sci USA 117:32557–32565. https://doi.org/10.1073/pnas.2013773117

Brandini S, Bergamaschi P, Cerna MF et al (2018) The Paleo-Indian entry into South America according to mitogenomes. Mol Biol Evol 35:299–311. https://doi.org/10.1093/molbev/msx267

Brochado JP (1984) An ecological model of the spread of pottery and agriculture into Eastern South America. Tese (Doutorado)-Urbana-Champaign: University of Illinois

Browning SR, Browning BL, Daviglus ML et al (2018) Ancestry-specific recent effective population size in the Americas. PLoS Genet 14:e1007385. https://doi.org/10.1371/journal.pgen.1007385

Bush MB, Nascimento MN, Åkesson CM et al (2021) Widespread reforestation before European influence on Amazonia. Science 372:484–487. https://doi.org/10.1126/science.abf3870

Campbell L, Grondona V (2012) The Indigenous languages of South America: a comprehensive guide. Walter de Gruyter

Campelo dos Santos AL, Owings A, Sullasi HSL et al (2022) Genomic evidence for ancient human migration routes along South America's Atlantic coast. Proc R Soc B Biol Sci 289:20221078. https://doi.org/10.1098/rspb.2022.1078

Capodiferro MR, Aram B, Raveane A et al (2021) Archaeogenomic distinctiveness of the Isthmo-Colombian area. Cell 184:1706–1723.e24. https://doi.org/10.1016/j.cell.2021.02.040

Caro-Consuegra R, Nieves-Colón MA, Rawls E et al (2022) Uncovering signals of positive selection in Peruvian populations from three ecological regions. Mol Biol Evol 39. https://doi.org/10.1093/molbev/msac158

Carvajal-Carmona LG, Soto ID, Pineda N et al (2000) Strong Amerind/white sex bias and a possible Sephardic contribution among the founders of a population in northwest Colombia. Am J Hum Genet 67:1287–1295. https://doi.org/10.1016/S0002-9297(07)62956-5

Castro E Silva MA, Hünemeier T (2023) A multidisciplinary overview on the Tupi-speaking people expansion. Am J Biol Anthropol. https://doi.org/10.1002/ajpa.24876

Castro E Silva MA, Nunes K, Lemes RB et al (2020) Genomic insight into the origins and dispersal of the Brazilian coastal natives. Proc Natl Acad Sci USA 117:2372–2377. https://doi.org/10.1073/pnas.190907511

Castro E Silva MA, Ferraz T, Couto-Silva CM et al (2022) Population histories and genomic diversity of South American Natives. Mol Biol Evol 39. https://doi.org/10.1093/molbev/msab339

Chmyz I (1983) Sétimo relatório de pesquisas realizadas na área de Itaipu (1981-83). Projeto Arqueológico Itaipu:1–106

Clement CR, Denevan WM, Heckenberger MJ et al (2015) The domestication of Amazonia before European conquest. Proc Biol Sci 282:20150813. https://doi.org/10.1098/rspb.2015.0813

Cooke CA, Abbott MB (2008) A paleolimnological perspective on industrial-era metal pollution in the central Andes, Peru. Sci Total Environ 393:262–272. https://doi.org/10.1016/j.scitotenv.2007.12.034

Cordaux R, Deepa E, Vishwanathan H, Stoneking M (2004) Genetic evidence for the demic diffusion of agriculture to India. Science 304:1125. https://doi.org/10.1126/science.109581

Corrêa ÂA (2015) Pindorama de mboîa e îakaré: continuidade e mudança na trajetória das populações Tupi. Universidade de São Paulo, Agência USP de Gestão da Informação Acadêmica (AGUIA)

Couto-Silva CM, Nunes K, Venturini G et al (2023) Indigenous people from Amazon show genetic signatures of pathogen-driven selection. Sci Adv 9:eabo0234. https://doi.org/10.1126/sciadv.abo023

Crawford JE, Amaru R, Song J et al (2017) Natural selection on genes related to cardiovascular health in high-altitude adapted Andeans. Am J Hum Genet 101:752–767. https://doi.org/10.1016/j.ajhg.2017.09.023

De Loma J, Vicente M, Tirado N et al (2022) Human adaptation to arsenic in Bolivians living in the Andes. Chemosphere 301:134764. https://doi.org/10.1016/j.chemosphere.2022.134764

Diamond J, Bellwood P (2003) Farmers and their languages: the first expansions. Science 300:597–603. https://doi.org/10.1126/science.107820

Dillehay TD (2009) Probing deeper into first American studies. Proc Natl Acad Sci USA 106:971–978. https://doi.org/10.1073/pnas.0808424106

Dillehay TD, Ocampo C, Saavedra J et al (2015) New archaeological evidence for an early human presence at Monte Verde, Chile. PLoS One 10:e0141923. https://doi.org/10.1371/journal.pone.0141923

Dixon RMW, Aikhenvald AY (2006) The Amazonian languages. Cambridge University Press

e Silva MAC, Ferraz T, Bortolini MC et al (2021) Deep genetic affinity between coastal Pacific and Amazonian natives evidenced by Australasian ancestry. Proc Natl Acad Sci 118:e2025739118. https://doi.org/10.1073/pnas.202573911

Eichstaedt CA, Antão T, Pagani L et al (2014) The Andean adaptive toolkit to counteract high altitude maladaptation: genome-wide and phenotypic analysis of the Collas. PLoS One 9:e93314. https://doi.org/10.1371/journal.pone.0093314

Eichstaedt CA, Antao T, Cardona A et al (2015) Positive selection of AS3MT to arsenic water in Andean populations. Mutat Res 780:97–102. https://doi.org/10.1016/j.mrfmmm.2015.07.007

Espinoza-Navarro O, Diaz J, Rodríguez H, Moreno A (2011) Effects of altitude on anthropometric and physiological patterns in Aymara and non-Aymara population between 18 and 65 years in the province of parinacota Chile (3.700 masl). Int J Morphol 29:34–40

Fernandes DM, Sirak KA, Ringbauer H et al (2021) A genetic history of the pre-contact Caribbean. Nature 590:103–110. https://doi.org/10.1038/s41586-020-03053-2

Ferraz T, Suarez Villagran X, Nägele K et al (2023) Genomic history of coastal societies from eastern South America. Nat Ecol Evol 7:1315–1330. https://doi.org/10.1038/s41559-023-02114-9

de Filippo C, Bostoen K, Stoneking M, Pakendorf B (2012) Bringing together linguistic and genetic evidence to test the Bantu expansion. Proc Biol Sci 279:3256–3263. https://doi.org/10.1098/rspb.2012.0318

Flegontov P, Altınışık NE, Changmai P et al (2019) Palaeo-Eskimo genetic ancestry and the peopling of Chukotka and North America. Nature 570:236–240. https://doi.org/10.1038/s41586-019-1251-y

de la Fuente C, Ávila-Arcos MC, Galimany J et al (2018) Genomic insights into the origin and diversification of late maritime hunter-gatherers from the Chilean Patagonia. Proc Natl Acad Sci 115:E4006–E4012. https://doi.org/10.1073/pnas.1715688111

Gaspar MD, DeBlasis P, Fish SK, Fish PR (2008) Sambaqui (Shell Mound) societies of Coastal Brazil. In: Silverman H, Isbell WH (eds) The handbook of South American Archaeology. Springer, New York, pp 319–335

Goldberg A, Mychajliw AM, Hadly EA (2016) Post-invasion demography of prehistoric humans in South America. Nature 532:232–235. https://doi.org/10.1038/nature17176

Gómez-Carballa A, Pardo-Seco J, Brandini S et al (2018) The peopling of South America and the trans-Andean gene flow of the first settlers. Genome Res 28:767–779. https://doi.org/10.1038/nature17176

González-José R, Bortolini MC, Santos FR, Bonatto SL (2008) The peopling of America: craniofacial shape variation on a continental scale and its interpretation from an interdisciplinary view. Am J Phys Anthropol 137:175–187. https://doi.org/10.1002/ajpa.20854

Graeber D, Wengrow D (2021) The dawn of everything: a new history of humanity. Allen Lane

Graf KE, Buvit I (2017) Human dispersal from Siberia to Beringia: assessing a Beringian standstill in light of the archaeological evidence. Curr Anthropol 58:S583–S603. https://doi.org/10.1086/693388

Gregorio de Souza J, Alcaina Mateos J, Madella M (2020) Archaeological expansions in tropical South America during the late Holocene: assessing the role of demic diffusion. PLoS One 15:e0232367. https://doi.org/10.1371/journal.pone.0232367

Guernier V, Hochberg ME, Guégan J-F (2004) Ecology drives the worldwide distribution of human diseases. PLoS Biol 2:e141. https://doi.org/10.1371/journal.pbio.0020141

Harris DN, Song W, Shetty AC et al (2018) Evolutionary genomic dynamics of Peruvians before, during, and after the Inca Empire. Proc Natl Acad Sci USA 115:E6526–E6535. https://doi.org/10.1073/pnas.1720798115

Headland TN, Bailey RC (1991) Introduction: Have hunter-gatherers ever lived in tropical rain forest independently of agriculture? Hum Ecol Interdiscip J 19:115–122. https://doi.org/10.1007/BF00888742

Heckenberger M (2013) The Arawak Diaspora: perspectives from South America. Oxford University Press

Heckenberger M, Neves EG (2009) Amazonian archaeology. Annu Rev Anthropol 38(1): 251–266. https://doi.org/10.1146/annurev-anthro-091908-164310

Heckenberger MJ et al (2003) Amazonia 1492: pristine forest or cultural parkland? Science 301(5640):1710–1714. https://doi.org/10.1126/science.1086112

Hill JD, Santos-Granero F (2002) Comparative Arawakan histories: rethinking language family and culture area in Amazonia. University of Illinois Press

Hoffecker JF, Elias SA, Scott GR et al (2023) Beringia and the peopling of the Western Hemisphere. Proc Biol Sci 290:20222246. https://doi.org/10.1098/rspb.2022.2246

Hoorn C, Wesselingh FP, ter Steege H et al (2010) Amazonia through time: Andean uplift, climate change, landscape evolution, and biodiversity. Science 330:927–931. https://doi.org/10.1126/science.1194585

Hünemeier T (2024) Biogeographic perspectives on human genetic diversification. Mol Biol Evol 41(3):msae029. https://doi.org/10.1093/molbev/msae029

Hünemeier T, Amorim CEG, Azevedo S et al (2012) Evolutionary responses to a constructed niche: ancient Mesoamericans as a model of gene-culture coevolution. PLoS One 7:e38862. https://doi.org/10.1371/journal.pone.0038862

Iriarte J, Smith RJ, Gregorio de Souza J et al (2017) Out of Amazonia: Late-Holocene climate change and the Tupi–Guarani trans-continental expansion. The Holocene 27:967–975. https://doi.org/10.1177/0959683616678461

Iriarte J, Elliott S, Maezumi SY et al (2020) The origins of Amazonian landscapes: plant cultivation, domestication and the spread of food production in tropical South America. Quat Sci Rev 248:106582. https://doi.org/10.1016/j.quascirev.2020.106582

Jacovas VC, Rovaris DL, Peréz O et al (2015) Genetic variations in the TP53 pathway in Native Americans strongly suggest adaptation to the high altitudes of the Andes. PLoS One 10:e0137823. https://doi.org/10.1371/journal.pone.0137823

Jacovas VC, Couto-Silva CM, Nunes K et al (2018) Selection scan reveals three new loci related to high altitude adaptation in Native Andeans. Sci Rep 8:12733. https://doi.org/10.1038/s41598-018-31100-6

Johnson RJ, Sánchez-Lozada LG, Nakagawa T et al (2022) Do thrifty genes exist? Revisiting uricase. Obesity 30:1917–1926. https://doi.org/10.1002/oby.23540

Jones TL, Schwitalla AW, Pilloud MA et al (2021) Historic and bioarchaeological evidence supports late onset of post-Columbian epidemics in Native California. Proc Natl Acad Sci USA 118. https://doi.org/10.1073/pnas.2024802118

Joseph SK, Migliore NR, Olivieri A et al (2023) Genomic evidence for adaptation to tuberculosis in the Andes before European contact. iScience 26:106034. https://doi.org/10.1016/j.isci.2023.106034

Keegan WF (1995) Modeling dispersal in the prehistoric West Indies. World Archaeol 26:400–420

Keegan WF, Hofman CL (2017) The early ceramic age. Oxford University Press

Kehdy FSG, Gouveia MH, Machado M et al (2015) Origin and dynamics of admixture in Brazilians and its effect on the pattern of deleterious mutations. Proc Natl Acad Sci USA 112:8696–8701. https://doi.org/10.1073/pnas.150444711

Lambeck K, Rouby H, Purcell A et al (2014) Sea level and global ice volumes from the Last Glacial Maximum to the Holocene. Proc Natl Acad Sci USA 111:15296–15303. https://doi.org/10.1073/pnas.141176211

Larson G, Piperno DR, Allaby RG et al (2014) Current perspectives and the future of domestication studies. Proc Natl Acad Sci USA 111:6139–6146. https://doi.org/10.1073/pnas.1323964111

Lathrap DW (1970) The upper Amazon. Thames and Hudson

Lawrence ES, Gu W, Bohlender RJ et al (2024) Functional/missense variant is associated with hematocrit in Andean highlanders. Sci Adv 10:eadj5661. https://doi.org/10.1126/sciadv.adj56
Lesnek AJ, Briner JP, Lindqvist C et al (2018) Deglaciation of the Pacific coastal corridor directly preceded the human colonization of the Americas. Sci Adv 4:eaar5040. https://doi.org/10.1126/sciadv.aar504
Lindo J, Haas R, Hofman C et al (2018) The genetic prehistory of the Andean highlands 7000 years BP though European contact. Sci Adv 4:eaau4921. https://doi.org/10.1126/sciadv.aau492
Loog L, Mirazón Lahr M, Kovacevic M et al (2017) Estimating mobility using sparse data: application to human genetic variation. Proc Natl Acad Sci USA 114:12213–12218. https://doi.org/10.1073/pnas.1703642114
Martinez VD, Vucic EA, Becker-Santos DD et al (2011) Arsenic exposure and the induction of human cancers. J Toxicol 2011:431287. https://doi.org/10.1155/2011/431287
Mas-Sandoval A, Mathieson S, Fumagalli M (2023) The genomic footprint of social stratification in admixing American populations. elife 12. https://doi.org/10.7554/eLife.84429
Meltzer DJ (2010) First peoples in a New World: Colonizing Ice Age America. University of California Press
Mendoza-Caamal EC, Barajas-Olmos F, García-Ortiz H et al (2020) Metabolic syndrome in indigenous communities in Mexico: a descriptive and cross-sectional study. BMC Public Health 20:339. https://doi.org/10.1186/s12889-020-8378-5
Mendoza-Revilla J, Chacón-Duque JC, Fuentes-Guajardo M et al (2022) Disentangling signatures of selection before and after European Colonization in Latin Americans. Mol Biol Evol 39. https://doi.org/10.1093/molbev/msac076
Meyer D, Aguiar VRC, Bitarello BD et al (2018) A genomic perspective on HLA evolution. Immunogenetics 70:5–27. https://doi.org/10.1007/s00251-017-1017-3
Moore LG (2001) Human genetic adaptation to high altitude. High Alt Med Biol 2:257–279. https://doi.org/10.3390/genes10020150
Moorjani P, Hellenthal G (2023) Methods for assessing population relationships and history using genomic data. Annu Rev Genomics Hum Genet 24:305–332. https://doi.org/10.1146/annurev-genom-111422-025117
Morello F, Borrero L, Massone M et al (2012) Hunter-gatherers, biogeographic barriers and the development of human settlement in Tierra del Fuego. Antiquity 86:71–87. https://doi.org/10.1017/S0003598X00062463
Moreno-Mayar JV, Potter BA, Vinner L et al (2018a) Terminal Pleistocene Alaskan genome reveals first founding population of Native Americans. Nature 553:203–207. https://doi.org/10.1038/nature25173
Moreno-Mayar JV, Vinner L, de Barros DP et al (2018b) Early human dispersals within the Americas. Science 362. https://doi.org/10.1126/science.aav2621
Nägele K, Posth C, Iraeta Orbegozo M et al (2020) Genomic insights into the early peopling of the Caribbean. Science 369:456–460. https://doi.org/10.1126/science.aba8697
Nakatsuka N, Luisi P, Motti JMB et al (2020) Ancient genomes in South Patagonia reveal population movements associated with technological shifts and geography. Nat Commun 11:3868. https://doi.org/10.1038/s41467-020-17656-w
Napolitano MF, DiNapoli RJ, Stone JH et al (2019) Reevaluating human colonization of the Caribbean using chronometric hygiene and Bayesian modeling. Sci Adv 5:eaar7806. https://doi.org/10.1126/sciadv.aar78
Neel JV (1962) Diabetes mellitus: a "thrifty" genotype rendered detrimental by "progress"? Am J Hum Genet 14:353–362
Neves EG (2011) Archaeological cultures and past identities in the pre-colonial Central Amazon. Ethnicity. In: Ancient Amazonian: reconstructing past identities from Archaeology, Linguistic and Etnohistory. University Press of Colorado, Boulder, pp 1–27
Neves EG (2013) Was agriculture a key productive activity in Pre-Colonial Amazonia? The stable productive basis for social equality in the Central Amazon. In: Brondízio ES, Moran EF (eds) Human-Environment interactions: current and future directions. Springer, Dordrecht, pp 371–388

Neves EG, Heckenberger MJ (2019) The call of the wild: rethinking food production in ancient Amazonia. Annu Rev Anthropol 48:371–388. https://doi.org/10.1146/annurev-anthro-102218-011057

Neves WA, Hubbe M (2005) Cranial morphology of early Americans from Lagoa Santa, Brazil: implications for the settlement of the New World. Proc Natl Acad Sci USA 102:18309–18314. https://doi.org/10.1073/pnas.0507185102

Noelli FS (1998) The Tupi: explaining origin and expansions in terms of archaeology and of historical linguistics. Antiquity 72:648–663

Noelli FS (2008) The Tupi expansion. In: Silverman H, Isbell WH (eds) The handbook of South American archaeology. Springer, New York, pp 659–670

Norris ET, Wang L, Conley AB et al (2018) Genetic ancestry, admixture and health determinants in Latin America. BMC Genomics 19(Suppl 8):861. https://doi.org/10.1186/s12864-018-5195-7

O'Brien MJ, Laland KN (2012) Genes, culture, and agriculture. Curr Anthropol 53:434–470. https://doi.org/10.1086/666585

O'Hagan Z, Chousou-Polydouri N, Michael L (2019) Phylogenetic classification supports a Northeastern Amazonian Proto-Tupí-Guaraní Homeland. LIAMES: Ling Indig Am 19:e019018. https://doi.org/10.20396/liames.v19i0.8655791

Ochoa-Guzmán A, Moreno-Macías H, Guillén-Quintero D et al (2020) R230C but not - 565C/T variant of the ABCA1 gene is associated with type 2 diabetes in Mexicans through an effect on lowering HDL-cholesterol levels. J Endocrinol Investig 43:1061–1071. https://doi.org/10.1007/s40618-020-01187-8

Palma-Silva C, Turchetto-Zolet AC, Fay MF, Vasconcelos T (2022) Drivers of exceptional Neotropical biodiversity: an updated view. Bot J Linn Soc 199:1–7. https://doi.org/10.1093/botlinnean/boac005

Pärssinen M, Gama T, Schaan DP (2005) Quando começou, realmente, a expansão guarani em direção às Serras Andinas Orientais? rsab 18:51–66. https://doi.org/10.24885/sab.v18i1.204

Pearce AJ, Beresford-Jones DG, Heggarty P (2020) Rethinking the Andes-Amazonia divide: a cross-disciplinary exploration. UCL Press

Pena SDJ (2002) Homo brasilis: aspectos genéticos, lingüísticos, históricos e socioantropológicos da formação do povo brasileiro, históricos e socioantropológicos da formação do …

Peña-Espinoza BI, Torre-Horta E, Ortiz-López MG, Menjivar M (2024) ABCA1 variant rs9282541 is associated with metabolic syndrome in Maya children. Ann Hum Genet. https://doi.org/10.1111/ahg.12546

Perego UA, Achilli A, Angerhofer N et al (2009) Distinctive Paleo-Indian migration routes from Beringia marked by two rare mtDNA haplogroups. Curr Biol 19:1–8. https://doi.org/10.1016/j.cub.2008.11.058

Perez SI, Postillone MB, Rindel D et al (2016) Peopling time, spatial occupation and demography of Late Pleistocene–Holocene human population from Patagonia. Quat Int 425:214–223. https://doi.org/10.1016/j.quaint.2016.05.004

Perez SI, Postillone MB, Rindel D (2017) Domestication and human demographic history in South America. Am J Phys Anthropol 163:44–52. https://doi.org/10.1002/ajpa.23176

Peripato V, Levis C, Moreira GA et al (2023) More than 10,000 pre-Columbian earthworks are still hidden throughout Amazonia. Science 382:103–109. https://doi.org/10.1126/science.ade25

Perry GH, Dominy NJ, Claw KG et al (2007) Diet and the evolution of human amylase gene copy number variation. Nat Genet 39:1256–1260. https://doi.org/10.1038/ng2123

Peyrégne S, Slon V, Kelso J (2023) More than a decade of genetic research on the Denisovans. Nat Rev Genet. https://doi.org/10.1038/s41576-023-00643-4

Pinotti T, Bergström A, Geppert M et al (2019) Y Chromosome sequences reveal a short Beringian standstill, rapid expansion, and early population structure of Native American Founders. Curr Biol 29:149–157.e3. https://doi.org/10.1016/j.cub.2018.11.029

Piperno DR, Mcmichael C, Bush M (2015) Amazonia and the anthropocene: what was the spatial extent and intensity of human landscape modification in the Amazon Basin at the end of prehistory?. The Holocene 25. https://doi.org/10.1177/0959683615588374

Pope KO, Pohl ME, Jones JG et al (2001) Origin and environmental setting of ancient agriculture in the lowlands of Mesoamerica. Science 292:1370–1373. https://doi.org/10.1126/science.292.5520.1370
Posth C, Nakatsuka N, Lazaridis I et al (2018) Reconstructing the deep population history of Central and South America. Cell 175:1185–1197.e22. https://doi.org/10.1016/j.cell.2018.10.027
Potter BA, Baichtal JF, Beaudoin AB et al (2018) Current evidence allows multiple models for the peopling of the Americas. Sci Adv 4:eaat5473. https://doi.org/10.1126/sciadv.aat547
Prates L, Politis GG, Perez SI (2020) Rapid radiation of humans in South America after the last glacial maximum: a radiocarbon-based study. PLoS One 15:e0236023. https://doi.org/10.1371/journal.pone.0236023
Qin P, Stoneking M (2015) Denisovan ancestry in East Eurasian and Native American populations. Mol Biol Evol 32:2665–2674. https://doi.org/10.1093/molbev/msv141
Rademaker K, Hodgins G, Moore K et al (2014) Paleoindian settlement of the high-altitude Peruvian Andes. Science 346:466–469. https://doi.org/10.1126/science.1258260
Raghavan M, Skoglund P, Graf KE et al (2014) Upper Palaeolithic Siberian genome reveals dual ancestry of Native Americans. Nature 505:87–91. https://doi.org/10.1038/nature12736
Raghavan M, Steinrücken M, Harris K et al (2015) Population genetics. Genomic evidence for the Pleistocene and recent population history of Native Americans. Science 349:aab3884. https://doi.org/10.1126/science.aab3884
Ramallo V, Bisso-Machado R, Bravi C et al (2013) Demographic expansions in South America: enlightening a complex scenario with genetic and linguistic data. Am J Phys Anthropol 150:453–463. https://doi.org/10.1002/ajpa.22219
Ranere AJ, Cooke RG (2021) Late glacial and Early Holocene migrations, and Middle Holocene settlement on the lower isthmian land-bridge. Quat Int 578:20–34. https://doi.org/10.1016/j.quaint.2020.06.002
Rasmussen M, Li Y, Lindgreen S et al (2010) Ancient human genome sequence of an extinct Palaeo-Eskimo. Nature 463:757–762. https://doi.org/10.1038/nature08835
Ratnam J, Bond WJ, Fensham RJ et al (2011) When is a "forest" a savanna, and why does it matter? Glob Ecol Biogeogr 20:653–660. https://doi.org/10.1111/j.1466-8238.2010.00634.x
Reich D, Patterson N, Campbell D et al (2012) Reconstructing Native American population history. Nature 488:370–374. https://doi.org/10.1038/nature11258
Roosevelt AC (2013) The Amazon and the anthropocene: 13,000 years of human influence in a tropical rainforest. Anthropocene 4:69–87. https://doi.org/10.1016/j.ancene.2014.05.001
Rostain S, Dorison A, de Saulieu G et al (2024) Two thousand years of garden urbanism in the Upper Amazon. Science 383:183–189. https://doi.org/10.1126/science.adi631
Rothhammer F, Dillehay TD (2009) The late Pleistocene colonization of South America: an interdisciplinary perspective. Ann Hum Genet 73:540–549. https://doi.org/10.1111/j.1469-1809.2009.00537.x
Rouse I (1997) From prehistory in Haiti, a study in method. In: Lyman RL, O'Brien MJ, Dunnell RC (eds) Americanist culture history: fundamentals of time, space, and form. Springer US, Boston, pp 201–217
Rull V (2008) Speciation timing and neotropical biodiversity: the Tertiary-Quaternary debate in the light of molecular phylogenetic evidence. Mol Ecol 17:2722–2729. https://doi.org/10.1111/j.1365-294X.2008.03789.x
Rull V (2011) Neotropical biodiversity: timing and potential drivers. Trends Ecol Evol 26:508–513. https://doi.org/10.1016/j.tree.2011.05.011
Sage AP, Minatel BC, Ng KW et al (2017) Oncogenomic disruptions in arsenic-induced carcinogenesis. Oncotarget 8:25736–25755. https://doi.org/10.18632/oncotarget.15106
dos Santos EJM, da Silva ALS, Ewerton PD et al (2015) Origins and demographic dynamics of Tupí expansion: a genetic tale. Bol Mus Para Emílio Goeldi Ciênc hum 10:217–228. https://doi.org/10.1590/1981-81222015000200003

Scheel-Ybert R, Macario K, Buarque A et al (2008) A new age to an old site: the earliest Tupiguarani settlement in Rio de Janeiro State? An Acad Bras Cienc 80:763–770. https://doi.org/10.1590/S0001-37652008000400015

Scheib CL, Li H, Desai T et al (2018) Ancient human parallel lineages within North America contributed to a coastal expansion. Science 360:1024–1027. https://doi.org/10.1126/science.aar6851

Schieber M, Chandel NS (2014) ROS function in redox signaling and oxidative stress. Curr Biol 24:R453–R462. https://doi.org/10.1016/j.cub.2014.03.034

Schlebusch CM, Gattepaille LM, Engström K et al (2015) Human adaptation to arsenic-rich environments. Mol Biol Evol 32:1544–1555. https://doi.org/10.1016/j.cub.2014.03.034

Schroeder H, Sikora M, Gopalakrishnan S et al (2018) Origins and genetic legacies of the Caribbean Taino. Proc Natl Acad Sci 115:2341–2346. https://doi.org/10.1073/pnas.1716839115

Sikora M, Pitulko VV, Sousa VC et al (2019) The population history of northeastern Siberia since the Pleistocene. Nature 570:182–188. https://doi.org/10.1038/s41586-019-1279-z

Sikora M, Canteri E, Fernandez-Guerra A et al (2023) The landscape of ancient human pathogens in Eurasia from the Stone Age to historical times. bioRxiv. https://doi.org/10.1101/2023.10.06.561165

Silva MACE, Ferraz T, Hünemeier T (2022) A genomic perspective on South American human history. Genet Mol Biol 45:e20220078. https://doi.org/10.1590/1678-4685-GMB-2022-0078

Skoglund P, Reich D (2016) A genomic view of the peopling of the Americas. Curr Opin Genet Dev 41:27–35. https://doi.org/10.1016/j.gde.2016.06.016

Skoglund P, Mallick S, Bortolini MC et al (2015) Genetic evidence for two founding populations of the Americas. Nature 525:104–108. https://doi.org/10.1038/nature14895

Smith AH, Marshall G, Yuan Y et al (2006) Increased mortality from lung cancer and bronchiectasis in young adults after exposure to arsenic in utero and in early childhood. Environ Health Perspect 114:1293–1296. https://doi.org/10.1289/ehp.8832

Sokal RR, Oden NL, Wilson C (1991) Genetic evidence for the spread of agriculture in Europe by demic diffusion. Nature 351:143–145. https://doi.org/10.1038/351143a0

Souza Mello AA, Kneip A (2017) New linguistic (and some archeological) evidence that point the origin of the tupi-guarani peoples in eastern amazon. Lit lingüíst:299–312

Stannard DE (1992) American Holocaust: the conquest of the new world. Oxford University Press

Stoneking M, Arias L, Liu D et al (2023) Genomic perspectives on human dispersals during the Holocene. Proc Natl Acad Sci USA 120:e2209475119. https://doi.org/10.1073/pnas.2209475119

Strauss A, Oliveira RE, Bernardo DV et al (2015) The oldest case of decapitation in the New World (Lapa do Santo, East-Central Brazil). PLoS One 10:e0137456. https://doi.org/10.1371/journal.pone.0137456

Strauss A, Mariano Rodrigues IM, Baeta A et al (2020) The archaeological record of Lagoa Santa (East-Central Brazil): from the Late Pleistocene to historical times. In: Lagoa Santa Karst: Brazil's Iconic Karst Region, pp 227–281. https://doi.org/10.1007/978-3-030-35940-9_12

Sutter RC (2021) The Pre-Columbian peopling and population dispersals of South America. J Archaeol Res 29:93–151. https://doi.org/10.1007/s10814-020-09146-w

Tamm E, Kivisild T, Reidla M et al (2007) Beringian standstill and spread of Native American founders. PLoS One 2:e829. https://doi.org/10.1371/journal.pone.0000829

Tarazona-Santos E, Carvalho-Silva DR, Pettener D et al (2001) Genetic differentiation in South Amerindians is related to environmental and cultural diversity: evidence from the Y chromosome. Am J Hum Genet 68:1485–1496. https://doi.org/10.1086/320601

Thornton R (1987) American Indian Holocaust and survival: a population history since 1492. University of Oklahoma Press

Tishkoff SA, Reed FA, Ranciaro A et al (2007) Convergent adaptation of human lactase persistence in Africa and Europe. Nat Genet 39:31–40. https://doi.org/10.1038/ng1946

Van Den Bergh K, Du Laing G, Montoya JC et al (2010) Arsenic in drinking water wells on the Bolivian high plain: field monitoring and effect of salinity on removal efficiency of iron-oxides-containing filters. J Environ Sci Health A Tox Hazard Subst Environ Eng 45:1741–1749. https://doi.org/10.1080/10934529.2010.513262

Villarreal-Molina MT, Aguilar-Salinas CA, Rodríguez-Cruz M et al (2007) The ATP-binding cassette transporter A1 R230C variant affects HDL cholesterol levels and BMI in the Mexican population: association with obesity and obesity-related comorbidities. Diabetes 56:1881–1887

Voight BF, Kudaravalli S, Wen X, Pritchard JK (2006) A map of recent positive selection in the human genome. PLoS Biol 4:e72. https://doi.org/10.1371/journal.pbio.0040072

Walker RS, Wichmann S, Mailund T, Atkisson CJ (2012) Cultural phylogenetics of the Tupi language family in lowland South America. PLoS One 7:e35025. https://doi.org/10.1371/journal.pone.0035025

Webster TH, Wilson Sayres MA (2016) Genomic signatures of sex-biased demography: progress and prospects. Curr Opin Genet Dev 41:62–71. https://doi.org/10.1016/j.gde.2016.08.002

Wells JCK, Stock JT (2020) Life history transitions at the origins of agriculture: a model for understanding how niche construction impacts human growth, demography and health. Front Endocrinol 11:325. https://doi.org/10.3389/fendo.2020.00325

Wen B, Li H, Lu D et al (2004) Genetic evidence supports demic diffusion of Han culture. Nature 431:302–305. https://doi.org/10.1038/nature02878

Willerslev E, Meltzer DJ (2021) Peopling of the Americas as inferred from ancient genomics. Nature 594:356–364. https://doi.org/10.1038/s41586-021-03499-y

Yang S, Ye K (2021) Recent advances in understanding the adaptive evolution of metabolic genes and traits. Curr Opin Clin Nutr Metab Care 24:308–314. https://doi.org/10.1097/MCO.0000000000000770

Zhou D, Udpa N, Ronen R et al (2013) Whole-genome sequencing uncovers the genetic basis of chronic mountain sickness in Andean highlanders. Am J Hum Genet 93:452–462. https://doi.org/10.1016/j.ajhg.2013.07.011

Zizumbo-Villarreal D, Colunga-GarcíaMarín P (2010) Origin of agriculture and plant domestication in West Mesoamerica. Genet Resour Crop Evol 57:813–825. https://doi.org/10.1007/s10722-009-9521-4

The manufacturer's authorised representative in the EU is Springer
Nature Customer Service Centre GmbH, Europaplatz 3, 69115 Heidelberg,
Germany. If you have any concerns regarding our products, please
contact ProductSafety@springernature.com

Printed and bound by CPI Group (UK) Ltd, Croydon, CR0 4YY
21/09/2026
02219973-0011